SAP PRESS e-books

Print or e-book, Kindle or iPad, workplace or airplane: Choose where and how to read your SAP PRESS books! You can now get all our titles as e-books, too:

- By download and online access
- For all popular devices
- And, of course, DRM-free

Convinced? Then go to **www.sap-press.com** and get your e-book today.

Configuring SAP® Plant Maintenance

SAP PRESS

SAP PRESS is a joint initiative of SAP and Galileo Press. The know-how offered by SAP specialists combined with the expertise of the Galileo Press publishing house offers the reader expert books in the field. SAP PRESS features first-hand information and expert advice, and provides useful skills for professional decision-making.

SAP PRESS offers a variety of books on technical and business-related topics for the SAP user. For further information, please visit our website: *www.sap-press.com*.

Ashish Saxena
SAP Extended Warehouse Management—Configuration, Use, and Processes
2015, approx. 800 pp., hardcover
ISBN 978-1-59229-996-6

Jawad Akhtar
Production Planning and Control with SAP ERP
2013, 1033 pp., hardcover
ISBN 978-1-59229-868-6

Karl Liebstückel
Plant Maintenance with SAP—Practical Guide (3rd Edition)
2014, 669 pp., hardcover
ISBN 978-1-59229-929-4

Martin Murray
Warehouse Management with SAP ERP—Functionality and Technical Configuration (2nd Edition)
2011, 581 pp., hardcover
ISBN 978-1-59229-409-1

Karl Liebstückel

Configuring SAP® Plant Maintenance

Galileo Press

Bonn • Boston

Galileo Press is named after the Italian physicist, mathematician, and philosopher Galileo Galilei (1564–1642). He is known as one of the founders of modern science and an advocate of our contemporary, heliocentric worldview. His words *Eppur si muove* (And yet it moves) have become legendary. The Galileo Press logo depicts Jupiter orbited by the four Galilean moons, which were discovered by Galileo in 1610.

Editor Laura Korslund
Acquisitions Editor Katy Spencer
German Edition Editor Kerstin Billen
Translation Lemoine International, Inc., Salt Lake City, UT
Copyeditor Julie McNamee
Cover Design Mai Loan Nguyen Duy, Graham Geary
Photo Credit Shutterstock.com/44947069/© Christian Lagerek
Layout Design Vera Brauner
Production Graham Geary
Typesetting Publishers' Design and Production Services, Inc.
Printed and bound in the United States of America, on paper from sustainable sources

ISBN 978-1-59229-932-4

1st edition 2014
1st German edition published 2014 by Galileo Press, Bonn, Germany

Library of Congress Cataloging-in-Publication Data
Liebstückel, Karl.
[Instandhaltung mit SAP-Customizing. English]
Configuring SAP plant maintenance / Karl Liebstückel. — 1st edition.
pages cm
ISBN 978-1-59229-932-4 (print) — ISBN 1-59229-932-6 (print) — ISBN 978-1-59229-933-1 (e-book) —
ISBN 978-1-59229-934-8 (print and e-book) 1. Plant maintenance—Data processing. 2. SAP ERP. I. Title.
TS192.L55 2014
658.2'02—dc23
2014016173

Contents at a Glance

1 About This Book 13

2 SAP Projects in Plant Maintenance 21

3 Organizational Structures 73

4 Generic Functions 105

5 Structuring of Technical Systems 153

6 Maintenance Processing 243

7 Preventive Maintenance 419

8 Additional Business Processes 437

9 Usability 547

A Project Plans and Overviews 609

Dear Reader,

You know the expression: no two snowflakes are the same. And if something as small as a snowflake never has a twin, you certainly can't expect two manufacturing plants, and therefore their maintenance, to be the same! With this thought in mind, I present to you your next best friend: a comprehensive resource to making those tweaks to SAP Enterprise Asset Management (SAP EAM) that will make your plant help your business process alignment.

Expert author Karl Liebstückel has already received accolades from readers like you on his expert plant maintenance and configuration knowledge. I'm confident that the skills he's built in his accomplished career will manifest in this title, as well. From setting up master data in the system, to customizing order handling and everything in between, this book presents the configuration knowledge you need to work in an easy, no-frills manner.

As always, we appreciate your business and welcome your feedback. What did you think about *Configuring SAP Plant Maintenance*? Your comments and suggestions are the most useful tools to help us improve our books for you, the reader. We encourage you to visit our website at *www.sap-press.com* and share your feedback.

Laura Korslund
Editor, SAP PRESS

Galileo Press
Boston, MA

laura.korslund@galileo-press.com
www.sap-press.com

Contents

Acknowledgments 11

1 About This Book 13

1.1 Target Audience 14
1.2 What This Book Can and Cannot Do 15
1.3 Structure of This Book 16
1.4 How to Read This Book 17

2 SAP Projects in Plant Maintenance 21

2.1 A Possible Process for Your SAP Plant Maintenance Project 21
2.1.1 Implementation Strategy 21
2.1.2 Methodology 23
2.2 General Risk Factors and Success Factors in SAP Projects: An Empirical Survey 28
2.2.1 Risk Sources 28
2.2.2 Success Factors 31
2.3 Tips for Your Plant Maintenance Project 32
2.3.1 Phase 1: Project Preparation 33
2.3.2 Phase 2: Business Blueprint 60
2.3.3 Phase 3: Realization 67
2.3.4 Phase 4: Final Preparation 69
2.3.5 Phase 5: Go-Live and Support 70
2.3.6 Phase 6: Operate 70

3 Organizational Structures 73

3.1 General SAP Organizational Units 74
3.1.1 Company Code 75
3.1.2 Controlling Area 78
3.1.3 Plant General 82
3.2 The Plant from a Maintenance Perspective 85
3.3 Maintenance-Specific Organizational Units 89
3.4 Work Centers 93

4 Generic Functions 105

4.1 Object Information 106
4.2 Status Management 108
4.3 Number Assignment 114
4.4 Warranties 117
4.5 Measuring Points and Counters 121
4.6 Permits 125
4.7 Partners 127
4.8 Documents 133
4.9 Field Selection 142
4.10 List Variants 146
4.11 Multilevel Lists 150

5 Structuring of Technical Systems 153

5.1 Structuring Elements 153
5.2 Functional Locations and Reference Functional Locations 155
5.3 Equipment 167
5.4 Fleet Management 177
5.5 Links and Networks 186
5.6 Linear Asset Management 189
5.7 Materials and Assemblies 201
5.7.1 Material Types 202
5.7.2 Field Selection 218
5.8 Serial Numbers 225
5.9 Bill of Material 232

6 Maintenance Processing 243

6.1 Notifications 245
6.1.1 Notification Types 247
6.1.2 Catalogs and Catalog Profiles 261
6.1.3 Other Functions 269
6.2 Orders 279
6.2.1 Order Types 284
6.2.2 Material Planning 302
6.2.3 Scheduling 330

6.2.4 Capacity Planning 344
6.2.5 Printing 353
6.2.6 Estimated Costs and Costing 363
6.2.7 Order Settlement 380
6.2.8 Project Management and Investment Management 388
6.2.9 Other Functions 397
6.3 Completion Confirmations 408
6.3.1 SAP EAM Completion Confirmations 408
6.3.2 CATS Confirmations 414

7 Preventive Maintenance 419

7.1 Task Lists 421
7.2 Maintenance Plans 431

8 Additional Business Processes 437

8.1 External Processing 438
8.1.1 External Processing as an Individual Purchase Order 439
8.1.2 External Processing with Service Specifications 446
8.2 Refurbishment 454
8.2.1 Material Master 456
8.2.2 Refurbishment Notifications 462
8.2.3 Refurbishment Orders 463
8.2.4 Material Requirements Planning 469
8.3 Subcontracting 471
8.4 Calibration of Test/Measurement Equipment 477
8.4.1 Customizing Functions for Plant Maintenance 479
8.4.2 Customizing Functions for Quality Management 494
8.5 Pool Asset Management 514
8.6 Maintenance Event Builder 533
8.7 Shift Reports and Shift Notes 535

9 Usability 547

9.1 Category 1: The User's Own Options 549
9.1.1 General User Parameters 549
9.1.2 Maintenance-Specific User Parameters 551

9.1.3 Roles and Favorites ... 556
9.1.4 List Variants ... 557
9.1.5 Personalizing Input Help ... 564
9.1.6 Buttons and Key Combinations ... 565
9.1.7 Table Controls ... 565
9.2 Category 2: The IT Department's Nonprogramming Options ... 568
9.2.1 Transaction Variants ... 568
9.2.2 Customizing ... 575
9.2.3 Action Box ... 578
9.2.4 GuiXT ... 579
9.2.5 SAP Screen Personas ... 586
9.3 Category 3: The IT Department's Programming Options ... 589
9.3.1 Upstream Transactions ... 589
9.3.2 BAPIs ... 591
9.3.3 Web Interface ... 593
9.3.4 Customer Exit ... 596
9.3.5 BAdIs and Enhancement Points ... 600
9.3.6 Workflow ... 602

Appendices ... 609

A Project Plans and Overviews ... 609
A.1 Functional Scope of SAP EAM and Assignment of Priorities ... 609
A.2 Authorization Objects in SAP EAM ... 614
A.3 BAPIs for Plant Maintenance ... 616
A.4 Customer Exits for Plant Maintenance ... 623
B The Author ... 629

Index ... 631

Acknowledgments

I would like to thank everyone who contributed to the success of this book:

- My editor at SAP PRESS, **Kerstin Billen**, for being so patient with me, for her meticulous work and commitment to this project, and for her many suggestions on how the content could be improved.
- My English edition editors, **Laura Korslund** and **Graham Geary**, for spending so much time and effort on making the book available in English.
- My former colleague at SAP, **Dr. Achim Krüger**, for providing me with information and images in relation to SAP Screen Personas, including previously unpublished material.
- The employees at **Symantec Inc.** for providing me with images and software in relation to SAP GuiXT.
- Three research associates at the Institute for Design and Information Systems (IDIS), **Daniel Schwarz**, **Tobias Schlereth**, and **Florian Wolf**, for developing the EAM app, and three student assistants at the same institute, **Johannes Bähr**, **Sebastian Büchse**, and **Michael Landwehr**, for enhancing the app and adapting it to higher education systems.
- My former boss, **Rolf-Peter Westhues**, for giving me every opportunity during my time at SAP to get the best from my role. Without the freedom he gave me, I would not have acquired the wealth of experience that laid the foundations for this book.
- My sons, **Justin** and **Jonas**, who now work within the SAP environment and were able to use their professional judgment to provide constructive feedback.
- And, last but not least, my wife **Brigitta**, for doing without my presence on many occasions while I wrote this book, and for her loving and moral support.

If you plan for a year, plant a seed. If for ten years, plant a tree. If for a hundred years, teach the people. When you sow a seed once, you will reap a single harvest. When you teach the people, you will reap a hundred harvests.
—KUAN CHUNG TZU, Chinese philosopher, Book of Master Kuan

1 About This Book

The book entitled *Plant Maintenance with SAP—Practical Guide*[1] introduced the functional side of SAP's plant maintenance solution, namely SAP Enterprise Asset Management (SAP EAM), which is part of SAP ERP release 6.0. There, the focus was on the system's options and limitations in terms of accomplishing day-to-day tasks. Because this practical guide addressed business topics in particular, it adopted an application-specific viewpoint.

You are now reading *Configuring SAP Plant Maintenance*, which adopts a technical viewpoint and focuses on the *how*, for example:

- *How* do I set up my organizational structures?
- *How* should I configure the Customizing settings so the asset structures can be mapped to the users' requirements?
- *How* should I configure Customizing so that my business processes can be handled in the manner discussed with the users?
- *How* should I configure the usability tools so that there is greater user acceptance of the system?

This book provides an overview of the configuration options currently available. Using more than 25 years of experience working in SAP plant maintenance on more than 70 customer projects, we'll not only show you *how* you can adapt SAP EAM to the needs of your enterprise and your users, but we'll also highlight some essential do's and don'ts.

This book uses customer examples to illustrate what other companies have done. We also provide numerous practical tips that are equally useful whether you are a beginner or an advanced user of a previous release of the system.

---

1 K. Liebstückel, *Plant Maintenance with SAP—Practical Guide* (3rd edition, SAP PRESS, 2014).

1.1 Target Audience

This book always addresses *you* directly. So, who are you? And what can you expect from this book?

- If you are a *project lead* responsible for an SAP plant maintenance project, this book gives you a lot of advice about project management and the configuration and adjustment options available within the system.
- If you are a *member of a project team* that has been assigned to an SAP implementation project and are responsible for adjusting the system to the needs of the enterprise and users, this book gives you plenty of tips in relation to the configuration and adjustment options available within the system.
- If you are an *external consultant* and, as a specialist consultant, you are looking for some background information, explanations in relation to the configuration and adjustment options associated with the system, and recommendations for best practices, which you can then impart to your customers, this is the book for you.
- If you are an *internal consultant* (often known as a *business consultant*), and you are to take over all IT-related support for the user department from the IT or organizational department, this book also gives you many tips in relation to the configuration and adjustment options available within the system.
- If you are a *key user*, which means that you help your colleagues in their daily work and therefore need more background information on the system than your colleagues, this book provides a great deal of background information in relation to system behavior and how this behavior manifests itself. This book enables you to have more targeted and skilled discussions about the needs of your users with the relevant project team members, project lead, or consultants.
- If you are a *developer* who is looking for help with programming (e.g., for interfaces or add-ons), this book is, to some extent, for you. At various points throughout the book, we'll give you information about technical implementations (e.g., customer exits, BAPIs, and BAdIs) without actually discussing the implementation or programming itself.

Who is this book unsuitable for? And what will you not find here?

- If you are an *end user* who is looking for user documentation in relation to the SAP system, this book is not for you.

- If you are a *manager* and have to decide whether or not to implement SAP EAM, this book is not for you because it contains only system-oriented information. In other words, it doesn't provide any specific information about whether or not the SAP system is suitable for you.
- If you're interested in SAP plant maintenance on a general level, this book is not for you. Instead of adding to your basic understanding of plant maintenance or providing an overview, the many details contained in this book will most likely cause confusion. Please consult the practical guide instead.

1.2 What This Book Can and Cannot Do

You can expect the following from this book:

- Show you how to implement the SAP system. In particular, you will learn what you need to be aware of during implementation (success factors) and what could endanger your implementation project (risk factors).
- Provide you with a basic understanding of the philosophy associated with configuring SAP systems, especially their options and limitations.
- Use Customizing settings to not only demonstrate how you can adjust the system to the needs of your enterprise, but also to highlight the limitations associated with some settings.
- Recommend which settings to configure and which to avoid.
- Introduce you to technologies that you can use to adjust the system interface to the needs of your users and to identify the system's limitations.
- Give you many useful tips and tricks for your SAP EAM system.

However, this book doesn't describe any applications or functions. It provides some programming information, but it doesn't provide any source code and is not end-user documentation.[2]

A lesson learned from experiences with past projects is that every enterprise has its own idea of how the system should be used. This means, for example, that every enterprise maps its technical assets differently, every enterprise sets up its business processes differently, and every enterprise needs to connect to different systems, and so on and so forth. Therefore, you should regard the information presented

2 This is usually not the case with SAP documentation.

in this book as a resource for adapting the system to your enterprise's individual needs and thereby creating your own SAP plant maintenance system.

1.3 Structure of This Book

This book is divided into nine chapters, as briefly outlined next.

Chapter 2, SAP Projects in Plant Maintenance, discusses the findings of an empirical study in which other enterprises were asked to state what they believe are the reasons behind the success factors and risk factors of their SAP projects. You'll learn how you can approach an SAP plant maintenance project in a methodical way and what you should watch out for in this kind of project.

Organizational structures are the basis of everything in an SAP system. **Chapter 3**, Organizational Structures, explains the generic SAP organizational units. It also shows you which plant maintenance-specific organizational units and work centers are required for other procedures and how you should configure these.

The SAP system has Customizing functions that are used by several objects (e.g., field selection, status control, and lists). To avoid having to describe these for every single object, **Chapter 4**, Generic Functions, describes the various configuration and adjustment options and, in each case, references which object uses which function.

SAP provides various elements (e.g., equipment and functional locations) for mapping a custom asset structure. **Chapter 5**, Structuring of Technical Systems, describes the settings available to you within the SAP system so that you can adjust certain aspects of structuring your technical system to your requirements.

Chapter 6, Maintenance Processing, shows you all possible settings for a basic process, which you can use to adjust processing of maintenance tasks to your requirements. These range from notification creation, order planning, and order control through to order completion. Recommendations are also provided in terms of which settings you should configure and which settings don't make any sense.

In **Chapter 7**, Preventive Maintenance, you'll learn special system settings in relation to tasks lists and maintenance plans as well as recommendations in terms of how you should configure these settings.

In **Chapter 8**, Additional Business Processes, we'll examine business processes that go beyond normal maintenance processing (e.g., refurbishment, subcontracting, or Pool Asset Management). The settings described in Chapter 6 will be used here, but only special, additional configuration options will be described. For example, the issue of special Customizing functions for refurbishment is addressed here as well as other things that you need to be aware of. The same applies to the other business processes mentioned here.

In **Chapter 9**, Usability, you'll learn how to improve the user interface of the SAP system and therefore increase acceptance among your users. This will be examined from the following three perspectives: What can end users do themselves? What can IT do without the need for programming? At which points can IT intervene with programming? This will be discussed without describing the actual programming itself.

The **appendix** contains additional useful information such as prioritizing the expansion phases when implementing SAP EAM, as well as a complete list of customer exits.

1.4 How to Read This Book

To make it easier for you to use this book, we've attached great importance to presenting the Customizing functions in a clear and concise manner. Because you likely won't read this book as you would a piece of fiction (i.e., from start to finish) but rather use it as a reference guide, we've provided a standard structure for each Customizing function. An example of the layout and content follows.

Customizing Function (per the IMG)

The heading here contains the name of the Customizing function as it appears in the SAP Implementation Guide (IMG) so that you can find it there. A definition and general description of this function is provided in the first section.

Prerequisites

The second section explains those prerequisites that you must fulfill so that you can configure the Customizing function. These may be other Customizing functions that you need to configure in advance or organizational prerequisites.

Customizing Path
This section contains the exact Customizing path for the Customizing function in the IMG.

Transaction
Alternatively, you can use one or more transactions to call many of the Customizing functions contained in the system. This is practical because transactions generally enable you to access the function much faster than using the path. Unfortunately, however, not every function is assigned an SAP transaction.

Settings or Recommended Settings
This section introduces you to the various settings within the relevant Customizing function. These may be entries that you need to make or indicators that you need to set. We'll also describe the effects associated with such settings.

For each Customizing function, you'll also receive at least one figure as a visual aid for understanding the verbal explanations provided in the book (see Figure 1.1).

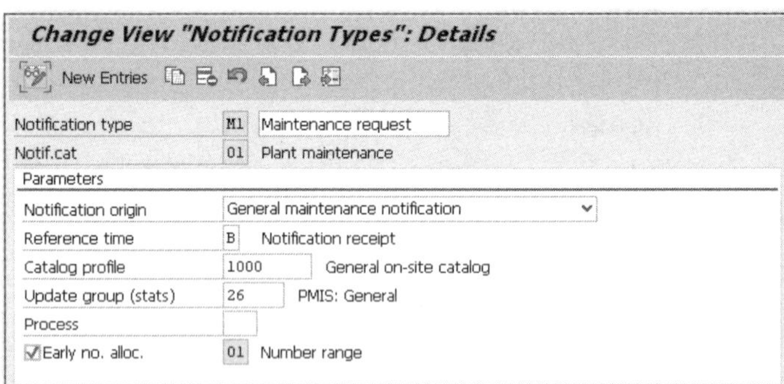

Figure 1.1 Introduction: Sample Figure

[+] **Recommendations**

Normally, at the end of the general explanations, you'll find recommendations in relation to which parameters to set, how to configure the Customizing function, and what to bear in mind.

We'll also provide recommendations in relation to what you should avoid. Generally, these recommendations will be contained in boxes similar to this one.

Special Icons Used in this Book

To make it easier for you to use this book, particular information is highlighted using the following special icons:

Warning
Boxes marked with this icon provide particularly important information about the topic being discussed. We'll also use this icon to warn you about potential sources of error. **[!]**

Practical Tip
Boxes marked with this icon give you recommendations in relation to the Customizing settings or tips from professional experience. **[+]**

Now, all that remains is the hope that you'll garner many ideas and a great deal of inspiration for your own enterprise from reading this book.

In the spirit of a well-known Chuang Tzu quotation, I hope that the information contained in this book falls on fertile ground and yields lasting results for you and your users.

Sincerely,
Karl Liebstückel

This chapter provides a road map for your plant maintenance project by discussing an empirical survey of the risk factors and success factors in SAP projects. This chapter also arms you with many useful tips regarding what you need to watch out for while implementing your plant maintenance project.

2 SAP Projects in Plant Maintenance

In every SAP implementation project, you lay the foundations that determine whether your users will be able to successfully use the SAP system and achieve the desired goals, or whether the SAP system will turn out to be a disappointment for you, your company's top management, and your users. In an SAP implementation project, there are many things that you need to ensure are done (the *Do's*); these are your success factors. However, there are at least as many potential sources of error that you need to avoid (the *Don'ts*); these are your risk factors. This chapter will give you several useful tips to ensure that your SAP plant maintenance is a success.

2.1 A Possible Process for Your SAP Plant Maintenance Project

In the basic discussion of a possible process for your SAP plant maintenance project, you should differentiate between the implementation strategy and its accompanying methodological approach.

2.1.1 Implementation Strategy

Most enterprises implement SAP plant maintenance in a series of stages rather than in a single step (i.e., the full range of functions in all plants). Therefore, during project planning, you need to make a fundamental decision in relation to the stages in which you'll implement SAP plant maintenance. There are two aspects to consider:

- What are the functional stages of the implementation?
- What are the spatial stages of the implementation?

The functional stages in your SAP PM project could be subdivided as follows:

- **Stage 1**
  Implement the structuring of technical systems. (You can omit this first stage if you can carry over asset structures from a legacy system.)
- **Stage 2**
  Implement notification processing and order processing.
- **Stage 3**
  Implement preventive maintenance.
- **Stage 4**
  Expand the system by adding potential enhancements (e.g., Mobile Asset Management, a CAD connection, SAP NetWeaver Portal with roles, test equipment processing, and plant maintenance projects).

The spatial stages, for their part, can be subdivided as follows:

- **Stage 1**
  Implementation in a pilot system.
- **Stage 2**
  Implementation in one plant only.
- **Stage 3**
  Roll-out to other plants.

This is the basis on which you create a plan that specifies which functions are to be implemented in which operation or plant and at what point in time. You then need to make a fundamental decision about which of the following strategies you deem to be the most useful for your project:

- **Horizontal strategy**
  Order processing implementation in all plants simultaneously.
- **Vertical strategy**
  Full implementation in one plant, followed by a roll-out to other plants.
- **Combined strategy**
  Full range of functions implementation in one plant, followed by a roll-out of order processing to all plants.

Figure 2.1 shows a summary of the functional and spatial stages.

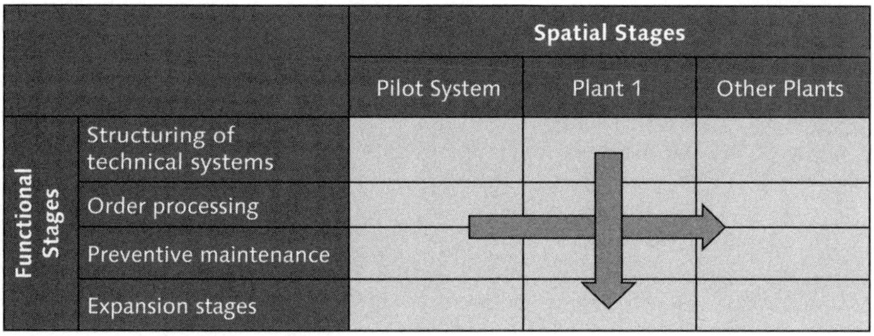

Figure 2.1 Implementation Strategy

2.1.2 Methodology

As is the case in all SAP projects, it's recommended that in SAP implementation projects for plant maintenance, you use the ASAP Roadmap for orientation purposes (see Figure 2.2). *ASAP* (Accelerated SAP) is a methodology developed by SAP for implementing SAP systems. The first version of ASAP was launched in 1996. Since then, numerous versions have been launched. Most recently, Version 8 was launched in 2012. There are two variants of ASAP8 (see *https://service.sap.com/asap*):

- Standard ASAP
- Agile ASAP

Agile ASAP integrates agile project management and development methods such as Scrum. It works in short planning and development cycles to keep the planning phase as short as possible and to obtain tangible and verifiable results as quickly as possible. In terms of their content design, there is little difference between both ASAP variants. Therefore, this won't be discussed in great detail here.

ASAP8 differentiates between the following six phases in project implementation (Figure 2.2):

1. Project Preparation
2. Business Blueprint
3. Realization
4. Final Preparation

5. Go-Live and Support
6. Operate

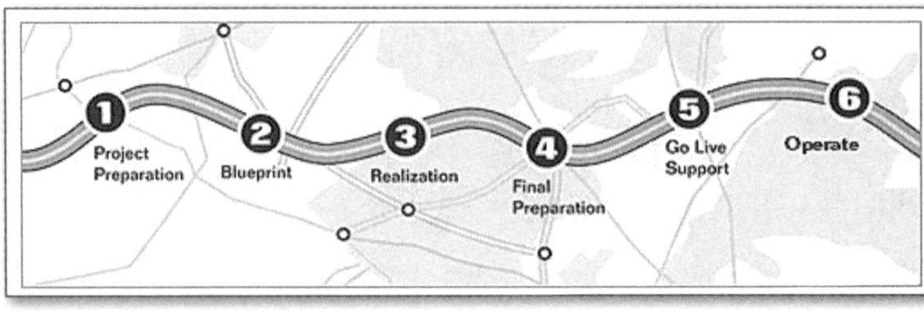

Figure 2.2 ASAP Roadmap

Phase 1: Project Preparation

In phase ❶, *Project Preparation*, you define the organizational aspects of the project and formulate an implementation concept. You carry out the project preparation work either as the first activity within the implementation project or in the form of a feasibility study in advance. The following are the main activities in this phase:

- Define the goals of implementing an SAP system. (Important: Also define what you *don't* want to happen when you implement an SAP system.)
- Define your implementation strategy.
- Define the organizational aspects of the project structure.
- Define project standards and a procedure.
- Create a preliminary project plan.
- Plan your internal marketing.
- Create a project application and get it approved.
- Decide on and adapt the documentation tools.
- Decide on and adapt the monitoring instruments.
- Define the change management process.
- Set up the system environment.
- Train the project team.
- Decide on a training strategy for key users and end users.
- Carry out an as-is analysis of planned processes.
- Design the process map and draft the Business Process Master List (BPML).

- Draw up a test strategy for the go-live.
- Sort through existing master data material.
- Define a data migration strategy.
- Perform an interface inventory.
- Define which project tools are to be used.

Phase 2: Business Blueprint

Phase ❷, *Business Blueprint*, comprises all activities needed to create a detailed business blueprint for using the SAP PM system and for building a prototype. In this phase, you perform the following activities:

- Develop a business blueprint for organizational structures.
- Develop a business blueprint for master data (in particular, the structuring of technical systems).
- Develop a business blueprint for the business processes and functions in SAP PM.
- Document and visualize the various business processes (e.g., using event-driven process chains, which are described in more detail in Section 2.3.1).
- Design a business blueprint for transferring legacy data, for transferring new master data, or both.
- Design a business blueprint for reporting and analyses.
- Draft and establish a communication plan for communicating with stakeholders.
- Create a fit-gap list.
- Create a statement of work and a requirements specification for interfaces.
- Create a statement of work and a requirements specification for enhancements (Gap-List).
- Create an archiving concept.
- Design an authorization concept.
- Create a concept for end-user training.
- Perform Customizing for prototypes.
- Build prototypes based on sample master data.
- Refine the test strategy.

Phase 3: Realization

Phase ❸, *Realization*, involves implementing and mapping the business blueprint. The following are the main activities in this phase:

- Perform final Customizing.
- Adapt the layout of screens and lists.
- Develop, implement, and test the data transfer programs.

  Alternatively: Develop SAP tools (e.g., Legacy System Migration Workbench, LSMW) for the transfer of data.
- Enter master data (if this isn't a separate stage in the implementation).
- Develop enhancements (e.g., print programs, customer exits, and analyses).
- Set up roles and authorizations.
- Create test plans and test cases.
- Perform interim and final tests on the application.
- Perform system and performance tests.
- Perform user acceptance tests.
- Configure settings for archiving and other jobs.
- Train the key users.
- Develop training courses for the end users (e.g., documentation and training examples).
- Develop user documentation.
- Plan for the transfer from the project phase to live operation.

Phase 4: Final Preparation

Phase ❹, *Final Preparation*, comprises all the activities that are required shortly before and in preparation for the system go-live. These include the following:

- Train the end users.
- Transfer the legacy data to the live system.
- Perform the going live check.
- Develop a cut-over plan.

Phase 5: Go-Live and Support

You then go live on the planned key date (Phase ❺) and start processing the first notifications and orders in your new system. The following are typical activities performed at and subsequent to this point:

- Provide support during live operation (i.e., support during the start-up phase).
- Provide additional training for end users.
- Hold the kick-out meeting to signal official project completion.

Phase 6: Operate

Phase ❻, *Operate*, mainly involves improving the operations of the SAP system and identifying any potential for optimization. These improvements include the following:

- Optimized documentation
- Optimized training
- Optimized implementation
- Optimized testing
- Optimized technical operations
- Optimized business processes
- Optimized system upgrades
- Optimized system maintenance

Like every project, your plant maintenance implementation project involves success factors and risk factors. It can only serve to help your project if you benefit from the experiences of other enterprises rather than making every mistake yourself. For this reason, the results of an empirical survey are provided next. Then, based on this survey and personal experiences, this chapter will provide you with several tips for your plant maintenance project.

2.2 General Risk Factors and Success Factors in SAP Projects: An Empirical Survey

Learn from the experiences of others was the principle on which an empirical survey was conducted at the Würzburg-Schweinfurt University of Applied Sciences in Germany, the content and results of which will be discussed in this section.[1]

For the survey, a catalog of 33 potential risk sources and 27 potential success factors was developed and then sent to SAP user companies. The companies were asked to rank each risk source, in accordance with their own experiences, as *Completely true*, *True*, *Not completely true*, and *Not true*. To rank the success factors, a scale with the values *Very important*, *Important*, *Neutral*, and *Not important* was used. A total of 148 companies participated in the survey, and we analyzed the results.

2.2.1 Risk Sources

The following potential risk sources were assigned to various categories:

Risk Sources in the Project Environment

- R01: insufficient documentation on business processes
- R02: no experience with software implementation projects of this type and scope
- R03: insufficiently high status of the SAP implementation project
- R04: end-user requirements not adequately taken into account
- R05: resistance by parties involved
- R06: requirements of the SAP system not prioritized
- R07: change requests made continually throughout duration of project

Risk Sources Associated with the Project Goals

- R08: insufficient communication of goals that were set for the implementation of the SAP system
- R09: hidden goals that were set for the implementation

1 I would like to thank Mr. Andreas Weber who conducted this survey as part of his thesis entitled "Risk Management in Business Software Projects – Concept and Empirical Study."

- R10: unrealistic goals
- R11: unclear requirements

Risk Sources in Project Management

- R12: unstructured projects procedures, milestones unknown
- R13: insufficient monitoring of deadlines and costs
- R14: insufficient adaptation of schedule and budget planning when change requests are made
- R15: poor planning leading to retroactive change requests
- R16: insufficient control of project on part of project managers
- R17: insufficient inclusion of project team members in planning
- R18: poor information flow in project

Risk Sources Associated with Project Organization

- R19: lack of clarity regarding competencies, contact persons, and responsibilities
- R20: work overload; project team members not made free for project work
- R21: inadequate equipment (such as PCs and software)

Risk Sources Associated with the Project Team

- R22: insufficient management of and specialist IT knowledge among project team members
- R23: insufficient know-how on the part of consultants
- R24: carelessness in selection of consultants
- R25: lack of staff, not enough project team members
- R26: poor motivation among project team members

Risk Sources in Project Flow

- R27: inadequate project budget
- R28: not enough time for project planning
- R29: unrealistic completion date, not enough buffer time

- R30: unexpected delays, deadline postponements
- R31: not enough time overall
- R32: problems with hardware configuration and SAP competence
- R33: insufficient quality assurance

Analysis

If we group together the first two categories, *Completely true* and *True*, we get what is known as a risk radar, as shown in Figure 2.3.

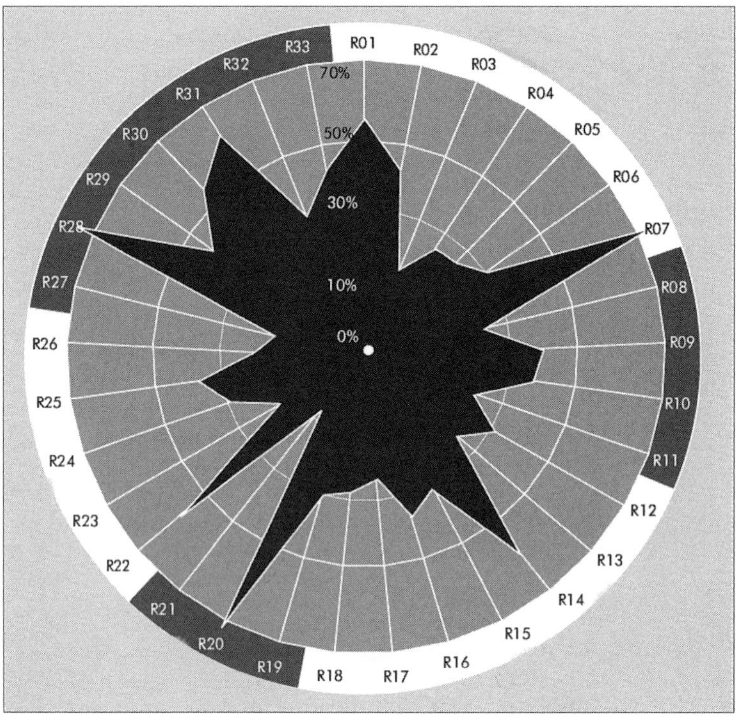

Figure 2.3 Risk Radar

The farther away the points are from the center of the grid, the greater the likelihood that an implementation project will be confronted with the risk in question in the course of the project.

- In the project environment, a particularly high likelihood of occurrence was recorded for *insufficient documentation* and *change requests* during the project.

- In terms of project management, it became clear that *plans were not adjusted* when changes occurred.
- When it came to project organization, staff work *overload* and insufficient specialist knowledge stood out.
- Finally, under project flow, there was a high likelihood of occurrence for *not enough time for planning*, unexpected *delays*, and *insufficient time*.

2.2.2 Success Factors

Let's now look at the success factors as shown in Table 2.1.[2]

Success Factor	Ranking	Very Important
E12: support from top management	1	82.2%
E02: competent project management	2	78.1%
E15: good collaboration	3	67.1%
E01: structured procedures	4	67.1%
E10: inclusion of all parties involved	5	65.8%
E05: technical competence of staff	6	64.4%
E21: clear goals	7	61.6%
E16: realistic scheduling	8	57.5%
E25: identification of project team members with project	9	58.9%
E19: early detection of problems	10	54.8%
E03: appropriate project organization	11	54.8%
E14: open information policy	12	52.1%
E26: appropriate reduction of daily operations tasks in project team members' workload	13	47.9%

Table 2.1 Success Factors of an SAP Project

2 Note: Weighting factors were used to calculate the rankings (*Very important* = 4, *Important* = 3, *Neutral* = 2, *Not important* = 1). Therefore, the ranking does not correspond to the percentage value under *Very Important*.

Success Factor	Ranking	Very Important
E07: appropriate number of project team members	14	39.7%
E11: appropriate budget parameters	15	38.4%
E13: availability of tools	16	37.0%
E24: clear competence guidelines	17	45.2%
E27: quality assurance during project	18	39.7%
E04: experience from other projects	19	41.1%
E22: high priority of implementation project	20	34.2%
E06: appropriate milestone planning	21	26.0%
E17: early training	22	26.0%
E18: appropriate size of work packages	23	19.4%
E23: small number of requirements changes	24	23.3%
E20: detailed project documentation	25	26.0%
E08: use of external consultants	26	27.4%
E09: small project team	27	8.2%

Table 2.1 Success Factors of an SAP Project (Cont.)

According to those surveyed, the following were particularly important for the success of the project:

- Support from top management
- Competent project management
- Team work
- Structured procedures
- Technically competent staff

2.3 Tips for Your Plant Maintenance Project

Based on the information about the general ASAP methodology, the results of the empirical survey, and our personal experiences in plant maintenance projects, let's consider some tips and tricks in relation to your own projects. This section adheres to the time sequence of the ASAP Roadmap, as described in Section 2.1.2.

2.3.1 Phase 1: Project Preparation

In the Project Preparation phase, you set the project goals, create a detailed schedule and budget plan, and perform other tasks that lay the foundations for the actual content of the project.

General Information on Project Preparation

The importance of good project preparation is often underestimated.

Errors in the Project Preparation Stage Have Serious Repercussions [+]

A basic rule of thumb is as follows: The earlier in the project a mistake is made, the greater the effect it will have on the project as a whole. You should therefore proceed with particular caution during the Project Preparation stage. Leave enough time for project planning, and devote all your energy into getting the most from this phase.

As shown in the empirical survey, those surveyed in the user companies believed that, in retrospect, there had been too little time for preparation in their project.

Schedule Sufficient Time for Project Preparation [+]

After you've completed all the tasks relating to project preparation, and only then, start with the business blueprint.

The poorer the quality of the planning, the more likely it is that improvements will have to be made later on. Similarly, the better the project planning, the better you'll be able to handle another of the top three problems: change requests throughout the duration of the project. Experience has shown that poor advanced planning means that several retroactive improvements will have to be made later on in the process. Nonetheless, for the following reasons, you won't be able to completely avoid changes during the course of your project:

- External reasons (e.g., SAP brings out a new release, or top management decides to restructure).
- Project-internal reasons (e.g., feedback from the reference group means that corrections have to be made, errors are detected, or the concept has to be refined).

Because changes within a project are unavoidable, you should define what form these should take.

[+]

Implement a Change Management Policy

Implement a change management policy, and define the following guidelines:

- Which changes are permitted and which aren't
- What types of changes you anticipate
- Which documents are required for each change type
- What approval procedure applies to each change type

Implementation Goal

The goal represents the starting point for each SAP implementation (in other words, what you want to achieve with this implementation).

[+]

Define the Goal of Your Plant Maintenance Project

As soon as you've defined a concrete goal, do the following:

- Formulate it in writing and publish it (e.g., a poster in the project room, an email to employees, an announcement on the intranet, etc.).
- Orientate all project activities toward this goal.
- Define what you *don't* want to happen when you implement the project (e.g., downsizing). In this way, you avoid any unfounded fears and unfulfilled hopes.

Together with your team and intended users, visualize what you want to achieve by implementing the SAP Enterprise Asset Management (SAP EAM) system. Principles that seem self-evident are nonetheless frequently violated:

- A goal may be formulated, but it isn't followed.
- There is a goal, but it is neither verbalized nor communicated.

The following question naturally arises: Which goals do you want to achieve by implementing an SAP EAM system in your enterprise? As a guide, the following is a short list of goals that customers often set in relation to plant maintenance with SAP EAM:

- We want to *reduce breakdowns* to *x*%.
- We want to *increase system availability* to *y*%.

- Our customers want *comprehensive documentation* for our maintenance tasks so that we can continue to deliver high-quality products to them.
- Our vendors want *comprehensive documentation* for our maintenance tasks so that we can continue to use their warranty measures.
- We want to reduce our technicians' nonpurposeful *travel times and wait times* by *x*%.
- We need *complete asset master data*.
- We need *cost transparency*.
- We don't want to invest *any more monitoring resources* in maintenance dates, and the maintenance dates are permitted to vary by *z* days only.
- We want to make more *targeted use of our employees' qualifications and availability*.
- We have to wind down our previous system *Y* by a particular key date (December 31) because it's no longer being maintained (the goal is to *wind down a legacy system*).
- We need an *up-to-date overview* of our spare parts stock.
- We want to *reduce working capital* by *x*%.
- We want to improve the *delivery behavior of our vendors and service providers*.

You'll achieve the most success if you specify a concrete goal and work towards it. A good example is a customer in the automotive component supplier industry that was not satisfied with the downtime rate of his assets, which was 6.2% at the time. Thus, *reducing downtime* was the top priority in his plant maintenance project. Therefore, the first phase of the implementation project was devoted to breakdown management. Breakdowns were recorded and analyzed, and countermeasures were put in place. It quickly became obvious that delayed reporting of breakdowns, especially at the ends of shifts, was causing an order bottleneck in the following shift and, consequently, unnecessarily long downtimes. To counteract this situation, floating shifts were introduced in the plant maintenance department, and as a result, the downtime rate was reduced to 2.8% within six months.

Competencies and Responsibilities

As the empirical survey and personal experiences repeatedly show, *support from top management* is the most important success factor in an SAP project.

[+]

Secure Support from Top Management

A must for project professionals is to secure the active support of top management and publicize this support.

Don't settle for a simple "That's acceptable" or "Go ahead." Instead, ensure that top management plays an active role in the project. What exactly could this role consist of? For example, top management could perform the following tasks:

- Stress the importance of the project and confirm its full support publicly at a company meeting.
- Inform employees about plans for the project in a company-wide email.
- Get regular briefings on the progress of the project.
- Make decisions as a member of the steering committee.

[+]

Demand the Necessary Competencies

Another must for project professionals is to ensure that the project has the necessary resources and that the project leader has the necessary competencies.

In many cases, the project is assigned to an employee who works in the line structure of a subdepartment and, as such, can at best write concepts but doesn't have the authority to make decisions about them. In this case, a major error is being committed.

The project must occupy as a high a position as possible within the enterprise organization; for example, at one of the following levels:

- Technical management
- Top management
- Plant management
- Area management
- Or similar

Make sure that the project leader can make technical and project-related decisions (possibly subject to approval by the steering committee).

Make the Decision-Making Structure as Compact as Possible [+]

An excessive number of decision-making instances is one of the main causes of overly long SAP projects.

You want to avoid the situation that arose within one company in the transport sector as described here. Three years into the project, the company had just about reached the stage where the structuring of technical systems had been agreed on by all decision makers.

Attach Importance to the Right Competencies [+]

Avoid the following situation: The people with the technical competence can't decide anything, and the people who have to make decisions on disciplinary grounds lack the technical knowledge.

You should also note the following points when putting together the project team:

Assemble the Right Project Team [+]

Ensure that your plant maintenance project has all the right qualitative and quantitative components. In your project, you require the following:

- The technical competence from the user department (mechanical engineering, electronics, measurement and control technology, building services, work scheduling, and similar areas)
- The technical competence from IT or the organization
- A knowledgeable and experienced consultant on your project team

The question of how big your project team should be isn't an easy one to answer. Small plant maintenance projects in SAP sometimes consist of a single person, whereas very large ones often have project teams of more than 25 people. Although this is a broad range, consider the following tip.

Have Fully Dedicated Staff on Your Team [+]

Make sure that at least one of the project team members is 100% assigned to the project.

In many projects, there is the mistaken belief, which is often put into practice, that the same results can be achieved with three team members at 50% or six team members at 30%. This is a big mistake. From a project management viewpoint, the following equation applies:

$1 * 100 > 3 * 50 > 6 * 30$

Although this isn't mathematically correct, what it means is that an employee who can devote 100% of his time to the project is more productive than three employees at 50%; and three employees at 50% are more productive than six employees at 30%.

Why? For one thing, the larger the project team, the more time is required for consultation and communication. Furthermore, 50% can very easily turn into 30%, then 20%, then 10%, and then 5%, until finally the team member in question has 0% availability for your project.

Although you may be thinking that this is just a theory or that it will never work here, you should ensure the following just to be safe.

[+]

Set Aside Specific Project Days

If you don't succeed in assigning employees 100% to the project, make sure that all project team members attend scheduled meetings.

Example: "Every Thursday and Friday are devoted to working on this project." However, you should also ensure that your employees can work in peace on certain project days.

[+]

Obtain a Separate Project Room

Employees have too many distractions at their own desks (the telephone, answering colleagues' questions, etc.) to enable them to fully commit themselves to working on the project. Furthermore, a project room has its own dynamic and energy, which enables everyone to concentrate solely on the project.

Closely linked to the above statements is the following tip.

[+]

Don't Overwork Your Employees

An employee who is already working at 120% capacity on daily tasks is incapable of working on an SAP project as well.

Despite this truth, the prevailing impression is that many companies assume that employees can work on the SAP project "alongside" their regular work.

Another issue that you should clarify at the outset is the input required for the project (e.g., to get your project budget authorized by top management).

Input Planning

Project input depends greatly on several factors that you have to consider, including the following:

- The spatial distribution of the implementation (see Section 2.1.1)
- The used and targeted range and depth of functions (also see the book entitled *Plant Maintenance with SAP—Practical Guide* [3rd edition, SAP PRESS, 2014])
- Your company's internal decision-making and approval structures
- Your project competence or the project lead's authority to decide
- The quality of existing master data
- The number of new master data records required
- The number and complexity of the interfaces to be created
- The quantity and scope of new development work (e.g., reports, print programs, customer exits, and a web user interface)
- Integration with other SAP applications
- The scope of end-user training
- Whether it concerns winding down a legacy system or a completely new implementation

With the last point, in particular, many enterprises mistakenly believe that if a legacy system is already in use, it should take even less time to implement an SAP system because, for example, the master data is already available in electronic form. Unfortunately, this isn't the case.

Additional Effort If a Legacy System Exists [!]

It takes *considerably* more effort to wind down a legacy system than to implement a new system.

There are several reasons for this:

- First, an IT system always revolves around a specific organizational structure and its individual business processes. Furthermore, it's always necessary to adjust this organizational structure when you change the IT system currently in use. This is always the case when you wind down a legacy system (e.g., replacing Maximo with SAP).
- Second, each IT system has a specific range of functions, its own terminology, and an individual layout, all of which the users have become accustomed to over the years. If you install a new IT system, everything is alien at first, looks strange, and just doesn't seem suitable. Concerns such as, "But the field must be on the top left of the screen, not in the middle right of the screen," are voiced.
- Third, every familiarization process is difficult for all parties involved (the end users, the application consultants, IT, and the decision makers). Reservations creep in and obstacles arise, all of which have to be broken down, and this all happens repeatedly in the implementation phase.

In SAP EAM projects, project input usually fluctuates between the following values:

- Less than 50 days for small, new implementation projects with a high decision-making competence, existing master data, no interfaces, and very little additional programming
- Significantly more than 10,000 days for the winding down of legacy systems in large international projects with multilevel approval structures, manual entry of new master data, a lot of additional programming, and new connections to non-SAP systems

As an additional indicator for estimating project input, an input distribution over the project phases may be useful, as follows:

- Project Preparation: 5–10%
- Business Blueprint: 25–30%
- Realization: 45–50%
- Final Preparation: 20–30%

For your own input planning, consider combining two procedures.

Top-Down Estimate and Bottom-Up Planning [+]

The procedure is as follows:

1. Carry out a top-down estimate (i.e., a rough input estimate based on the goals and framework conditions).
2. Break it down into the individual project phases.
3. Then, use the planning for individual work packages to perform bottom-up planning.
4. If the two plans are very different, examine your assumptions very closely.

Another aspect that you need to specify in the Project Preparation phase is the type of documentation you want to create.

Documentation

One of the most important risk factors mentioned by the companies who were surveyed was insufficient documentation. Therefore, note the following important tip.

Create a Documentation Concept ("Document of Documents") [+]

Here, define the following:

- What documents are to be created and for what purpose?
- What documentation tool will be used?
- What names will be assigned to the documents?
- Where will they be stored?
- Who is responsible for the documentation?

During the course of your project, you'll require the following documents, among others:

- **Project plan**
  General plan, focus on milestones and work packages, parties involved in the project, distribution of responsibilities.
- **Business case**
  Document for calculating the cost effectiveness of the project, which is usually required for budget requests.
- **Monitoring documents**
  Documents that you, as the project lead, can use to monitor milestones, activities, compliance with deadlines, and so on.

- **As-is analysis**
  Inventory of existing master data and business processes.
- **BPML (Business Process Master List)**
  Overview of new business processes.
- **BPP (Business Process Procedures)**
  Detailed concept of new business processes.
- **Customizing**
  Documents that contain all Customizing settings.
- **GAP list**
  List of all points not covered in the standard system and for which an organizational or technical solution for the system must be found.
- **WRICEF list**
  List of all workflows, reports, interfaces, conversions, enhancements, and forms (WRICEF); that is, a list of all programming tasks, practically an overview of the solutions for the GAP list.
- **User roles**
  Definition of roles, including their menus and authorizations.
- **Test plan and test documents**
  Documents for planning and conducting tests as well as recording the results of such tests.
- **Data transfer**
  Document that contains details about the objects that are to be transferred from the legacy system to the SAP system.
- **Program development**
  Detailed programming requirements.

  Depending on the variant, this may also be divided into a *detailed specification* (on the business side) and a *technical specification* (on the technical side).
- **Training plan**
  Overview of when, where, and from whom users will receive training and what topics will be covered.
- **User training material**
  For holding the requisite training courses before the system goes live.

- **User documentation**
  Reference material.
- **Cut-over plan**
  Details of the transfer to a live system and a go-live.
- **Feedback forms**
  For problem messages.
- **Sprint backlog**
  If you apply Scrum to your project.
- **Project completion report**
  Document that is transferred to the user department involved in the go-live, basis for final presentation.
- **Document of documents**
  If you're using a large number of document types, the central document in which you describe the documents available as well as their purpose and content.

You may not require so many documents. Furthermore, you may choose to combine many of these documents into one document, or you may even need to use other documents not listed here.

However, don't worry. You don't have to reinvent the wheel for the umpteenth time.

SAP Templates [+]

On SAP Service Marketplace (*http://service.sap.com/asap*) and in SAP Solution Manager, SAP provides templates for all of the documents listed here as well as other documents (e.g., a costing sheet for the business case or capacity calculations, presentations, checklists, report templates, and many more).

Training

Both the empirical survey and personal experiences have shown that good training of project team members is a prerequisite for high-quality project work.

Provide Comprehensive Training for Project Team Members [+]

Another must for project professionals is to train your project team members fully, either through attendance at SAP's standard courses or an in-house training workshop.

It isn't enough to simply give the project team members a brief introduction. If you want them to successfully perform their conceptual work, they need to have comprehensive knowledge of SAP Enterprise Asset Management (SAP EAM).

SAP currently offers the following training courses (shown in Figure 2.4).

[+] **PLM300 and PLM315 for Project Team Members**

At the very minimum, your project team members should complete courses PLM300 and PLM315, as well as others as required.

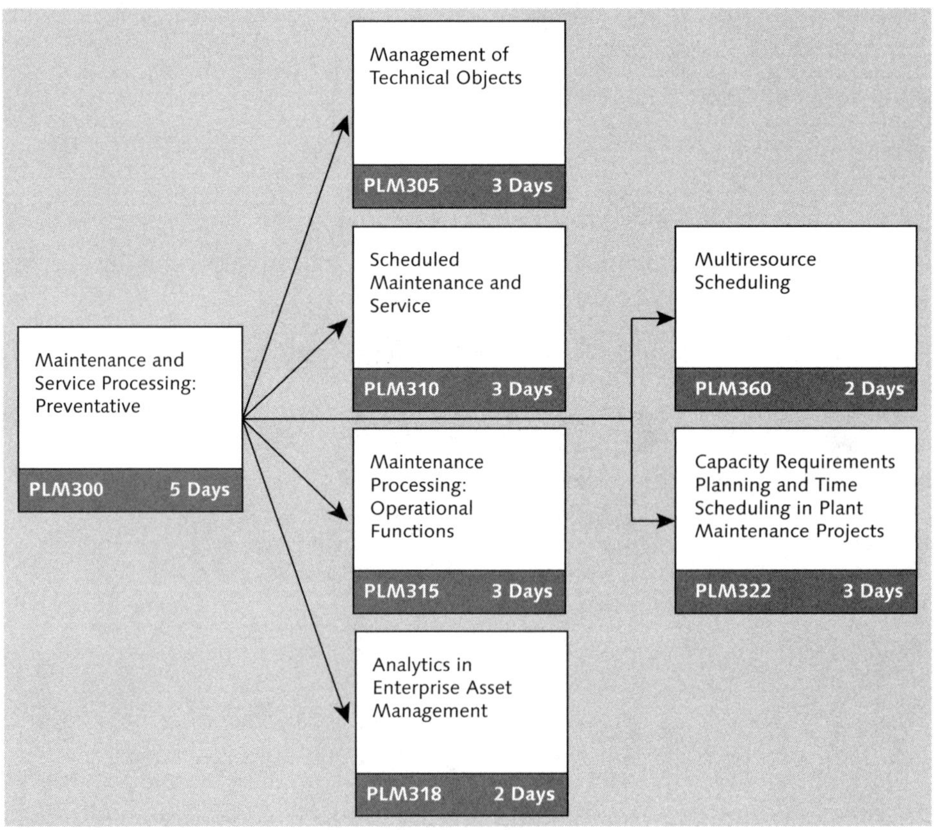

Figure 2.4 SAP Courses on SAP EAM

Another aspect of the Project Preparation phase is one that was already explained in Section 2.1.1, but will be mentioned here again for the sake of completeness.

Create an Implementation Strategy [+]

Create a plan that specifies which functions are to be implemented in which operation or plant and at what point in time. You must choose one of the following strategies:

- **Horizontal strategy**
  Order processing in all plants.
- **Vertical strategy**
  Full implementation in one plant followed by a roll-out to other plants.
- **Combined strategy**
  Full range of functions in one plant followed by a roll-out of order processing to all plants.

Furthermore, remember the following tip, which was also given in the book *Plant Maintenance with SAP—Practical Guide* mentioned earlier.

Have the Courage to Leave Some Gaps [+]

The full range of functions in the SAP system doesn't have to be (and should not be) implemented all at once.

Marketing

Your project will have a greater chance of success if it has the full support of all employees, especially intended users, right from the start, and if you can create a "power base" for the project. A first step in this direction (there will be others later on) is publicity work.

Do Something Good and Talk about It [+]

Make your project known within your company (in-house marketing).

There are many ways to do this. Here are some examples of what other companies have done:

- A transportation company designed a flyer and distributed it among its staff.
- An energy provider had a presentation playing on a screen in its main foyer.

- An automotive supplier set up a home page to report on the current status of the project.
- A chemicals company sent a monthly newsletter to staff—in particular, intended users—and top management.
- A transportation company produced a CD containing a presentation and description of the project as well as the project plan.
- Many companies convene official kick-off meetings for staff and the project team.

There are many inexpensive ways to announce your project. Be creative!

As-Is Analysis and Feasibility Study

The as-is analysis (i.e., the analysis of the current situation) is an aspect that should be a given because the project can't proceed without it. Nonetheless, it's often neglected—or omitted altogether—in implementation projects.

[+]

Perform an As-Is Analysis (Feasibility Study)

Before you do anything with your system, you need to carry out a careful as-is analysis of the framework conditions (in particular, your business processes). The time and effort required to perform a full and correct feasibility study will definitely be worth it.

A feasibility study should pursue several goals:

- To analyze, document, and visualize the master data, plant maintenance processes, and materials management processes used previously as well as the reporting requirements.
- To define the range of plant maintenance functions available with the SAP software and to assign them to potential expansion phases.
- To serve as a basis for further planning of the project details (deadlines, effort, staff, costs), serve as a basis for establishing the training workshops in the Business Blueprint phase, and serve as a basis for using SAP software to draft to-be processes (see Figure 2.5).

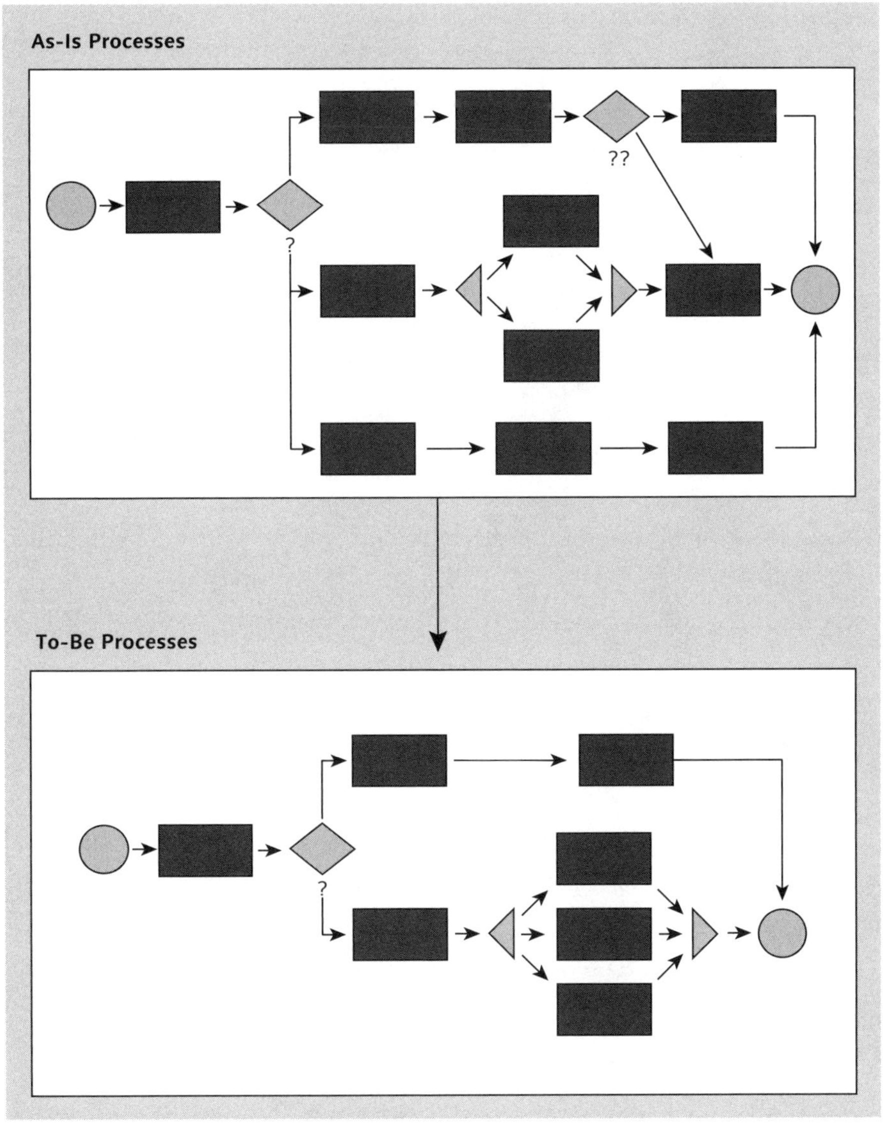

Figure 2.5 As-Is and To-Be Processes

To avoid starting from scratch, consider the following list of questions that should be discussed and answered as part of such a feasibility study.

Key Questions in Relation to Asset Management

1. What types of objects are to be managed?
    - Buildings
    - Assets
    - Machines
    - Fleet objects
    - Industrial trucks
    - Operating resources
    - Tools
    - Test equipment
    - Other
2. Is there a distinction between fixed assets and inventories with changing installation locations?
3. How large is the quantity structure?

Detailed Questions in Relation to Asset Management

1. Which information is to be defined for which objects?
2. Are inventories to be kept in stock?
3. Are several labels to be displayed in parallel for the assets?
4. Are the inventories and assets to be classified (e.g., according to e-class or a separate classification procedure)?
5. Are bills of materials (BOMs) to be managed for spare parts?
    - Are they single-level or multilevel BOMs?
    - Do they concern a single aggregate or category?
    - Are material numbers assigned to spare parts in the SAP system?
6. Are changes to the asset structure to be documented?
7. Are documents (e.g., images, drawings, and process instructions) to be assigned to the assets and/or inventories?
    - Are these documents stored electronically?
    - If so, where are they stored and how?

- If so, are there master records for the documents?

8. Are warranties to be managed?
   - Are they to be assigned to the assets and/or inventories?
   - Are they time-based or performance-based warranties?
9. Are the assets and/or inventories mapped as asset master records in the SAP system?

Key Questions in Relation to Plant Maintenance Processes

1. Which plant maintenance processes are to be mapped in the SAP system in future?
   - Breakdown resolution
   - Repair
   - Maintenance and inspection
     - Time-based
     - Performance-based
     - Time-based and performance-based
     - Condition-based
   - Calibration processing or inspection processing
   - Use of external companies
   - Investment measures
   - Construction of new assets
   - Manufacture of spare parts
   - Refurbishment
   - Shift log
   - Tool and fixture construction
2. How are the core processes currently handled?
3. Who is involved in the planning and execution, and at what point and to what extent do they become involved?
4. What do you consider to be the weaknesses and functional deficits in the current plant maintenance processes? What is to be or can be improved?

Detailed Questions in Relation to Plant Maintenance Processes

1. Which workshops need to be mapped (electronics, mechanical engineering, measurement and control technology, etc.)?
2. Are there such things as catalogs for damage codes or cause codes?
3. Which workshop documents are required?
4. Is there a need to perform capacity planning (available capacity, capacity load, and capacity leveling)?
5. Is it necessary to perform resource planning (i.e., which employees are available to work when, for how long, and on which order)?
6. Are budgets to be assigned to tasks?
7. Are technical or business approval procedures to be mapped?
8. What types of completion confirmations are to be mapped?
   - Time confirmations
   - Technical completion confirmations
   - Material confirmations
   - Shift notes
9. How are the activities allocated to the asset cost centers?
10. Are task lists to be managed?
    - Do they concern a single aggregate or category?
    - What information do task lists contain?
    - Are spare parts also to be defined there?
11. How large is the quantity structure for task lists?
12. If you want time-based maintenance and inspection:
    - Are the tasks subject to a cycle?
    - Are there different cycles within a task list? If so, what are they?
13. If you want performance-based maintenance and inspection:
    - Which types of counters (OH, mls, unit, etc.) are available?
    - How are the counter readings recorded?
      - In the SAP system
      - Mobile

    - By data transfer
- Are the tasks subject to a cycle?
- Are there different cycles within a task list?
    - If so, what are they?
- Can you assign several counters to one maintenance plan?

14. If you want to perform condition-based maintenance:
    - Which types of measuring points (temperature, pressure, rotational speed, flow rate, etc.) are available?
    - How are the counter readings recorded?
        - In the SAP system
        - Mobile
        - By data transfer
15. How large is the quantity structure for maintenance plans?

Key Questions in Relation to Purchase and Warehouse Processes

1. Which materials management processes are to be mapped in the SAP system in the future?
    - Procurement of direct materials (free-text purchase orders)
    - Procurement of spare parts in stock
    - Use of external companies
        - Individual commissioning
        - Outline agreements
        - Service specifications
        - Work centers in external companies
2. How are these processes currently handled?
3. Who is involved in the planning and execution, and at what point, and to what extent, do they become involved?
4. What do you consider to be the weaknesses and functional deficits in the current materials management processes? What is to be or can be improved?

Detailed Questions in Relation to Purchase and Warehouse Processes

1. Are the vendors and service providers mapped as vendor master records in the SAP system?
2. Are outline agreements with vendors and service providers to be mapped?
   - If so, how is the contract release order to be triggered?
3. Are there companies for which separate work centers are to be managed?
   - If so, what is the commissioning and settlement procedure for the work centers?
4. Are the services to be created as service master data or when commissioning service specifications?
5. Are spare parts to be mapped as material master data in the SAP system?
6. Are spare parts to be planned in an order?
   - If so, will spare parts catalogs be available on the intranet?
   - Are automatic reservations to be triggered?
   - Are spare parts to be subject to inventory management?
     - If so, what is the withdrawal process?
     - If so, are material requirements to be planned for spare parts?
     - If so, is reorder point planning to be used? Or something else?
   - Are spare parts to be procured for direct consumption only?
7. Are there vendor consignment stores?
   - If so, how are these managed?
   - Who is responsible for replenishment?
   - How are withdrawals settled?
8. Are there approval procedures for the procurement processes?
9. How large is the quantity structure for spare parts, vendors, and outline agreements?

Key Questions in Relation to Reporting and Key Figures

1. Which lists and reports are currently in use, who uses them, and when are they made available?

    - Which of these lists and reports must be available as soon as the SAP system goes live?
    - Which lists could be made available later?
2. Which key figures are currently in use, who uses them, and when are they made available?
    - Which of these key figures must be available as soon as the SAP system goes live?
    - Which key figures could be made available later?
3. What do you consider to be the weaknesses and functional deficits in the current lists, reports, and key figures? What must be or should be improved?

Information Acquisition

One question arises in this context: Who do you approach to get as full and correct a picture as possible of existing processes? From experience of carrying out as-is analyses, the following are reliable information sources:

- People who are involved in executing, controlling, and monitoring the business processes
- Users of identical or similar IT systems
- Customers who are communicative and often creative knowledge-holders
- Business partners who work with the process (e.g., vendors)
- Subject-matter experts (SMEs), who should be asked to give critical feedback
- Company management, which ultimately has to approve the processes

Another question arises here: What techniques should you use to acquire the information needed for a full and correct as-is analysis? The following techniques have proven themselves in practice to be most effective:

- Establishing user workshops
- Observing the relevant employees at work
- Working with the business processes under analysis
- Taking on the role of an outsider (e.g., the production director acts as a customer)
- Using questionnaires

- Holding interviews
- Brainstorming with the parties involved
- Discussion with SMEs
- Looking through existing forms, documentation, descriptions, manuals, and other resources
- Verbally describing the organizational structure and process structure (e.g., organigrams)

Of course, your company may also have other ways of conducting information research.

[+]

Results of the Feasibility Study

The feasibility study should deliver the following results:

- You know which master data, plant maintenance processes, materials management processes, and reports are to be mapped in the SAP system.
- You know which plant maintenance functions are to be implemented in the SAP system and have assigned them to three potential expansion phases (A/B/C, see Appendix A.1).
- You now also have a basis for further planning of project details such as deadlines, effort, staff, and costs.
- You know which user workshops are to be established in the business blueprint.
- You have a basis for using SAP software to draft to-be processes.

Business Process Modeling Methods

Now let's think about the external form in which the business processes will be documented and presented; that is, what business process modeling methods you'll use.

In principle, because of their superior visualization capabilities, graphical and tabular formats are better than a purely textual format.

Value chain diagrams (VCD, see Figure 2.6) are suitable for depicting the flow of complex processes, such as plant maintenance projects, on a general level. VCDs represent predecessor-successor relationships (including multilevel ones), such as those in work packaging and sequencing.

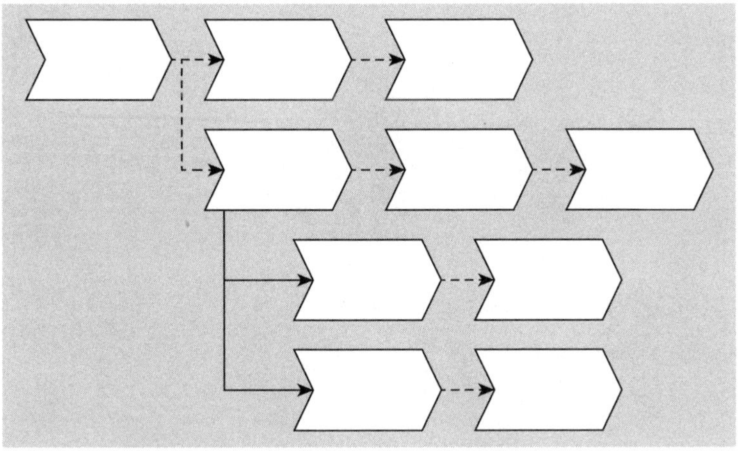

Figure 2.6 Value Chain Diagram

Event-driven process chains (EDPC) are suitable for processes of low and medium complexity. Simple and advanced versions of these are available.

The simple version of an EDPC represents a process flow on a more detailed level than a VCD and consists of two object categories:

- Functions (e.g., print the order, check material availability)
- Events (e.g., malfunction report received, order confirmed)

Advanced EDPCs (aEDPC) are used when additional object categories have to be modeled:

- Documents (e.g., a job ticket, confirmation slip, object list)
- Organizational units (e.g., work scheduling, mechanical engineering workshop)
- Process interfaces (e.g., purchase order handling, final costing)
- Information systems (e.g., SAP ERP Transaction IW31)
- Files or databases (e.g., AFRU [order confirmations] and BANF [purchase requisitions])

It's up to you whether you use object categories in your business process model, and if so, which ones. Figure 2.7 shows a sample extract from such a model.

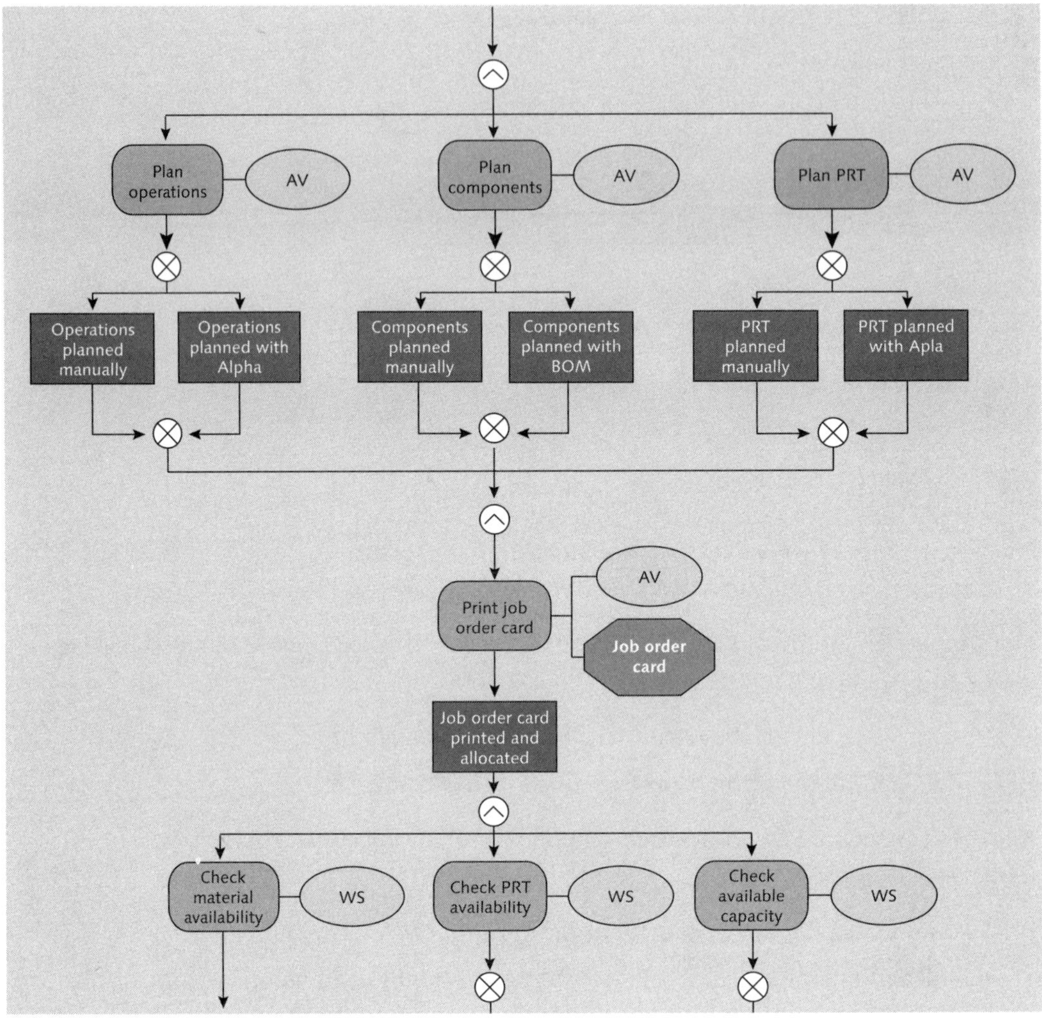

Figure 2.7 Advanced Event-Driven Process Chain

Transaction/event chain diagrams (TECD) are suitable for simple processes but not for more complex processes. A central characteristic of TECDs is the fact that they assign object categories to fixed columns. TECDs are available in a purely graphical format or in a verbal tabular format. In its graphical format, they have the same object categories as an aEDPC (see Figure 2.8).

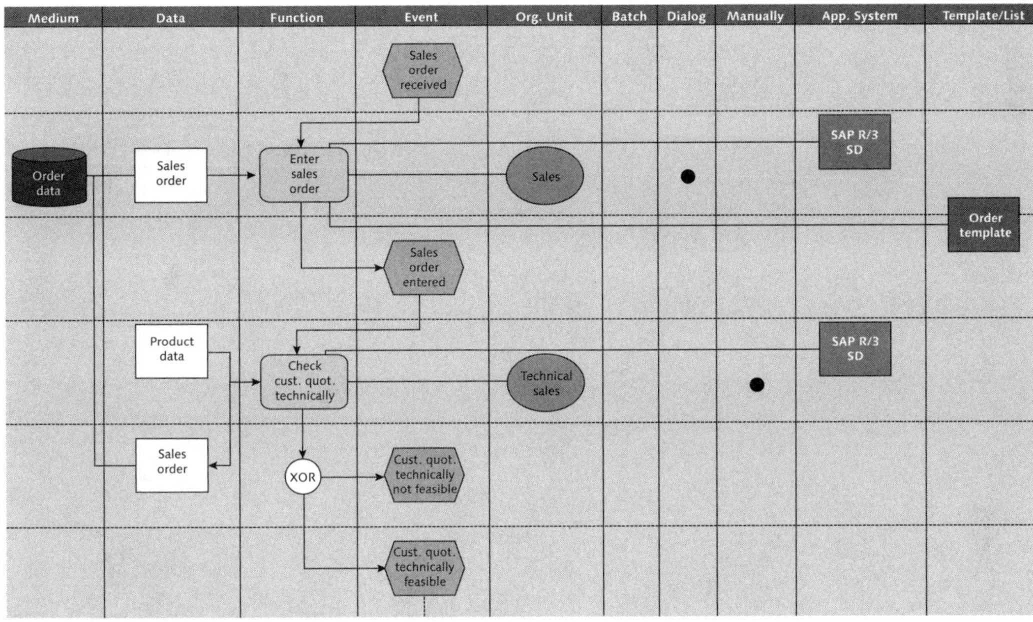

Figure 2.8 Graphical Transaction/Event Chain Diagram

Table 2.2 shows a TECD in a verbal tabular format.[3]

No.	Event/Result	Function/ Activity	Input (Data)	Output (Data)	Parties Involved/ Org.	System Function
1.	Maintenance task completed	Confirm single entry or, in the operation list, continue with item 3.	Order number, personnel number, time	Confirmation data record	Staff WI	SAP: IW41
2.	Confirmation complete	Continue with item 4.		Confirmation data record		

Table 2.2 Tabular Transaction/Event Chain Diagram

3 See also M. Kempchen, "Processing of Maintenance Tasks: Evaluation Report." In SAP R/3 PM (Plant Maintenance), Munich 2007.

No.	Event/Result	Function/ Activity	Input (Data)	Output (Data)	Parties Involved/ Org.	System Function
3.	Operation list created	Select PM operations to be confirmed and branches in single record entry.	Selection data, operation list, personnel number, time	Confirmation list	Staff WI	SAP: IW48
4.	Confirmations complete	Check confirmation data.	Selection data, confirmation list	Modified data records	Staff WI	SAP: IW47
5.	Difference detected	Make correction (cancel or follow-up entry).	PM order number, correction data		Staff WI	SAP: IW48 or IW41

Table 2.2 Tabular Transaction/Event Chain Diagram (Cont.)

Business Process Model and Notation (BPMN) has become very popular in IT projects. In terms of its outward appearance, BPMN looks completely different from an aEDPC (see Figure 2.9), but it uses a similar set of symbols:

- An *activity* describes a task that must be completed in a business process (e.g., record the demand for spare parts).
- A *gateway* represents a decision point (split/fork, e.g., in stock or not in stock) or a point at which different control flows converge (join/merge, e.g., courier booked and material packaged).
- An *event* is something that occurs in a business process (e.g., a message is received, a certain date is reached, such as three days after sending the purchase order, or an exception situation arises).
- *Sequence flows* connect activities, gateways, and events. They represent the sequence in which activities are executed.
- A *pool* represents a participant, that is, a user, user role, or system (e.g., a buyer).

- A *lane* is a subdivision of a pool that spans the complete length of the pool (e.g., the process line of the buyer).
- A *message flow* shows that two lanes or pools in a business process diagram or two elements of these exchange messages (e.g., transfer of requirement from the process flow of the maintenance planner to the process flow of the buyer).
- An *annotation* is a comment that can be assigned to an element of a business process.
- A *data object* represents an artifact that is processed by the business process. Data objects can be used to represent electronic objects such as documents or data records (e.g., the EAM order) as well as physical objects such as spare parts or documents (e.g., the pull list).

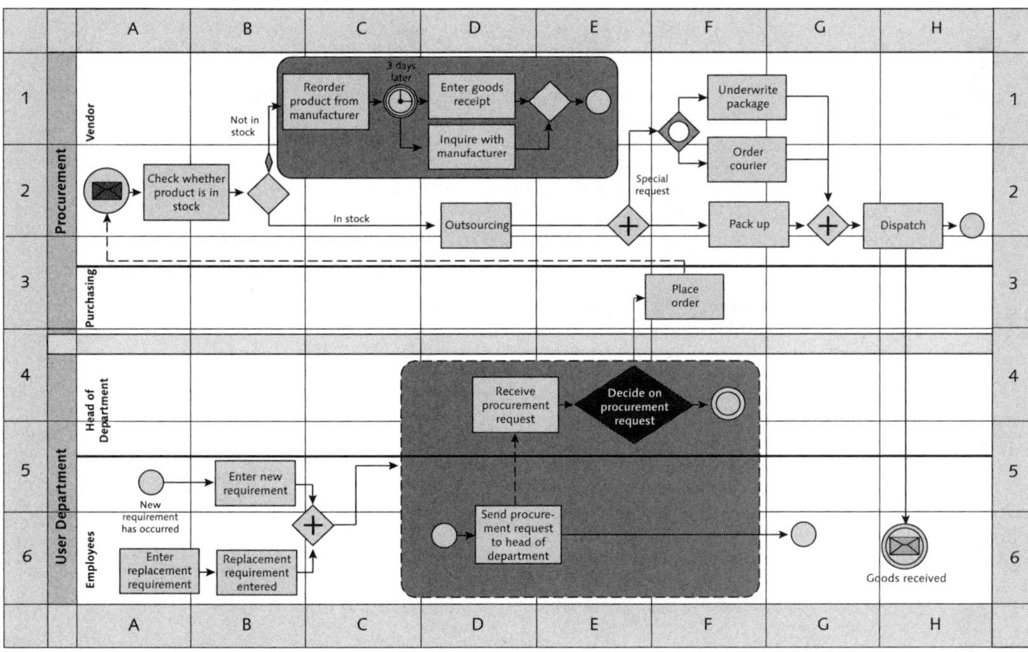

Figure 2.9 Business Process Model and Notation

[+]

Visualize the Business Processes

Create business process models of the as-is and to-be processes. Use different model types to do this:

- VCD for an overview of complex business processes
- EDPC or BPMN for a detailed representation of complex processes or to represent less complex processes
- TECD in a tabular or graphical format for simple processes

System Landscape

Another error that many companies make in this phase is to make the SAP system landscape available too late.

[+]

Set Up a Test System Early On

At the very latest, a test SAP system must be available after your staff has finished their SAP training. If they don't have access to a test system at this stage, they won't be able to practice what they've learned and will quickly forget most of it. In other words, plan to set up a test system early on.

2.3.2 Phase 2: Business Blueprint

A large number of tips were provided in the book *Plant Maintenance with SAP—Practical Guide* that you can apply during the Business Blueprint phase, in particular. The following is a brief list of some common questions:

- **Organizational units**
  What needs to be done in cross-cost accounting maintenance processing?
- **Structuring of technical systems**
  To what extent should you perform structuring of technical systems? What elements are to be used?
- **Business processes**
  How are external companies mapped? How can the layout of orders be determined?
- **Integration**
  How can equipment and assets be aligned with each other? Which material requirements planning (MRP) procedure is best for spare parts?

- **Maintenance controlling**
  How can active availability control be set up in budgeting? How are dynamic date calculations performed?
- **New SAP technology**
  What is the difference between SAP NetWeaver Portal and SAP NetWeaver Business Client (NWBC)? Which follow-up entry options are available to you?
- **Usability**
  How do you configure maintenance-specific user parameters? Or how can you adjust table controls to the relevant user requirements?

In the Business Blueprint phase, all of these questions are reflected in your detailed concept. More tips are provided next that you should observe in the Business Blueprint phase, especially in relation to organizational issues.

The most important thing is perhaps for the employees to support the project and, in particular, the future system. You should therefore secure their support as early as possible.

[+]

Involve Your Intended Users in the Design Phase

Only a system that is accepted by its users is a genuinely useful one. Therefore, it's important that you involve your intended users in the design phase of the project.

Here are some suggestions for involving your users in the design phase:

- Start a *reference group*, that is, a group of selected users who will provide technical input and feedback to the project group.
- *Present your interim results* at regular intervals (live presentation, newsletter, home page).
- Conduct individual surveys and get *feedback* from these.

Usability and User Acceptance

The book *Plant Maintenance with SAP—Practical Guide* devotes an entire chapter to usability and user acceptance, and it discusses various organizational and user-specific measures. Furthermore, Chapter 9 of this book provides various Customizing and programming options. Therefore, this topic is not discussed in detail here, but its importance is underlined by providing some general tips.

[+]

Keep the System Design as Simple as Possible

Irrespective of whether it concerns the structuring of technical systems or business process handling, user acceptance increases directly in line with the simplicity of the system. It's better to have 80% of the system you want with 100% user acceptance than 100% of the system you want with 20% user acceptance.

The SAP system isn't exactly known for its usability. We'll look at the question of whether this is really true and what you can do about it in your own project.

[+]

Do Everything You Can to Make the System as User Friendly as Possible

Be particularly proactive in terms of addressing the topic of the *unusability* of the SAP system. During the implementation phase, be attentive to the concerns of your employees regarding the usability of the SAP system. Don't dismiss the topic; rather, accept users' fears and try to alleviate them.

The following is another tip on how you can increase user acceptance right from the start.

[+]

Low-Hanging Fruits Are Easy to Reach and Entice Employees

Use the as-is analysis to identify aspects with which the users have problems or aspects that are perceived to be suboptimal. Offer them a solution to these specific aspects that represents an improvement on the current situation. Your users will then be more prepared to support the new system.

Because these aspects are very individual, here are some examples of what other companies have done:

- Created a list of information that can only be obtained following extensive research
- Created messages in the production system about today's document movements
- Automatically calculated maintenance dates instead of searching through folders
- Printed the pull list directly in the warehouse for advance picking, instead of having the warehouse worker wait a long time for the good issue
- Sent breakdown notifications via a paging system rather than having the worker walk back to the support desk to pick up messages

- Automatically issued maintenance lists to production by email rather than having the worker unable to gain access to the facility and therefore unable to carry out his maintenance tasks, simply because production was completely unaware of the maintenance, and therefore the facility was still up and running

If you're observant, you'll notice issues like these in your own company. And if you're a little creative, you'll be able to provide your users with improvements.

Now let's deal with an aspect that is sometimes a little neglected at the start of a project.

Authorization Concept

The authorization concept controls which user in which organizational unit may execute which functions on which objects. This is a general topic, not a maintenance-specific one. As such, responsibility for the authorization concept lies with the IT department, the coordination department, or similar. However, you should ensure, in good time, that an authorization concept is created to control the use of SAP EAM.

[+]

Authorization Concept: A Project within a Project

Don't forget to design an authorization concept alongside the detailed technical concept.

Section A.2 of Appendix A contains a summary of the authorization objects that are available in SAP EAM and the organizational units, functions, and fields that are checked in each case.

This topic won't be covered in any further detail here; instead, refer to the relevant specialist literature.[4] The most important settings will be explained in detail here, for example:

- How to define authorizations on the basis of authorization objects
- How to combine several authorizations to form an authorization profile
- How to create single roles using Transaction PFCG (see Figure 2.10) and combine them to form composite roles
- How to define users with the requisite authorizations

4 Example: Volker Lehnert and Katharina Stelzner, *Authorizations in SAP Software: Design and Configuration.* (SAP PRESS, 2010).

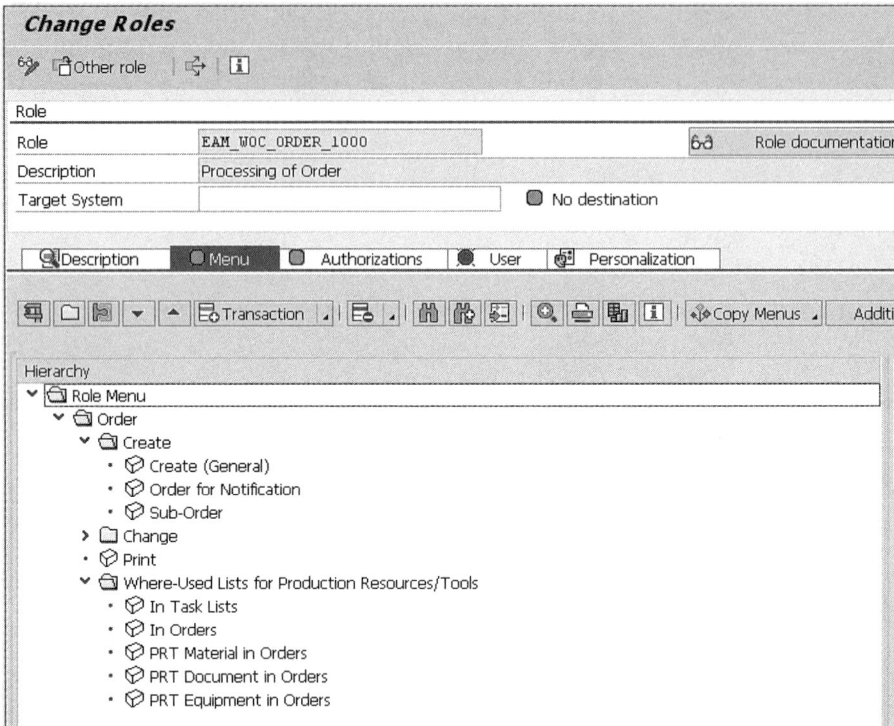

Figure 2.10 Using Transaction PFCG to Maintain Single Roles

Legacy Data Transfer and Master Data Maintenance

There are always two ways to transfer your master data to the SAP system: manually or automatically. If you're changing from another maintenance planning and control system to the SAP system, or if the asset data is in another electronic format (e.g., in a CAD system), you should always try to transfer this data to the SAP system automatically.

[+] **Use SAP Tools**

SAP provides the following two standard tools for transferring data from upstream or legacy systems, namely a maintenance-specific data transfer (Transaction IBIP) or the general data transfer workbench (Transaction LSMW). Both tools enable you to transfer the maintenance objects and their data to the SAP system.

Plant maintenance batch input (IBIP), as shown in Figure 2.11, is a data transfer program that is tailored to the special database objects in plant maintenance.

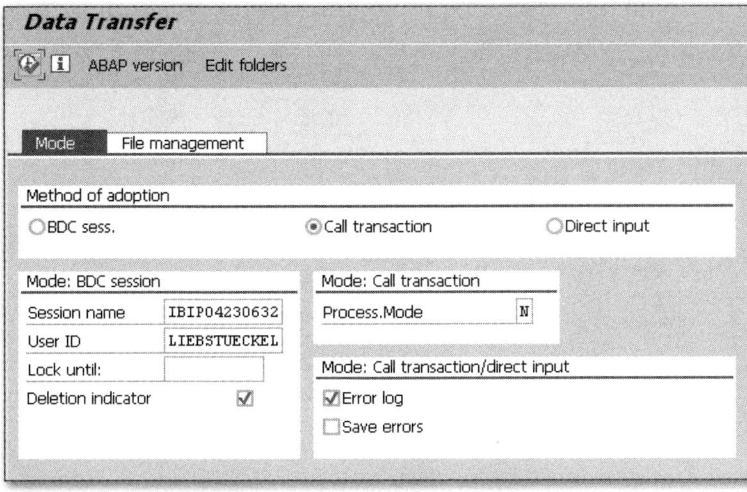

Figure 2.11 SAP Plant Maintenance Batch Input: IBIP

In contrast, the Legacy System Migration Workbench (LSMW), shown in Figure 2.12, is a general data transfer workbench that can transfer not only maintenance objects but also objects from many SAP applications. The main difference between the systems is as follows:

- **IBIP**
  IBIP requires a source file whose field sequence has to correspond to a predefined structure. You can find out the names of these structures (e.g., IBIPEQUI for the equipment master record) in the IBIP documentation, and you can find out the field structure in the Data Dictionary (Transaction SE11). No further prerequisites apply if the structure is adhered to.
- **LSMW**
  LSMW doesn't require any fixed structure. Instead, it uses field mapping, which assigns the fields of the source structure to the fields of the SAP object. The source structure is described on a per-project basis in the SAP system.

This makes LSMW more flexible but also more complex than IBIP.

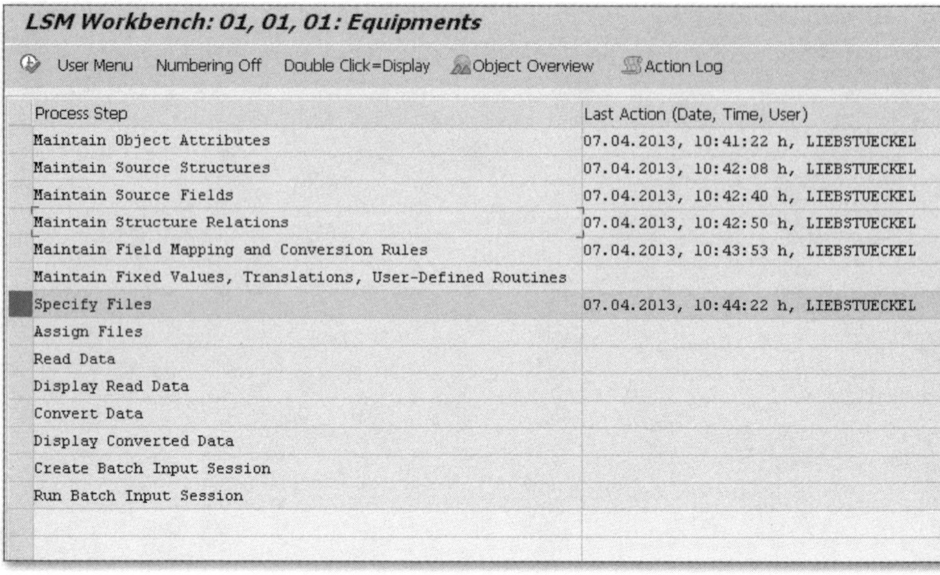

Figure 2.12 Legacy System Migration Workbench

If you're setting up IT support for your plant maintenance system from scratch (i.e., you don't have any master data in electronic form), you'll have to enter the data manually.

[+]

Office Tools for Data Entry

Besides the standard approach of creating data directly in the SAP system, you can also use the RIACCESS program to export the SAP table structures to Microsoft Access as a local database, enter the data locally on your PC, and then transfer it to the SAP system using IBIP or LSMW.

Regardless of what procedure you choose, you need to plan.

[+]

Plan the Data Entry

Create a detailed data retrieval plan. In this plan, you should specify the following points in particular:

- Which employees are involved in the data transfer
- What kind of templates (index cards, folders, drawings, etc.) are used
- How the data should be formatted and prepared

- The dates by which the data should be entered into the system
- How the quality of the data is to be monitored
- Who is responsible for quality control
- How the master data is to be released
- How long the individual activities last

Customizing for Prototypes

Last but not least, there is an extremely important task in the Business Blueprint phase: building a prototype.

Build a Prototype [+]

During the Business Blueprint phase, you make a large number of decisions that affect Customizing. Don't delay Customizing until the Realization phase. Instead, configure all of the Customizing settings in the Business Blueprint phase and build a prototype in this phase.

The following data belongs to this prototype:

- All controlling Customizing settings for the organizational structures, the structuring of technical systems, the business processes, and integration with other components
- Representative settings in noncontrolling Customizing settings, such as maintenance planner groups, purchasing groups, MRP controllers, locations, plant sections, and similar tables, which don't have to be fully maintained in the prototype
- Representative master data (work centers, functional location structure, equipment, task lists, etc.)

In such a prototype, reproduce the future live system approximately so that your users have an opportunity to see the system in advance.

2.3.3 Phase 3: Realization

In the Realization phase, the plans and concepts that were created in the Business Blueprint phase must be turned into action. To do this, you need to complete the following tasks:

- Finalize the Customizing settings. You already made some controlling Customizing settings in the business blueprint so that you could build a prototype. However, in the noncontrolling Customizing settings (e.g., maintenance planner groups, purchasing groups, MRP controllers, locations, plant sections, and similar tables), you made representative entries only. You must now complete these settings.
- Develop the required enhancements and interface programs.
- Prepare the legacy data for transfer to the new system and set up the transfer tools, namely LSMW or IBIP.
- Enter the master data directly into the SAP system.
- Develop the end-user training courses and documentation.
- Arrange for the end-user training courses to be held.
- Set up the users, roles, and authorizations.

These are defined tasks that don't require any further explanation here. Instead, let's discuss the important topic of testing. Before going live, you should test your implemented concept extensively.

[+]

Conduct Sufficient Testing

Extensive testing in the implementation phase is a prerequisite for a successful go-live. Therefore, schedule testing in time and conduct all tests in full.

There are various types of tests:

- **Functional tests**
  This is the most basic level of testing. It examines programs and transactions for errors and evaluates them. These tests focus on the program's internal functions. The main aim is to become familiar with the system in terms of one's own requirements.
- **Scenario tests**
  During the Customizing process, it's necessary to test transaction chains that are dependent on one another and reflect central business processes and scenarios. These tests focus on complete business processes.
- **Integration tests**
  A final test examines integration aspects by using predefined business processes or scenarios that simulate the processes in your system. These business processes

already use the migrated data and are executed in the actual IT infrastructure, which comprises the SAP system, third-party software, system interfaces, and various hardware and software components.

2.3.4 Phase 4: Final Preparation

The Final Preparation phase comprises all activities that are required shortly before and in preparation for the system go-live. This includes transferring legacy data to the production system. However, because the data transfer process rarely runs smoothly and in full, you'll need some time to correct and complete the data. How much time you'll need for this depends on the volume of changes required.

Allow One Day to One Week for Data Maintenance [+]

Schedule sufficient time (one day or several days) to correct errors and complete the master data.

Another activity that you need to complete in this phase is end-user training. Although SAP training courses are recommended for project team members, this doesn't apply to end users because SAP training is generally unsuitable for end users.

Conduct In-House End-User Training [+]

End-user training should be provided by employees in your company, ideally by the representatives of the user department in the project team. For this training, use the system environment as well as examples of the technical objects that the end users will encounter in their work later on. It's also important to provide end-user training close to the go-live date. Otherwise, the users will have forgotten much of what they learned by the time the system goes live.

Cut-Over Plan

At the very least, you should have created a complete cut-over plan by the time you conduct end-user training.

Develop a Cut-Over Plan [+]

A cut-over plan contains all the activities that are to be carried out either directly before, during, or shortly after the go-live. The cut-over plan thus serves the function of a "string around your finger," that is, a reminder to ensure that you don't forget anything.

This plan often contains what appear to be minor issues. However, it's precisely the minor things that are easily forgotten in something big like a system go-live. This plan can contain the following items:

- Any master data that still has to be created manually because it isn't worthwhile to transfer it manually
- The settings that have to be made to the system because they can't be transported
- Any selection and display variants that need to be set
- What batch jobs need to be triggered
- What number ranges have to be either created manually or whose transport has to be triggered manually

2.3.5 Phase 5: Go-Live and Support

Your new system then goes live on the planned key date and you start to process the first notifications and orders there. The most important thing in this phase is to ensure that your employees support the systems.

[+]

Arrange for Project Team Members to Attend the Workshop

Organize end-user support for the first few days after the go-live. Ideally, this should be on-site user support by members of the project team. Furthermore, publicize a central hotline number and an email address so that problems can be resolved quickly.

Provided that the tests were carried out in a targeted and intensive manner, you should not experience any design difficulties or system problems in this phase. Experience shows that the issues that arise in this phase tend to relate to working with the software or concern detailed questions, and even these issues taper off after the initial familiarization phase.

2.3.6 Phase 6: Operate

The final phase (*Operate*) mainly involves improving the operations of the SAP system in plant maintenance and identifying which measures in the project phase offer potential for optimization in future projects. These include the following:

- Optimized documentation
- Optimized training

- Optimized implementation
- Optimized testing
- Optimized technical operations
- Optimized business processes
- Optimized system upgrades
- Optimized system maintenance

Therefore, this chapter concludes the project-related questions and tips. In the next chapter, we'll turn our attention to the first set of system-related questions, namely the organizational structures that you need to set up.

This chapter introduces you to the indispensable elements of maintenance processing within the SAP system, namely the general organizational units, the maintenance-specific organizational units, and the work center.

3 Organizational Structures

SAP organizational units define your enterprise's organizational structure in general, and your plant maintenance in particular. Consequently, they form the basis for all master data and business processes.

SAP provides a range of organizational units that you can use to map your enterprise structure:

- Some are also used in other applications (e.g., the company code, controlling area, or plant).
- Some are used in plant maintenance whereby plant maintenance has its own perspective (e.g., the plant as a maintenance planning plant and as a maintenance plant).
- Finally, some organizational units concern plant maintenance only (e.g., the plant section or the maintenance planner group).

Organizational Units as Authorization Objects [+]

Each organizational unit has at least one authorization object. Generally, however, it has more than one. Consequently, organizational units are perfectly suited to assigning or restricting authorizations. For example, you can use the maintenance planner group to specifically assign authorization for processing maintenance orders.

Even though it's possible to cautiously enhance the enterprise structure during live operation, deleting organizational units already in use in live systems is a very delicate matter. Changing the organizational structure involves the following tasks, among others:

- Identifying and adjusting affected Customizing tables
- Identifying and adjusting affected selection variants
- Checking custom developments because they may have been developed on the basis of certain organizational structures
- Creating new master data
- Checking the authorization concept
- "Archiving data so that the document history isn't lost"[1]
- Checking and correcting all potentially affected documents (e.g., you can no longer continue to process an order if the planner group has been deleted)
- Checking and correcting all potentially affected master data (e.g., you can no longer continue to process a piece of equipment if the work center has been deleted)

From a plant maintenance perspective, defining organizational structures also concerns the following areas:

- The general SAP organizational units (e.g., the controlling area, company code, plant, and storage location)
- A definition of the maintenance-specific organizational units (e.g., location or plant section)
- Customizing for the work centers in plant maintenance (e.g., mechanical engineering, electronics, and measurement and control technology)

3.1 General SAP Organizational Units

Organizational units are the basis of all master data and business processes in SAP ERP. In the following sections, you'll get to know the most important organizational units from the perspective of plant maintenance and learn how you should use the associated Customizing functions. Figure 3.1 shows some typical general organizational units.

---

1 Martin Murray, *Materials Management with SAP ERP: Functionality and Technical Configuration* (3rd edition, SAP PRESS, 2011).

[+]

General Organizational Units Are Usually Predefined

If you implement SAP Enterprise Asset Management (SAP EAM), the general organizational units in the SAP system (e.g., the company code, controlling area, and plant) are usually predefined because you defined them when you implemented other components, such as Controlling (CO) and Materials Management (MM). Therefore, you can only influence the design if SAP EAM is implemented from the outset, or if you define separate organizational units from a pure maintenance perspective.

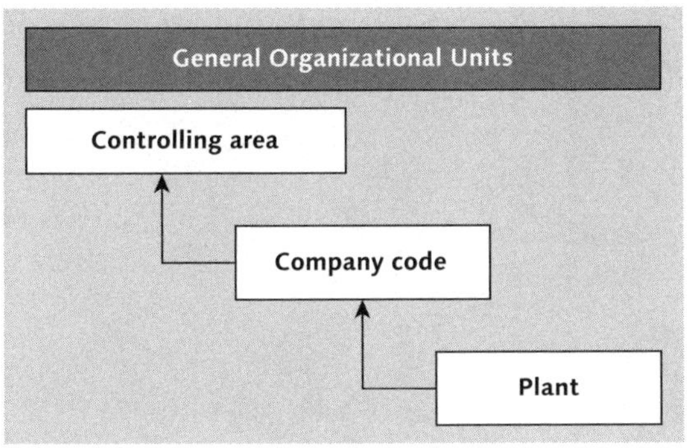

Figure 3.1 General Organizational Units

3.1.1 Company Code

In SAP ERP, the company code maps an independent accounting unit and is therefore the smallest organizational unit in Financial Accounting (FI). You therefore use a company code to map, within your corporate group (client), a company that is required to create a balance sheet, profit and loss account, and other year-end documents.

Edit Company Code

The company code provides the basis for many other applications, including SAP EAM. You use this Customizing function to create new company codes and to make changes to existing company codes.

Customizing Path

ENTERPRISE STRUCTURE • DEFINITION • FINANCIAL ACCOUNTING • EDIT, COPY, DELETE, AND CHECK COMPANY CODE

Prerequisites
You must maintain the countries, languages, and currencies beforehand.

Settings
Choose the subfunction EDIT COMPANY CODE DATA. The system then displays a table of defined company codes (see Figure 3.2).

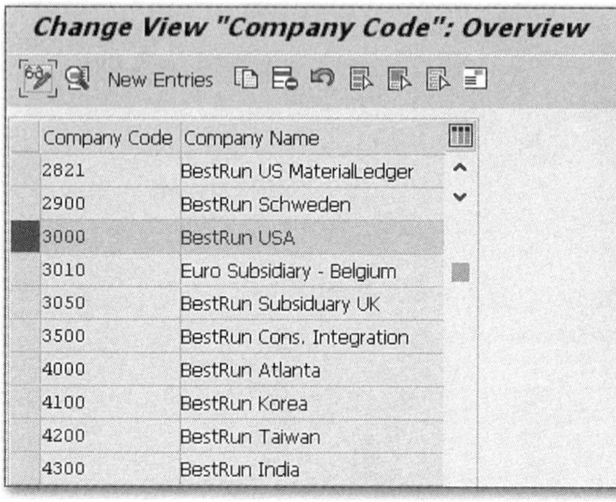

Figure 3.2 Company Codes: Overview

On the detail screen, you maintain key parameters such as COUNTRY or CURRENCY (see Figure 3.3).

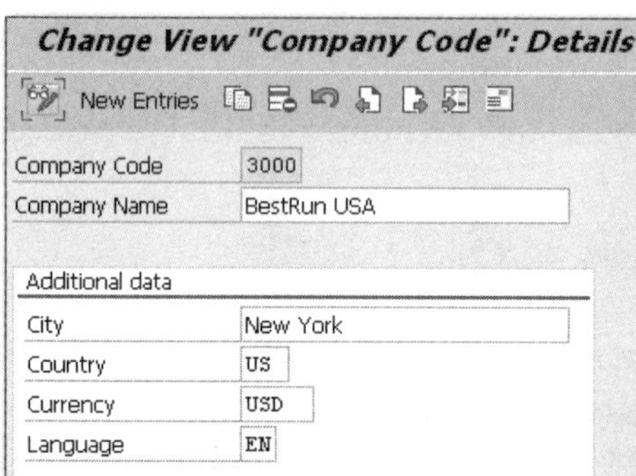

Figure 3.3 Company Code: Basic Data

You can then use the icon to specify the address data for the company code.

Global Parameters for Company Code

You can use this Customizing function to make key specifications in relation to the company code. These include, among other things, the chart of accounts and the fiscal year variant.

Customizing Path

Financial Accounting (New) • Financial Accounting Global Settings (New) • Global Parameters for Company Code • Enter Global Parameters

Prerequisites

You must define these specifications in advance, either as master data (e.g. Chart of Accts) or in other Customizing functions (e.g., Company, Credit control area, and Fiscal Year Variant).

Settings

From a plant maintenance perspective, the following fields are most important: Chart of Accts and Fiscal Year Variant (see Figure 3.4).

These fields represent the following:

- The *chart of accounts* is a directory of all general ledger (GL) accounts. It provides the basis for all business transactions in SAP ERP. For each business transaction, postings are made in the background to one or more GL accounts. For example, in the case of a material withdrawal to a maintenance order, at least one material stock account (e.g., Stock of Spare Parts) and a consumption account (e.g., Spare Parts Used) are used. "A chart of accounts must be assigned to each company code. The assigned chart of accounts is known as an operational chart of account; that is, the day-to-day postings are made to the accounts of the operational chart of accounts. The operational chart of accounts is used by Financial Accounting and Controlling."[2]
- The *fiscal year variant* defines the fiscal year and the posting periods. The fiscal year can correspond to the calendar year. However, this doesn't have to be the case; for example, if you don't want to perform year-end closing operations during the public holidays in December. The posting periods usually contain the number of posting periods (generally 12 months) and the number of special

2 Martin Munzel, and Renata Munzel, *SAP Controlling—Customizing* (Bonn: SAP PRESS, 2011, p. 25). This book is in German language only.

periods. A special period is a special posting period that subdivides the last regular posting period for closing operations. A fiscal year can have a maximum of 12 posting periods and 4 special periods.

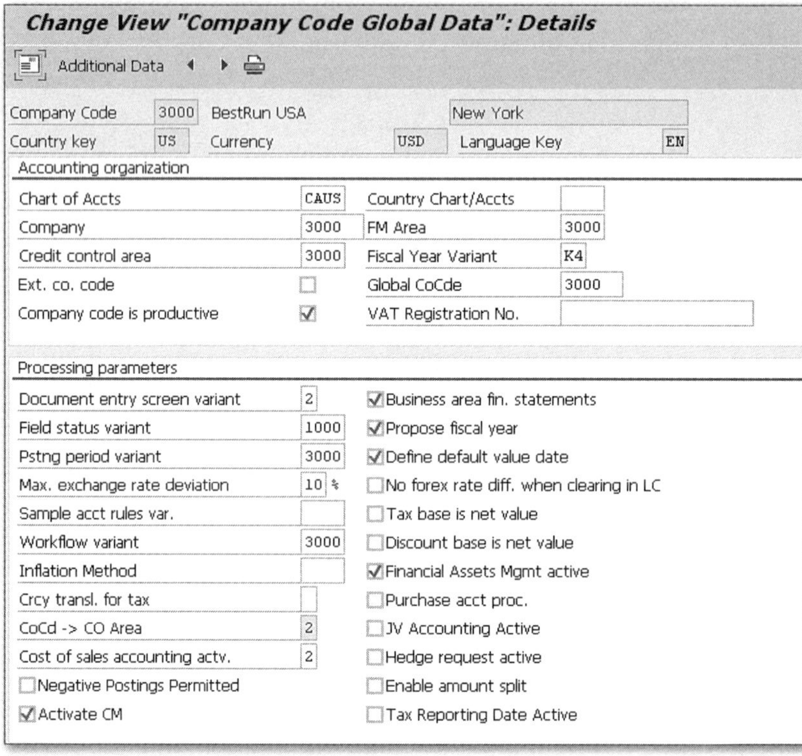

Figure 3.4 Company Code: Global Parameters

3.1.2 Controlling Area

The controlling area is the main organizational unit in CO and is the basis for all other master data (e.g., cost centers or activity types).

Maintain Controlling Area

A controlling area is defined as an organizational unit within a company that is used to represent a closed system for cost accounting purposes. A controlling area may include single or multiple company codes that may use different currencies.

These company codes must use the same operational chart of accounts and the same fiscal year variant.

Customizing Path

Enterprise Structure • Definition • Controlling • Maintain Controlling Area

Transaction

OX06

Prerequisites

You must define the currency, chart of accounts, and fiscal year variant beforehand.

Settings

Choose the subfunction Maintain Controlling Area to check or change the settings for a controlling area (see Figure 3.5).

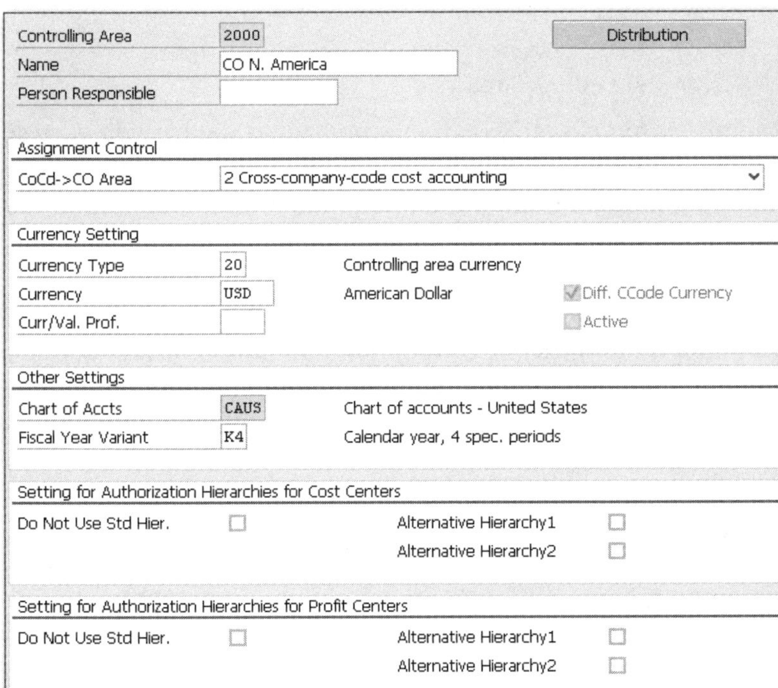

Figure 3.5 Controlling Area

A company code maps a company from an external perspective and is primarily created for tax reasons. In other words, company codes are created to optimize an enterprise's tax burden. Consequently, a company code doesn't necessarily reflect the actual set of services. In contrast, a controlling area maps a company from an internal perspective. Primarily, it's created for the actual business processes and the exchange of services between the various business units.

For example, Mr. Miller, a production planner, and Mr. Lawrence, a maintenance planner, work in the same office and once belonged to the same company (company code) before the decision was made to outsource plant maintenance. Since then, Mr. Lawrence has continued to plan the repair orders and maintenance tasks for Mr. Miller's machine facilities and still works in the same office as Mr. Miller. However, he now belongs to another company (company code). The day-to-day processes have not changed. Mr. Lawrence can continue to plan his orders in the same way as before because, from an internal perspective (controlling area), his company belongs to the same company as Mr. Miller's production, and the CROSS-COMPANY-CODE COST ACCOUNTING option is activated there.

Company Code—Assign Controlling Area

You use this Customizing function to assign the company code that you want to perform common cost center accounting to a controlling area. You only have to explicitly assign company codes to the controlling area if you want to perform cross-company-code cost accounting.

Customizing Path

ENTERPRISE STRUCTURE • ASSIGNMENT • CONTROLLING • ASSIGN COMPANY CODE TO CONTROLLING AREA

Transaction

OX19

Prerequisites

You must define the company code and controlling area beforehand. Also, both must have the same chart of accounts and fiscal year variant.

Settings

As shown in Figure 3.6, you call the affected controlling area (here: 2000) and assign the necessary company codes (here: 3000, among others).

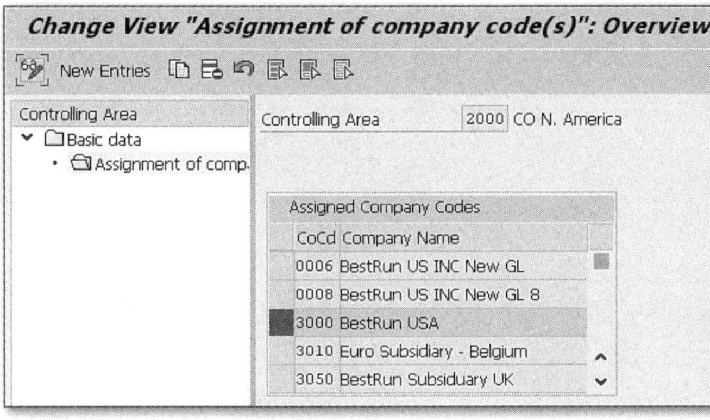

Figure 3.6 Company Code: Assign Controlling Area

Due to poor experiences with other constellations, we unreservedly agree with the following recommendation from Martin Munzel and Renata Munzel.

Creating the Organizational Structure [+]

SAP strongly recommends using only one controlling area for the entire enterprise. In other words, you assign all company codes to this one controlling area. The advantages of this are as follows:

- A company code is permitted to be assigned to a controlling area only if both use the same fiscal year variant and chart of accounts. This is the only way in which you can ensure that all company codes use the same chart of accounts and thus ensure a basic level of standardization.
- You must create master data such as cost elements, cost centers, and profit centers for each controlling area. If you have only one controlling area, this considerably reduces the time and effort needed for master data maintenance.
- You can run cross-company-code analyses in CO. In SAP ERP, it's not possible to display data from different controlling areas in one report.
- Basic controlling functions such as cross-company-code sales, transfer prices, allocations in CO or cross-plant costing, and, of course, cross-plant and cross-company code plant maintenance, are only possible within a controlling area. [3]

3 Martin Munzel, and Renata Martin, *SAP Controlling—Customizing* (Bonn: SAP PRESS, 2011, p. 32). This book is in German Language only.

3.1.3 Plant General

The plant is the central and most important organizational unit in logistics.

Define, Copy, Delete, Check Plant

You require a plant for the following areas, for example:

- In Production Planning (PP) for production order processing and capacity requirements planning
- In Purchasing as a purchasing or supplying plant
- In Shipping as a delivering plant
- In SAP Inventory Management (IM) as an organizational unit in which you manage stocks and values
- In SAP Quality Management (QM) as a unit in which you edit inspection lots
- In Materials Management (MM) for forecasts, material requirements planning (MRP), and product costing
- For the management of all Logistics master data (work centers, task lists, bills of material [BOMs], and materials)

Customizing Path

Enterprise Structure • Definition • Logistics – General • Define, Copy, Delete, Check Plant

Prerequisites

You must maintain the language keys, country keys, regions, and factory calendar beforehand.

Settings

Choose the subfunction Define Plant to access an overview of existing plants (see Figure 3.7).

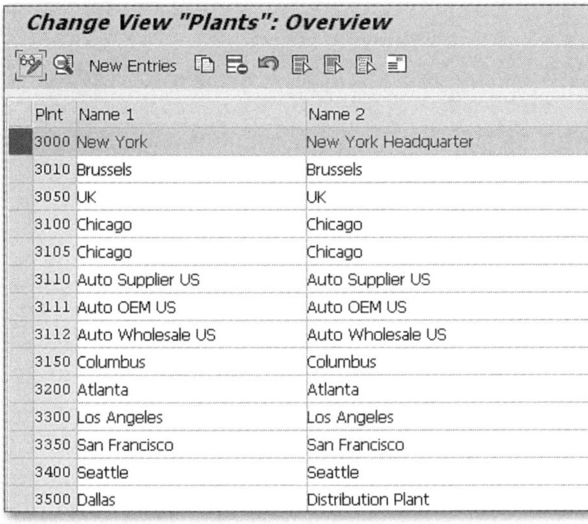

Figure 3.7 Plants: Overview

On the detail screen for plants, you can maintain plant addresses in particular (see Figure 3.8). These addresses are later required as object addresses for technical objects or delivery addresses for spare parts.

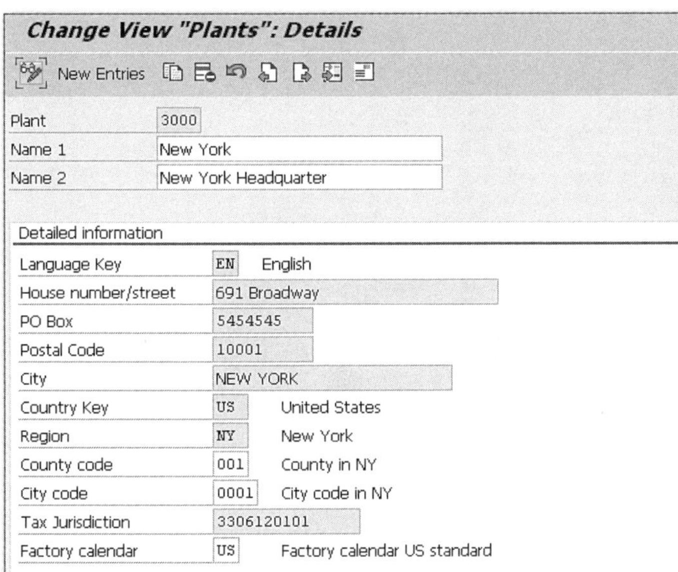

Figure 3.8 Plant: Address

Plant—Assign Company Code

You use the Customizing function PLANT • ASSIGN COMPANY CODE to assign a plant to its company code. At any one time, a plant can only ever be assigned to one company code. Conversely, this means that, over time, you can assign a plant to another company code. This is necessary, for example, if you outsource your maintenance area, and it becomes a separate company.

Customizing Path

ENTERPRISE STRUCTURE • ASSIGNMENT • LOGISTICS – GENERAL • ASSIGN PLANT TO COMPANY CODE

Transaction

OX19

Prerequisites

You must define the plant and company code beforehand.

Settings

As shown in Figure 3.9, you simply define a new entry when you assign the plant (here: 3000) to a company code (here: 3000).

Change View "Assignment Plant - Company Code": Overview

New Entries

Assignment Plant - Company Code

CoCd	Plnt	Name of Plant	Company Name	Status
3000	3000	New York	BestRun USA	
3010	3010	Brussels	Euro Subsidiary - Belgium	
3050	3050	UK	BestRun Subsiduary UK	
3000	3100	Chicago	BestRun USA	
3000	3105	Chicago	BestRun USA	
3000	3110	Auto Supplier US	BestRun USA	
3000	3111	Auto OEM US	BestRun USA	
3000	3112	Auto Wholesale US	BestRun USA	
3000	3150	Columbus	BestRun USA	
3000	3200	Atlanta	BestRun USA	
3000	3300	Los Angeles	BestRun USA	
3000	3350	San Francisco	BestRun USA	
3000	3400	Seattle	BestRun USA	
3000	3500	Dallas	BestRun USA	

Figure 3.9 Assignment Plant: Company Code

3.2 The Plant from a Maintenance Perspective

The plant is, without doubt, the most important organizational unit for plant maintenance. It fulfills several maintenance functions as both a maintenance planning plant and as a maintenance plant.

An overview of the various organizational units in SAP EAM is provided in Figure 3.10.

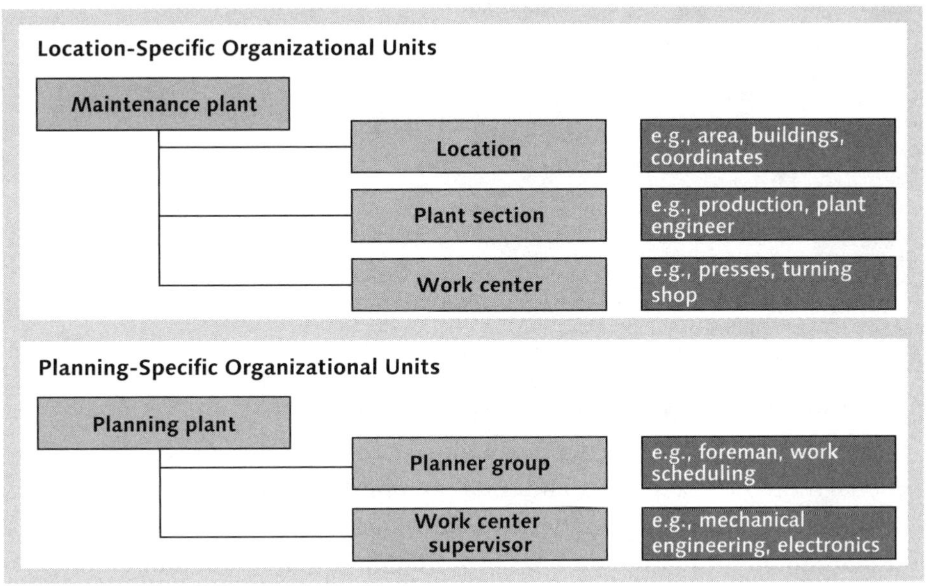

Figure 3.10 Organizational Units in SAP EAM

Maintain Maintenance Planning Plant

A plant is responsible for planning maintenance activities. This type of plant is called a *maintenance planning plant* (or *planning plant* for short). You use the Customizing function MAINTAIN MAINTENANCE PLANNING PLANT to convert a plant to a planning plant.

Customizing Path

ENTERPRISE STRUCTURE • DEFINITION • PLANT MAINTENANCE • MAINTAIN MAINTENANCE PLANNING PLANT

Prerequisites

You must define the maintenance planning plant as a normal logistics plant beforehand (see Section 3.1.3).

Settings

You obtain an overview of existing maintenance planning plants (see Figure 3.11), and you can add new plants to this overview.

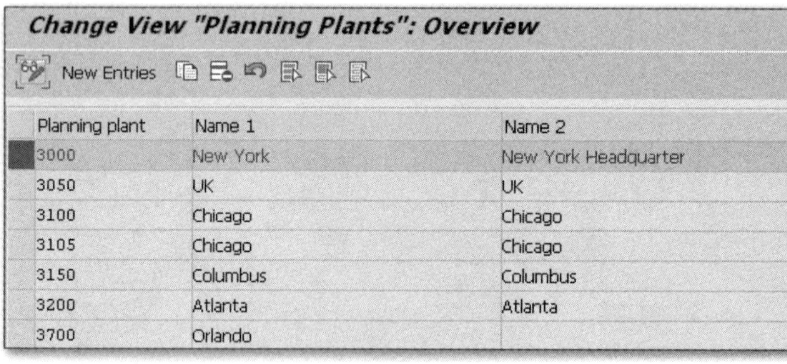

Change View "Planning Plants": Overview

New Entries

Planning plant	Name 1	Name 2
3000	New York	New York Headquarter
3050	UK	UK
3100	Chicago	Chicago
3105	Chicago	Chicago
3150	Columbus	Columbus
3200	Atlanta	Atlanta
3700	Orlando	

Figure 3.11 Maintenance Planning Plant

Maintenance Plant—Assign Maintenance Planning Plant

All of the technical objects to be maintained are physically present in a plant (functional location, equipment, and serial number). Here, this plant is known as a *maintenance plant*. A plant becomes a maintenance plant if you create a technical object there. Maintenance plants are not stored in a separate Customizing table.

Every plant in which maintenance processes are to be planned or executed must have an assignment from the maintenance plant to the planning plant. You use the Customizing function ASSIGN MAINTENANCE PLANNING PLANT to perform this assignment (see Figure 3.12).

Customizing Path

ENTERPRISE STRUCTURE • ASSIGNMENT • PLANT MAINTENANCE • ASSIGN MAINTENANCE PLANNING PLANT TO MAINTENANCE PLANT

Change View "Allocation of PlanPlants to MaintPlants":

Plnt	Name 1	PlPl	Name 1
3000	New York	3000	New York
3010	Brussels		
3050	UK	3200	Atlanta
3100	Chicago		
3105	Chicago	3100	Chicago
3110	Auto Supplier US		
3111	Auto OEM US	3000	New York
3112	Auto Wholesale US	3000	New York
3150	Columbus	3000	New York
3200	Atlanta	3200	Atlanta
3300	Los Angeles		
3350	San Francisco		
3400	Seattle		
3500	Dallas		

Figure 3.12 Assignment of Maintenance Plant: Planning Plant

Prerequisites

You must define both plants (maintenance plant and planning plant) as normal logistics plants beforehand (see Section 3.1.3). If a plant isn't assigned to a planning plant (e.g., plant 1100), you can't create any technical objects there (maintenance plant), nor can you plan or execute maintenance tasks (planning plant).

Settings

In the PlPl column, assign the general logistics plant to a planning plant if it contains assets, or if you want to plan or perform maintenance work.

For business processes in SAP EAM, you need to differentiate between order planning and execution in only one plant and order planning and execution in different plants.

Plant-Specific Maintenance

In practice, you most frequently encounter a situation where the maintenance requirement is planned in the plant in which it originates, the orders are fulfilled by workshops in the same plant, and the spare parts are stored within the same plant. In Figure 3.13, this plant is known as Plant 1000. Here, the following applies: maintenance plant = planning plant = spare parts storage plant.

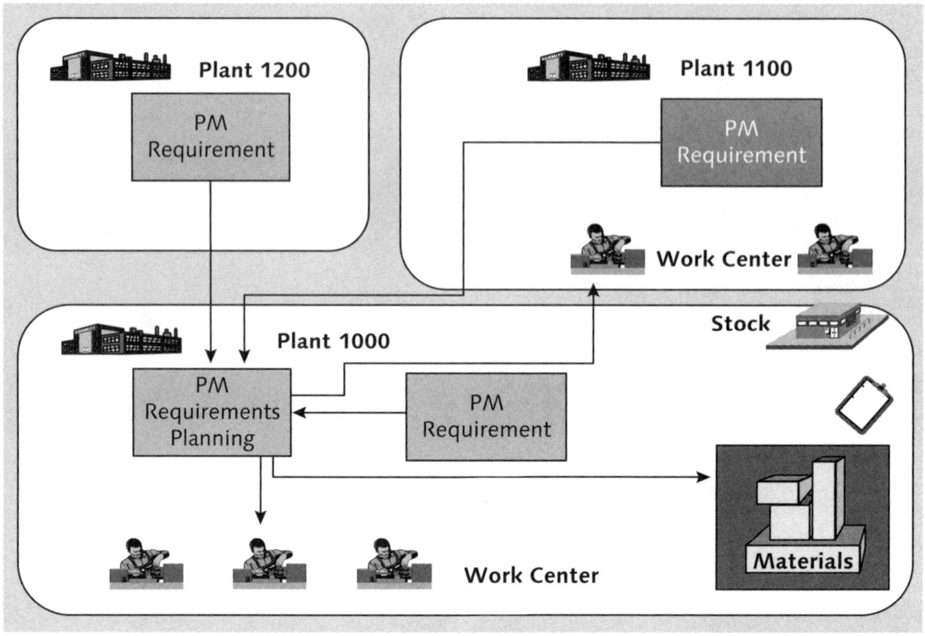

Figure 3.13 Plant and Plant Maintenance

Cross-Plant Maintenance

There are other constellations in addition to plant-specific maintenance:

- There is a requirement in a plant (here, 1200) because an asset is to be maintained there (i.e., in the maintenance plant), but all other functions (planning, order execution, and spare parts storage) are the responsibility of another plant (here, 1000).
- There is a requirement in a plant (here, 1100), and additional partial functions (order execution) are also the responsibility of this plant, but other partial functions (order planning and spare parts storage) are the responsibility of other plants (here, 1000).

Cross-plant maintenance isn't a problem if the maintenance plant of the technical object and the plant of the executing work center are in the same company code. The same applies if the plants are in different company codes but belong to the same controlling area. This is also a standard scenario.

However, a problem occurs if the plants belong to different controlling areas. Here, there is no standard scenario but rather a customer-vendor relationship. Therefore, in this case, the maintenance plant (customer) has to trigger purchase orders, and the plant of the work center (vendor) triggers a sales order and its associated invoice. The invoice is, in turn, recorded as an incoming invoice in the maintenance plant. All in all, this is a very cumbersome procedure, but it can be simplified as described next.

Plants in Different Controlling Areas

If you implement cross-plant maintenance and your plants are in different controlling areas, the following approach is recommended:

- In the work center plant, create a cost center for the actual maintenance plant.
- Assign all of the technical objects to the work center plant (as a maintenance plant) and to this cost center.
- Process all maintenance orders in the work center plant.
- Manually issue periodic invoices (e.g., monthly) from the work center plant whereby the customer maintenance plant is debited the amount and the cost center is credited the same amount.

This procedure saves you from having to create purchase orders, sales orders, and individual invoices, as well as posting individual incoming invoices.

3.3 Maintenance-Specific Organizational Units

Additional, maintenance-specific organizational units (either maintenance plant-specific or planning plant-specific) play an important role within a plant (refer back to Figure 3.10). Technical objects (functional location and equipment) also contain all of the maintenance plant-specific and planning plant-specific data, which is then copied to notifications and orders. This data is explained in more detail here.

Work centers either perform or are responsible for maintenance tasks. They reference a planning plant or a maintenance plant (see Section 3.4).

Define Planner Groups

A maintenance planner group is responsible for planning maintenance tasks. It also references a planning plant. You use the Customizing function DEFINE PLANNER GROUPS to maintain maintenance planner groups.

Customizing Path

PLANT MAINTENANCE AND CUSTOMER SERVICE • MAINTENANCE AND SERVICE PROCESSING • BASIC SETTINGS • GENERAL DATA • DEFINE PLANNER GROUPS

Prerequisites

Because maintenance planner groups are always created for a specific planning plant, you must create the plant as both a general logistics plant and a maintenance planning plant beforehand.

Settings or Recommended Settings

Assign a three-digit key, description, and, if necessary, a telephone number to each maintenance planner group (see Figure 3.14).

[+] **Recommendation for Using Maintenance Planner Groups**

You set up maintenance planner groups, for example, if you want to map work scheduling or individual maintenance planners known by name.

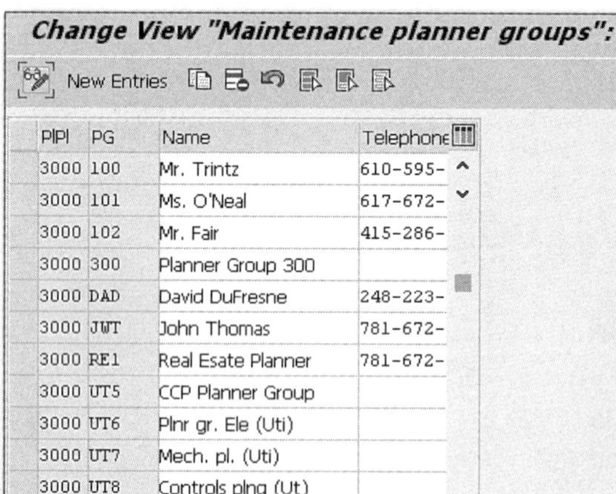

PlPl	PG	Name	Telephone
3000	100	Mr. Trintz	610-595-
3000	101	Ms. O'Neal	617-672-
3000	102	Mr. Fair	415-286-
3000	300	Planner Group 300	
3000	DAD	David DuFresne	248-223-
3000	JWT	John Thomas	781-672-
3000	RE1	Real Esate Planner	781-672-
3000	UT5	CCP Planner Group	
3000	UT6	Plnr gr. Ele (Uti)	
3000	UT7	Mech. pl. (Uti)	
3000	UT8	Controls plng (Ut)	

Figure 3.14 Maintenance Planner Groups

Define Location

You use a label to indicate the physical location of a technical object. You always define a location with reference to a maintenance plant. Furthermore, you use the Customizing function DEFINE LOCATION to maintain locations.

Customizing Path

ENTERPRISE STRUCTURE • DEFINITION • LOGISTICS – GENERAL • DEFINE LOCATION

Prerequisites

Because locations are always created for a specific plant, you must define the maintenance plant as a normal logistics plant beforehand. Agree on this with your colleagues from PP and Asset Accounting because they will also use the location table for work centers and assets.

Settings or Recommended Settings

Assign a 10-digit number (maximum) and name to each location (see Figure 3.15).

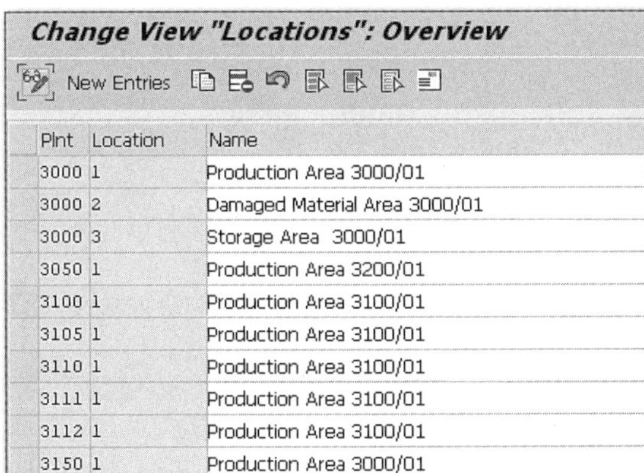

Figure 3.15 Location

[+]

Recommendation for Assigning Names to Locations

In practice, either building numbers (e.g., F141 or WDF21) or, if they exist, plant coordinates (e.g., A01 or K15) have become commonly used locations.

Define Plant Sections

The *plant section* enables you to divide a maintenance plant into production areas. The person responsible for the plant section is the contact person for all coordination between production and plant maintenance. A plant section can be assigned to each piece of equipment and each functional location. You can also specify a plant section when processing notifications and orders.

You define the responsibilities associated with operating a (production) facility as a plant section. To this end, you use the Customizing function DEFINE PLANT SECTIONS to maintain plant sections.

Customizing Path

PLANT MAINTENANCE AND CUSTOMER SERVICE • MAINTENANCE AND SERVICE PROCESSING • BASIC SETTINGS • GENERAL DATA • DEFINE PLANT SECTIONS

Transaction

OIAB

Prerequisites

Because plant sections are always assigned to a plant, you must define the plant as a general logistics plant beforehand.

Settings or Recommended Settings

Assign a three-digit number, a name and, if necessary, a telephone number, to each plant section (see Figure 3.16).

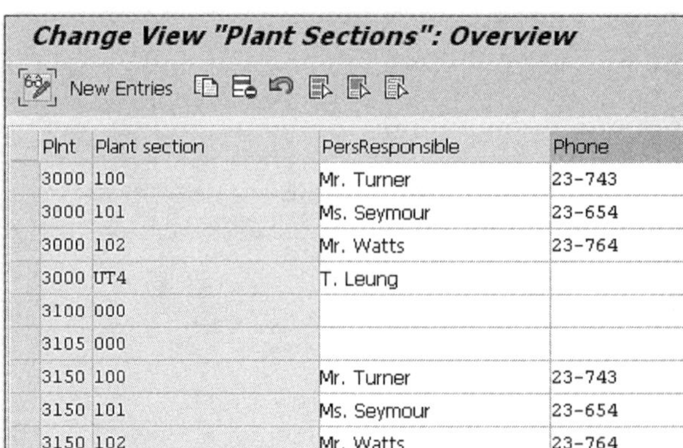

Plnt	Plant section	PersResponsible	Phone
3000	100	Mr. Turner	23-743
3000	101	Ms. Seymour	23-654
3000	102	Mr. Watts	23-764
3000	UT4	T. Leung	
3100	000		
3105	000		
3150	100	Mr. Turner	23-743
3150	101	Ms. Seymour	23-654
3150	102	Mr. Watts	23-764

Figure 3.16 Plant Sections

Recommendation for the Plant Section [+]

In practice, either the plant engineer responsible for the asset or the production area belonging to the asset have proven themselves as plant sections.

3.4 Work Centers

From a maintenance perspective, a work center represents either an individual person (e.g., Mr. John Smith, a technician) or a workshop; that is, a group of people. In practice, the following workshops are commonly used:

- Mechanical workshop
- Electrical workshop
- Measurement and control
- Machine center
- Welding workshop
- Paint shop
- Cleaning line
- Building services engineering

In SAP EAM, work centers are used in many locations:

- Responsible work center in the equipment master record and functional location master record
- Responsible work center in a maintenance item
- Responsible work center in the header of a task list
- Performing work center in the operations of a task list
- Responsible work center in the notification
- Responsible work center in the order header
- Performing work center in the operations of an order

Next up are the most important Customizing functions for creating and maintaining work centers.

Define Work Center Types and Link to Task List Application

You use this Customizing function to define the work center types. When you create a work center, you must always assign a work center type (see Figure 3.17). The work center type determines the following:

- The permitted *task list application* (e.g., SAP PM and PP)
- The screen sequence (e.g., Basic Data and Capacities)
- The field selection
- The use of change documents to document changes

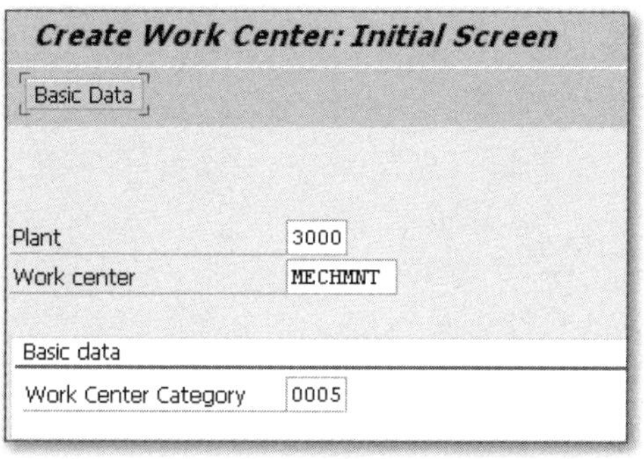

Figure 3.17 Create Work Center: Initial Screen

Prerequisites

There are no prerequisites.

Customizing Path

Plant Maintenance and Customer Service • Maintenance Plans, Work Centers, Task Lists and PRTs • Work Centers • General Data • Define Work Center Types and Link to Task List Application

Transaction

OIZA

Settings or Recommended Settings

SAP delivers work center type 0005 (Plant maintenance) (see Figure 3.18). Usually, this work center type is sufficient. Therefore, you only need to create your own work center type if you require other control options or advanced control options.

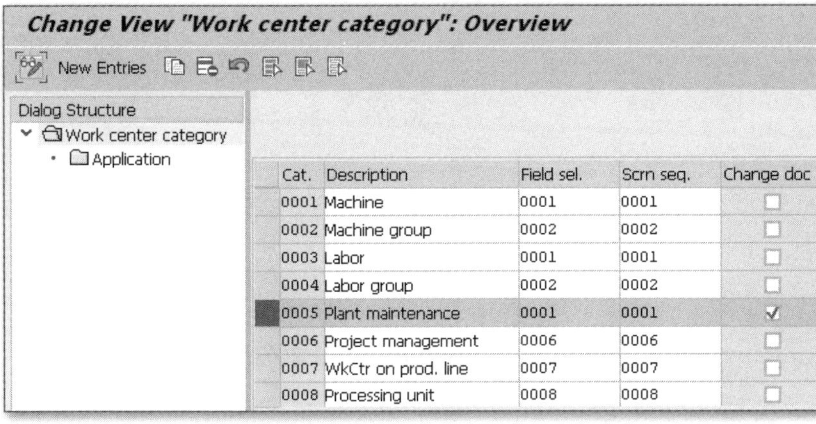

Figure 3.18 Work Center: Work Center Type

To enable you to use the work center type in SAP EAM, you must assign it to a task list application (APPLICATION) (see Figure 3.19). Make sure that your work center type is assigned to task list application I for plant maintenance. If you also want to use your work centers in other applications (e.g., in production orders as a production aid), you need to add other task list applications.

Work Center Type 0005 Should Suffice [+]

Usually, work center type 0005 is sufficient for assignments to SAP EAM, and you don't have to define any other work center types.

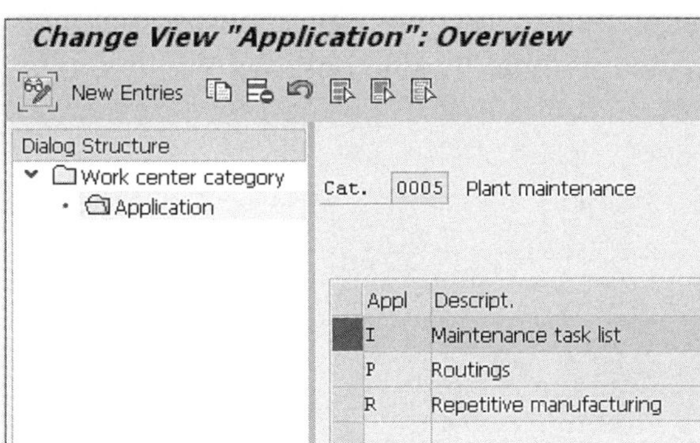

Figure 3.19 Work Center: Task List Application

Define Field Selection

You use this Customizing function to define field attributes when maintaining work centers. You can choose from the following options:

- Required entry field
- Hidden field
- Display field
- Normal entry field

These fields are divided into *screen groups* (see Figure 3.20).

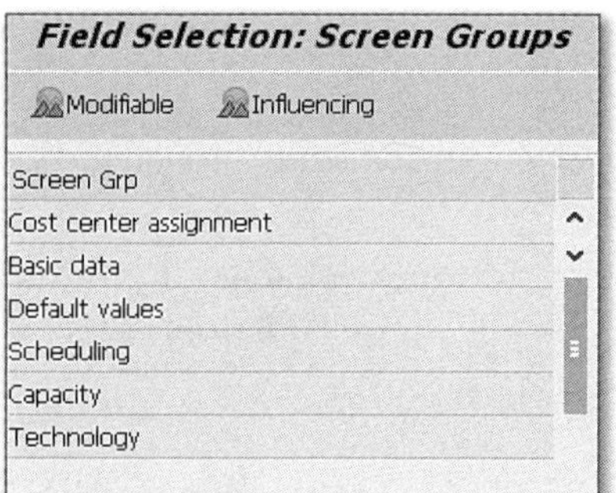

Figure 3.20 Work Center: Screen Groups

Prerequisites

Because we recommend that you don't configure a general field selection for all work center types but instead configure a specific field selection for each work center type, you must maintain the work center type beforehand.

Customizing Path

Plant Maintenance and Customer Service • Maintenance Plans, Work Centers, Task Lists and PRTs • Work Centers • General Data • Define Field Selection

Transaction

OBFA

Settings or Recommended Settings

In the overview screen, choose Influencing to configure the field selection on the detail screen on the basis of the relevant work center type. Figure 3.21 shows an example of the field selection for the BASIC DATA screen group.

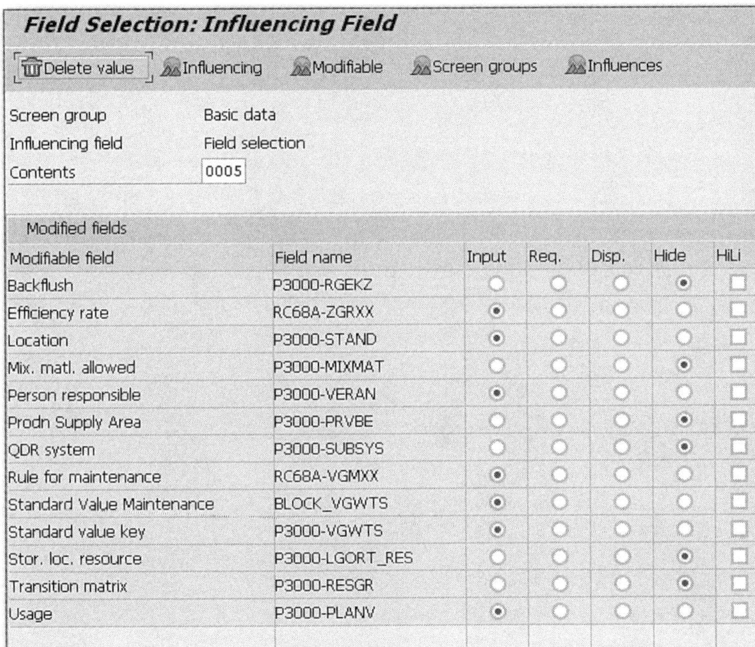

Figure 3.21 Work Center: Field Selection

Define Standard Value Keys

You use this Customizing function to define *standard value keys*. The standard values are plan values for determining the execution time for operations in the task list and order.

Prerequisites

There are no prerequisites.

Customizing Path

Plant Maintenance and Customer Service • Maintenance Plans, Work Centers, Task Lists and PRTs • Work Centers • General Data • Define Standard Value Keys

Transaction

OIZ2

Settings or Recommended Settings

The task lists and orders in SAP EAM always take into account the duration of an operation for scheduling orders and the work associated with costing and capacity requirements planning. In the case of work centers used solely by SAP EAM, use the standard value key for which no standard values are defined. SAP delivers the standard value key SAP0 for this purpose (see Figure 3.22). Assign this key to each of your work centers in SAP EAM.

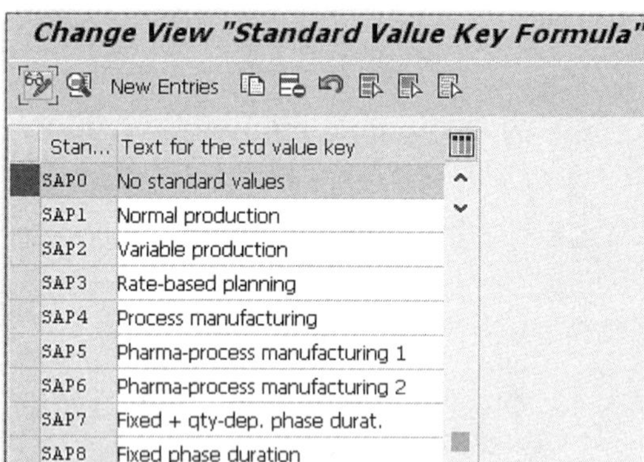

Change View "Standard Value Key Formula"

New Entries

Stan...	Text for the std value key
SAP0	No standard values
SAP1	Normal production
SAP2	Variable production
SAP3	Rate-based planning
SAP4	Process manufacturing
SAP5	Pharma-process manufacturing 1
SAP6	Pharma-process manufacturing 2
SAP7	Fixed + qty-dep. phase durat.
SAP8	Fixed phase duration

Figure 3.22 Work Center: Standard Value Key

[+]

Assign Standard Value Key SAP0

The standard value key in the standard SAP system is sufficient. You don't have to define any other standard value keys. However, make sure to assign it to each of your SAP work centers.

Define Task List Usage Keys

You use this Customizing function to specify the task lists and order categories in which a *work center* is permitted to be used. For example, you could differentiate between whether you want to use work centers in maintenance processing or inspection processing.

Prerequisites

There are no special prerequisites.

Customizing Path

PLANT MAINTENANCE AND CUSTOMER SERVICE • MAINTENANCE PLANS, WORK CENTERS, TASK LISTS AND PRTS • WORK CENTERS • TASK LIST DATA • DEFINE TASK LIST USAGE KEYS

Transaction

OIZD

Settings or Recommended Settings

For plant maintenance purposes, SAP delivers task list usage 004 (ONLY MAINTENANCE TASK LISTS) and task list usage 009 (ALL TASK LIST TYPES) (see also Figure 3.23).

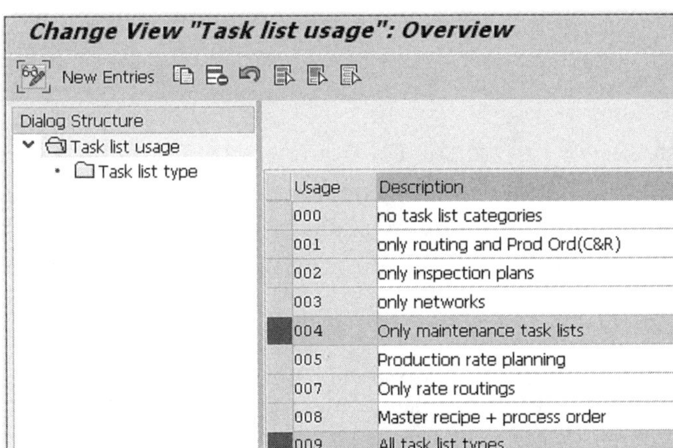

Figure 3.23 Work Center: Task List Usage

You should assign one of these task list usages to each of your work centers in SAP EAM. Specify the relevant task list types on the detail screen (e.g., E = equipment task lists, A = general maintenance task lists, and T = functional location task lists) as shown in Figure 3.24.

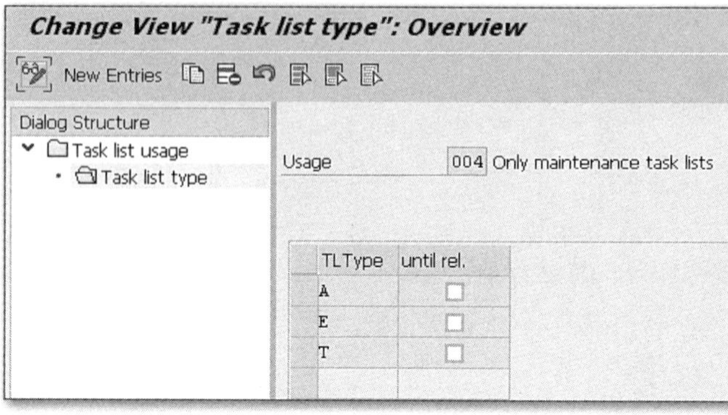

Figure 3.24 Work Center: Task List Types

[+]

Assign Task List Usage Key 004 or 009

The task list usage keys in the standard SAP system are sufficient. You don't have to define any other task list usage keys. However, make sure to assign one of these task list usage keys to each of your work centers.

Maintain Control Keys

On the DEFAULT VALUES screen, you can assign a control key to a work center. From here, the control key is transferred as a default value to the operations associated with a task list or order. You can use the *control key* of an operation to define which business functions you want to execute or how you want to handle the *operation*.

Prerequisites

There are no technical prerequisites. Because, however, the control key is used not only in SAP EAM but also in PP, QM, and project processing, you should consult with your colleagues from these areas on the naming convention for the control key (see Figure 3.25).

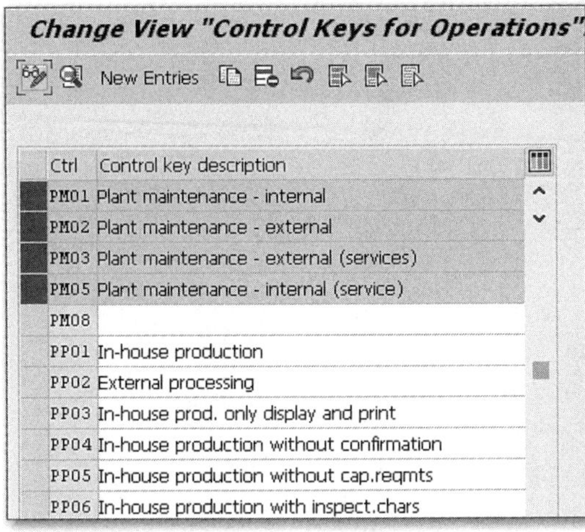

Figure 3.25 Work Center: Overview of Control Keys

Customizing Path

Plant Maintenance and Customer Service • Maintenance Plans, Work Centers, Task Lists and PRTs • Work Centers • Task List Data • Maintain Control Keys

Settings or Recommended Settings

You use the control key to define which of the following business functions are to be executed in the order at a later stage (see Figure 3.26):

- Whether the operation is to be part of costing
- Whether the operation is to be scheduled
- Whether the operation is to generate capacity requirements
- Whether completion confirmations are expected for the operation
- Whether the operation is to be assigned to an external company
- Whether an externally processed operation is to be taken into account in scheduling
- Whether service specifications are to be set up in the operation
- Whether the operation is to be printed
- Whether time tickets or completion confirmation slips are also to be printed for the operation
- Whether inspection characteristics are to be assigned to the operation

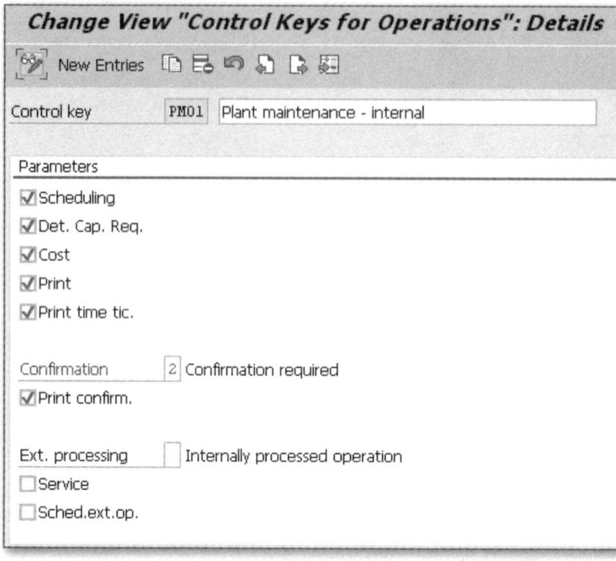

Figure 3.26 Work Center: Control Keys Detail Screen

[+]

Using the Control Key

You can use the control key to control, in detail, the business functions that an operation is to have (cost, print, confirm, assign externally, schedule, etc.).

You require at least two control keys: a key for internal processing and a key for external processing. You can use other control keys as required.

You should always define the control key in the work center as a default value so that you don't always have to manually enter it in the task list and order.

Configure Screen Sequence for Work Center

You use this Customizing function to define which screens are to appear when you maintain a work center, as well as the order in which they are to appear.

Prerequisites

Because the work center type determines the order in which the screens are displayed, you must define the work center type beforehand.

Customizing Path

Plant Maintenance and Customer Service • Maintenance Plans, Work Centers, Task Lists and PRTs • Work Centers • Configure Screen Sequence for Work Center

Transaction

OIZU

Settings or Recommended Settings

Figure 3.27 shows an overview of the screens most commonly used for the work center. Choose New Entries to insert screens that you want to maintain, and choose to delete those screens that you don't need. The following screens are available:

- Basic Data
- Defaults
- Scheduling
- Capacities
- Costing

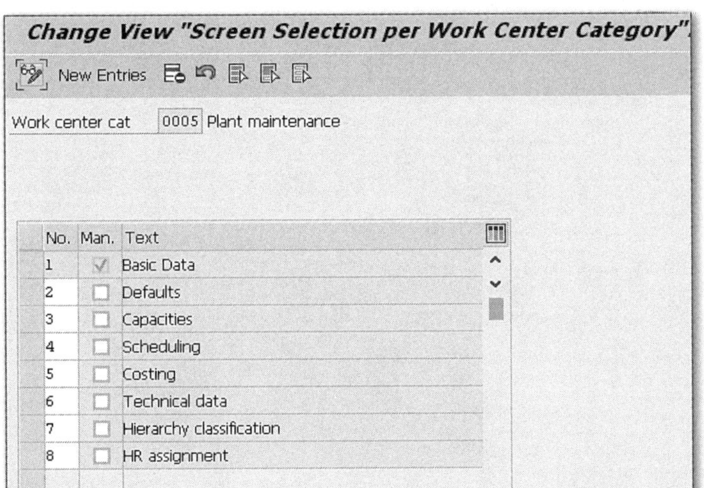

Figure 3.27 Work Center: Screen Selection

- Technical data
- Hierarchy classification
- HR assignment
- Res.NetworkRelation.

Use the No. column to define the order in which these screens are displayed.

This chapter introduces you to generic functions, such as status management and partners; both of which are used in several locations within SAP Enterprise Asset Management. You'll learn how to configure your tools as well as obtain numerous tips in relation to configuring Customizing settings for these functions.

4 Generic Functions

In SAP Enterprise Asset Management (SAP EAM), some functions are used in several locations or by several objects. For example, you can use status management not only to document the condition of technical objects (functional location, equipment, and serial number), but also to manage notification and order processes. The field selection control that you use to hide fields or to force entries can also be used with each SAP object.

Table 4.1 shows which generic functions can be used for which SAP object.

Function	Functional Location	Equipment	Material Number	Notification	Order	Task List	Maintenance Plan
Object information	X	X		X	X		
Status management	X	X		X	X		
Number assignment		X	X	X	X	X	X
Warranties	X	X		X	X	X	
Measuring points/counters	X	X			X	X	X
Permits	X	X			X		
Partners	X	X		X	X		

Table 4.1 Generic Functions: Overview

Function	Functional Location	Equipment	Material Number	Notification	Order	Task List	Maintenance Plan
Documents	X	X	X	X	X	X	
Field selection	X	X	X	X	X	X	X
List variants	X	X	X	X	X	X	X
Multilevel lists	X	X		X	X		

Table 4.1 Generic Functions: Overview (Cont.)

4.1 Object Information

When you receive a new notification from a requester and need to decide whether the maintenance task is to be performed or not, it's very useful to find out concise information about the object by using *object information*.

Define Object Information Keys

You can use this Customizing function to configure a dialog box that contains concise information about the reference object (known as object information).

You can call object information in the following locations within the system:

- Equipment and serial numbers
- Functional locations
- Notifications
- Orders

Prerequisites
There are no special prerequisites.

Customizing Path
PLANT MAINTENANCE AND CUSTOMER SERVICE • MASTER DATA IN PLANT MAINTENANCE AND CUSTOMER SERVICE • BASIC SETTINGS • DEFINE OBJECT INFORMATION KEYS

Transaction
OIMD

Settings
You use the object information key to define which technical object data is displayed in the object information window (see Figure 4.1). The object information also contains historical data from the information system as well as data from the notifications and orders processed in the system.

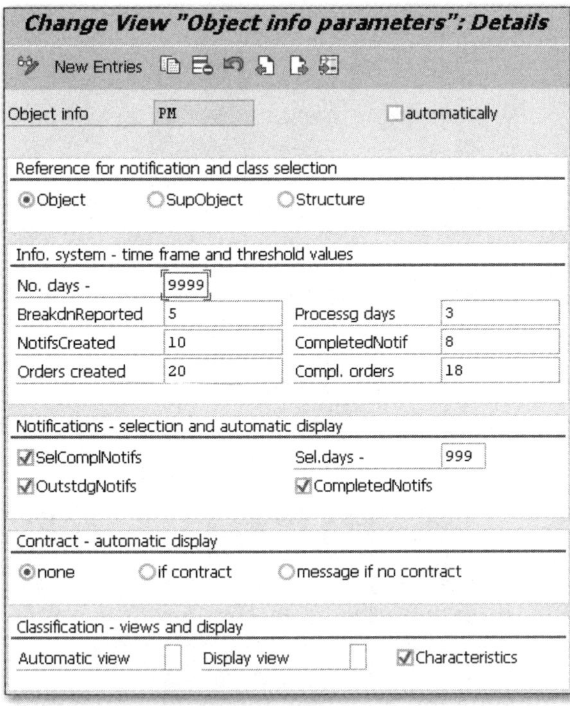

Figure 4.1 Object Information

In the REFERENCE FOR NOTIFICATION AND CLASS SELECTION area, you define which notification and order information is to be displayed:

- For a technical object (OBJECT)
- For the superior object (SUPOBJECT)
- For the entire structure (STRUCTURE)

If you've selected object information for the object specified, you use the AUTOMATICALLY indicator to define whether this object information is displayed automatically (e.g., immediately after you call a notification or order). If you don't set this indicator, you can use the [i] button to call the object information.

In the INFO. SYSTEM – TIME FRAME AND THRESHOLD VALUES area, you can define the following:

- The period to be selected in the information system (highest value is 9,999 days; that is, approximately 27 years)
- Threshold values, which, when exceeded, are indicated by [icon] in the OBJECT INFORMATION window

In the NOTIFICATIONS – SELECTION AND AUTOMATIC DISPLAY area, you can define the following:

- The period to be selected in the notification list (highest value is 9,999 days; that is, approximately 27 years)
- The notification status to be considered in the selection (outstanding and/or completed)

In the CLASSIFICATION – VIEWS AND DISPLAY area, you essentially define whether the characteristics of the object are to be displayed (by setting the CHARACTERISTICS indicator).

[+]

Object Information Can Be Very Useful

You should define object information keys in your SAP system because they will enable you to obtain a clear and concise overview of the technical object's situation in master data and in processing. However, you should only display this object information when you actually need it. In other words, don't set the AUTOMATICALLY indicator.

4.2 Status Management

Both the technical objects themselves, as well as complete notification and order processing, are associated with general SAP status management. Here, you have to distinguish between the system status and the user status.

For certain business processes, the system sets the *system statuses* both internally and automatically as part of its general status management. This happens, for example, in the equipment master INST when the equipment is installed at a functional location or in the order CONF when an order receives a final confirmation. The system status is displayed on the left-hand side of the status screen (see Figure 4.2).

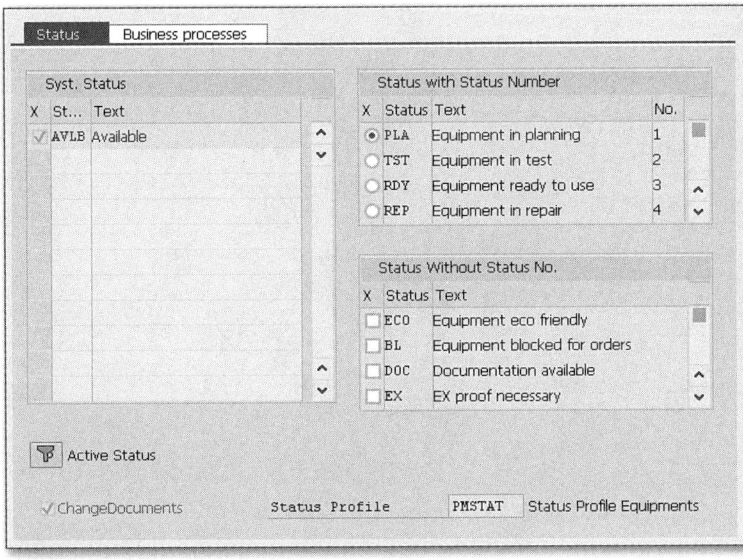

Figure 4.2 System Status and User Status

Define User Status

In addition to the predefined system statuses, you can freely define *user statuses* that fulfill your requirements. The user status is set and displayed on the right-hand side of the status screen (see Figure 4.2). SAP distinguishes between two types of user statuses:

- STATUS WITH STATUS NUMBER (upper half of the screen)
  You can select only one radio button here.
- STATUS WITHOUT STATUS NO. (lower half of the screen)
  You can select any number of checkboxes here.

The business processes that you can execute for each of these object statuses is also defined in status management. If, for example, you set a piece of equipment to inactive, the status display informs you which business processes are still allowed (green traffic light), which will trigger a warning message (yellow traffic light), and which are forbidden (red traffic light) (see Figure 4.3).

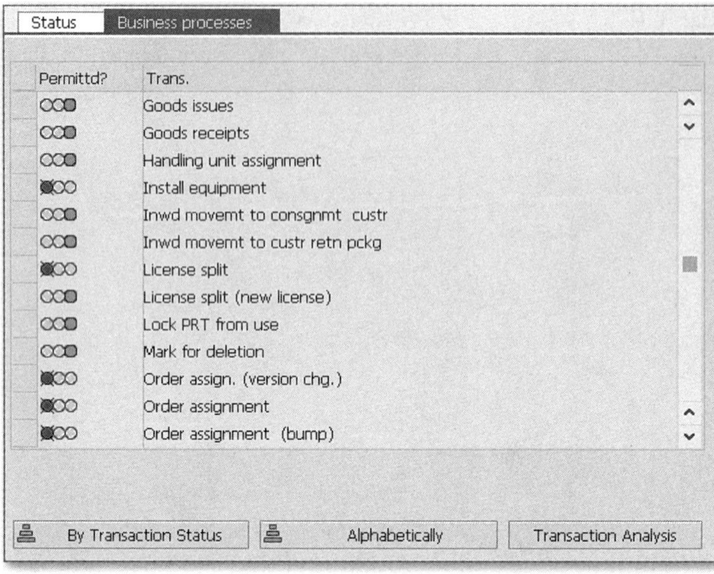

Figure 4.3 Business Processes

You can define a user status for the following objects:

- Equipment and serial numbers
- Functional locations
- Notifications
- Orders and order operations

For the sake of simplicity, let's consider the example of a piece of equipment to illustrate status management. However, status management can be used in a similar manner for all other objects.

Prerequisites

There are no special prerequisites.

Customizing Paths

Plant Maintenance and Customer Service • Master Data in Plant Maintenance and Customer Service • Basic Settings • Define User Status

Plant Maintenance and Customer Service • Maintenance and Service Processing • Maintenance and Service Orders • General Data • User Status for Orders • Define Status Profile

Plant Maintenance and Customer Service • Maintenance and Service Processing • Maintenance and Service Notifications • Notification Processing • User Status for Notifications • Define Status Profile

Transaction
OIBS

You always access the same Customizing function irrespective of which Customizing path you call or whether you use the transaction.

Settings
When you call the Customizing function, you first obtain an overview of all status profiles irrespective of which application or object they are valid for. In other words, you see not only the status profiles for maintenance objects, but also all defined status profiles. If you now create a new status profile, you define each status in succession (see Figure 4.4).

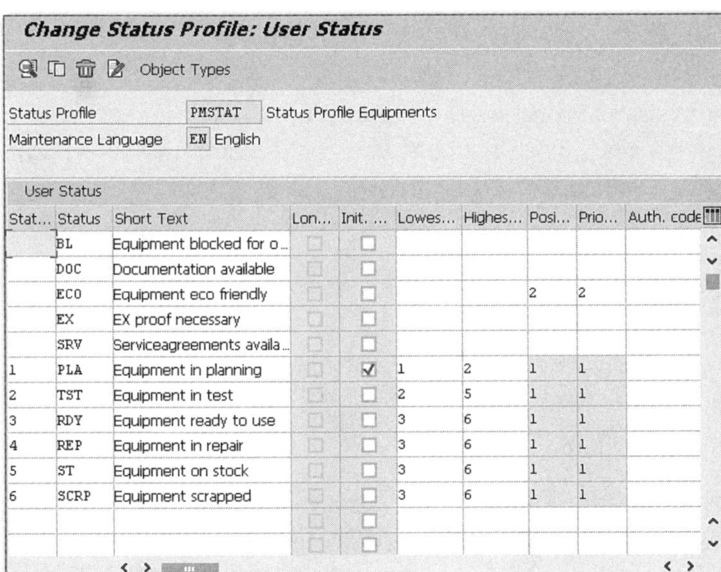

Figure 4.4 Status Profile: Status

The statuses *without* a status number are sorted on the upper half of the screen, while the statuses *with* a status number are sorted on the lower half of the screen (according to the status number).

Status numbers are used for the following reasons:

- Statuses with a status number are mutually exclusive. In the example shown, a piece of equipment can be ready to use (status number 3) or in repair (status number 4).
- You generally use the statuses with a status number to map a process (e.g., execution of a planned repair order). In the example shown, virtually the entire life cycle of a piece of equipment, that is, from planning and operation through to scrapping, is mapped.
- One of these statuses (generally the status with the lowest status number) is selected as the initial status and later automatically set in the object. In this case, PLA is the initial status.
- If you set a specific status with a status number, you can only change to another status if it's within the upper and lower limits defined for the status number. If, in the example shown, the piece of equipment is assigned the status Equipment in repair (status number 4), the next status must lie between 3 (Equipment ready to use) and 6 (Equipment scrapped). It's not possible to return to 2 (Equipment in test).

If you want to assign business processes to a status, double-click the status, and choose [icon] to add new entries. For each status, you can define how the relevant business process is to be handled (see Figure 4.5):

- Allowed
  The business process is explicitly allowed.
- Warning
  The system issues a warning when you execute the business process.
- Forbidd.
  The business process is forbidden.

Finally, you must ensure that the status profile can also be used by the relevant object category. Choose Object TYPES to display a list of all object types associated with status management (see Figure 4.6), and then select the relevant object type. You can also permit the same status profile for several object types. For example, you can make a status profile available for both equipment and functional locations. However, you can also use a different status profile for notifications, orders, and order operations.

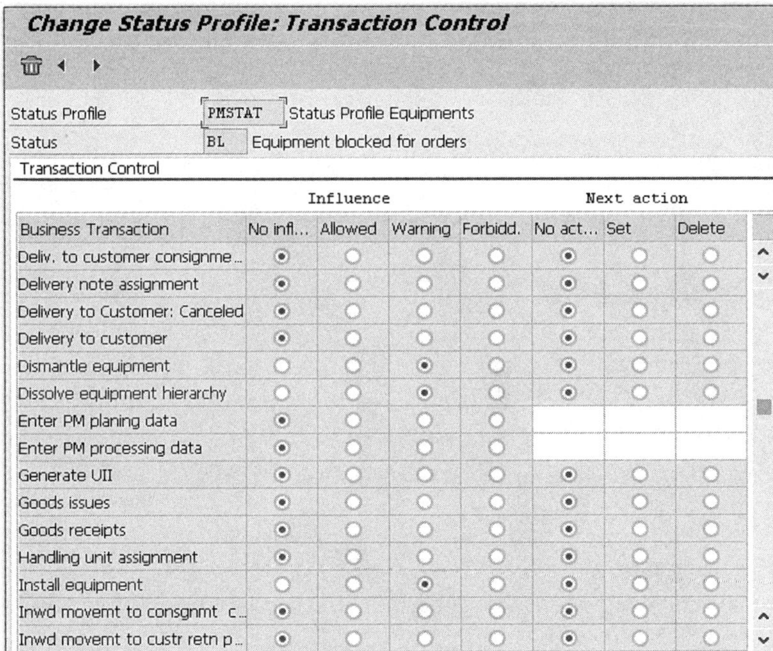

Figure 4.5 Status: Business Processes

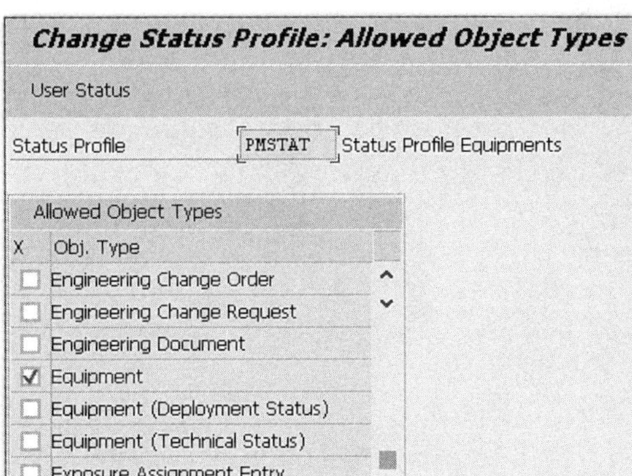

Figure 4.6 Status Profile: Object Categories

You then use special object-based Customizing functions to assign a status profile to an object (e.g., Assign User Status Profile to Equipment Category or Define Functional Location Category).

[+]

Make Good Use of the Status Profile Options

A purposeful status definition enables you to control in detail which business functions are to be allowed or forbidden for which object and in which situation.

It also enables you to search for objects (e.g., to display all equipment currently being planned).

4.3 Number Assignment

A number must be assigned to each data record. Apart from a few exceptions (e.g., functional locations), you adopt a similar approach to number assignment for all objects in the SAP system. You must manually assign a unique number to each object, or you must allow the system to assign a unique number. The following number assignment options are available:

- **Internal number assignment**
  The system automatically assigns consecutive numbers within a predefined number range.
- **External number assignment**
  The user assigns the numbers manually.

Define Number Ranges

You use this Customizing function to define the type of number assignment for each object. The procedure described here is used for the following maintenance objects:

- Equipment
- Serial numbers
- Object links
- Material numbers
- Notifications
- Orders
- Measuring points and counters

- Measurement documents
- Master warranties
- Task lists
- Maintenance plans

For the sake of simplicity, let's continue to consider the example of a piece of equipment to illustrate number assignment. However, a similar approach to number assignment is adopted for all other objects.

Groups are always used in conjunction with number assignment. Here, several object types (e.g., several equipment categories, notification types, or order types) are combined together to form a group. You can then define an internal and external number range for the group. In the example shown, several equipment categories can belong to the same group and therefore have the same number range (see Figure 4.7).

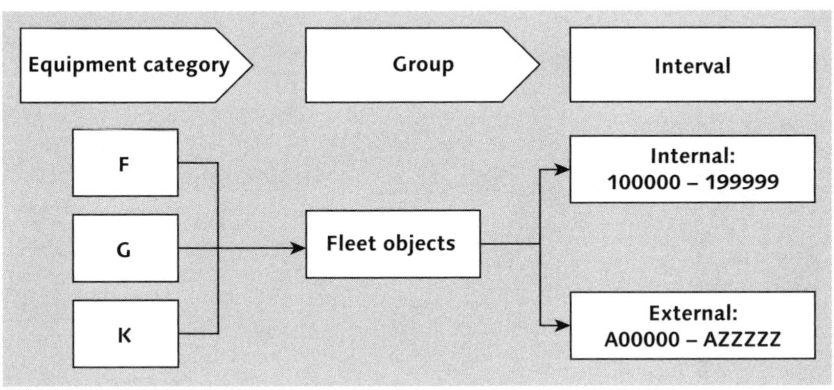

Figure 4.7 Number Assignment: Groups

Prerequisites

You must define the object types (e.g., equipment categories, order types, and notification types) beforehand.

Customizing Path

Plant Maintenance and Customer Service • Master Data in Plant Maintenance and Customer Service • Technical Objects • Equipment • Equipment Categories • Define Number Ranges

Please note that different objects have different Customizing paths.

Transaction
OIEN

Different objects also have different transactions (e.g., Transaction IW20 (notifications), Transaction OION (orders), Transaction IN20 (object links), Transaction BG20 (warranties), Transaction OIL4 (general maintenance task lists), and Transaction IP20 (maintenance plans).

Settings or Recommended Settings
If you want to define a number range for a group, proceed as follows:

1. After you call the Customizing function, choose Groups.

   An overview appears of which groups exist and which object types (in this case, equipment categories) are assigned to which groups or which object types aren't assigned yet (see Figure 4.8).

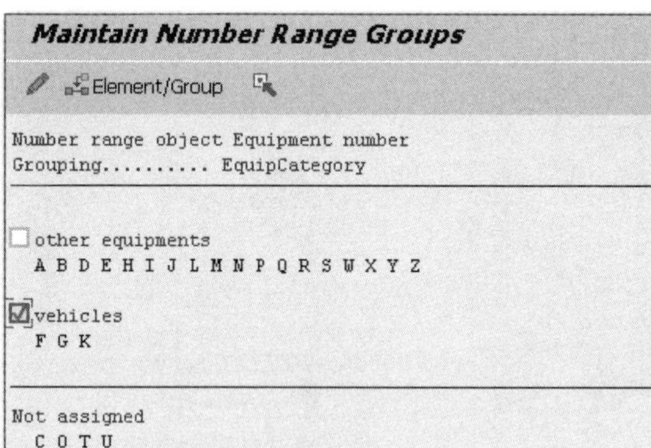

Figure 4.8 Number Range Groups

2. Use the menu option GROUP • INSERT GROUP to create a new group. At the same time, the system prompts you to define some number ranges, namely an internal number range and/or an external number range (see Figure 4.9).
3. If you now want to assign an equipment category to the group, double-click the equipment category, select the group, and choose Element/Group.

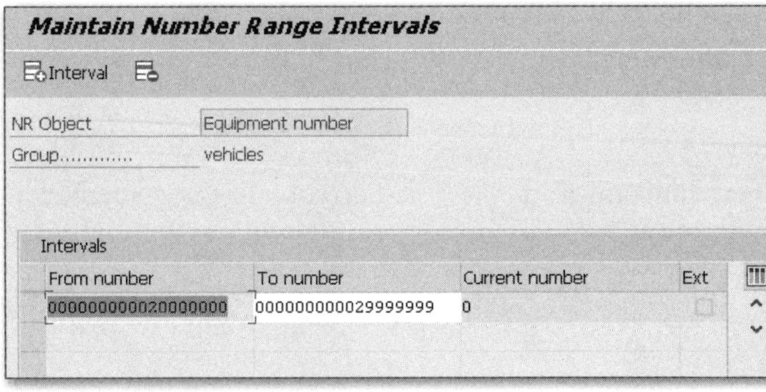

Figure 4.9 Number Ranges

4. Use the menu option INTERVALS • MAINTAIN to make subsequent changes to the number range intervals.
5. Use the menu option GROUP • MAINTAIN TEXT to subsequently change the name of the group.

[+]

Define Short Numbers

For usability purposes, you should define numbers that are as short as possible. Your users will thank you for only having to enter a 6-digit number (e.g., 800123) rather than a 12-digit number (e.g., 800121232343).

4.4 Warranties

A warranty is a commitment from a manufacturer, vendor, or salesperson to a customer that services are guaranteed for a particular period of time, without the customer being billed. A warranty always refers to a technical object (functional location, equipment, or serial number).

Warranties are used in the following areas:

- Equipment and serial numbers
- Functional locations
- Notifications
- Orders

If you want to assign warranties to technical objects and then use these warranties in maintenance processing, the following Customizing functions are available to you.

Check Warranty Categories

You use this Customizing function to define whether you are the guarantor or warrantee.

Prerequisites
There are no special prerequisites.

Customizing Path
Plant Maintenance and Customer Service • Master Data in Plant Maintenance and Customer Service • Basic Settings • Warranties • Check Warranty Categories

Transaction
GM01

Settings or Recommended Settings
In the standard SAP delivery, a distinction is made between the following warranty categories (see Figure 4.10):

- **Warranty category I (inbound)**
  This warranty category exists if you are the warrantee.
- **Warranty category O (outbound)**
  This warranty category exists if you are the guarantor.

Figure 4.10 Warranty Categories

Delete Customer Warranties [+]

Delete warranty category O because, in SAP EAM, you are generally only the warrantee and never have to manage warranties granted to customers.

Define Warranty Types

You use this Customizing function to define whether you want to manage one or more warranty types for a warranty category. For example, for warranty category I (inbound), you can define one warranty type for a manufacturer warranty and another warranty type for a vendor warranty.

Prerequisites

You must define the warranty categories beforehand. If you want to assign a partner determination procedure and status profile, you must also define them beforehand.

Customizing Path

PLANT MAINTENANCE AND CUSTOMER SERVICE • MASTER DATA IN PLANT MAINTENANCE AND CUSTOMER SERVICE • BASIC SETTINGS • WARRANTIES • DEFINE WARRANTY TYPES

Settings or Recommended Settings

If you want to distinguish between vendor warranties and manufacturer warranties, you can define two warranty types for this purpose (see Figure 4.11).

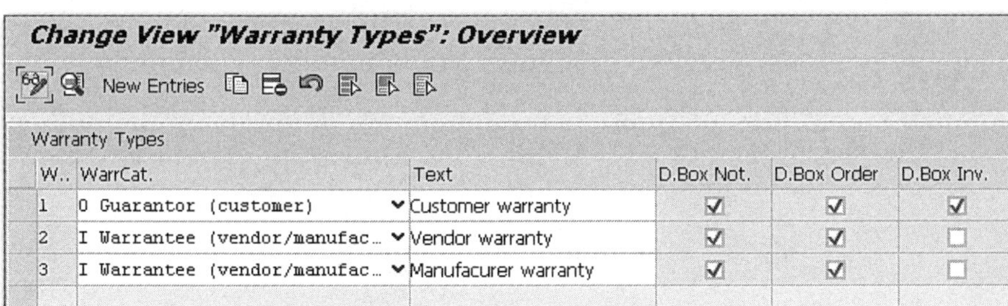

W..	WarrCat.	Text	D.Box Not.	D.Box Order	D.Box Inv.
1	O Guarantor (customer)	Customer warranty	☑	☑	☑
2	I Warrantee (vendor/manufac...	Vendor warranty	☑	☑	☐
3	I Warrantee (vendor/manufac...	Manufacurer warranty	☑	☑	☐

Figure 4.11 Warranty Types

For each warranty type, define the following:

- Do you want the system to display a dialog box when you open a notification?
- Do you want the system to display a dialog box when you open an order?
- Do you want to assign a status profile?
- Do you want to assign a partner determination procedure?

[+]

Vendor Warranties and Manufacturer Warranties

As a precaution, you should always create two warranty types for the inbound warranty category:

- One for the manufacturer warranty
- One for the vendor warranty

Define Warranty Counters

You use this Customizing function to define counters for use in warranty processing. However, this Customizing function is only necessary if you work with master warranties or performance-based warranties. You don't require counters if you enter only basic warranty data (start and end dates) directly in the technical object (piece of equipment or functional location).

Prerequisites

You must define the counters as characteristics in the classification system beforehand.

Customizing Path

PLANT MAINTENANCE AND CUSTOMER SERVICE • MASTER DATA IN PLANT MAINTENANCE AND CUSTOMER SERVICE • BASIC SETTINGS • WARRANTIES • DEFINE WARRANTY COUNTERS

Transaction

GM04

Settings or Recommended Settings

When you create a master warranty, the system automatically proposes the counters marked as DEFAULT. Furthermore, if you want to measure time, you should always specify a counter that you've defined as a time-dependent counter (see Figure 4.12).

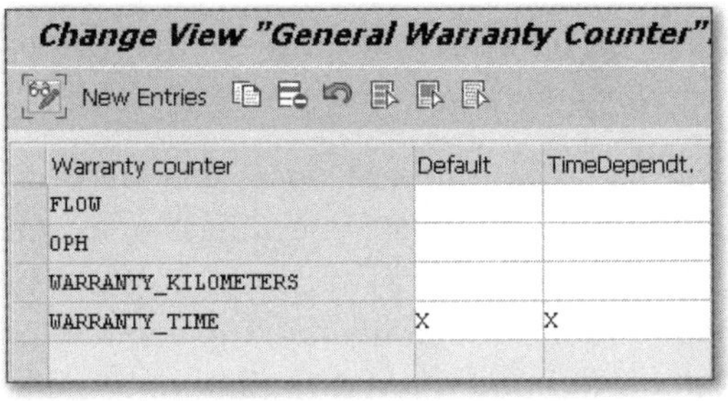

Figure 4.12 Warranty Counters

4.5 Measuring Points and Counters

Measuring points mark those locations at which the current condition of an asset is described (e.g., temperature, number of revolutions, pressure, level of contamination, and viscosity). At individual measuring points, you can specify target values as well as upper- and/or lower limits. Measuring points are a basic prerequisite for condition-based maintenance.

Counters are the tools you can use in the SAP system to represent the wear and tear of an object, consumption, or the reduction of an object's useful life (e.g., an odometer, operation hours counter, numbers of pieces, and output in tons). Counters are a basic prerequisite for performance-based plant maintenance.

Measuring points and counters are used in the following locations within the system:

- Pieces of equipment and serial numbers
- Functional locations
- Task lists
- Maintenance plans
- Order confirmations

If you want to assign measuring points and counters to technical objects and then use them in maintenance processing, several Customizing functions are available to you.

Make System Settings for Measuring Points and Measurement Documents

You use this Customizing function to define global system settings. In the current SAP version, however, there is only one—the generation of interval documents.

What are interval documents? If you use the measurement document transfer, difference measurement documents are transferred to the subordinate pieces of equipment when you enter a measurement document for a functional location. In the example shown in Figure 4.13, 14 measurement documents are generated for the pieces of equipment installed at functional location K1-M each time you enter a measurement document for the functional location. Therefore, in the case of larger assets, the system may easily generate several hundred measurement documents each time you enter one measurement document.

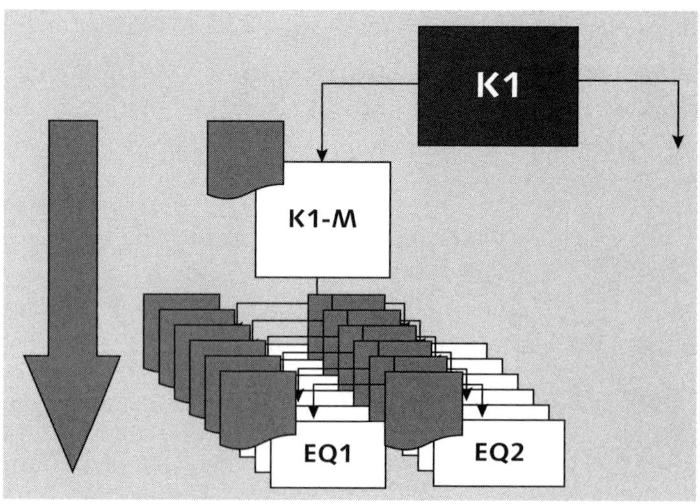

Figure 4.13 Measurement Document Transfer

If you don't want this to happen, you can activate interval documents. Then, a single *interval document* is created (and updated repeatedly) for the length of time that the piece of equipment is installed at the functional location. Therefore, the interval document represents a recurring document that is repeatedly overwritten with the latest data. A new interval document is only generated if the installation location of the piece of equipment changes.

Prerequisites

There are no special prerequisites.

Customizing Path

PLANT MAINTENANCE AND CUSTOMER SERVICE • MASTER DATA IN PLANT MAINTENANCE AND CUSTOMER SERVICE • BASIC SETTINGS • MEASURING POINTS, COUNTERS AND MEASUREMENT DOCUMENTS • MAKE SYSTEM SETTINGS FOR MEASURING POINTS AND MEASUREMENT DOCUMENTS

Settings or Recommended Settings

Set the TRANSFER GENERATES INTERVAL DOCUMENTS indicator (see Figure 4.14).

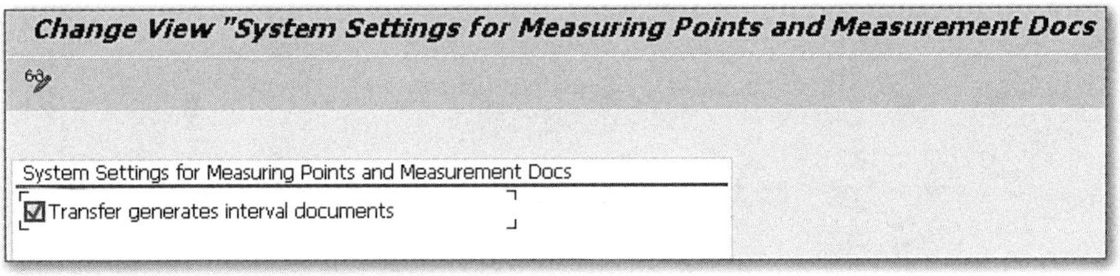

Figure 4.14 Interval Documents

Define Measuring Point Categories

You must assign a *measuring point category* to each measuring point and counter. The measuring point category, in turn, assigns certain controlling attributes to the measuring point or counter. You use this Customizing function to define the measuring point categories and therefore generic attributes that should apply to measuring points and counters.

Prerequisites

If you want to assign catalogs, you must maintain them beforehand.

Customizing Path

Plant Maintenance and Customer Service • Master Data in Plant Maintenance and Customer Service • Basic Settings • Measuring Points, Counters and Measurement Documents • Define Measuring Point Categories

Settings

You can use the measuring point category to assign the following attributes to a measuring point or counter (see Figure 4.15):

- MeasPosUniqnss
  This determines whether you want the measurement position to be *unique*. You use the measurement position to define the installation position of the measuring point associated with the measuring point object, for example, measurement position 1 (measurement of inside temperature) or measurement position 2 (measurement of outside temperature). You can use the uniqueness check for measurement positions to prevent a measurement position from being defined several times for each object. The following settings are available to you: 0 = no check, 1 = check at object level, and 2 = check at client level.
- Catalog type
  This is used when entering the measurement reading to record the inspection results (e.g., C = damage codes).
- MeasRge message
  This determines the system response that occurs when the measurement range limits are exceeded: blank = no message, W = warning message, or E = error message.
- TolPeriod (sec)
  This specifies which tolerance period is permitted when entering future measurement readings if, in the event of a data transfer from an external system, the external server has a different system time than the SAP system. The tolerance period is specified in seconds.
- Linear Asset
  This determines whether you can enter linear data for a measuring point.

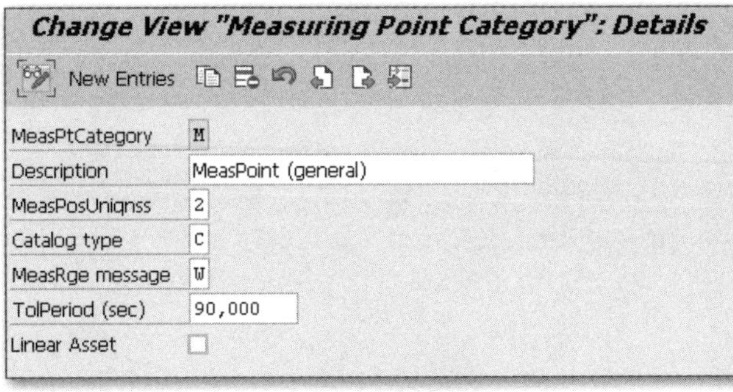

Figure 4.15 Measuring Point Category

Other Customizing functions are available for controlling the measuring points and counters. However, they have already been described previously as generic functions:

- Create Number Ranges for Measuring Points
  Number assignment is explained in detail in Section 4.3. Because measuring point numbers are only assigned internally, create only one number range, namely 01.
- Create Number Ranges for Measurement Documents
  Number assignment is explained in detail in Section 4.3. Because measurement document numbers are only assigned internally, create only one number range, namely 01.
- Define Field Selection for Measuring Points and Measurement Documents
  Field selection is explained in detail in Section 4.9.
- Set List Editing for Measuring Point Lists
  For more information about the technique used to design lists, see Section 4.10.
- Set List Editing for Measurement Document Lists
  For more information about the technique used to design lists, see Section 4.10.

4.6 Permits

For some technical objects, certain regulations or conditions must be taken into account when using them or performing maintenance work. You can define such regulations for a technical object as permits. These permits must be granted for the orders during maintenance processing. Important permits in plant maintenance include the following:

- Fire permits
- Notifications of environmental protection
- Welding permits
- Drivers' licenses
- Fire protection permits
- Vat access permits
- Activation authorizations
- Technical inspection certificates
- Explosion protection zones

Permits are used in the following locations:

- Pieces of equipment and serial numbers
- Functional locations
- Orders

The Customizing functions available to you for use with permits are described next.

Define Permit Categories

You use this Customizing function to define permit categories. You can use the permit category to group permits together. Therefore, to facilitate this, you must assign each permit to a permit category. However, the permit categories aren't used for controlling; rather, they are used only for grouping and selection purposes.

Prerequisites

There are no specific prerequisites.

Customizing Path

Plant Maintenance and Customer Service • Master Data in Plant Maintenance and Customer Service • Basic Settings • Permits • Define Permit Categories

Settings
Defines the groups according to their individual requirements (see Figure 4.16).

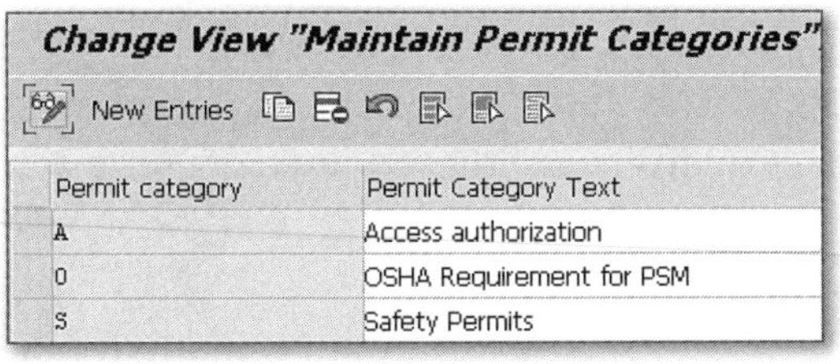

Figure 4.16 Permit Categories

4.7 Partners

By default, the SAP system knows only a few organizational units that you can assign to a technical object or order (e.g., maintenance planner group and responsible work center).

By defining *partners*, you can considerably expand these competencies and responsibilities and specify them in greater detail. A partner can be an individual or legal entity, and is either an internal organizational unit (e.g., departments, cost centers, or persons) or an external organizational unit (e.g., vendors, manufacturers, or service providers).

Partners are used in the following locations:

- Pieces of equipment and serial numbers
- Functional locations
- Notifications
- Orders

Figure 4.17 shows the following procedure for assigning partners to technical objects or processing objects:

- The *partner type* is predefined by SAP and always contains a database table (customer, contact person, vendor, user, personnel number, organizational unit, and position).
- In Customizing (Define Partner Determination Procedure and Partner Function), you can freely define *partner functions* with reference to a partner

type. For example, you can define partner functions such as the manufacturer, plant vendor, and service provider, and refer all functions to the *Vendor* database table.

- You can freely define a *partner determination procedure*. This is a grouping of partner functions that specifies which partner functions are permitted or must always be specified. For example, you can determine that the manufacturer and vendor must always be specified for a piece of equipment, but that the service provider is an optional specification.

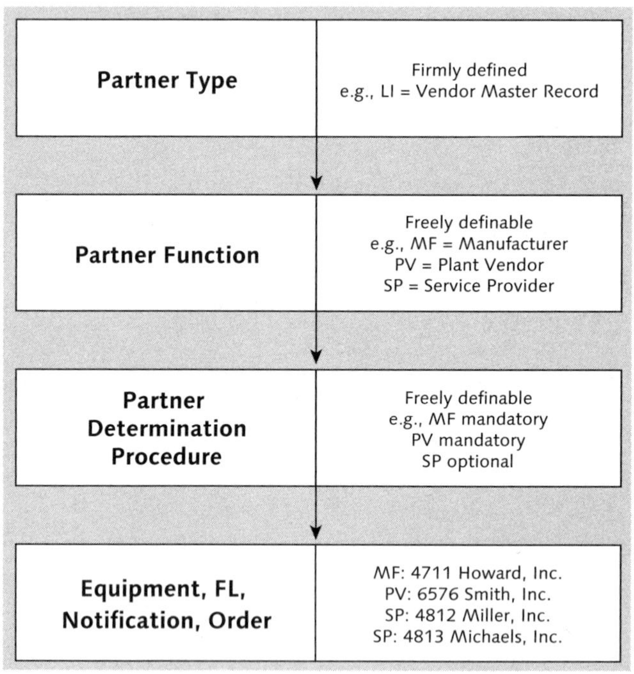

Figure 4.17 Partner

Define Partner Determination Procedure and Partner Function

You use this Customizing function to define partner functions. In particular, you use it to define the partner type to which a partner function is to belong. Furthermore, you can also use this Customizing function to define the partner determination procedures and to assign the corresponding partner functions to a partner determination procedure.

Prerequisites
There are no specific prerequisites.

Customizing Paths
Plant Maintenance and Customer Service • Master Data in Plant Maintenance and Customer Service • Basic Settings • Partners • Define Partner Determination Procedure and Partner Function

Plant Maintenance and Customer Service • Maintenance and Service Processing • Maintenance and Service Notifications • Notification Creation • Partners • Define Partner Determination Procedure and Partner Function

Plant Maintenance and Customer Service • Maintenance and Service Processing • Maintenance and Service Orders • Partners • Define Partner Determination Procedure and Partner Function

Transaction
VOP2

Choose Plant Maintenance, and then click Change Partner.

Settings
You use the Partner Functions function (see Figure 4.18) to define the partner functions that you require in technical objects or processing objects.

Take Care When Editing Partner Functions [!]

You use the Customizing function Define Partner Functions to maintain not only partner functions in SAP EAM, but also all partner functions used in the SAP system (e.g., in QM or customer service).

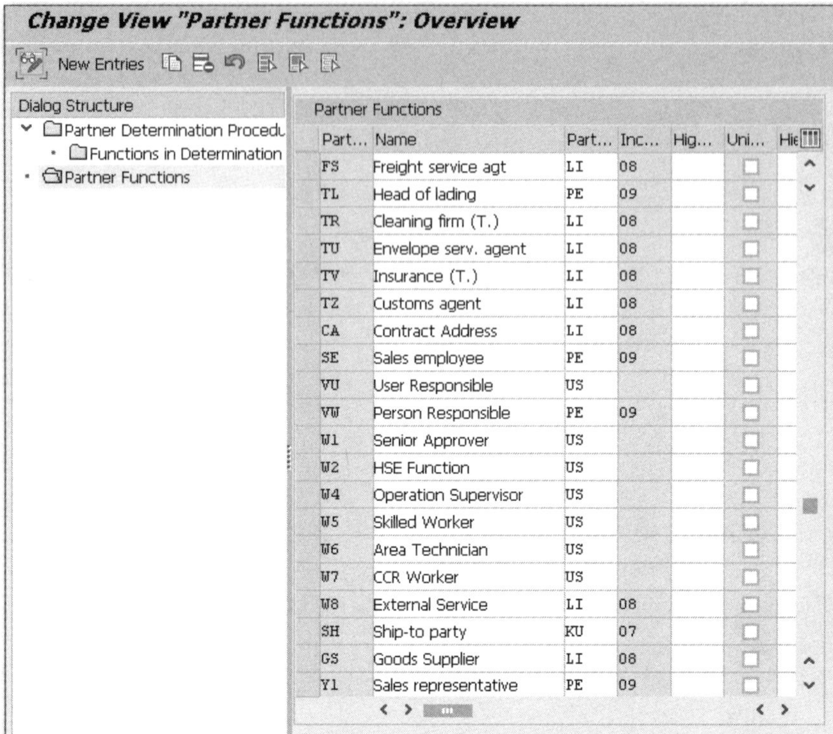

Part...	Name	Part...	Inc...	Hig...	Uni...	Hie
FS	Freight service agt	LI	08		☐	
TL	Head of lading	PE	09		☐	
TR	Cleaning firm (T.)	LI	08		☐	
TU	Envelope serv. agent	LI	08		☐	
TV	Insurance (T.)	LI	08		☐	
TZ	Customs agent	LI	08		☐	
CA	Contract Address	LI	08		☐	
SE	Sales employee	PE	09		☐	
VU	User Responsible	US			☐	
VW	Person Responsible	PE	09		☐	
W1	Senior Approver	US			☐	
W2	HSE Function	US			☐	
W4	Operation Supervisor	US			☐	
W5	Skilled Worker	US			☐	
W6	Area Technician	US			☐	
W7	CCR Worker	US			☐	
W8	External Service	LI	08		☐	
SH	Ship-to party	KU	07		☐	
GS	Goods Supplier	LI	08		☐	
Y1	Sales representative	PE	09		☐	

Figure 4.18 Partner Functions

You assign each partner function to a partner type. From an SAP EAM perspective, the following partner types are relevant:

- PE = person
- US = SAP system user
- LI = vendor
- S = position
- O = organizational unit

You use the PARTNER DETERMINATION PROCEDURES function to maintain the partner determination procedures that you require, and you use the FUNCTIONS IN DETERMINATION function to assign the partner functions required in each case (see Figure 4.19). You also define the following attributes here (the four indicators from left to right):

- No Change possible
  Whether it will be possible to change the partner retroactively.
- Partner mandatory
  Whether the partner function is to be mandatory.
- Unique
  Whether the partner function is unique (i.e., the partner function isn't permitted to be assigned to the same object repeatedly).
- Appointments
  Whether an appointment is to be entered in the employee's calendar.

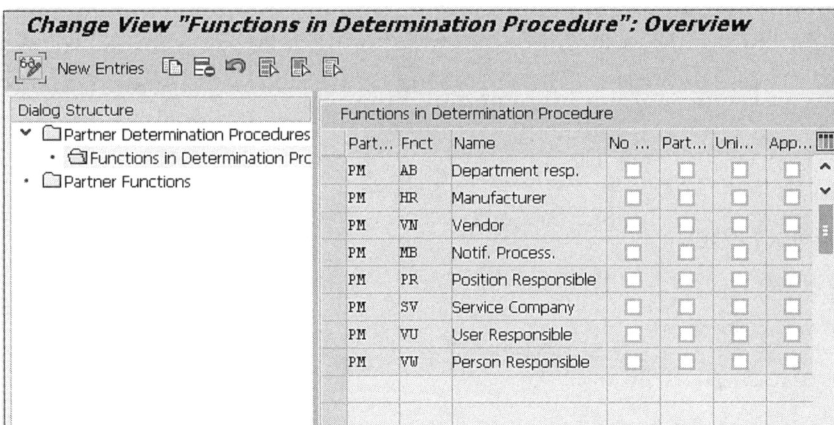

Figure 4.19 Partner Determination Procedure

Customizing Functions for the Field Selection for Partners

You use these Customizing functions to define which fields are to be displayed for which partner type at their place of use (e.g., in the equipment master or order).

Prerequisites

You must define partner functions for each partner type beforehand.

Customizing Paths

Plant Maintenance and Customer Service • Master Data in Plant Maintenance and Customer Service • Basic Settings • Partners

Plant Maintenance and Customer Service • Maintenance and Service Processing • Maintenance and Service Notifications • Notification Creation • Partners Field Selection for List Display of Partner Data

Then, the following Customizing functions are available to you:

- Define Field Selection for List Display of Vendor Data
- Define Field Selection for HR Data List Display
- Define Field Selection for Organizational Unit List Display
- Define Field Selection for Position List Display
- Define Field Selection for List Display of User Data
- Define Field Selection for Address Data List Display

Transactions

- OIR1 – Customer Data
- OIR2 – Vendor Data
- OIR3 – HR Data
- OIR4 – Contact Persons
- OIR5 – Organizational Units
- OIR6 – Positions
- OIR7 – User Data
- OIR8 – Address Data

Settings

Figure 4.20 shows the settings for the vendor field selection. However, these are representative of all other settings for field selections. Here, you define which fields are to be invisible as well as the display positions for those fields that you want to display.

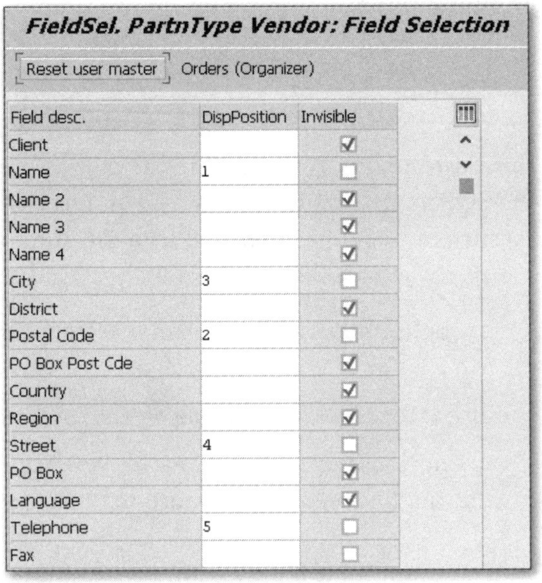

FieldSel. PartnType Vendor: Field Selection

Reset user master | Orders (Organizer)

Field desc.	DispPosition	Invisible
Client		☑
Name	1	☐
Name 2		☑
Name 3		☑
Name 4		☑
City	3	☐
District		☑
Postal Code	2	☐
PO Box Post Cde		☑
Country		☑
Region		☑
Street	4	☐
PO Box		☑
Language		☑
Telephone	5	☐
Fax		☐

Figure 4.20 Partner Type: Field Selection

4.8 Documents

Many companies want to link the technical objects and/or processing objects with documents such as design drawings, work instructions, checklists, images, inspection instructions, exploded drawings, measuring and control technology profiles, 3D models, and so on.

In the SAP system, you can manage your documents as document master records (Transactions CV01N to CV04N). Within these document master records, you define a link to the original documents (e.g., a link to the storage location of the CAD file or a link to the file server for Microsoft Office documents) whereby the link is retained or the original documents are uploaded to an SAP database.

In SAP EAM, documents are used in the following objects:

- Pieces of equipment and serial numbers
- Functional locations
- Bills of material (BOMs)

- Notifications
- Orders
- Task lists

The next section doesn't describe the Customizing required for the full scope of functions in the SAP document management system because, for pure plant maintenance purposes, it's generally sufficient to define a reduced scenario and to configure the necessary Customizing. This scenario looks as follows:

- To differentiate this scenario from other application areas, you want this scenario to have its own *document type*.
- Furthermore, you want to be able to use it for *plant maintenance objects* only.
- To prevent other enterprise areas from encroaching on this scenario, you want to create a separate *secure storage area* in the SAP system and store the original documents there.
- All other functions in the SAP document management system (e.g., status profiles and versioning) don't play any role here.

We'll discuss how to map this scenario in Customizing next.

Define Document Types

You use this Customizing function to configure the basic settings for a document type (e.g., number assignment or object links).

Prerequisites

You must define the number ranges (external and internal) beforehand.

Customizing Path

Cross-Application Components • Document Management • Control Data • Define Document Types

Settings

Figure 4.21 shows the basic settings for a document type.

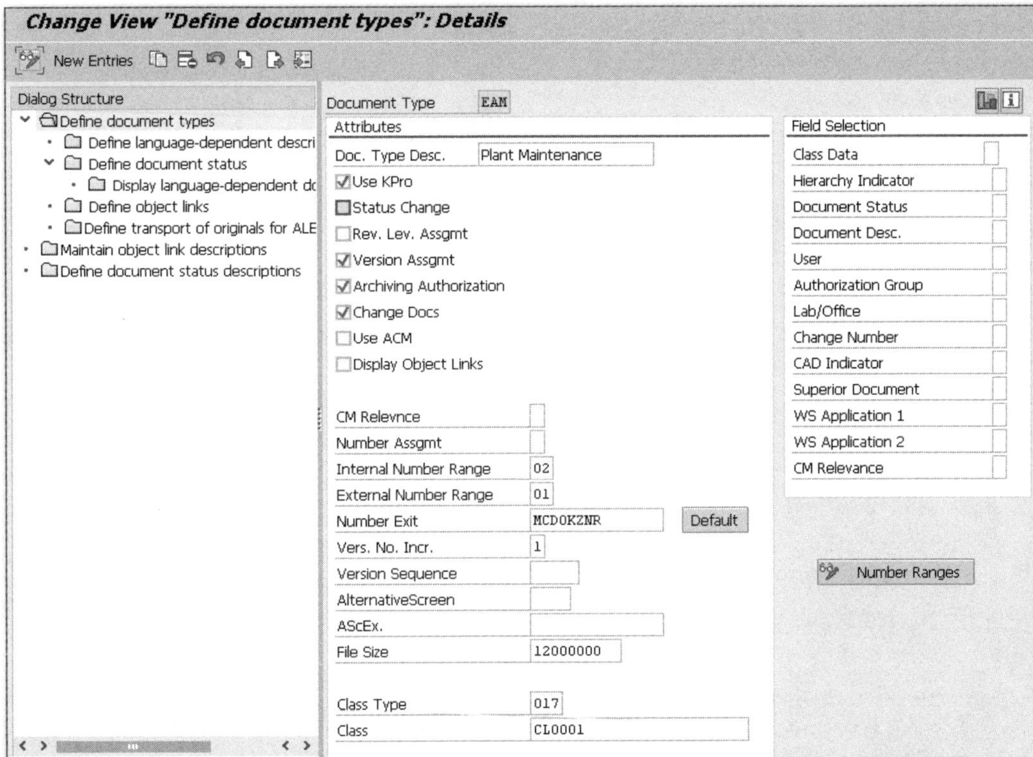

Figure 4.21 Document Type: Basic Settings

The following describes how to configure these settings from an SAP EAM perspective:

- Use KPro
  This indicator controls the use of Knowledge Provider to store original files in defined storage systems (known as content repositories). If you define a new document type, you should use only Knowledge Provider to control file storage. Don't set this indicator retroactively for a document type. If original files are already maintained for a document type, and Knowledge Provider wasn't used previously to control file storage, the system can no longer manage these original files.

- Version Assgmt
  If you set this indicator, the system assigns the new version number automatically. Because version numbers are rarely used in plant maintenance, version 00 is automatically assigned for the document.

- ARCHIVING AUTHORIZATION
  This indicator controls whether the original files can be archived.
- CHANGE DOCS
  Set this indicator if you want to use change documents to log changes to the document info record.
- INTERNAL NUMBER RANGE/EXTERNAL NUMBER RANGE
  You use the number range controls to assign the external and internal number ranges.
- FILE SIZE
  The specifications in relation to the maximum file size are relevant only if you don't use KPro to store the files.

You use the subfunction DEFINE DOCUMENT STATUS (see Figure 4.22) to define the status to be used. In SAP EAM, you generally don't need to use a multilevel status profile such as those used in product development. You require only a status that has activated the RELEASE FLAG indicator so that the document can be used in other business processes.

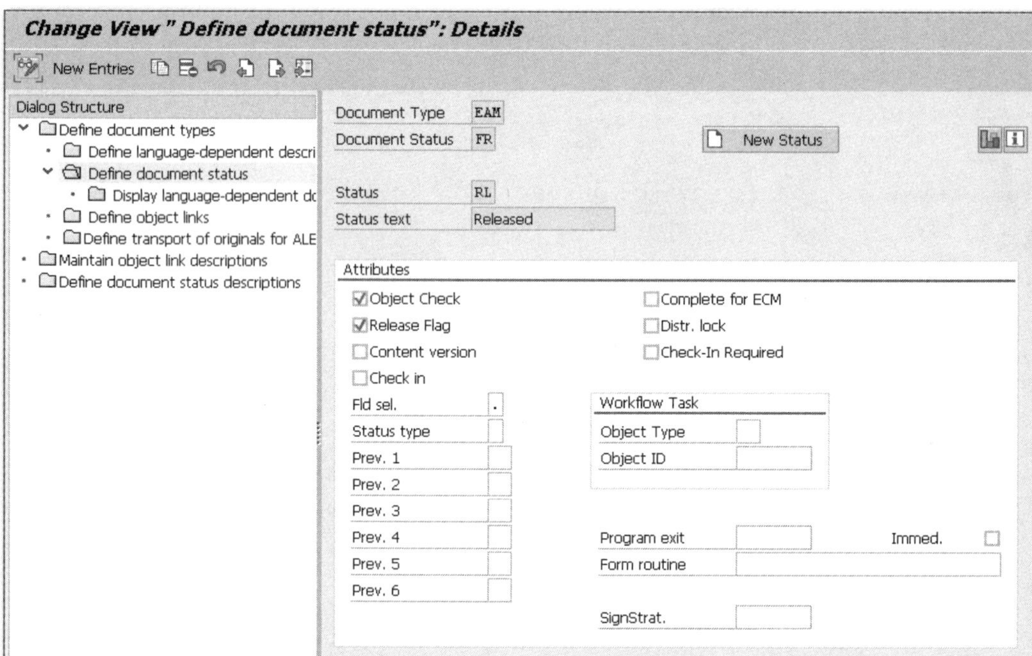

Figure 4.22 Document Type: Status

You use the subfunction DEFINE OBJECT LINKS (see Figure 4.23) to define the SAP objects to which documents of this document type can be assigned.

From an SAP EAM perspective, these are the following:

- Equipment (EQUI)
- Functional locations (IFLOT)
- Reference functional locations (IRLOT)
- Object links (INET)
- Materials (MARA)
- Plant materials (MARC)
- Measuring points (IMPTT)
- Notifications (PMQMEL)
- Tasks (IMAV)
- Orders (PMAUFK)
- Order operations (PMAFVC)
- BOM headers (STKO_DOC)
- BOM items (STPO_DOC)
- Task lists (PMPLKO)

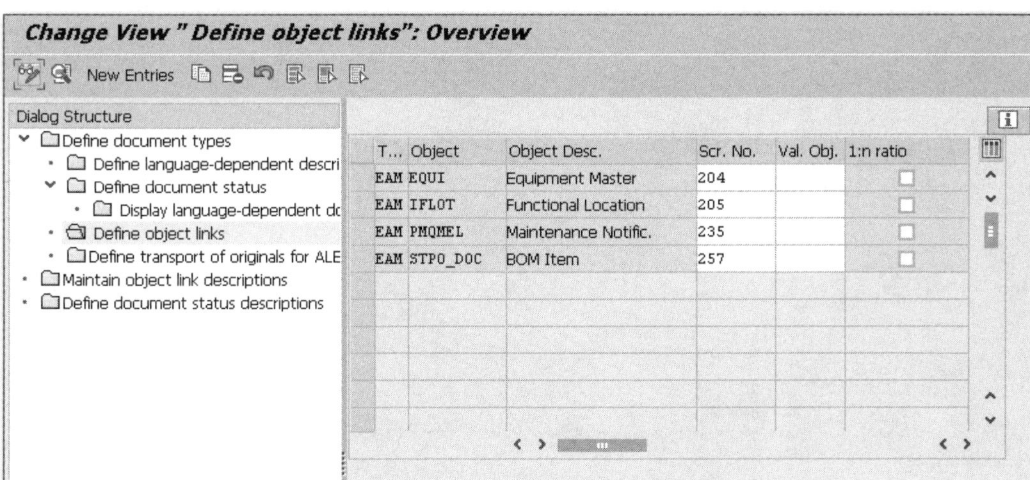

Figure 4.23 Document Type: Object Links

Don't set the 1:N RATIO indicator. If you do, this means that you can only assign a single document of this document type to the object. Generally, this doesn't make sense.

Maintain Storage System

You use this Customizing function to define the storage location (known as the content repository) for the original files. The system uses these storage systems for documents whose document type allows for storage via Knowledge Provider.

Prerequisites

If you want to use your own database tables to physically store documents, you must define these beforehand in Transaction SE11. Furthermore, their structure must correspond to the structure of the database Table SDOKCONT1.

Customizing Path

CROSS-APPLICATION COMPONENTS • DOCUMENT MANAGEMENT • GENERAL DATA • SETTINGS FOR STORAGE SYSTEMS • MAINTAIN STORAGE SYSTEM

Transaction

OAC0

Settings

Figure 4.24 shows the settings for a storage system.

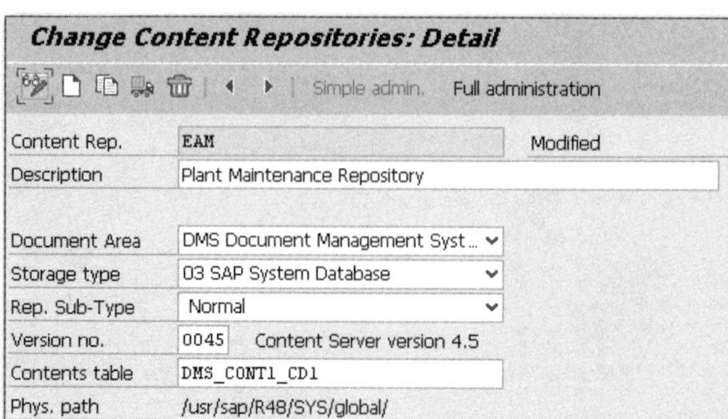

Figure 4.24 Storage System

The following describes how to configure these settings from a plant maintenance perspective:

- The database of the SAP system should be the Storage type.
- You should always select the latest Version no. for the database interface (currently 0045).
- The Contents table is the database table in which the documents are stored. You can define this table as a client-specific or cross-client table. You must define your own database tables in Transaction SE11 beforehand.

Maintain Storage Category

You use this Customizing function to set up one or more logical views (storage categories) of a storage system (content repository). Having several views of the same storage system enables you to store the original files in a highly structured manner.

Prerequisites

You must define the content repository (storage system) beforehand.

Customizing Path

Cross-Application Components • Document Management • General Data • Settings for Storage Systems • Maintain Storage Category

Transaction

OACT

Settings

Figure 4.25 shows you how to assign a category to a content repository.

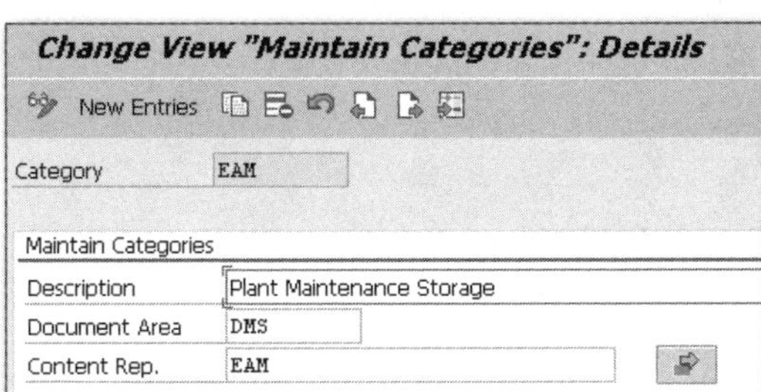

Figure 4.25 Storage Category

Define Workstation Application

For all file types that you want to use in SAP EAM (e.g., PDF, JPG, DWG, and DOC), you must use this Customizing function to specify the application to which they belong as well as how you want to control these file types.

Prerequisites

You must install the necessary programs for starting the relevant file types beforehand.

Customizing Path

Cross-Application Components • Document Management • General Data • Define Workstation Application

Settings

Figure 4.26 shows an overview table of all applications and associated file types. The standard SAP delivery already contains a large number of applications and file types, which you should check for accuracy and completeness. If the necessary file types don't exist, you must add the corresponding new entries to this table.

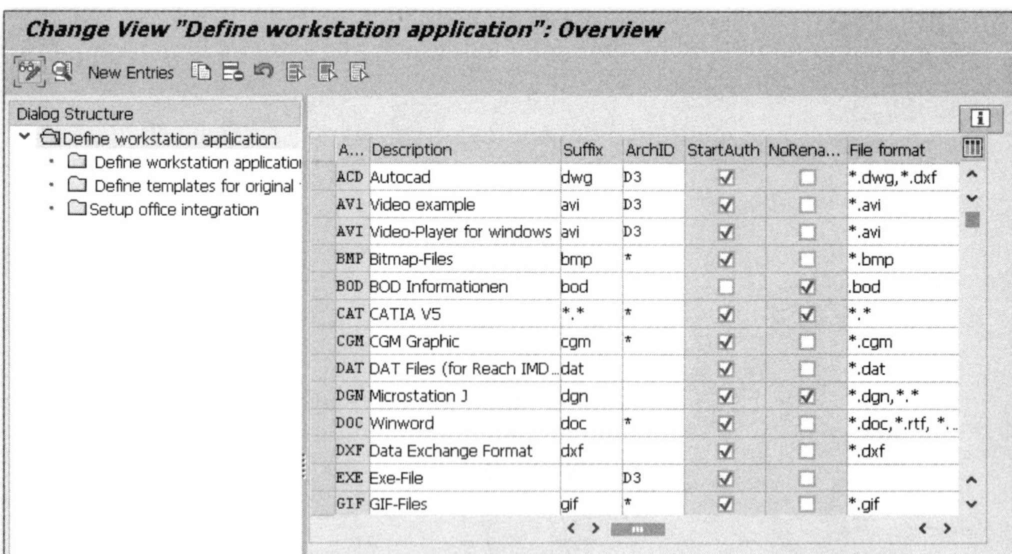

Figure 4.26 Work Station Applications: Overview

As an example, Figure 4.27 shows the detail screen for the application JPG. Here, the following is important:

- FILE FORMAT
  Be sure to assign *all* potential file formats (e.g., DOC application with the file formats DOC, DOCX, and RTF, among others).
- START AUTHORIZATION
  Select this indicator so that the application starts immediately after you double-click the file (e.g., a PDF in the equipment master) and is permitted to display the document.

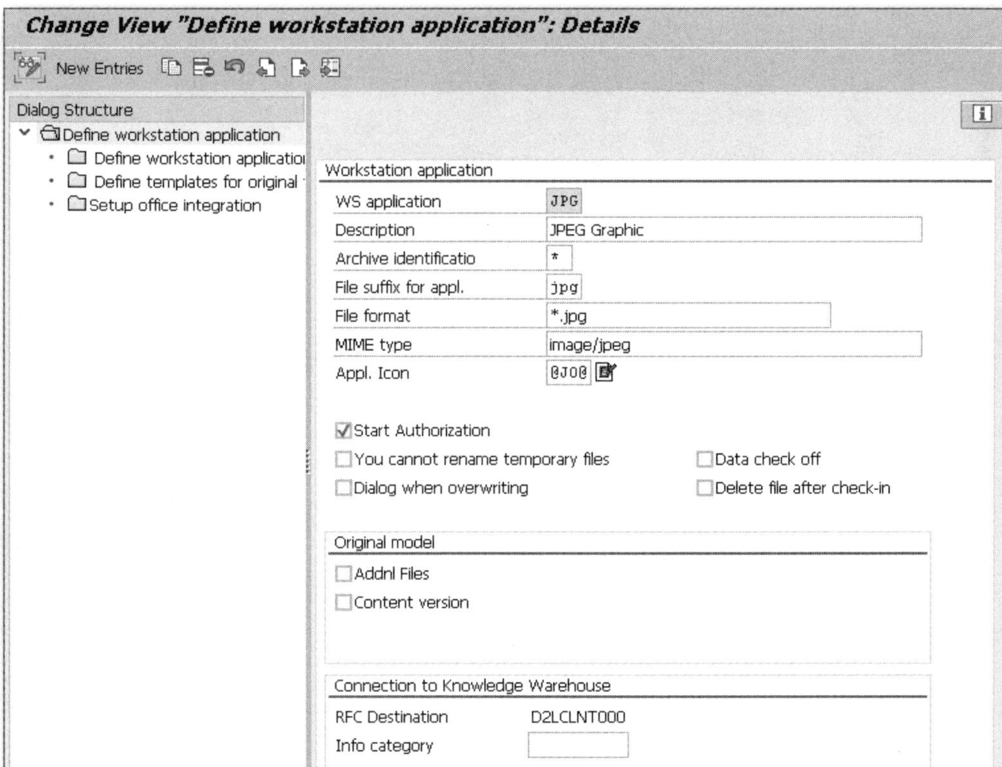

Figure 4.27 Workstation Applications: Detail Screen

You use the DEFINE WORKSTATION APPLICATION IN NETWORK subfunction to define which program is to start when a certain file type is called. In Figure 4.28, this is shown for JPEG files.

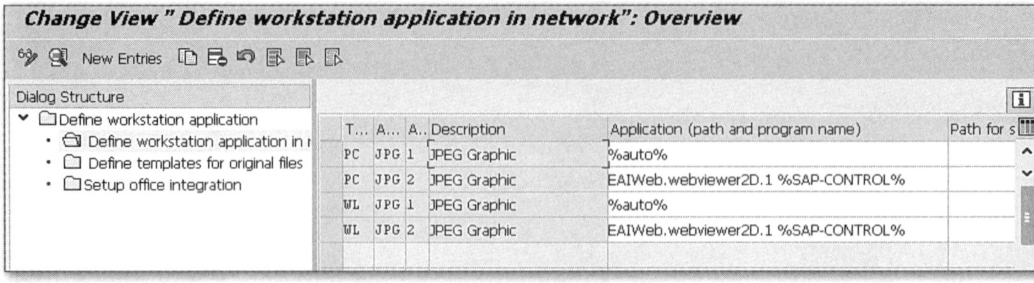

Figure 4.28 Define Workstation Application in Network

Note the following regarding this subfunction:

- It always makes sense to specify "%auto%" as the first alternative so that the workstation then automatically uses the general MIME technology to search for the correct program.
- As a second alternative, enter a specific path for the program that you want to start.

For certain purposes, SAP also provides integrated viewers that can be called using a predefined string (e.g., 2D CAD drawings via *EAIWeb.webviewer2D.1 %SAP-CONTROL%*, 3D CAD drawings via *EAIWeb.webviewer3D.1 %SAP-CONTROL%*, or right hemisphere files via *%VIEWER-CONTROL% %SAPPROVIS%*).

4.9 Field Selection

You use field selections to define attributes for each object type. For example, you can determine the following:

- Which field must be maintained
- Whether the field is merely a display field
- Whether it's a normal entry field
- Whether the field is to be hidden

Furthermore, you can and should base the field selection on *influencing fields*.

The field selection technique is available to you for almost all SAP objects. From an SAP EAM perspective, they are, in particular:

- Equipment and serial numbers
- Functional locations and reference functional locations
- Materials
- Notifications
- Orders, order operations, and components
- Completion confirmations
- Measuring points and measurement documents
- Maintenance plans
- Task lists
- Work centers

Customizing Functions for Field Selection

For each of the preceding objects, the system has separate Customizing functions for controlling the field selection. However, the technique is always the same. Next we'll discuss this technique using the example of the field selection for functional locations.

Prerequisites

If you want to base the field selection on influencing fields (e.g., the field selection for a functional location is based on the *functional location category* or the field selection for an order is based on the *order type*), you must define these influencing factors beforehand.

Customizing Path

Plant Maintenance and Customer Service • Master Data in Plant Maintenance and Customer Service • Technical Objects • Functional Locations • Define Field Selection for Functional Locations

Transactions

- OIAE for functional location-specific data
- OIAF for fields that are common to both the functional location and the piece of equipment

Furthermore, for the field selection, separate transactions are assigned to all other objects; for example, Transaction OIAD (Equipment), Transaction OIW0 (Maintenance Plans), Transaction OP5A – OP5H (Task Lists), Transaction OIAL (Notifications), Transaction OIAN (Order Header Data), Transaction OIOPD (Order Operations), or Transaction OIZN (Completion Confirmations).

Settings

You use the Field Selection for Special Functional Location Fields subfunction to define the field selection for fields that concern the functional location only.

You use the Field Selection for Functional Location (Common Equipment/Location Fields) subfunction to define those fields that are common to both the functional location and the piece of equipment.

After you select the function and choose Influencing, you access an overview of all influencing fields. In the case of the functional location, the following fields are suitable:

- The activity category (Add, Change, and/or Display)
- The object type
- The functional location category

In particular, the functional location category is frequently selected so that the field selection is based on this category. You then have the following options for each field (see Figure 4.29):

- Input
  The field is a normal entry field.
- Req.
  The field is a required entry field.
- Disp.
  The field is a display only field.
- Hide
  The field is hidden.
- HiLi
  The field is highlighted (in a different color).

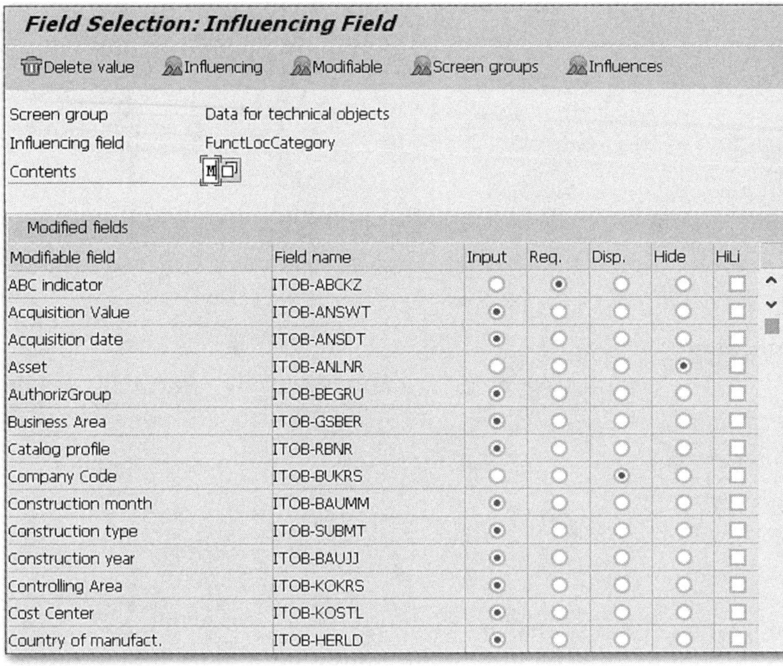

Figure 4.29 Field Selection: Functional Location

If you want to base the field selection on several influencing fields (e.g., on the functional location category and object type), two influencing factors then apply to a field. SAP has clearly defined priority rules for such cases (see Figure 4.30).

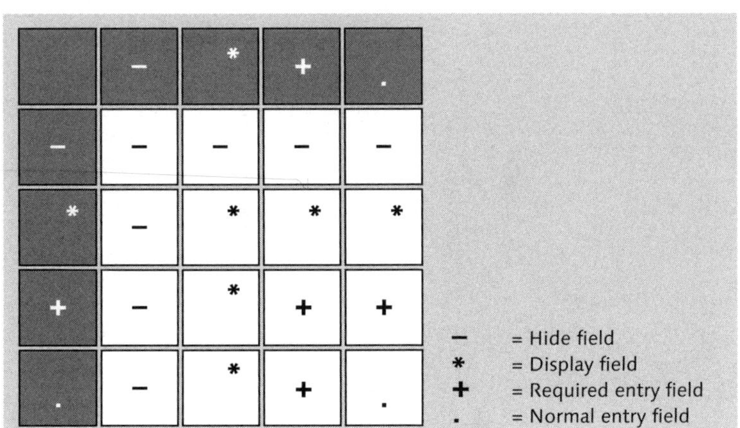

Figure 4.30 Field Selection: Priority Rules

The priority rules are explained here:

- **Priority 1: Hide**
  If one of the influencing factors defines a field as a field that is to be hidden, it's hidden irrespective of the field selection control for other influencing fields.
- **Priority 2: Display**
  If it's not hidden, and an influencing factor defines it as a display field, it becomes a display field, even if other influencing factors define it as a required entry or normal entry field.
- **Priority 3: Required**
  If it's not a hidden field or display field, and an influencing factor defines it as a required entry field, it becomes a required entry field, even if other influencing factors define it as a normal entry field.
- **Priority 4: Normal**
  If no influencing factor defines it elsewhere, this field remains a normal entry field.

Highlighting fields in color isn't subject to any priority rules. If an influencing factor defines a field as highlighted, it's flagged with another color.

4.10 List Variants

In SAP EAM, a large number of lists are available to you. SAP provides lists for the following objects:

- Equipment (IE05, IH08)
- Fleet objects (IE36, IE37)
- Serial numbers (IQ08, IQ09)
- Functional locations (IL05, IH06)
- Reference functional locations (IL15, IH07)
- Object links for functional locations (IN15, IN16)
- Object links for equipment (IN18, IN19)
- Materials (IH09)
- Notifications (IW28, IW29)

- Notification items (IW68, IW69)
- Activities (IW64, IW65)
- Tasks (IW66, IW67)
- Orders (IW38, IW39)
- Order operations (IW37, IW49)
- Completion confirmations (IW47)
- Measuring points (IK07, IK08)
- Measurement documents (IK17, IK18)
- Maintenance plans (IP15, IP16)
- Maintenance items (IP17, IP18)
- Task lists (IA08, IA09)

In the case of all lists, a large number of Customizing functions for selecting and determining the layout of the lists are available to you, each of which is described next.

Customizing Functions for Configuring Lists

You can use these Customizing functions to define the following presettings for all users:

- Determine all potential selection fields.
- Set specific values for the selection fields.
- Perform the field selection for the list display.

These settings apply to all users who haven't created their own user-specific variants or don't use other general variants.

Similarly, list editing, which is introduced here, applies to all aforementioned objects. For the sake of simplicity, we'll use the example of a functional location to illustrate the list editing technique.

Prerequisites

Generally, there are no prerequisites. If, however, you want to define a selection or display variant that makes reference to other Customizing functions (e.g., functional location category), you must define it beforehand.

Customizing Path

Plant Maintenance and Customer Service • Master Data in Plant Maintenance and Customer Service • Technical Objects • Functional Locations • Set List Editing for Functional Locations

Transaction

OIW6

Furthermore, separate transactions are generally assigned to all other Customizing functions for list editing; for example, Transaction OIYC (Equipment), Transaction OIYH (Serial Numbers), Transaction OIWW (Maintenance Plans), Transaction OIW5 (Maintenance Dates), Transaction OIWO (Task Lists), Transaction OIWI (Notifications), Transaction OIWL (Orders), Transaction OIWU (Operations), or Transaction OIW3 (Completion Confirmations).

Settings

When you execute the Customizing function, you obtain the following three subfunctions:

- List for selection screen: Display mode
- List for selection screen: Change mode
- Maintain list for field selection

You should first maintain the selection screen. Then, depending on your selection, you'll obtain a list of all potential selection fields. You can then choose Variant Attributes to access the detail screen. Here, you can decide how to proceed for each field. The settings on this screen have the following meaning (see Figure 4.31):

- Protect field
  If selected, the relevant selection criterion is write protected at runtime (e.g., if a user starts the variant).
- Hide field
  If selected, the relevant selection criterion is hidden when a user starts a variant or changes the values.
- Hide field 'BIS'
  If selected, the same applies here as in Hide field, and then it's not possible to specify an interval.

- REQUIRED FIELD
  If selected, a value must be entered for the relevant selection criterion when the user starts a variant.

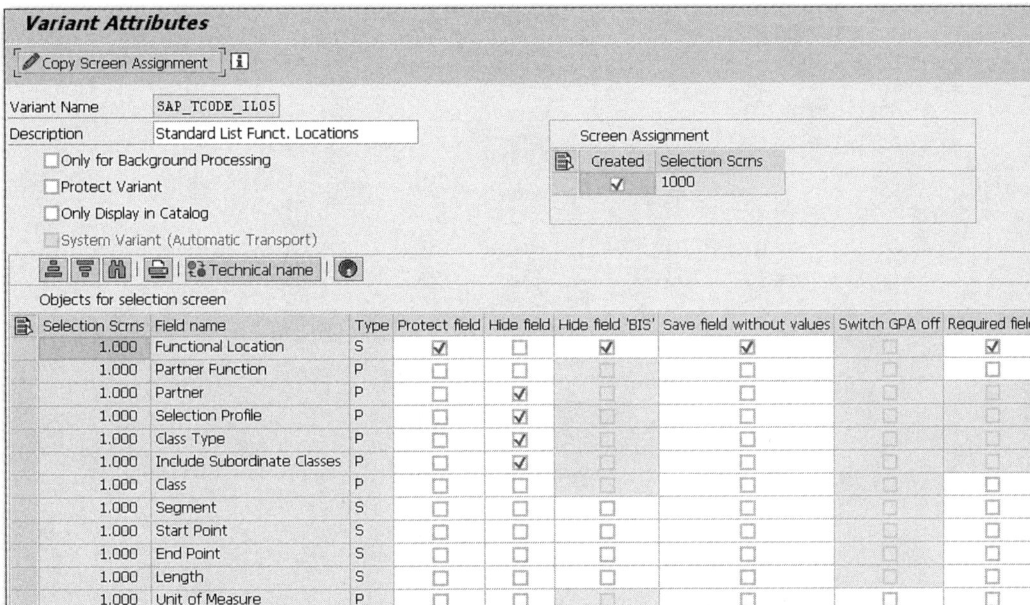

Figure 4.31 Functional Location: Selection Variant

You use the subfunction MAINTAIN FIELD SELECTION LIST to determine the layout of the standard list. You can then choose to access a dialog box in which you can configure these settings (see Figure 4.32):

- DISPLAYED COLUMNS
  Which fields are to be displayed?
- SORT ORDER
  According to which criteria is the list to be sorted?
- FILTER
  Is a filter to be used?
- VIEW
  Do you want to output an SAP list or an Excel table?

- DISPLAY
  For example, do you want a striped pattern, or do you want to optimize the column width?

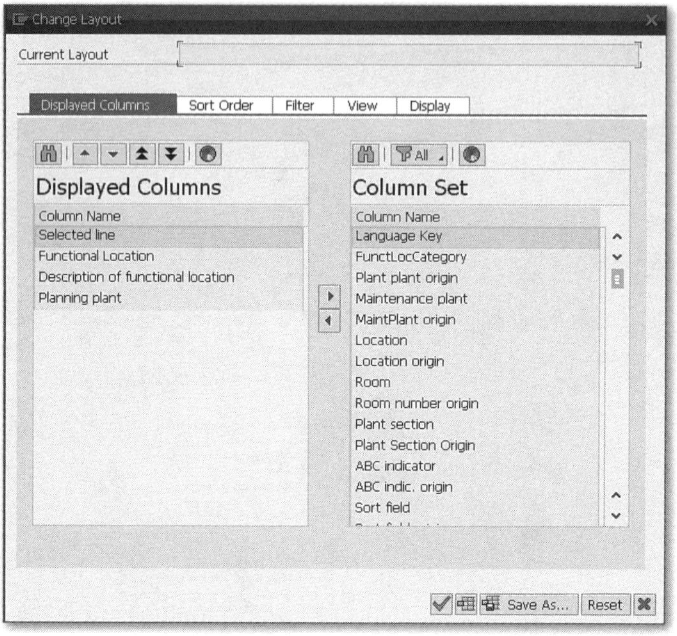

Figure 4.32 Functional Location: List Variant

4.11 Multilevel Lists

From an SAP EAM perspective, multilevel lists are available for the following:

- Equipment (IE07)
- Functional locations (IL07)
- Reference functional locations (IH18)
- Notifications (IW30)
- Orders (IW40)
- Task lists (IA10)

In the case of a list object (e.g., a functional location), multilevel lists are used to display other objects (e.g., orders, notifications, and documents) for a functional location.

Customizing Functions for the Field Selection for Multilevel List Displays

You use this Customizing function to adjust the presettings for multilevel lists. Here, you define which fields are to be displayed for which objects. These settings apply to all users who haven't created their own user-specific variants or don't use other general variants.

The multilevel list technique is available to you for all of the preceding SAP objects. For the sake of simplicity, we'll use the example of a functional location to illustrate the multilevel list technique.

Prerequisites

There are no specific prerequisites.

Customizing Path

Plant Maintenance and Customer Service • Master Data in Plant Maintenance and Customer Service • Technical Objects • Functional Locations • Field Selection for Multilevel List Displays of Functional Location

The following subfunctions are available for the individual objects:

- Define Field Selection for Functional Location Fields
- Define Field Selection for Equipment Usage Data Fields
- Define Field Selection for Equipment Master Data Fields
- Define Field Selection for Partner Data Fields
- Define Field Selection for Notification Data Fields
- Define Field Selection for Order Data Fields
- Define Field Selection for Class Data Fields
- Define Field Selection for Characteristic Data Fields
- Define Field Selection for Document Management Fields
- Define Field Selection for Object Link Data Fields
- Define Field Selection for Measuring Point and Counter Fields

- Define Field Selection for Measurement Document and Counter Reading Fields
- Define Field Selection for Permit Fields

Settings

Figure 4.33 shows the settings for the order segment in the multilevel list for functional locations. However, these are representative of all other objects.

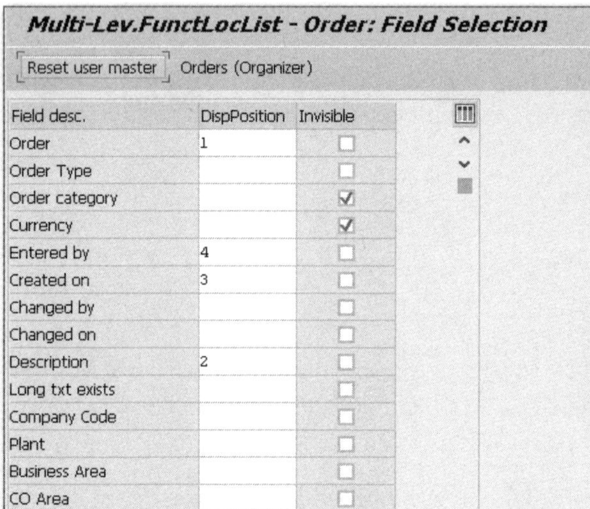

Figure 4.33 Field Selection: Multilevel List for Functional Locations (Order Segment)

- DispPosition
  You define, as a presetting, the position at which the field is to appear. By default, fields without an item number aren't displayed. When designing their individual layouts, users can change or delete the item number as well as enter an item number for other fields so that they are displayed.
- Invisible
  You define whether the field is to be invisible for user-specific list designs. If a field is invisible, a user can't select it for his variant.

This chapter introduces you to the structuring elements of SAP EAM. You'll learn how to configure these elements, and find tips that deal with configuring the Customizing settings for asset structures.

5 Structuring of Technical Systems

A requirements-oriented approach to the structuring of technical systems is the basis for being able to use SAP Enterprise Asset Management (SAP EAM) to map business processes in plant maintenance, and subsequently process them. The following issues are discussed in the book *Plant Maintenance with SAP—Pracitcal Guide*:

- Structuring elements to use
- Depth of the structure
- Structure levels for particular elements
- Information to store
- Functions to use
- Levels to map assets to

5.1 Structuring Elements

SAP EAM provides you with a broad range of potential structuring elements; namely functional locations, reference functional locations, equipment, object links, serial numbers, maintenance assemblies, materials, and different types of bills of material (BOMs).

Functional locations represent a complex, generally multilevel asset structure. You must create each element of the asset structure as a functional location. Therefore, functional locations are used to establish a vertical asset structure. Functional locations usually represent immovable, functional units. Examples of functional locations include process plants in the chemical and pharmaceutical industries, power plants, production lines, buildings, pipeline systems, infrastructures, and computer networks.

Reference functional locations are solely templates for generating *real* functional location structures or for subsequently passing on data to real functional locations.

Equipment represents movable, individual aggregates (inventories). Examples include machines, pumps, engines, production resources/tools (PRTs), fleet objects (cars, trucks, forklift trucks, and industrial trucks), and IT inventories (PCs, printers, monitors, notebooks, and projectors). Equipment categories that are moved infrequently (e.g., pumps) are intended to be installed on functional locations. Other equipment categories (e.g., fleet objects) are not installed on functional locations, due to their constant movement.

You establish *object links* between different technical objects (e.g., pieces of equipment or functional locations). Such links exist, for example, between individual production units, between production plants and supply systems, and between supply systems and disposal systems. You use object links to form an *object network*. In this way, your assets can have a horizontal structure. Object links can't be the subject of business processes (e.g., malfunction reports). Rather, they are used for information and visualization purposes only.

Linear assets (Linear Asset Management) are technical systems with a linear infrastructure whose properties and conditions may change from stage to stage (dynamic segmentation). Examples of linear assets include pipelines, road networks, railway networks, overhead cables, pipes, and so on. You can create linear assets as technical objects (e.g., functional locations or pieces of equipment) and store linear data there.

Unlike a piece of equipment, a *material* doesn't represent an individual item but rather an object category (e.g., the category *pump normal 400–100* or the category *three-phase normal engine SM/I, 220/380 V, 50 Hz, 0.18 kW*). A material includes a specific number of objects of the corresponding category. You require materials for spare parts and for pieces of equipment and maintenance assemblies that are suitable for storage.

A *maintenance assembly* serves to provide a functional location or piece of equipment with a deeper structure. For example, a forklift truck could comprise the following maintenance assemblies: *lift type*, *chassis*, *brake system*, and *drive assembly*. You can also assign a maintenance assembly to a notification, order, or maintenance plan to specify the location of any damage that occurs.

You create *serial numbers* for a material number. Here, you can specify as many serial numbers as you want for one material number. A serial number is an individual item and corresponds to a piece of equipment. The serial number function enables you to place the equipment in storage. In terms of usage, a material serial number corresponds to a piece of equipment.

An *equipment bill of material* generally comprises a list of spare parts, and it's assigned directly to a piece of equipment. This means that only this piece of equipment can use this BOM.

A *BOM for a functional location* generally comprises a list of spare parts, and it's assigned directly to a functional location. This means that only this functional location can use this BOM.

A *material BOM* also comprises a list of spare parts. However, you can make the material BOM available for any number of pieces of equipment or functional locations by means of an *indirect assignment*.

In the following sections, you'll learn about the main settings for the individual technical objects.

5.2 Functional Locations and Reference Functional Locations

Functional locations represent a complex, generally multilevel, immovable asset structure. Therefore, functional locations are used to establish a vertical asset structure.

Reference functional locations are solely templates for generating *real* functional location structures or for subsequently passing on data to "real" functional locations. Reference functional locations can't be the subject of business processes (e.g., malfunction reports). Rather, they are used as a reference template.

Practical examples include the following:

- Process plants in the chemical, pharmaceutical, and food industries
- Power plants such as coal-fired, hydroelectric, and nuclear power plants
- Production lines in discrete manufacturing
- Complex machines such as automated systems or flexible production cells

- Real estate
- Power systems, gas systems, water systems, and heating systems
- Pipeline systems
- Infrastructure such as roads, public places, tracks, tunnels, and bridges
- Computer networks
- Complex fleet objects such as locomotives, high-speed trains, and tractors
- Aircraft

Figure 5.1 provides an overview of the Customizing functions for functional locations.

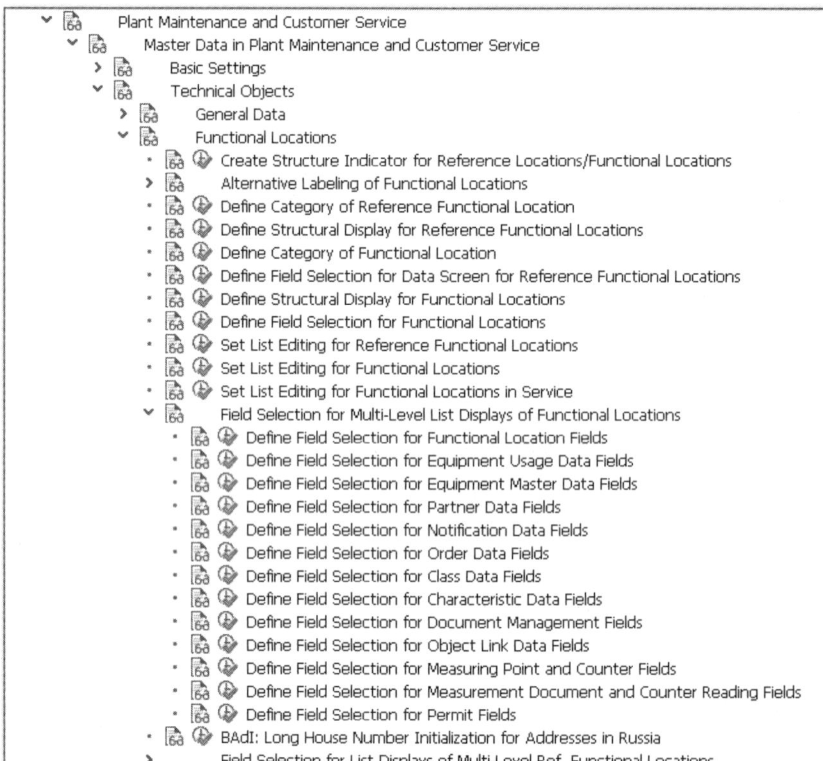

Figure 5.1 Functional Locations: Customizing Functions

Next, we'll explain the most important of these Customizing functions.

Create Structure Indicator for Reference Locations/Functional Locations

You always assign an external number to a functional location and reference functional location in accordance with the *structure indicator*. You manually assign this functional location number according to the specifications made using the structure indicator.

Prerequisites

There are no special prerequisites.

Customizing Path

Plant Maintenance and Customer Service • Master Data in Plant Maintenance and Customer Service • Technical Objects • Functional Locations • Create Structure Indicator for Reference Locations/Functional Locations

Transaction

OIPK

Settings

Figure 5.2 provides an overview of existing structure indicators. If you create a new structure indicator, you define the name and description of the structure indicator there.

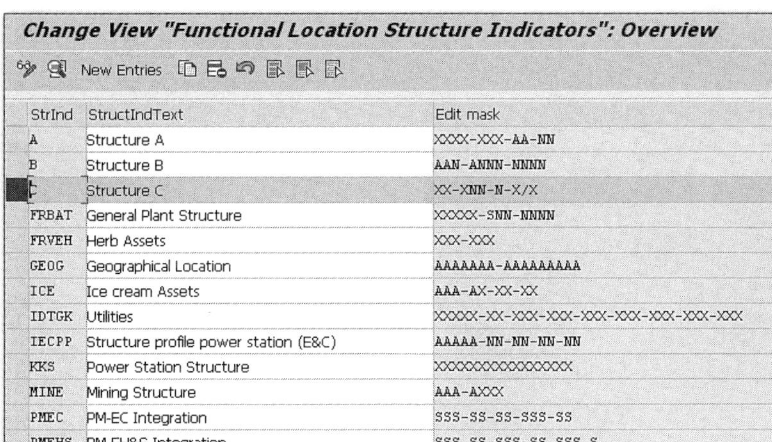
Change View "Functional Location Structure Indicators": Overview

New Entries

StrInd	StructIndText	Edit mask
A	Structure A	XXXX-XXX-AA-NN
B	Structure B	AAN-ANNN-NNNN
C	Structure C	XX-XNN-N-X/X
FRBAT	General Plant Structure	XXXXX-SNN-NNNN
FRVEH	Herb Assets	XXX-XXX
GEOG	Geographical Location	AAAAAAA-AAAAAAAAA
ICE	Ice cream Assets	AAA-AX-XX-XX
IDTGK	Utilities	XXXXX-XX-XXX-XXX-XXX-XXX-XXX-XXX-XXX
IECPP	Structure profile power station (E&C)	AAAAA-NN-NN-NN-NN
KKS	Power Station Structure	XXXXXXXXXXXXXXX
MINE	Mining Structure	AAA-AXXX
PMEC	PM-EC Integration	SSS-SS-SS-SSS-SS
PMEHS	PM-EH&S Integration	SSS-SS-SSS-SS-SSS-S

Figure 5.2 Functional Location: Structure Indicator Overview

On the detail screen, you configure additional settings (see Figure 5.3):

- The length of the functional location number
- The number of levels for the asset structure
- The number of characters for the individual level numbers
- The rule used to establish the functional location number:
  - A: Alphabetic characters
  - N: Numeric characters
  - X: Alphanumeric characters
  - S: Special characters and alphanumeric characters

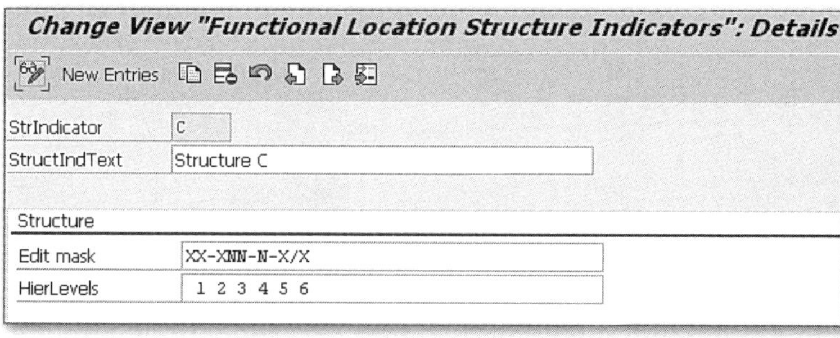

Figure 5.3 Functional Location – Structure Indicator: Detail Screen

Set View Profiles for Technical Objects

You can use this Customizing function to define the layout of screens associated with technical objects (e.g., equipment or functional locations). You determine the following:

- The number of tab pages
- Which screen groups to position on which tab page and where
- The name of the tab page
- Whether an icon is to be assigned to the tab page

Prerequisites

There are no special prerequisites. You assign defined view profiles to an equipment category or functional location category.

Customizing Path

Plant Maintenance and Customer Service • Master Data in Plant Maintenance and Customer Service • Technical Objects • General Data • Set View Profiles for Technical Objects

Settings

First, define a name for the view profile. Then, specify whether you want it to be valid for equipment or functional locations (see Figure 5.4).

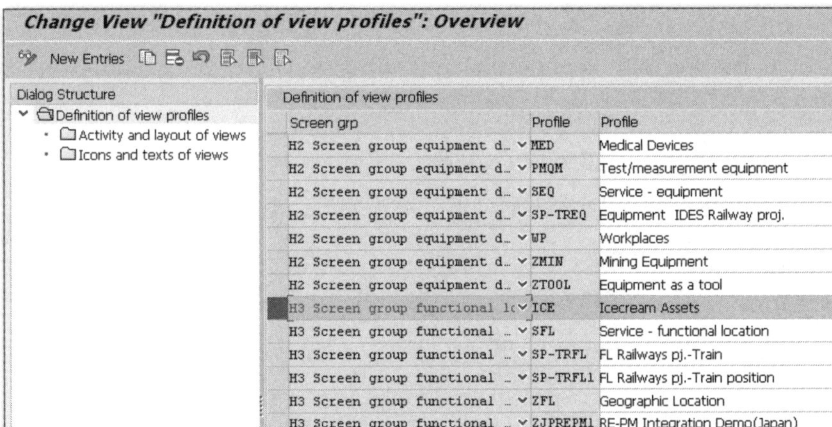

Figure 5.4 Functional Location: View Profile Definition

Then, use the Activity and layout of views subfunction to assign the relevant screen areas to the tab pages (see Figure 5.5).

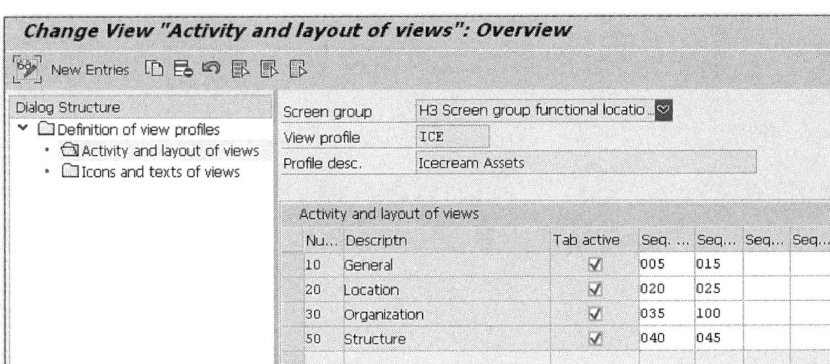

Figure 5.5 Functional Location: View Profile Layout

The following screen areas are available:

- 005 – General (e.g., size, weight, and inventory number)
- 010 – Reference (e.g., vendor, acquisition value, and acquisition date)
- 015 – Manufacturing (e.g., manufacturer, year of construction, and month of construction)
- 020 – Location (e.g., maintenance plant, location, and room)
- 025 – Address (e.g., zip code, town, telephone number, and fax number)
- 030 – Account assignment (e.g., cost center, asset, and company code)
- 035 – Responsibilities (e.g., planner group and responsible work center)
- 040 – Structuring (e.g., superior functional location and position)
- 045 – Equipment (e.g., installed equipment)
- 095/097 – Full-screen standard class/subscreen standard class
- 100 – Partner (e.g., partner function, name, and address)
- 105 – Long text
- 110 – Warranty (e.g., customer warranty and vendor warranty)
- 120/122 – Full-screen documents/subscreen documents
- 140/150 – User data with custom fields
- 141 – Log book data
- 300 – Linear data for functional locations
- 350 – Linear data for characteristics

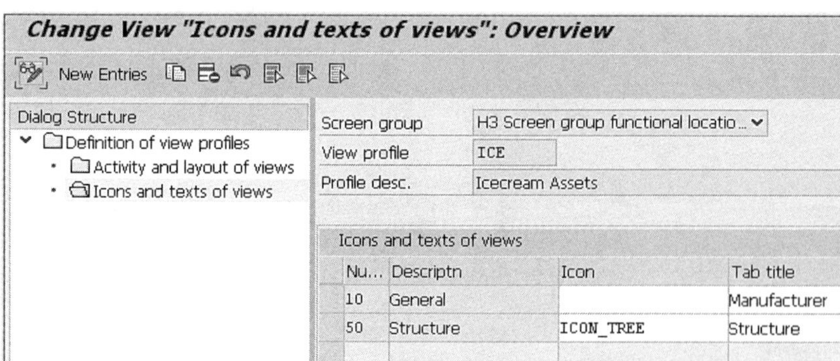

Figure 5.6 Functional Location: View Profile of Icons and Texts

Finally, you can use the ICONS AND TEXTS OF VIEWS subfunction to define your own titles for the tab pages and to assign an icon (see Figure 5.6). You can choose from more than 1,000 predefined icons.

Define Category of Functional Location

You can use this Customizing function to define different functional location categories.

Prerequisites

If you want to assign a status profile, partner determination procedure, measuring point category, object information key, and the LINEAR DATA category, you must define them beforehand.

Customizing Path

PLANT MAINTENANCE AND CUSTOMER SERVICE • MASTER DATA IN PLANT MAINTENANCE AND CUSTOMER SERVICE • TECHNICAL OBJECTS • FUNCTIONAL LOCATIONS • DEFINE CATEGORY OF FUNCTIONAL LOCATION

Settings

Figure 5.7 provides an overview of the functional location categories. Here, you can define the name and description of the category.

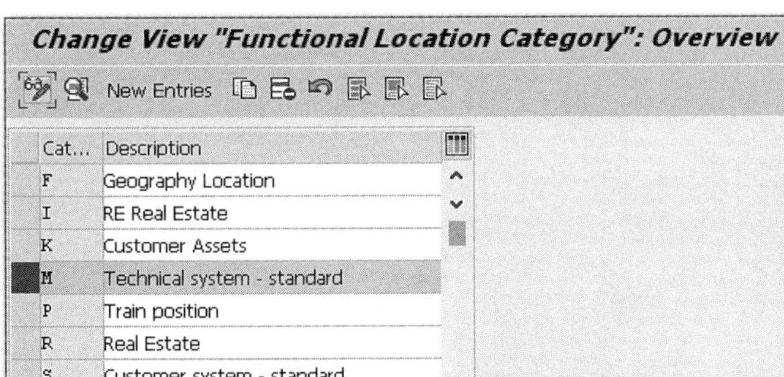

Figure 5.7 Overview of Functional Location Categories

On the detail screen, you can determine the following information for each category (see Figure 5.8):

- Status Profile
  The status profile to be assigned.
- PartnDet.Proc
  The partner determination procedure to apply.
- MeasPtCategory
  The measuring point category to use.
- View Profile
  The layout of the master record.
- Change Docs During Creation
  Change documents generated when creating or changing data.
- CustObject
  The category concerns customer assets.
- Object Info
  Information to be displayed as object information.

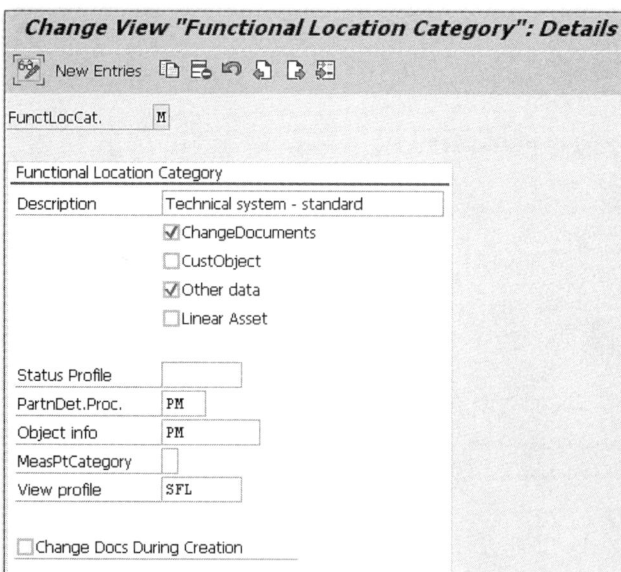

Figure 5.8 Functional Location Category: Detail Screen

- LINEAR ASSET
  Linear data to enter for the functional locations.

You can also base the field selection on the functional location category.

Define Structural Display for Functional Locations

You can use this Customizing function to define the layout of the structural display for functional locations in Transaction IH01.

The fields you select are used for thc field selection. If a user doesn't define his own field selection, the system displays the fields that you've activated.

Prerequisites

There are no special prerequisites.

Customizing Path

PLANT MAINTENANCE AND CUSTOMER SERVICE • MASTER DATA IN PLANT MAINTENANCE AND CUSTOMER SERVICE • TECHNICAL OBJECTS • FUNCTIONAL LOCATIONS • DEFINE STRUCTURAL DISPLAY FOR FUNCTIONAL LOCATIONS

Transaction

OIWP

Settings

You can use numbering to define the order in which the fields are displayed in the list. In other words, the field assigned the number 1 is the first field displayed in the list, the field assigned the number 2 is the second field displayed in the list, and so on (see Figure 5.9).

Each user can design his user-related list on the basis of the field list that you've defined. Here, however, the user can use only work with those fields that are visible. In other words, an individual user can't activate fields marked as INVISIBLE for his own user-specific field selection.

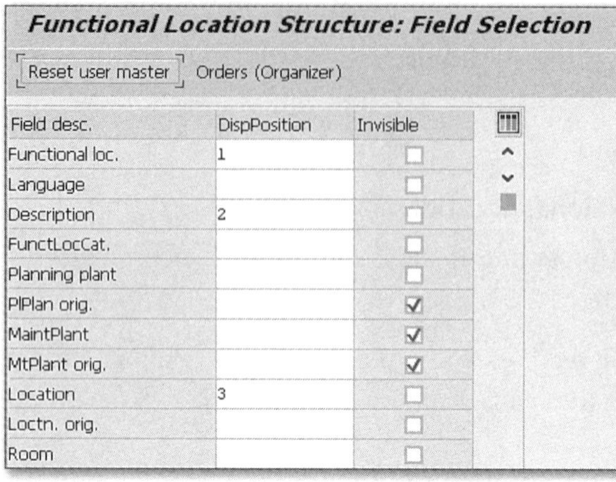

Figure 5.9 Functional Location: Structural Display

Alternative Labeling of Functional Locations

You should activate alternative labeling if you want to use one of the following functions later:

- You want to assign several numbers to a functional location structure or to each individual location for different perspectives (e.g., own naming conventions versus customer naming conventions).
- You want to create primary and secondary functional location structures (e.g., a primary structure for a power supply and a secondary structure for building services).
- You want to change the number assigned to a functional location (e.g., if you've assigned an incorrect number or you're assigning parts of an asset to another asset).

Prerequisites

There are no special prerequisites for simply activating alternative labeling. If, however, you want to use several labeling systems, you require the corresponding structure indicators (see the section discussing the structure indicator).

Customizing Path

Plant Maintenance and Customer Service • Master Data in Plant Maintenance and Customer Service • Technical Objects • Functional Locations •

Alternative Labeling of Functional Locations • Define Labeling Systems for Functional Locations

Transactions

OIPU and OIPV

Settings or Recommended Settings

To activate alternative labeling, set the Alt. label active (activate alternative labeling) indicator (see Figure 5.10).

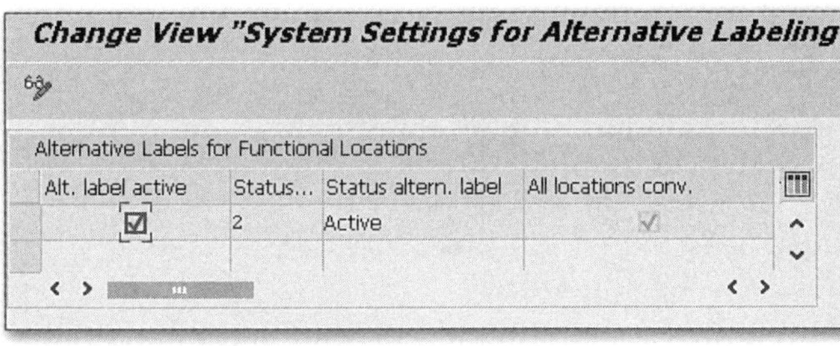

Figure 5.10 Activate Alternative Labeling

The following consequences are associated with activating alternative labeling:

- You can change the number of the functional location.
- Then, the following two database tables for managing the functional locations are activated: IFLOS (Labels for Functional Locations) and IFLOT (Root Table for Functional Locations).
- IFLOS contains the internal number of the functional location TPLNR and a reference to the number STRNO, which can be displayed on the interface and may change over time. Figure 5.11 depicts such a scenario: during processing, the functional location number ?0100000000000000093 was changed from ICE-FR-01 to ICE-FR-11.
- IFLOT contains the actual information about the functional location (e.g., the maintenance planning plant, maintenance planner group, description, etc.).
- If you activate alternative labeling retroactively (i.e., you've already entered functional locations), you must run Report RI_IFLOT2IFLOS. The program generates an entry in Table IFLOS for each entry in Table IFLOT. Each time you

enter a new functional location, the system automatically creates two entries—one in Table IFLOS and one in Table IFLOT.

- You can set up several labeling systems for functional locations.

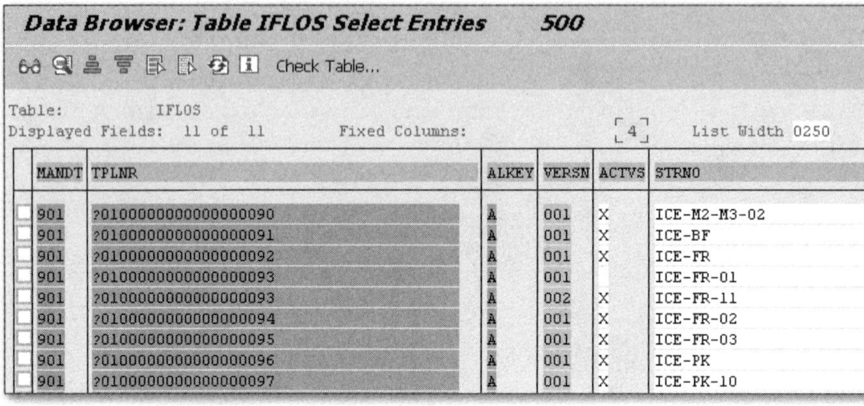

MANDT	TPLNR	ALKEY	VERSN	ACTVS	STRNO
901	?0100000000000000090	A	001	X	ICE-M2-M3-02
901	?0100000000000000091	A	001	X	ICE-BF
901	?0100000000000000092	A	001	X	ICE-FR
901	?0100000000000000093	A	001		ICE-FR-01
901	?0100000000000000093	A	002	X	ICE-FR-11
901	?0100000000000000094	A	001	X	ICE-FR-02
901	?0100000000000000095	A	001	X	ICE-FR-03
901	?0100000000000000096	A	001	X	ICE-PK
901	?0100000000000000097	A	001	X	ICE-PK-10

Figure 5.11 Table IFLOS

You use the Customizing function DEFINE LABELING SYSTEMS FOR FUNCTIONAL LOCATIONS to define alternative labeling systems (e.g., if you want to differentiate between internal structures and customer structures, or if you want to define parallel views—a mechanical view and electrical view—for the functional locations) (see Figure 5.12).

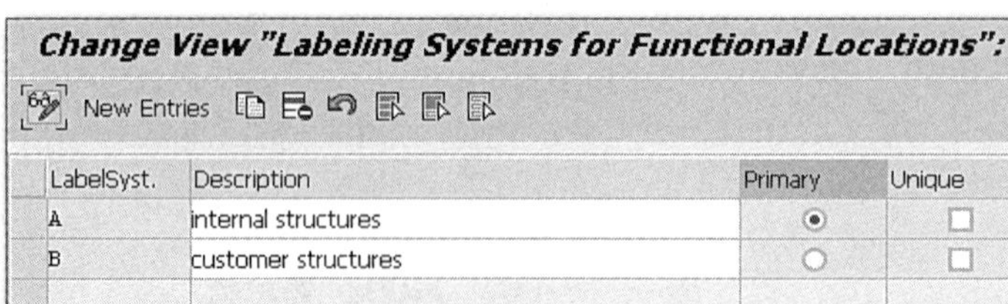

Figure 5.12 Labeling Systems for Functional Locations

You can also define whether you want to check the uniqueness of one labeling system against all other labeling systems. A uniqueness check is always performed within a labeling system.

One of the labeling systems must be selected as the main labeling system. This should be the system that is used by the majority of users.

5.3 Equipment

A piece of equipment represents a movable, individual aggregate (inventory). Example of equipment include the following:

- Machines
- Means of production (pumps and engines)
- Production resources/tools (PRTs)
- Test equipment (scales and gauges)
- Fleet objects (cars, trucks, forklift trucks, and industrial trucks)
- IT inventories (PCs, printers, monitors, notebooks, and projectors)
- Robots

You create a piece of equipment in the following scenarios:

- You install the devices on functional locations and want to provide evidence of a usage history.
- You want to store the devices.
- You want to manage individual data from a maintenance perspective (technical data such as performance data or organizational data such as the work center).
- You want to open notifications, orders, or maintenance plans.
- You have documentation commitments and must provide evidence of maintenance tasks performed.
- You want to collate and analyze technical data such as causes of damage, measurement readings, or counter readings.
- You want to verify costs.
- You want to refurbish a device.
- You want to subcontract an external supplier (subcontracting).
- You manage devices in a pool and want to support the borrowing process.
- You want to or need to perform calibrations or checks.

Figure 5.13 provides an overview of the Customizing functions for equipment.

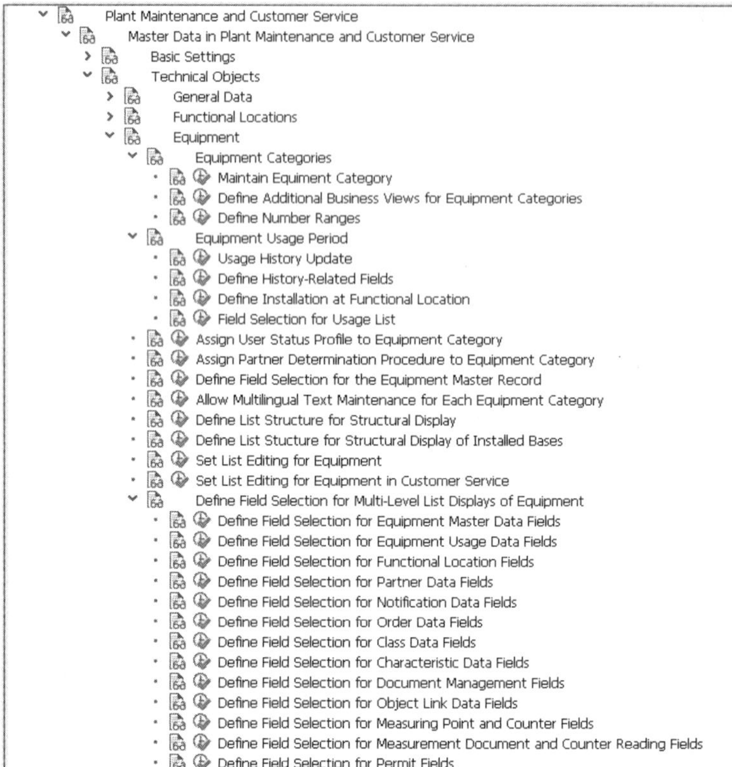

Figure 5.13 Equipment: Customizing Functions

Now, let's discuss the most important of these Customizing functions. The equipment category is at the heart of Customizing for equipment (see Figure 5.14).

You can use the equipment category to define the following:

- The number assignment or which equipment numbers can be assigned
- Whether change documents are written
- Which information is displayed as object information
- The layout of the equipment master (view profile) and which fields are to be maintained (field selection)
- Whether it concerns a linear object

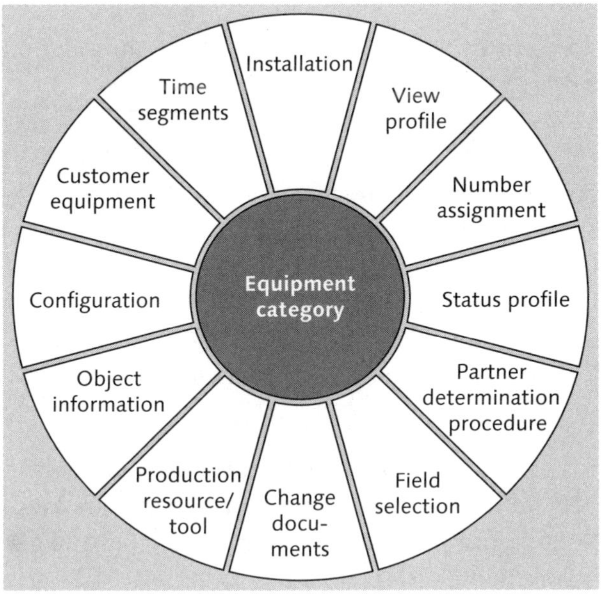

Figure 5.14 Equipment Category

- Which user status profile is used
- Whether the piece of equipment is a PRT
- Which partner determination procedure is used
- Whether it concerns a piece of customer equipment
- Whether the piece of equipment can be configured
- Whether time segments are written for changes and, if so, for which fields
- Whether the piece of equipment is permitted to be installed on functional locations

Maintain Equipment Category

You use this Customizing function to define equipment categories. This enables you to control which basic properties you want a corresponding equipment master record to have. You can use control properties to define not only how many equipment categories you require, but also which equipment categories you require.

Prerequisites

To assign equipment categories, you must define the relevant object information keys and view profile beforehand.

Customizing Path

Plant Maintenance and Customer Service • Master Data in Plant Maintenance and Customer Service • Technical Objects • Equipment • Equipment Categories • Maintain Equipment Category

Settings

The following settings are available to you in the columns shown in Figure 5.15:

- R
  You assign a reference category that is used to define history-relevant fields in an equipment master record. The following three reference categories are available: M Machines, S Customer equipment, and P Production Resources/Tools. M Machines is the default reference category.
- A
  In the case of external assignment, you can prohibit alphanumeric numbers.
- C
  You can activate change documents for verifying master data changes.
- CC
  You can define whether a change document is to be generated when creating a piece of equipment.
- W..
  You can trigger certain workflow events during master data maintenance.
- Object info
  You assign the Object info key.
- View Profile
  You assign the view profile.
- CTyp/MatNo
  You decide whether you want to check the synchronization between the material number and construction type. If so, you define whether you want the system to issue a warning message or error message if there is an inconsistency.

- Li...
  You determine whether you can enter linear data for a piece of equipment.

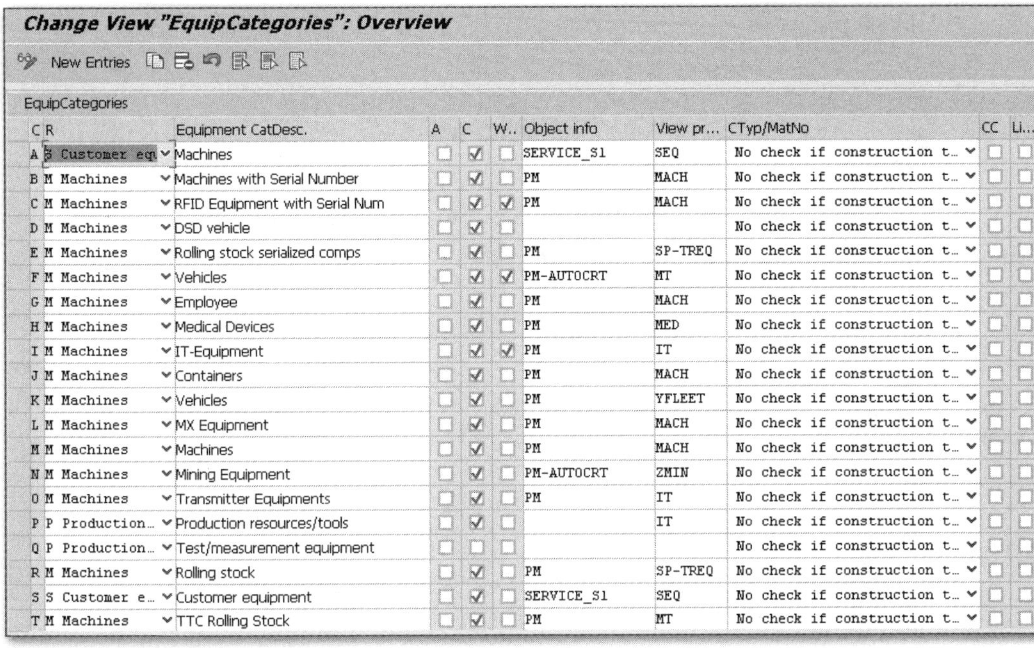

Change View "EquipCategories": Overview

New Entries

EquipCategories

C	R	Equipment CatDesc.	A	C	W..	Object info	View pr...	CTyp/MatNo	CC	Li...
A	S Customer eq	Machines	☐	☑	☐	SERVICE_S1	SEQ	No check if construction t...	☐	☐
B	M Machines	Machines with Serial Number	☐	☑	☐	PM	MACH	No check if construction t...	☐	☐
C	M Machines	RFID Equipment with Serial Num	☐	☑	☑	PM	MACH	No check if construction t...	☐	☐
D	M Machines	DSD vehicle	☐	☑	☐			No check if construction t...	☐	☐
E	M Machines	Rolling stock serialized comps	☐	☑	☐	PM	SP-TREQ	No check if construction t...	☐	☐
F	M Machines	Vehicles	☐	☑	☑	PM-AUTOCRT	MT	No check if construction t...	☐	☐
G	M Machines	Employee	☐	☑	☐	PM	MACH	No check if construction t...	☐	☐
H	M Machines	Medical Devices	☐	☑	☐	PM	MED	No check if construction t...	☐	☐
I	M Machines	IT-Equipment	☐	☑	☑	PM	IT	No check if construction t...	☐	☐
J	M Machines	Containers	☐	☑	☐	PM	MACH	No check if construction t...	☐	☐
K	M Machines	Vehicles	☐	☑	☐	PM	YFLEET	No check if construction t...	☐	☐
L	M Machines	MX Equipment	☐	☑	☐	PM	MACH	No check if construction t...	☐	☐
M	M Machines	Machines	☐	☑	☐	PM	MACH	No check if construction t...	☐	☐
N	M Machines	Mining Equipment	☐	☑	☐	PM-AUTOCRT	ZMIN	No check if construction t...	☐	☐
O	M Machines	Transmitter Equipments	☐	☑	☐	PM	IT	No check if construction t...	☐	☐
P	P Production...	Production resources/tools	☐	☑	☐		IT	No check if construction t...	☐	☐
Q	P Production...	Test/measurement equipment	☐	☐	☐			No check if construction t...	☐	☐
R	M Machines	Rolling stock	☐	☑	☐	PM	SP-TREQ	No check if construction t...	☐	☐
S	S Customer e...	Customer equipment	☐	☑	☐	SERVICE_S1	SEQ	No check if construction t...	☐	☐
T	M Machines	TTC Rolling Stock	☐	☑	☐	PM	MT	No check if construction t...	☐	☐

Figure 5.15 Equipment Categories

You can also base the functional location category on the field selection. (For more information, see Chapter 4, Section 4.9, which provides information about the field selection.)

Define Additional Business Views for Equipment Categories

In this work step, you can define additional views for each equipment category. The views represent additional tab pages within the equipment master record.

Prerequisites

You must define the equipment categories beforehand.

Customizing Path

PLANT MAINTENANCE AND CUSTOMER SERVICE • MASTER DATA IN PLANT MAINTENANCE AND CUSTOMER SERVICE • TECHNICAL OBJECTS • EQUIPMENT • EQUIPMENT CATEGORIES • DEFINE ADDITIONAL BUSINESS VIEWS FOR EQUIPMENT CATEGORIES

Settings

You can activate or deactivate the following views in these columns shown in Figure 5.16:

- S
  Activate this sales data view if you want to maintain fields such as SALES ORGANIZATION, DIVISION, SOLD-TO PARTY, and SHIP-TO PARTY.
- PRODUCTION RES/TOOLS
  Activate this PRT view if you want to perform the Calibration of Test and Measurement Equipment business process. For more information, see Chapter 8, Section 8.4.
- C
  Activate the equipment configuration view if you want to assign a configurable material to the piece of equipment.
- SD
  Activate the serial data view if you want to store serial data in addition to the normal equipment data (e.g., the material number, serial number, stock type, storage location, and stock batch), or if you want to generate a serial number history.
- OTHER DATA
  Activate the view for other data to then see an additional, freely configurable tab page.
- CC
  Activate the configuration control view.

When you activate a view, the assigned data fields and screens are automatically offered during maintenance. If the views are deactivated, you can maintain the data fields and screens. However, you must explicitly call them by means of a function key.

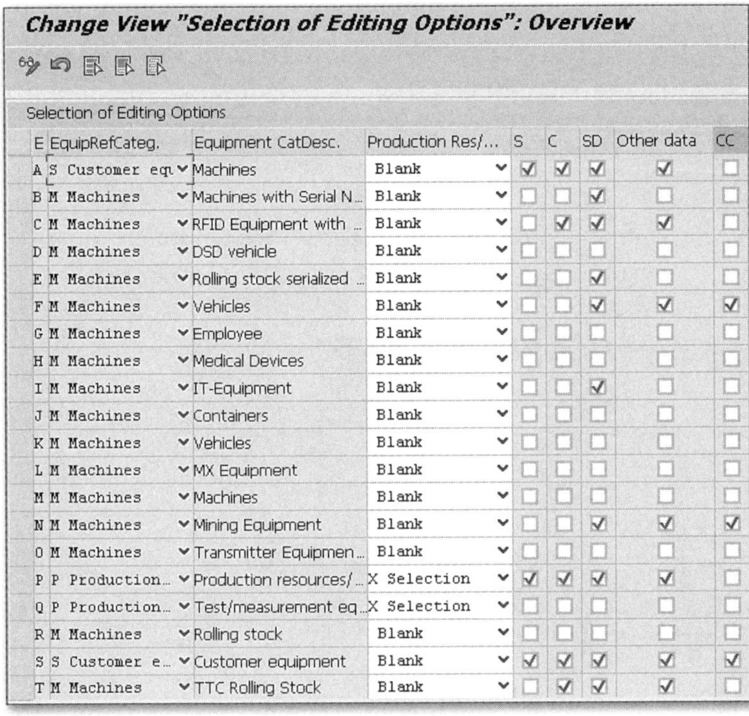

Change View "Selection of Editing Options": Overview

Selection of Editing Options

E	EquipRefCateg.	Equipment CatDesc.	Production Res/...	S	C	SD	Other data	CC
A	S Customer eq	Machines	Blank	☑	☑	☑	☑	☐
B	M Machines	Machines with Serial N...	Blank	☐	☐	☑	☐	☐
C	M Machines	RFID Equipment with ...	Blank	☐	☑	☑	☑	☐
D	M Machines	DSD vehicle	Blank	☐	☐	☐	☐	☐
E	M Machines	Rolling stock serialized ...	Blank	☐	☐	☑	☐	☐
F	M Machines	Vehicles	Blank	☐	☐	☑	☑	☑
G	M Machines	Employee	Blank	☐	☐	☐	☐	☐
H	M Machines	Medical Devices	Blank	☐	☐	☐	☐	☐
I	M Machines	IT-Equipment	Blank	☐	☐	☑	☐	☐
J	M Machines	Containers	Blank	☐	☐	☐	☐	☐
K	M Machines	Vehicles	Blank	☐	☐	☐	☐	☐
L	M Machines	MX Equipment	Blank	☐	☐	☐	☐	☐
M	M Machines	Machines	Blank	☐	☐	☐	☐	☐
N	M Machines	Mining Equipment	Blank	☐	☐	☑	☑	☑
O	M Machines	Transmitter Equipmen...	Blank	☐	☐	☐	☐	☐
P	P Production...	Production resources/...	X Selection	☑	☑	☑	☑	☐
Q	P Production...	Test/measurement eq...	X Selection	☐	☐	☐	☐	☐
R	M Machines	Rolling stock	Blank	☐	☐	☐	☐	☐
S	S Customer e...	Customer equipment	Blank	☑	☑	☑	☑	☑
T	M Machines	TTC Rolling Stock	Blank	☐	☑	☑	☑	☐

Figure 5.16 Equipment Categories: Other Data

Usage History Update

In this work step, you define whether a usage history is to be created for each equipment category. In contrast to the change documentation associated with change documents, the system not only stores the field contents of the fields both before and after the change, but also retains the time-dependent data fields for the entire equipment master record when the change is made. However, you can't base the usage history on all fields. Rather, you can base it only on the time-dependent data. The former includes, among others, the description, class, manufacturer, or warranty data. The latter includes, among others, the functional location, cost center, responsible work center, or maintenance planner group. In the Customizing function DEFINE HISTORY-RELATED FIELDS, you define those fields for which you want the usage history to be updated, for each reference category.

Prerequisites

You must define the equipment categories beforehand.

Customizing Path

PLANT MAINTENANCE AND CUSTOMER SERVICE • MASTER DATA IN PLANT MAINTENANCE AND CUSTOMER SERVICE • TECHNICAL OBJECTS • EQUIPMENT • EQUIPMENT USAGE PERIOD • USAGE HISTORY UPDATE

Settings

Select the TIME SEG. checkbox for those equipment categories for which you want the usage history to be updated (see Figure 5.17).

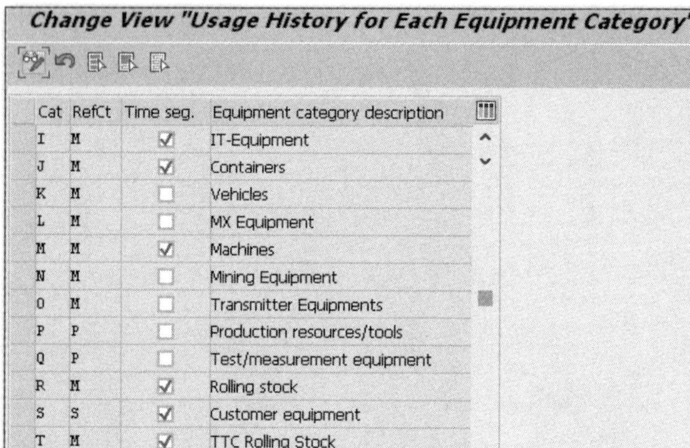

Change View "Usage History for Each Equipment Category"

Cat	RefCt	Time seg.	Equipment category description
I	M	☑	IT-Equipment
J	M	☑	Containers
K	M	☐	Vehicles
L	M	☐	MX Equipment
M	M	☑	Machines
N	M	☐	Mining Equipment
O	M	☐	Transmitter Equipments
P	P	☐	Production resources/tools
Q	P	☐	Test/measurement equipment
R	M	☑	Rolling stock
S	S	☑	Customer equipment
T	M	☑	TTC Rolling Stock

Figure 5.17 Equipment Category: Usage History

Define History-Related Fields

In this work step, you define the reference category for which you want the usage history to be updated when you change particular fields, on the basis of the Customizing function USAGE HISTORY UPDATE. Note that the usage history is always updated when you change the fields in the LOCATION and STRUCTURING screen areas if you've activated the update. These include the location, maintenance plant, functional location, or superior piece of equipment, for example.

Prerequisites

If you want to perform an update, you must activate the update for each equipment category beforehand.

Customizing Path

Plant Maintenance and Customer Service • Master Data in Plant Maintenance and Customer Service • Technical Objects • Equipment • Equipment Usage Period • Define History-Related Fields

Transaction

OIEZ

Settings

For each reference category, select those fields for which you want the usage history to be updated (see Figure 5.18).

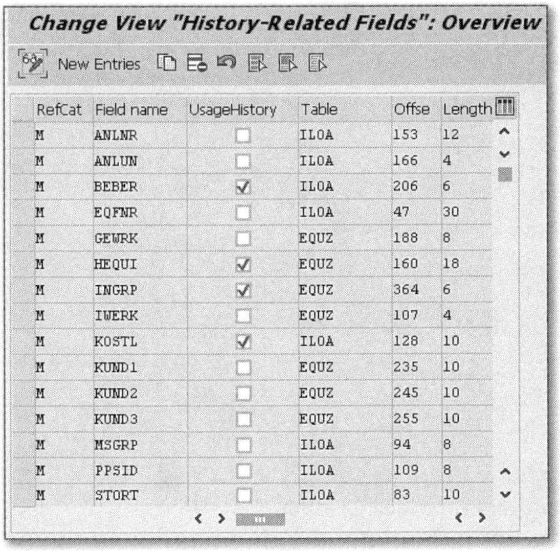

Change View "History-Related Fields": Overview

New Entries

RefCat	Field name	UsageHistory	Table	Offse	Length
M	ANLNR	☐	ILOA	153	12
M	ANLUN	☐	ILOA	166	4
M	BEBER	☑	ILOA	206	6
M	EQFNR	☐	ILOA	47	30
M	GEWRK	☐	EQUZ	188	8
M	HEQUI	☑	EQUZ	160	18
M	INGRP	☑	EQUZ	364	6
M	IWERK	☐	EQUZ	107	4
M	KOSTL	☑	ILOA	128	10
M	KUND1	☐	EQUZ	235	10
M	KUND2	☐	EQUZ	245	10
M	KUND3	☐	EQUZ	255	10
M	MSGRP	☐	ILOA	94	8
M	PPSID	☐	ILOA	109	8
M	STORT	☐	ILOA	83	10

Figure 5.18 Reference Category: History-Related Fields

Define Installation at Functional Location

In this work step, you define whether or not to permit an installation at a functional location, for each equipment category.

Prerequisites

You must define the equipment categories beforehand.

Customizing Path

Plant Maintenance and Customer Service • Master Data in Plant Maintenance and Customer Service • Technical Objects • Equipment • Equipment Usage Period • Define Installation at Functional Location

Settings

Select the equipment categories for which you want to permit an installation at functional locations (see Figure 5.19).

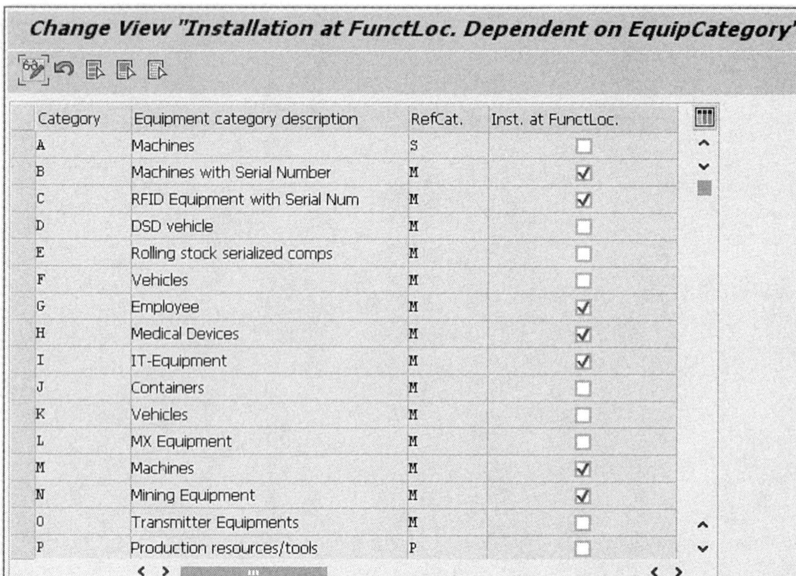
Change View "Installation at FunctLoc. Dependent on EquipCategory"

Category	Equipment category description	RefCat.	Inst. at FunctLoc.
A	Machines	S	☐
B	Machines with Serial Number	M	☑
C	RFID Equipment with Serial Num	M	☑
D	DSD vehicle	M	☐
E	Rolling stock serialized comps	M	☐
F	Vehicles	M	☐
G	Employee	M	☑
H	Medical Devices	M	☑
I	IT-Equipment	M	☑
J	Containers	M	☐
K	Vehicles	M	☐
L	MX Equipment	M	☐
M	Machines	M	☑
N	Mining Equipment	M	☑
O	Transmitter Equipments	M	☐
P	Production resources/tools	P	☐

Figure 5.19 Equipment Category: Installation at a Functional Location

Allow Multilingual Text Maintenance for Each Equipment Category

In this work step, you define whether you want to permit multilingual text maintenance for the equipment category, for each equipment category. This is necessary, for example, if your company is part of a group that operates internationally and stores its equipment on a cross-country and therefore cross-lingual basis.

Prerequisites

You must define the equipment categories beforehand.

Customizing Path

Plant Maintenance and Customer Service • Master Data in Plant Maintenance and Customer Service • Technical Objects • Equipment • Equipment Usage Period • Allow Multilingual Text Maintenance for Each Equipment Category

Settings

Select the equipment categories for which you want to permit multilingual text maintenance (see Figure 5.20).

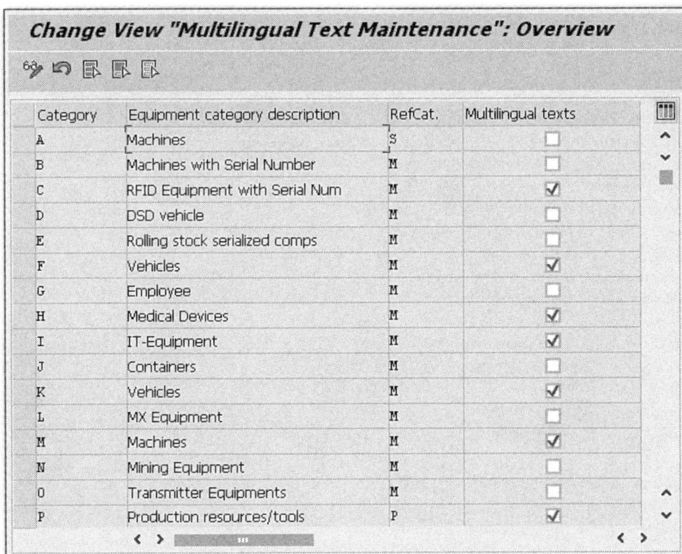

Change View "Multilingual Text Maintenance": Overview

Category	Equipment category description	RefCat.	Multilingual texts
A	Machines	S	
B	Machines with Serial Number	M	
C	RFID Equipment with Serial Num	M	X
D	DSD vehicle	M	
E	Rolling stock serialized comps	M	
F	Vehicles	M	X
G	Employee	M	
H	Medical Devices	M	X
I	IT-Equipment	M	X
J	Containers	M	
K	Vehicles	M	X
L	MX Equipment	M	
M	Machines	M	X
N	Mining Equipment	M	
O	Transmitter Equipments	M	
P	Production resources/tools	P	X

Figure 5.20 Equipment Category: Multilingual Texts

5.4 Fleet Management

In addition to the general Customizing settings, some special Customizing functions are available for managing equipment and/or functional locations as fleet objects. For example, you can configure the following:

- Fleet object-specific view profiles
- Consumable types
- Engine types
- Usage types

- Costing and account assignment of fuel consumption
- Service stations

Figure 5.21 provides an overview of the Customizing functions for fleet objects.

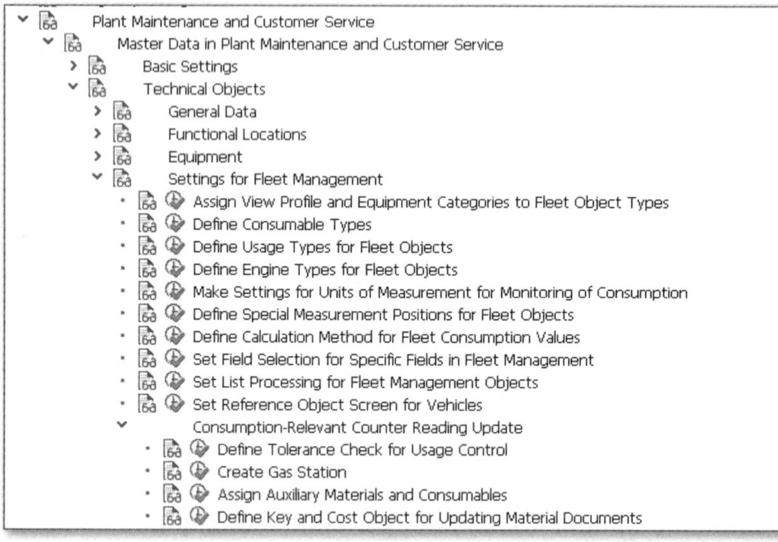

Figure 5.21 Fleet Objects: Customizing Functions

The most important of these Customizing functions are explained next.

Assign View Profile and Equipment Categories to Fleet Object Types

In this work step, you distinguish between different types of fleet objects (e.g., cars, trucks, mobile cranes, forklift trucks, industrial trucks, and wheel loaders). You can assign a custom view profile to each fleet object type so that you can use different layouts for forklift trucks and cars, for example. You also define default values for each equipment category.

Prerequisites

You must define the object types, view profiles, and equipment categories beforehand.

Customizing Path

PLANT MAINTENANCE AND CUSTOMER SERVICE • MASTER DATA IN PLANT MAINTENANCE AND CUSTOMER SERVICE • TECHNICAL OBJECTS • SETTINGS FOR FLEET MANAGEMENT • ASSIGN VIEW PROFILE AND EQUIPMENT CATEGORIES TO FLEET OBJECT TYPES

Settings or Recommended Settings

In the table, add object types that represent the fleet objects, and assign the relevant view profile and permitted equipment categories to each fleet object type (see Figure 5.22).

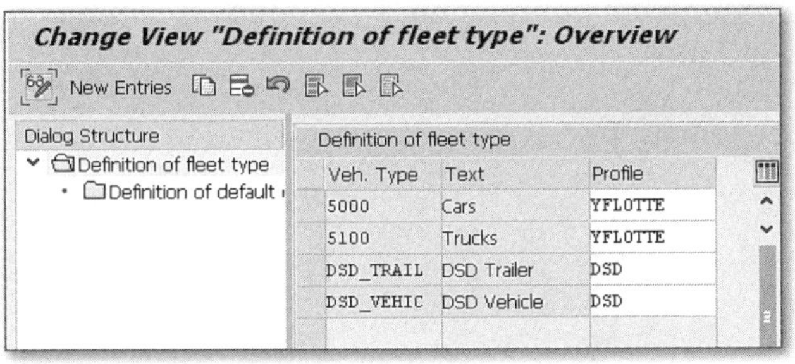

Figure 5.22 Fleet Objects: Fleet Object Type

Fleet Object Tab Pages Are Also Displayed [!]

The tab pages for the view profile specified here are displayed *in addition to* the tab pages for the view profile for the equipment category. If, for example, the equipment category has view profile A with tab pages 010 (General Data), 020 (Location Data), and 030 (Organizational Data), and the fleet object type has view profile B with tab pages 130 (Fleet Object ID/Dimensions) and 140 (Fleet Object – Technical Data), all five tab pages—010, 020, 030, 130, and 140—are displayed in the equipment master record.

Define Consumable Types

You can use this Customizing function to define consumable types for fleet objects. You can assign up to three consumable types to each fleet object.

Prerequisites

There are no special prerequisites.

Customizing Path

Plant Maintenance and Customer Service • Master Data in Plant Maintenance and Customer Service • Technical Objects • Settings for Fleet Management • Define Consumable Types

Settings

In the table, add the consumable types that are used to run your fleet objects and/or are required as a lubricant (see Figure 5.23).

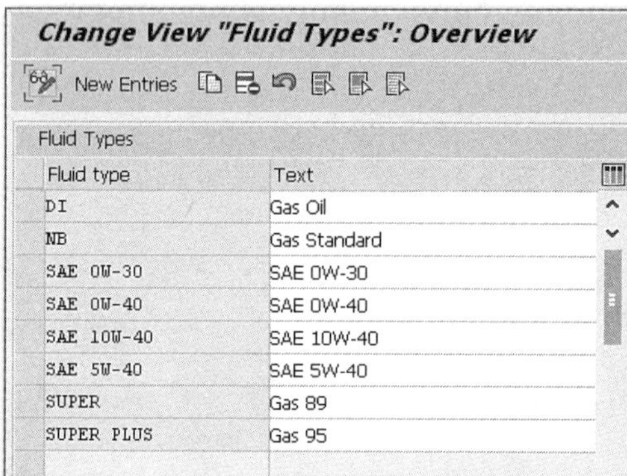

Figure 5.23 Fleet Objects: Consumables

Define Usage Types for Fleet Objects

You can use this Customizing function to define different usage types for fleet objects. For example, you can define whether the fleet objects are used primarily for business or transportation purposes.

Prerequisites

There are no special prerequisites.

Customizing Path

Plant Maintenance and Customer Service • Master Data in Plant Maintenance and Customer Service • Technical Objects • Settings for Fleet Management • Define Usage Types for Fleet Objects

Settings

In the table, add the usage types that you require for your fleet objects (see Figure 5.24).

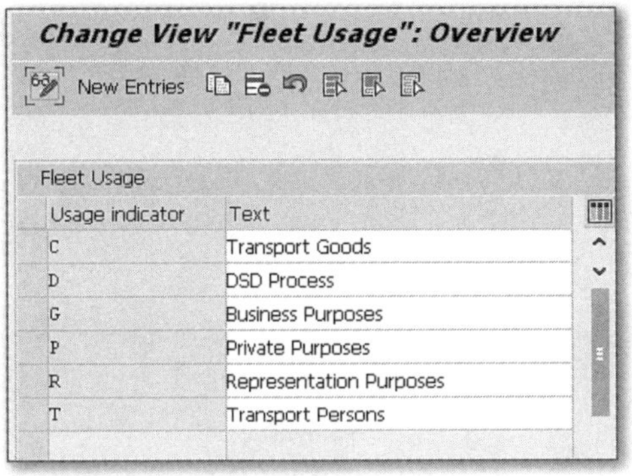

Figure 5.24 Fleet Objects: Usage Types

Define Engine Types for Fleet Objects

You can use this Customizing function to define different engine types for fleet objects (e.g., whether the fleet objects run on diesel, gas, or electricity).

Prerequisites

There are no special prerequisites.

Customizing Path

PLANT MAINTENANCE AND CUSTOMER SERVICE • MASTER DATA IN PLANT MAINTENANCE AND CUSTOMER SERVICE • TECHNICAL OBJECTS • SETTINGS FOR FLEET MANAGEMENT • DEFINE ENGINE TYPES FOR FLEET OBJECTS

Settings

In the table, add the engine types that you require for the different types of fleet objects in your fleet (see Figure 5.25).

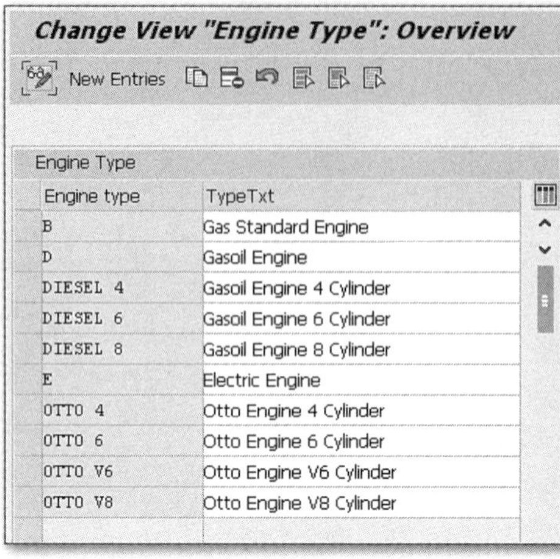

Engine type	TypeTxt
B	Gas Standard Engine
D	Gasoil Engine
DIESEL 4	Gasoil Engine 4 Cylinder
DIESEL 6	Gasoil Engine 6 Cylinder
DIESEL 8	Gasoil Engine 8 Cylinder
E	Electric Engine
OTTO 4	Otto Engine 4 Cylinder
OTTO 6	Otto Engine 6 Cylinder
OTTO V6	Otto Engine V6 Cylinder
OTTO V8	Otto Engine V8 Cylinder

Figure 5.25 Fleet Objects: Engine Type

Make Settings for Units of Measurement for Monitoring of Consumption

You can use this Customizing function to configure settings for determining the fuel consumption of your fleet objects. Here, fuel consumption is gauged using either a time-related calculation (e.g., liters per hour) or a distance-related calculation (e.g., liters per 100km).

Prerequisites

You must define the units of measurement beforehand (by using the Customizing function CHECK UNITS OF MEASUREMENT).

Customizing Path

PLANT MAINTENANCE AND CUSTOMER SERVICE • MASTER DATA IN PLANT MAINTENANCE AND CUSTOMER SERVICE • TECHNICAL OBJECTS • SETTINGS FOR FLEET MANAGEMENT • MAKE SETTINGS FOR UNITS OF MEASUREMENT FOR MONITORING OF CONSUMPTION

Settings

In the table, add the consumption units that you want to use to measure your fuel consumption (see Figure 5.26).

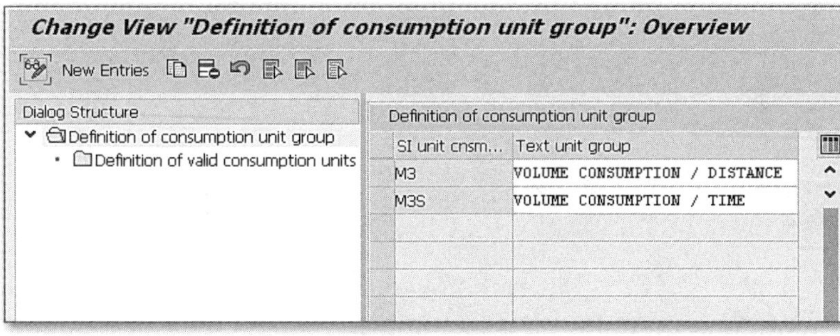

Figure 5.26 Fleet Objects: Consumption Units

You then use the subfunction DEFINITION OF VALID CONSUMPTION UNITS to define the corresponding units of measurement (e.g., liters per 100km or gallons per mile for the volume consumption per distance).

Note

You can define any number of units of measurement for each consumption unit group.

Define Special Measurement Positions for Fleet Objects

You can use this Customizing function to create special measurement positions that are filled with special fleet management functions. The measurement positions are used to perform fleet object-specific calculations (e.g., consumption) and are used as a basis for analyses by the Plant Maintenance Information System (PMIS) (Transaction MCIZ – Vehicle Consumption Analysis).

Prerequisites

You must define the units of measurement beforehand (by using the Customizing function CHECK UNITS OF MEASUREMENT).

Customizing Path

PLANT MAINTENANCE AND CUSTOMER SERVICE • MASTER DATA IN PLANT MAINTENANCE AND CUSTOMER SERVICE • TECHNICAL OBJECTS • SETTINGS FOR FLEET MANAGEMENT • DEFINE SPECIAL MEASUREMENT POSITIONS FOR FLEET OBJECTS

Settings

In the table, add the types of measurement positions according to which you want to calculate your fuel consumption in PMIS (see Figure 5.27).

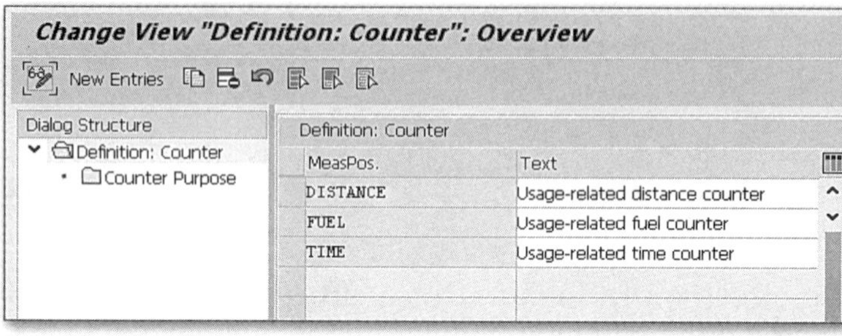

Figure 5.27 Fleet Objects: Consumption Counters

Consider the following example to understand the calculation method. You want to calculate the distance-related average consumption for a truck (e.g., diesel liters per 100km), so you create two counters in the piece of equipment (see Figure 5.28):

- You require a counter that enables you to measure distance. For this counter, you enter, for example, the counter definition "DISTANCE" in the MEASPOS. field, which, in turn, was defined in the COUNTER PURPOSE subfunction as 1 = USAGE-RELATED DISTANCE COUNTER.
- You require a second counter that enables you to measure consumption. For this counter, you enter, for example, the counter definition "FUEL" in the MEASPOS. field, which, in turn, was defined in the COUNTER PURPOSE subfunction as 3 = USAGE-RELATED FUEL COUNTER.

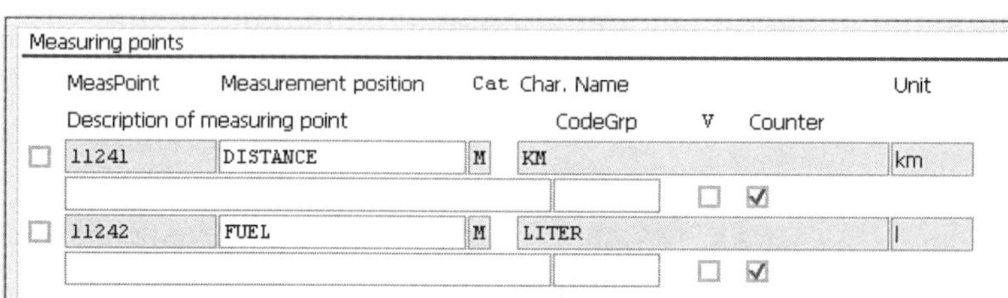

Figure 5.28 Fleet Objects: Measurement Positions

In your daily work, you'll now use the first counter with the measurement position DISTANCE to enter the distance data and the second counter FUEL to enter the consumption data.

PMIS can then use this information to calculate the average consumption. You define the method for calculating average consumption in the Customizing function DEFINE CALCULATION METHOD FOR FLEET CONSUMPTION VALUES.

Define Calculation Method for Fleet Consumption Values

You can use this Customizing function to define the methods for calculating the consumption values of fleet objects in PMIS (Transaction MCIZ – Vehicle Consumption Analysis).

Prerequisites
You must define the units of measurement (by using the Customizing function CHECK UNITS OF MEASUREMENT) and the measurement positions beforehand.

Customizing Path
PLANT MAINTENANCE AND CUSTOMER SERVICE • MASTER DATA IN PLANT MAINTENANCE AND CUSTOMER SERVICE • TECHNICAL OBJECTS • SETTINGS FOR FLEET MANAGEMENT • DEFINE CALCULATION METHOD FOR FLEET CONSUMPTION VALUES

Settings
You need to define the following (see Figure 5.29):

- Define the unit that is to be displayed (e.g., liter/100km, miles/gallon).
- To calculate consumption, define which fleet object counters you want to use in the consumption calculation. Enter these counters in the FUEL COUNTER column (here: "FUEL").
- Specify the primary counter in an additional column. The primary counter is the fleet object counter that you use as a basis for measuring the distance or operation time in relation to the consumption measured for the fleet object (here: "DISTANCE").

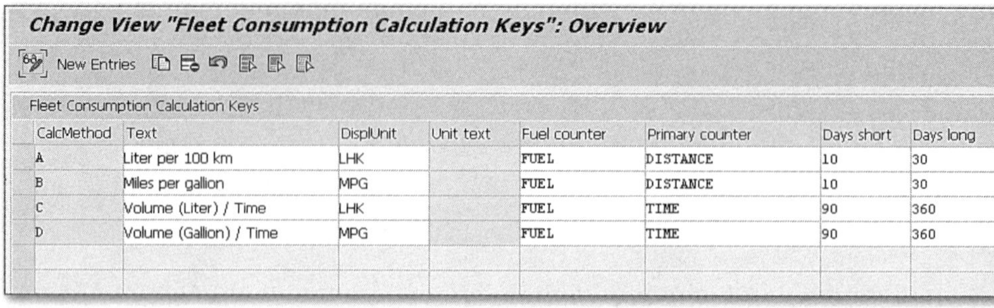

Change View "Fleet Consumption Calculation Keys": Overview

New Entries

Fleet Consumption Calculation Keys

CalcMethod	Text	DisplUnit	Unit text	Fuel counter	Primary counter	Days short	Days long
A	Liter per 100 km	LHK		FUEL	DISTANCE	10	30
B	Miles per gallion	MPG		FUEL	DISTANCE	10	30
C	Volume (Liter) / Time	LHK		FUEL	TIME	90	360
D	Volume (Gallion) / Time	MPG		FUEL	TIME	90	360

Figure 5.29 Fleet Objects: Consumption Calculation

Additional Customizing Functions

Although they aren't discussed in great detail here, the following are some additional Customizing functions in relation to fleet object consumption:

- Define Tolerance Check for Usage Control
  Define whether (and if so, to what extent) the consumption entered should differ from the short-term or long-term average consumption before the system issues warning or error messages.
- Create Gas Station
  Create internal gas stations with reference to the plant and storage location. When a consumption value is entered, the goods issue posting is automatically made from the relevant storage location.
- Assign Auxiliary Materials and Consumables
  Define the storage location in which a particular material number is to be posted for a particular consumable type.
- Define Key and Cost Object for Updating Material Documents
  Define the account assignment for the automatic issue posting (e.g., the cost center for a piece of equipment).

5.5 Links and Networks

SAP EAM provides functions that enable you to map links between different technical objects or systems (equipment or functional locations). For example, you can establish the following links:

- Between production units
- Between production plants and supply systems
- Between supply systems and disposal systems

Such object links form an *object network* that can give your assets a horizontal structure. Figure 5.30 provides an overview of the Customizing functions for object links.

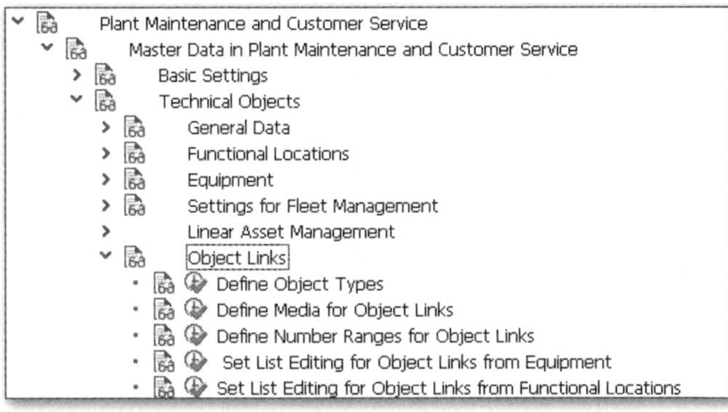

Figure 5.30 Object Links: Customizing Functions

The most important of these Customizing functions are explained next.

Define Object Types

You use this Customizing function to define object types for object links (similar to defining equipment categories for equipment or categories for functional locations).

Prerequisites

If you want to assign a status profile, you must define it beforehand.

Customizing Path

PLANT MAINTENANCE AND CUSTOMER SERVICE • MASTER DATA IN PLANT MAINTENANCE AND CUSTOMER SERVICE • TECHNICAL OBJECTS • OBJECT LINKS • DEFINE OBJECT TYPES

Settings

You can define the following (see Figure 5.31):

- STATPROF
  Assign a status profile.
- NO ALPHANU...
  Define whether alphanumeric characters are to be permitted.
- CHANGEDOCUMENTS
  Define whether change documents are to be generated.

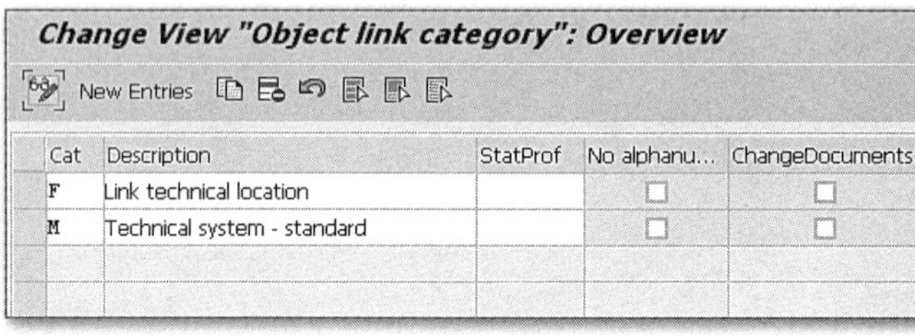

Change View "Object link category": Overview

New Entries

Cat	Description	StatProf	No alphanu...	ChangeDocuments
F	Link technical location		☐	☐
M	Technical system - standard		☐	☐

Figure 5.31 Object Links: Object Type

Define Media for Object Links

You use this Customizing function to define the media permitted for object links (e.g., water, gas, or granulate).

Prerequisites

If you want to assign a material number or color, you must define these beforehand.

Customizing Path

PLANT MAINTENANCE AND CUSTOMER SERVICE • MASTER DATA IN PLANT MAINTENANCE AND CUSTOMER SERVICE • TECHNICAL OBJECTS • OBJECT LINKS • DEFINE MEDIA FOR OBJECT LINKS

Transaction

OINM

Settings

You can define the following (see Figure 5.32):

- MEDIUM
  Whether the medium is managed under a material number.

- PN-RELEV.
  Whether the medium is relevant for production.
- COLOR
  Which color is assigned to the medium.

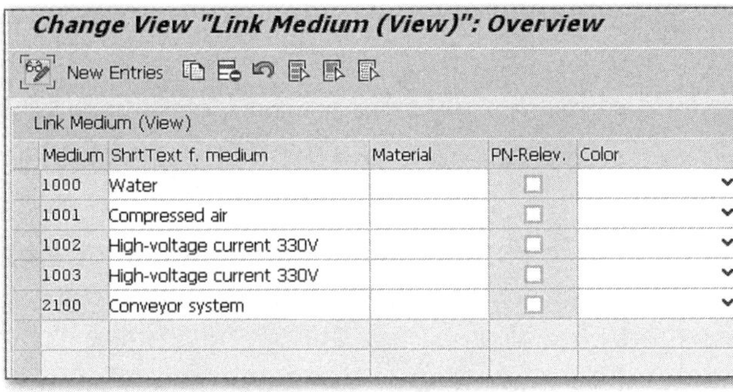

Figure 5.32 Object Links: Media

5.6 Linear Asset Management

Linear assets (Linear Asset Management) are technical systems with a linear infrastructure whose properties and conditions may change from stage to stage (dynamic segmentation). Examples of linear assets include the following:

- Pipelines
- Road networks
- Railway networks
- Overhead cables

As you can tell from these examples, this function is of particular interest to the oil and gas industry, utility companies (electricity, gas, and water), road maintenance companies, public authorities, and rail operators.

To use Linear Asset Management to its full potential, you must create the following conditions:

- In Customizing for the classification system, you use the Customizing function MAINTAIN OBJECT TYPES AND CLASS TYPES to create an organizational area (e.g., the organizational area L for linear objects).

- To enable you to enter linear data for measuring points and measurement documents, you use the Customizing function DEFINE MEASURING POINT CATEGORIES to define a measuring point category, and you set the LINEAR ASSET indicator (see the Customizing function DEFINE MEASURING POINT CATEGORIES and refer to Figure 5.16).
- To define a view profile for functional locations and a view profile for equipment, you use the Customizing function SET VIEW PROFILES FOR TECHNICAL OBJECTS. You also assign the field groups for linear data there (refer to the Customizing function SET VIEW PROFILES FOR TECHNICAL OBJECTS in Figure 5.4 and Figure 5.5).
- To define linear data for functional locations, you use the Customizing function DEFINE CATEGORY OF FUNCTIONAL LOCATION to define a functional location category. In particular, you assign the view profile there and set the LINEAR ASSET indicator (refer to the Customizing function DEFINE CATEGORY OF FUNCTIONAL LOCATION and Figure 5.8).
- To define linear data for equipment, you use the Customizing function MAINTAIN EQUIPMENT CATEGORY to define an equipment category. In particular, you assign the view profile here and set the LINEAR ASSET indicator (refer to the Customizing function MAINTAIN EQUIPMENT CATEGORY and Figure 5.15).

Additional Customizing functions are available within Linear Asset Management, an overview of which is provided in Figure 5.33.

Next, we'll explain the most important of these Customizing functions to you.

Define Offset Types

You use this Customizing function to define the necessary offset types for linear objects. Offset data is always required if, when defining a linear asset structure, you integrate objects that have a certain offset in relation to the actual linear object (e.g., if, for the linear object ROAD, you want to define a signaling device with a horizontal offset of 2 meters from the edge of the road and a vertical offset of 3.50 meters).

Prerequisites

You must define the units of measurement used for the offset type beforehand (by using the Customizing function CHECK UNITS OF MEASUREMENT).

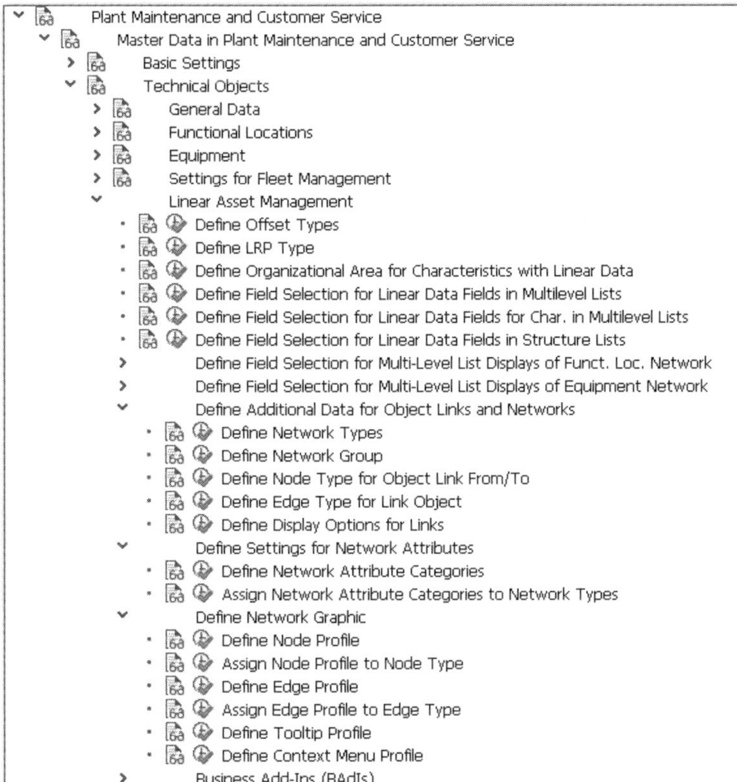

Figure 5.33 Linear Asset Management: Customizing Functions

Customizing Path

PLANT MAINTENANCE AND CUSTOMER SERVICE • MASTER DATA IN PLANT MAINTENANCE AND CUSTOMER SERVICE • TECHNICAL OBJECTS • LINEAR ASSET MANAGEMENT • DEFINE OFFSET TYPES

Settings

You can define the following (see Figure 5.34):

- OTC
  A two-digit code and description for the offset type.
- UoM
  The unit of measurement for the offset type.
- DEFAULT OFFSET
  Whether the offset type is to be proposed when you define linear objects.

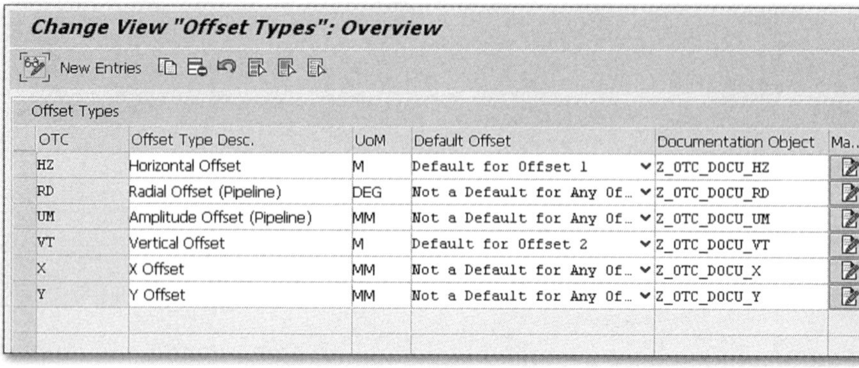

Figure 5.34 Linear Asset Management: Offset Types

Define LRP Type

Linear reference patterns (LRPs) are created so that you can assign a descriptive reference point to real technical objects or documents (orders, notifications, etc.). You can then specify these patterns in a notification (e.g., "The damage occurred 10km after the Kennebunk rest area and service station"). You assign a linear reference pattern type to each LRP.

You use this Customizing function to define the necessary LRP types for linear objects.

Prerequisites

There are no specific prerequisites.

Customizing Path

Plant Maintenance and Customer Service • Master Data in Plant Maintenance and Customer Service • Technical Objects • Linear Asset Management • Define LRP Type

Settings

You create an ID (with a maximum of four characters) and a description for the LRP type (see Figure 5.35).

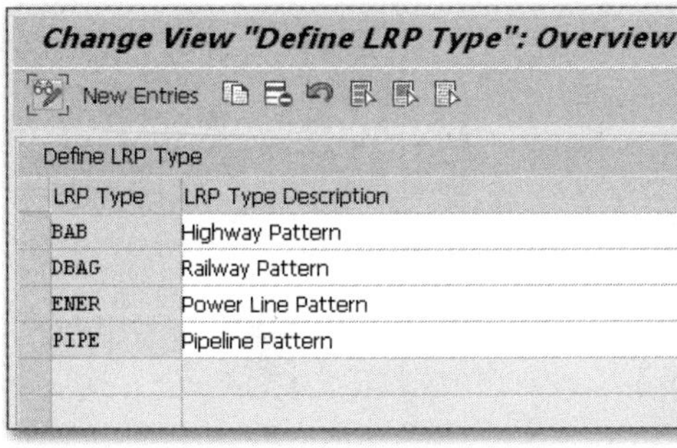

Figure 5.35 Linear Asset Management: Linear Reference Pattern Type

Define Organizational Area for Characteristics with Linear Data

You use this Customizing function to define which view you want to use in the classification system for the linear objects.

Prerequisites

You've used the Customizing function Cross-Application Components • Classification System • Classes • Maintain Object Categories and Class Types to define the corresponding application views for Class Type 002 (Linear Equipment) and 003 (Linear functional location) (see Figure 5.36).

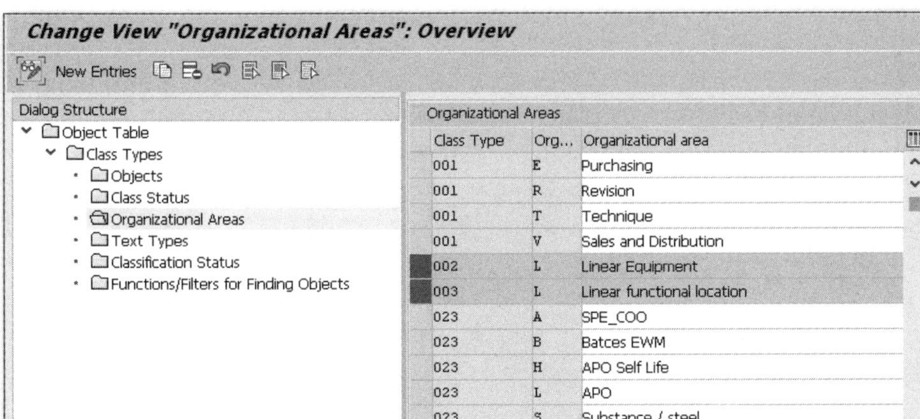

Figure 5.36 Classes: View for Linear Objects

Customizing Path

Plant Maintenance and Customer Service • Master Data in Plant Maintenance and Customer Service • Technical Objects • Linear Asset Management • Define Organizational Area for Characteristics with Linear Data

Settings

You assign the corresponding view to class types 002 (Linear Equipment) and 003 (Linear functional location) (see Figure 5.37). Other class types and objects aren't possible.

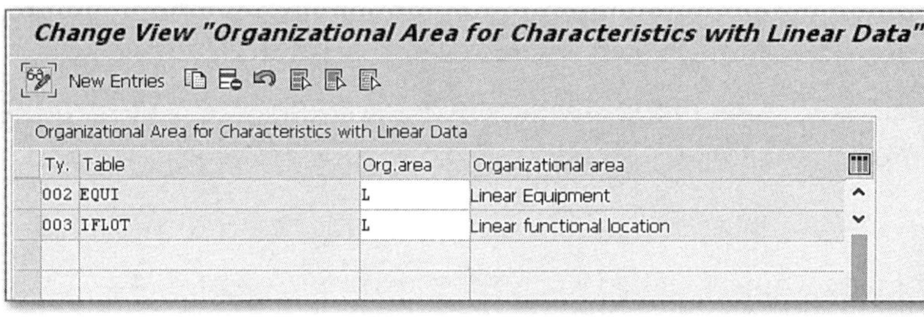

Figure 5.37 Linear Asset Management: Views

Object Networks

As of SAP ERP 6.0, EHP6, the business function `LOG_EAM_LINEAR_2` is available to you. When activated, you can map your linear assets as object networks that comprise several object links between functional locations or pieces of equipment. You can enter linear data at the network or object level, and you can display the object network in a graphical diagram. You can use network attribute values and properties to define additional information about your object networks (Transaction IN21FF). The following new single-level and multilevel list reports are available to you for analyzing your object network data: Transaction IN24 (Multilevel Functional Location Network List) and Transaction IN25 (Multilevel Equipment Network List). Furthermore, you can use some new Customizing functions that enable you to adjust this new function to your own requirements.

Define Network Types

You can use this Customizing activity to create network types. Then, in the Transaction IN21 (Create Object Network), you select one of these network types for the object network.

To enter network attribute values and any changes to these values, you use the Customizing function ASSIGN NETWORK ATTRIBUTE CATEGORIES TO NETWORK TYPES to assign your network attribute categories to the network types that you've created. You can assign one network attribute category to several network types.

Prerequisites

There are no special prerequisites.

Customizing Path

PLANT MAINTENANCE AND CUSTOMER SERVICE • MASTER DATA IN PLANT MAINTENANCE AND CUSTOMER SERVICE • TECHNICAL OBJECTS • LINEAR ASSET MANAGEMENT • DEFINE ADDITIONAL DATA FOR OBJECT LINKS AND NETWORKS • DEFINE NETWORK TYPES

Settings

Create the necessary (four-character) NETWORK TYPE with a corresponding NETWORK TYPE DESCRIPTION (see Figure 5.38). You can also determine whether change documents are to be generated when you change and/or create a network.

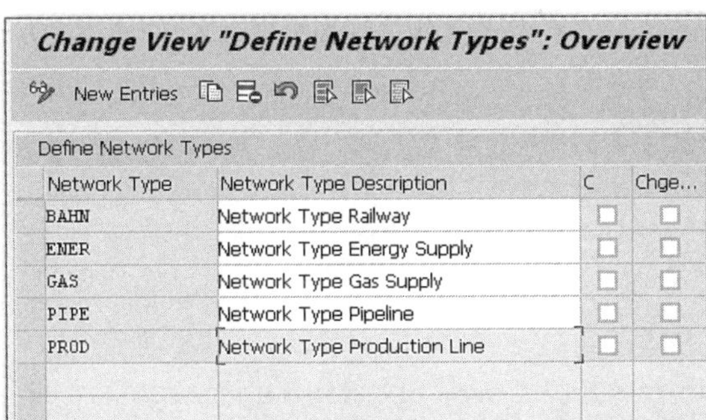

Figure 5.38 Object Networks: Define Network Types

Define Network Group

You can use this Customizing function to define network groups. If you create object networks, you can assign each object network to a network group. The network group is then used as a selection criterion in the list displays.

Prerequisites
There are no special prerequisites.

Customizing Path
Plant Maintenance and Customer Service • Master Data in Plant Maintenance and Customer Service • Technical Objects • Linear Asset Management • Define Additional Data for Object Links and Networks • Define Network Group

Settings
Create the necessary (10-character) network groups with a corresponding description (see Figure 5.39). Because the object networks do not contain a field for assignment to a plant, it makes sense to use a network group for this purpose. However, you can also set up network groups for all other usage purposes.

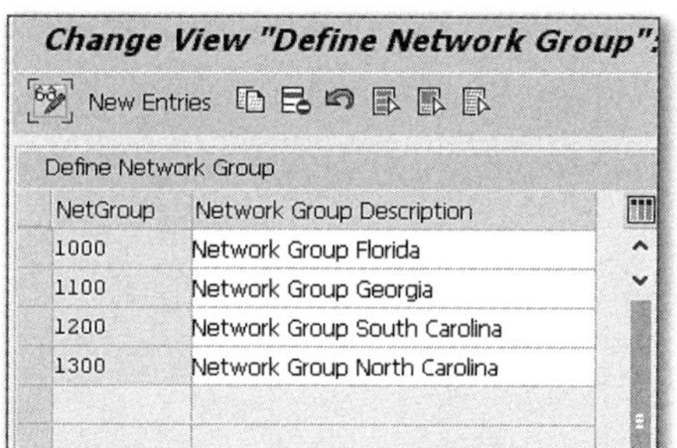

Figure 5.39 Object Networks: Define Network Group

Define Network Attribute Categories

You can use this Customizing function to define network attribute categories. You can then define network attributes and assign them to predefined categories.

Prerequisites
There are no special prerequisites.

Customizing Path
Plant Maintenance and Customer Service • Master Data in Plant Maintenance and Customer Service • Technical Objects • Linear Asset Management • Define Settings for Network Attributes • Define Network Attribute Categories

Settings

This Customizing function comprises several subfunctions:

- DEFINE NETWORK ATTRIBUTE CATEGORIES
  Defines the necessary (12-character) network groups with a corresponding description (see Figure 5.40). Here, you can specify whether this category and its attributes are linear assets by selecting the LIN. CAT. checkbox. If you select this field, the END POINT field for the lower-level attributes becomes ready for input. If you don't select this field, the system assumes that this category and its attributes aren't linear, and the END POINT field isn't ready for input.

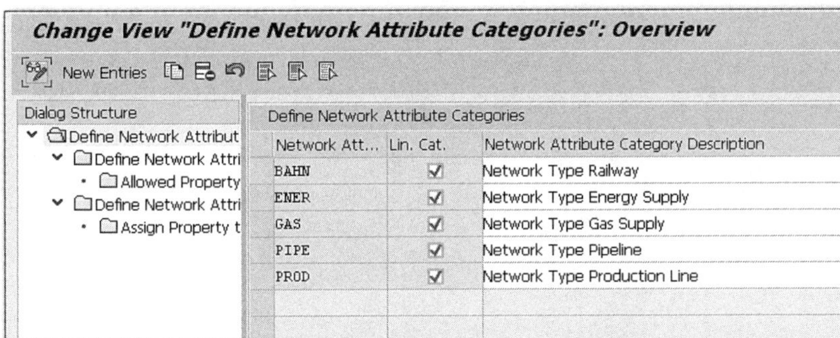

Figure 5.40 Object Networks: Network Attribute Categories

- DEFINE NETWORK ATTRIBUTE PROPERTIES
  Defines which technical properties or other properties you want to assign to a network attribute category. Figure 5.41 shows that you can assign track, sleeper, or line type attributes to a railway infrastructure. In each case, you specify the format (e.g., CHARACTER FORMAT or NUMBER FORMAT) and the field length.

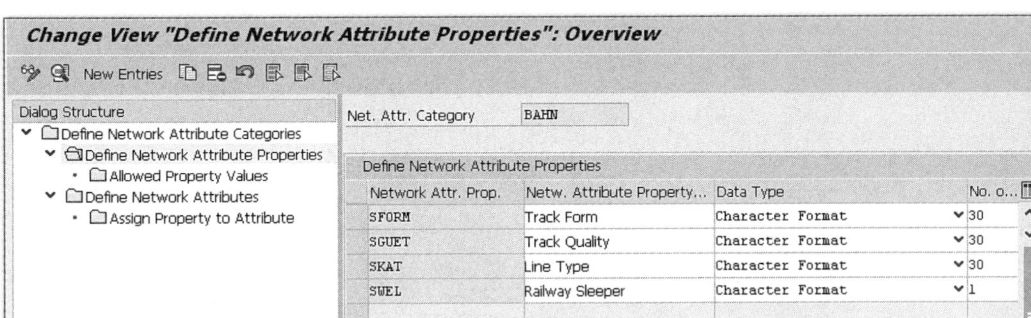

Figure 5.41 Object Networks: Network Attributes

- Allowed Property Values
  Defines the values permitted for each network attribute. Figure 5.42 shows, for example, the attributes permitted for the Line type property (e.g., regional route or long-distance route, passenger route, freight route, or mixed route). If you don't define a value table, you can freely define the relevant property in the master record for the object network.

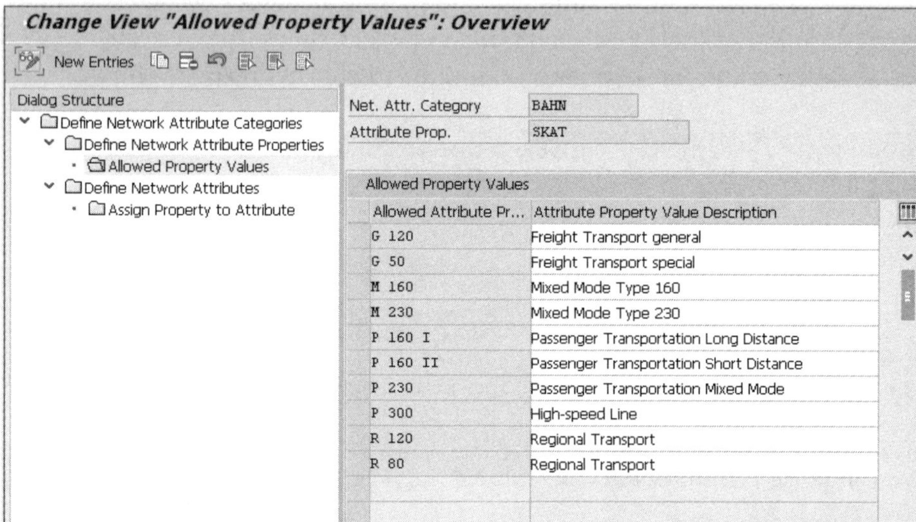

Figure 5.42 Object Networks: Properties of Network Attributes

Assign Network Attribute Categories to Network Types

You use this Customizing function to define which network attribute categories are used for which network type.

Prerequisites

You must define the network types and network attribute categories beforehand.

Customizing Path

Plant Maintenance and Customer Service • Master Data in Plant Maintenance and Customer Service • Technical Objects • Linear Asset Management • Define

Settings for Network Attributes • Assign Network Attribute Categories to Network Types

Settings

You assign the network attribute category to the network type. Figure 5.43 shows a 1:1 assignment. However, you can also assign several network attribute categories to a network type and the same network attribute category to different network types. In other words, an N:M assignment is also possible.

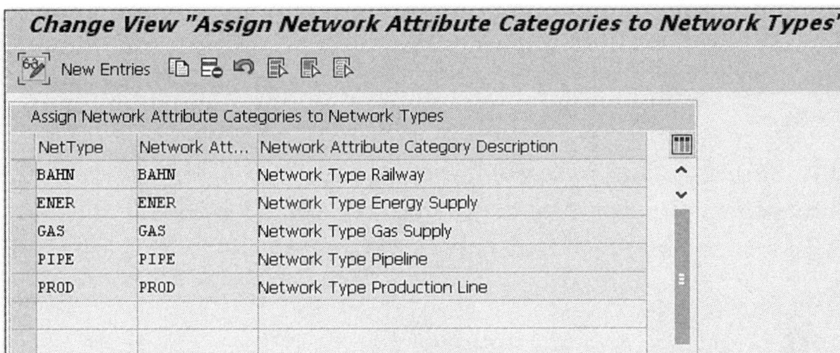

NetType	Network Att...	Network Attribute Category Description
BAHN	BAHN	Network Type Railway
ENER	ENER	Network Type Energy Supply
GAS	GAS	Network Type Gas Supply
PIPE	PIPE	Network Type Pipeline
PROD	PROD	Network Type Production Line

Figure 5.43 Object Networks: Network Type and Network Attribute Category

Customizing Functions for List Displays

Linear Asset Management has a range of Customizing functions for field selections:

- Define Field Selection for Linear Data Fields in Multilevel Lists (i.e., for Transactions IE07, IL07, IW30, and IW40)
- Define Field Selection for Linear Data Fields for Char. in Multilevel Lists (i.e., for Transactions IE07 and IL07)
- Define Field Selection for Linear Data Fields in Structure Lists (i.e., for Transactions IH01 and IH03)
- Define Field Selection for Multilevel List Displays of Funct. Loc. Network (i.e., for Transaction IN24)
- Define Field Selection for Multilevel List Displays of Equipment Network (i.e., for Transaction IN25)

The last two functions in the preceding list have the following subfunctions:

- Define Field Selection for Object Network Fields
- Define Field Selection for Network Attribute Values
- Define Field Selection for Object Links
- Define Field Selection for Functional Locations
- Define Field Selection for Notification Data Fields
- Define Field Selection for Order Data Fields
- Define Field Selection for Measuring Point and Counter Fields
- Define Field Selection for Measurement Document and Counter Reading Fields

How to make these field definitions was described in Chapter 4, Section 4.11. Figure 5.44 shows, for example, which data fields are selected for linear objects in the structural display for functional locations (Transaction IH01).

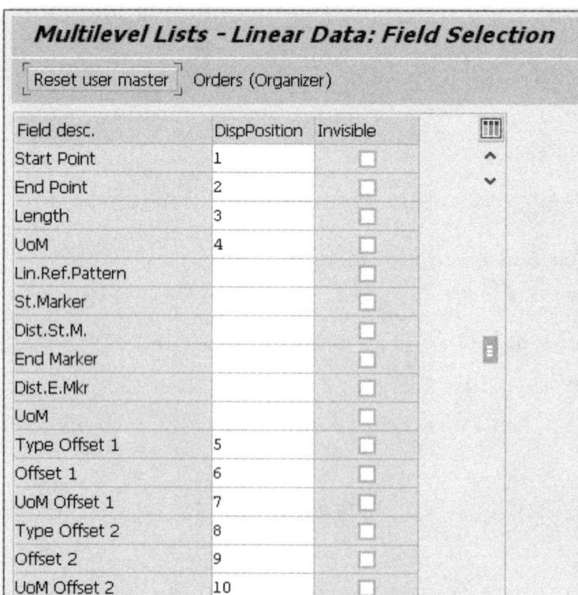

Figure 5.44 List Display with Linear Data

5.7 Materials and Assemblies

The material master contains information about materials that a company designs, procures, produces, stores, and sells. It integrates data from various areas within a company (e.g., purchasing or accounting). Unlike a piece of equipment or a functional location, a material master record doesn't describe just one thing, but rather similar items. In other words, a material master record generally represents several similar spare parts, assemblies, raw materials, and so on.

From a maintenance perspective, material master records are used for the following purposes:

- Some material master records are purchased as spare parts and then stored.
- Some material master records are purchased as equipment and then placed in storage as material/serial numbers.
- Some material master records are construction types that solely encompass a group of similar pieces of equipment or functional locations. This makes it possible to execute shared functions for the entire group (e.g., management of shared BOMs or general maintenance task lists).
- Some material master records serve as maintenance assemblies for substructures of pieces of equipment or functional locations.
- Some material master records fulfill the same function as spare parts. The only difference is that they aren't placed in storage, but instead are procured each time as nonstock items (e.g., because they are too expensive, too large, or rarely used).
- Some material master records are used as operating resources for maintenance tasks, but not as spare parts (e.g., tools or protective clothing).

Figure 5.45 provides an overview of the Customizing functions for the material master.

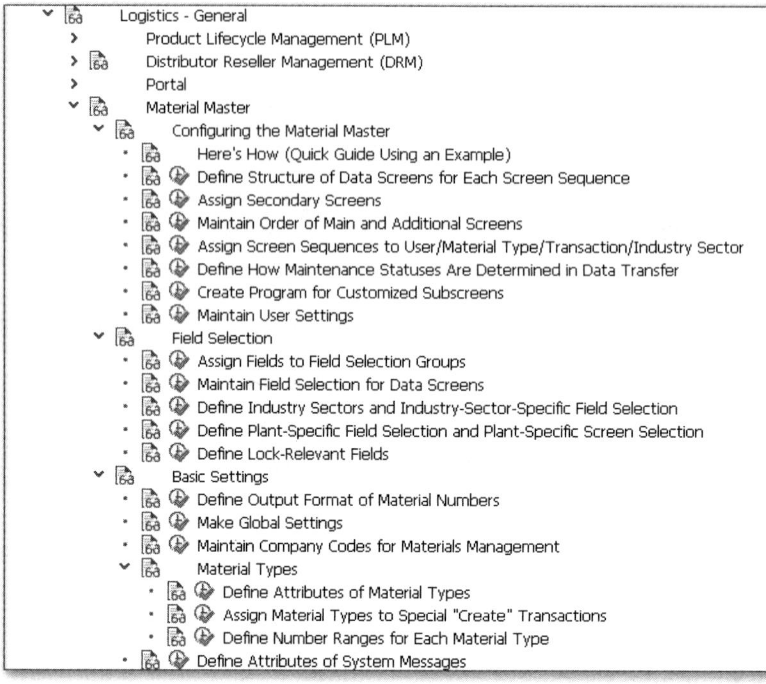

Figure 5.45 Materials: Customizing Functions

Next, we'll explain the most important of these Customizing functions to you.

5.7.1 Material Types

You group materials that have the same basic characteristics together by assigning them to one common material type. This enables you to manage different materials in a uniform manner, according to your company's requirements. An overview of potential material types is provided in Figure 5.46.

From a maintenance perspective, the following material types, which have a darker background in Figure 5.46, are of practical relevance:

- **Auxiliary materials**
  Auxiliary materials form part of a finished product but are insignificant and barely seen in the finished product (e.g., screws, adhesive, and welded joints).

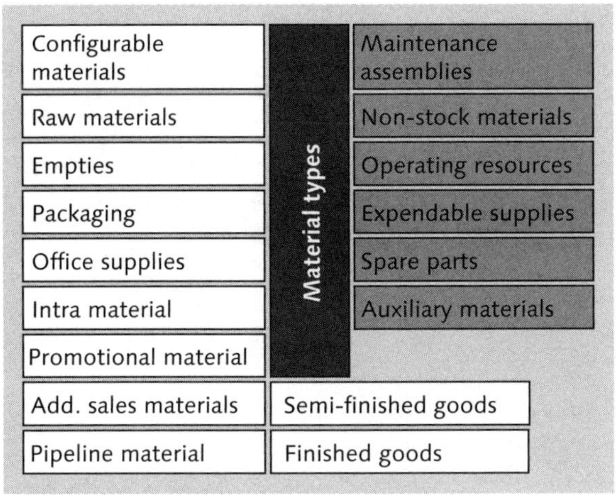

Figure 5.46 Material Types: Overview

- **Expendable supplies**
  Expendable supplies aren't part of the finished product but are required for the production process (e.g., lubricant, power, grease, oil, etc.).
- **Spare parts**
  Spare parts are used to replace defective parts. They can be purchased and placed in storage.
- **Maintenance assemblies**
  Maintenance assemblies aren't standalone objects but are rather logical elements that divide the technical objects in plant maintenance into clearly defined units. A forklift truck, for example, can be a technical object, while the lifting plant, gear shift, chassis, and so on, can be the associated maintenance assemblies.
- **Operating resources**
  Operating resources are required for maintenance tasks (e.g., tools and devices, test equipment, protective clothing, and safety devices). Unlike expendable supplies, they don't perish but wear out with use over time.
- **Nonstock materials**
  Nonstock materials aren't held in storage but are procured as required and used immediately. This may be the case, for example, because the parts are too expensive, are very rarely required, or are difficult to store (too large, too heavy, or too bulky). The master record of nonstock materials comprises purchasing data only.

Customizing for the Material Type

In Customizing, the material type is the main control element for the material master. You define the following for the material type (see Figure 5.47):

- The views that comprise the material master record (purchasing, storage, MRP, accounting, etc.)
- Whether the material number can be assigned internally or externally
- The number range interval from which the material number originates
- Which screens appear and in which order
- Which are main screens and which are secondary screens
- Which fields are to be maintained (field selection)
- Whether quantity and value changes are updated
- How the material prices are to be defined (standard price, moving average price, etc.)
- Whether the material is to be produced in-house or procured externally
- The accounts to which material consumption and material stock are posted if a material is placed in storage or removed from storage

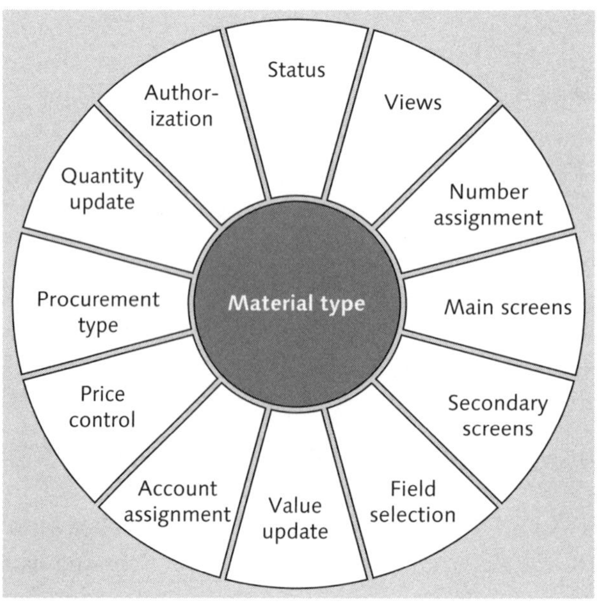

Figure 5.47 Material Type: Customizing

- The authorization check for access to the material master
- Which statuses can be assigned to the material master

Define Attributes of Material Types

You can use this Customizing function to define the basic definitions for a material type (e.g., the user departments, the reference to the screen sequence, or the statuses permitted).

Prerequisites

You must have configured the following Customizing settings:

- Field reference for the field selection
- Screen reference for the screen sequence
- Account category reference for automatic account assignment
- Valuation areas for the quantity and value update

If configured beforehand, you can also use the following Customizing settings:

- Authorization group
- Cross-plant material status

Customizing Path

Logistics - General • Material Master • Basic Settings • Material Types • Define Attributes of Material Types

Settings or Recommended Settings

You define the following settings for a material type. Figure 5.48 shows an example of a material type for spare parts.

- Field reference
  Establishes a link to the Customizing functions for field selections. Together with other influencing factors, they control the fields to be maintained for the material type.
- Screen reference (SRef material Type)
  Establishes a link to the Customizing functions for the screen control (Configuring the Material Master). Together with other influencing factors, they control the screens (main and secondary) for the material type.

- X-PLANT MATL STATUS
  Defines whether an initial material status is to be proposed for the materials of this material type. In this way, you can control, for example, that materials are first locked for certain uses (such as purchasing or BOMs) and then need to be explicitly released.

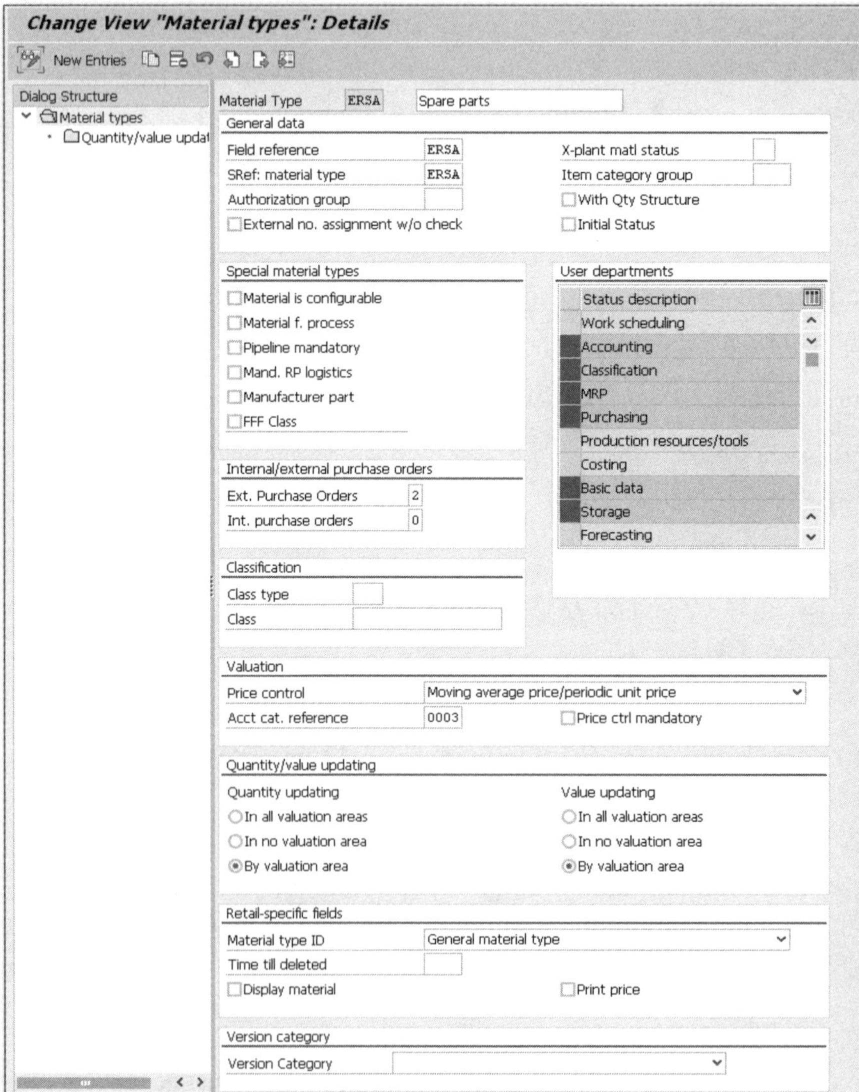

Figure 5.48 Material Type: Basic Settings

- USER DEPARTMENTS
  Defines which user departments are permitted to maintain material masters. In the case of maintenance materials, these are usually purchasing, MRP, storage, accounting, and classification, as well as plant stock and warehouse stock. You therefore also indirectly determine the screen sequence.
- INTERNAL/EXTERNAL PURCHASE ORDERS
  Defines the procurement type for materials of this material type: internal and/or external. You control whether the procurement type is prohibited (0), permitted but a warning message is issued (1), or always permitted (2). Maintenance materials are primarily procured externally.
- PRICE CONTROL
  Defines which price control procedure is to be proposed for materials of this material type: standard price or moving average price. In the case of maintenance materials, both procedures are used.

  If you don't want it to be possible to change this default value in the material master, set the PRICE CTRL MANDATORY indicator.
- ACCT CAT. REFERENCE
  Defines (together with other factors such as the company code, chart of accounts, or business process) the automatic account assignment for materials of this material type. Consult with your colleagues from accounting or ask them which account category reference is to be used for maintenance materials.
- QUANTITY/VALUE UPDATING
  Defines whether updates are to be handled in the same manner in all valuation areas (which usually correspond to the plants) or handled differently for each valuation area. In the case of maintenance materials, you generally base the quantity and price update on the valuation area and therefore on the plant.

  If you want to control quantity and value updating on the basis of the plant, use the QUANTITY UPDATING/VALUE UPDATING subfunctions to define which update is to occur for which plant (see Figure 5.49).

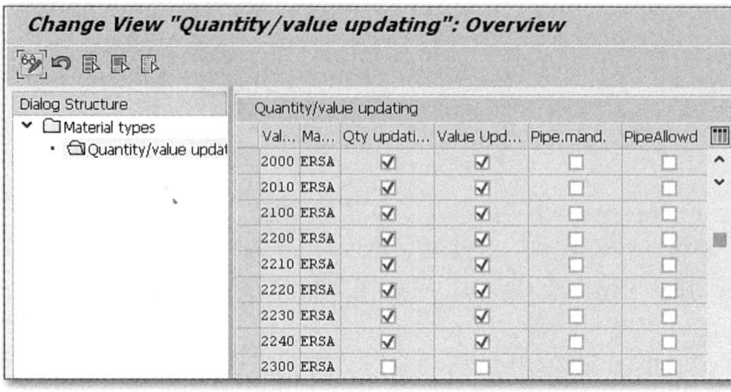

Change View "Quantity/value updating": Overview

Dialog Structure
- Material types
  - Quantity/value updat

Quantity/value updating

Val...	Ma...	Qty updati...	Value Upd...	Pipe.mand.	PipeAllowd
2000	ERSA	☑	☑	☐	☐
2010	ERSA	☑	☑	☐	☐
2100	ERSA	☑	☑	☐	☐
2200	ERSA	☑	☑	☐	☐
2210	ERSA	☑	☑	☐	☐
2220	ERSA	☑	☑	☐	☐
2230	ERSA	☑	☑	☐	☐
2240	ERSA	☑	☑	☐	☐
2300	ERSA	☐	☐	☐	☐

Figure 5.49 Material Type: Quantity and Value Updating

Configuring the Material Master

You can use this Customizing function to define your own material master layout. You can base your layout on the following influencing factors (see Figure 5.50):

- Transaction
- Material type
- User
- Industry sector

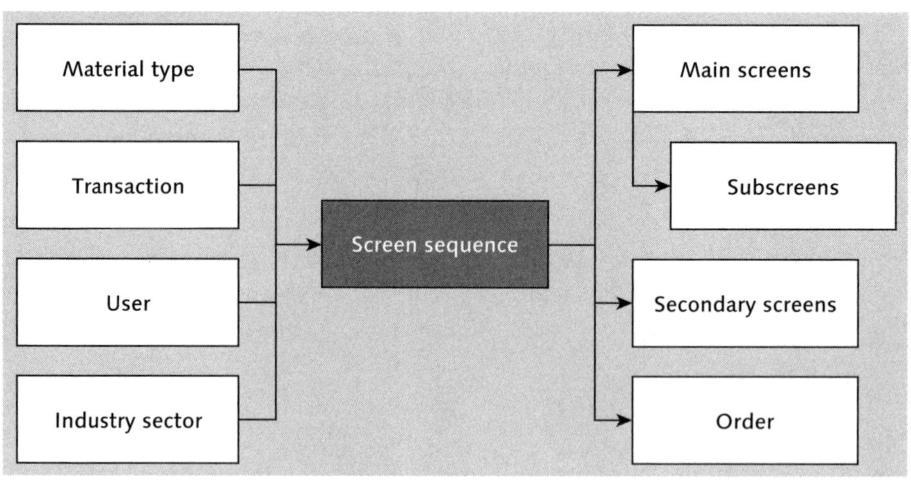

Figure 5.50 Screen Sequence: Overview

The influencing factors determine the *screen sequence*. For each screen sequence, you define the following (see Figure 5.50):

- Number and type of main screens
- Number and type of subscreens (screen groups) for each main screen
- Number and type of secondary screens
- Order of main screens and secondary screens

Customizing Path

Logistics - General • Material Master • Configuring the Material Master

The following example shows you the screen content definition for a typical material type; namely, spare parts (material type ERSA). The material master will contain the following main screens only (one full screen in each case) and in the following order:

- Basic Data
- Classification
- Documents
- Purchasing
- MRP
- Plant/Storage Data
- Accounting

Each of the main screens will contain a maximum of three screen groups, which provide typical maintenance information.

The following secondary screens are required in the following order:

- Multilingual Texts
- Units of Measure
- Basic Data Text

Define Structure of Data Screens for Each Screen Sequence

You use this Customizing function to specify a screen sequence and to assign the necessary main screens and their subscreens.

Prerequisites

There are no specific prerequisites.

Customizing Path

Logistics – General • Material Master • Configuring the Material Master • Define Structure of Data Screens for Each Screen Sequence

Transaction

OMT3B

Settings or Recommended Settings

The easiest way to generate a new screen sequence is to copy and adjust (i.e., reduce) screen sequence 01, which is delivered in the standard SAP system. For our example, we copied screen sequence ZE for Spare Parts (see Figure 5.51).

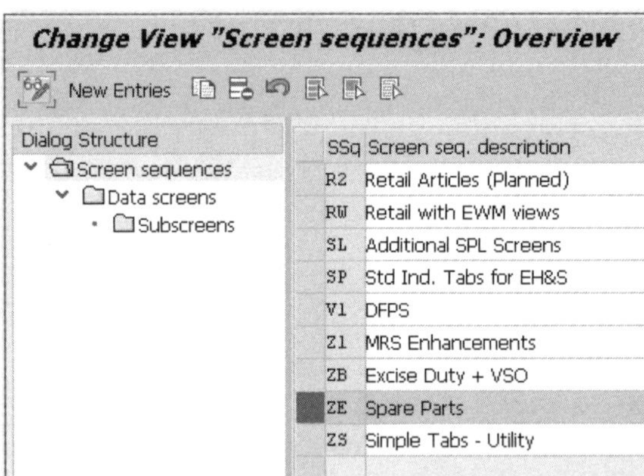

Figure 5.51 Screen Sequence: Overview

You should then use the Data screens subfunction to delete those data screens that you copied from screen sequence 01, but that don't require plant maintenance.

For our example, we need all of the screens listed in Figure 5.52. Here, you need to list all required screens (i.e., not only the main screens but also the secondary screens). The following settings are available:

- SCREEN DESCRIPTION
  Choose your own screen description for each screen.
- T
  Use the screen type to define whether the screen is a main screen (1) or a secondary screen (2).
- SCON
  Use the container screen to define the screen structure. In addition, you can use the F4 help to choose from approximately 50 different variants. For example, you can select whether the screen is a general container screen and, if so, with how many modules (any number between 6 and 20 modules), or whether the screen is a special screen (e.g., for a description or classification).
- MAINT. STATUS
  Define which maintenance status is to be set if you maintain this screen. In Figure 5.52, you can obtain the maintenance status associated with the relevant screen (e.g., E for purchasing or B for accounting).
- GUI STATUS
  Refine the screen types for main screens and secondary screens. The standard screen type for main screens is DATE, while the standard screen type for secondary screens is ADD0001.

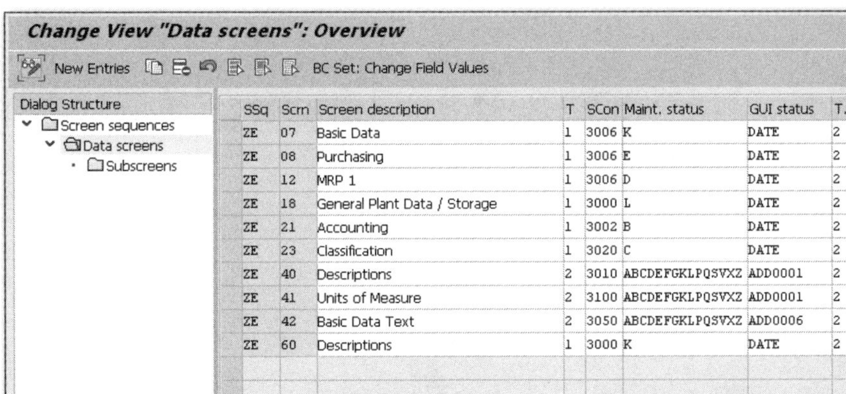

Figure 5.52 Screen Sequence: Overview of Data Screens

You use the SUBSCREENS subfunction to determine the screen areas required for each screen. Here, you can choose from a selection of almost 300 screen areas for the various views. The following is a representative selection of screen areas for the views in our example.

Figure 5.53 shows a collection of screen areas for BASIC DATA: 1002 = Material Numbers, 2001 = General Data, 2007 = Dimensions and EAN, 2003 = Document without Document Master Record, and 2004 = Document with Document Master Record.

[!]

Important Notes on the Definition

- Screen area 0001 represents an empty placeholder. If, for example, you select a container screen with 15 screen areas (as is the case here) but you require only 5, you must enter "0001" in the remaining screen areas.
- You can choose the VIEW SUBSCREEN button to preview the layout of the screen areas to be selected.
- You can choose the VIEW DATA SCREEN button to fully display the finished data screen.

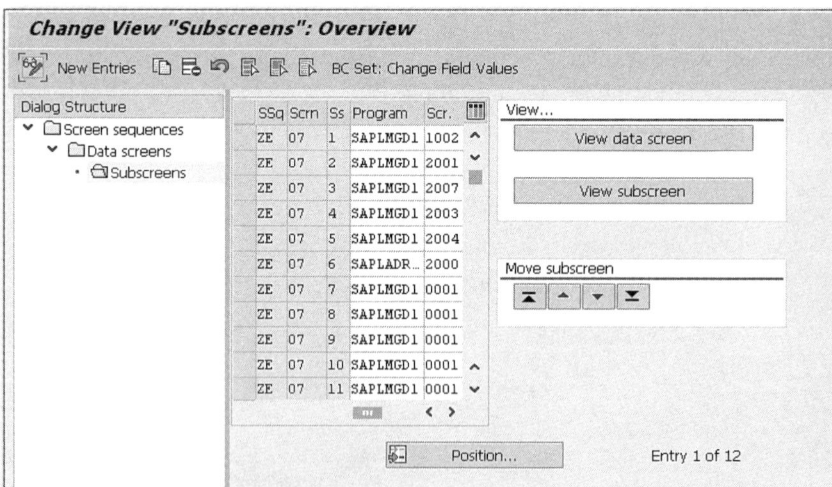

Figure 5.53 Screen Sequence: Subscreens for General Data

Figure 5.54 shows a representative collection of screen areas for PURCHASING: 1001 = Material Numbers with Plant, 2301 = Purchasing Data – General Data, 2302 = Purchasing Data – Purchasing Values, and 2303 = Purchasing Data – Other Data.

SSq	Scrn	Ss	Program	Scr.
ZE	08	1	SAPLMGD1	1001
ZE	08	2	SAPLMGD1	2301
ZE	08	3	SAPLMGD1	2302
ZE	08	4	SAPLMGD1	2303

Figure 5.54 Screen Sequence: Subscreens for Purchasing

Figure 5.55 shows a representative collection of data for MRP: 1001 = Material Numbers with Plant, 2482 = MRP Procedure, 2483 = Lot-Size Data, 2484 = Procurement, 2485 = Scheduling, and 2493 = Availability Check.

SSq	Scrn	Ss	Program	Scr.
ZE	12	1	SAPLMGD1	1001
ZE	12	2	SAPLMGD1	2482
ZE	12	3	SAPLMGD1	2483
ZE	12	4	SAPLMGD1	2484
ZE	12	5	SAPLMGD1	2485
ZE	12	6	SAPLMGD1	2493

Figure 5.55 Screen Sequence: Subscreens for MRP

Figure 5.56 shows the most important screen areas for PLANT/STORAGE DATA: 1005 = Material Numbers with Plant and Storage Location, 2701 = Storage Data – General Data, and 2702 = Storage Data – Shelf Life Data.

SSq	Scrn	Ss	Program	Scr.
ZE	18	1	SAPLMGD1	1005
ZE	18	2	SAPLMGD1	2701
ZE	18	3	SAPLMGD1	2702

Figure 5.56 Screen Sequence: Subscreens for Plant/Storage Data

Figure 5.57 shows the most important screen areas for ACCOUNTING: 1008 = Material Numbers with Plant and Valuation Type, 2801 = Valuation Data for Current Period – General Data, and 2802 = Valuation Data for Current Period – Current Valuation.

SSq	Scrn	Ss	Program	Scr.
ZE	21	1	SAPLMGD1	1008
ZE	21	2	SAPLMGD1	2801
ZE	21	3	SAPLMGD1	2802

Figure 5.57 Screen Sequence: Subscreens for Accounting

Assign Secondary Screens

You use this Customizing function to define whether a secondary screen is to be displayed as an additional screen if you choose ⇨ Additional Data to call it, or whether it's to be displayed as a secondary screen when you use the relevant button to call it directly.

Prerequisites

You've used the Customizing function Define Structure of Data Screens for Each Screen Sequence to define the basic layout of the screen sequence, and, in particular, you've assigned screens of screen type 2 (secondary screen). If no screens of screen type 2 exist, you can't assign any secondary screens.

Customizing Path

Logistics - General • Material Master • Configuring the Material Master • Assign Secondary Screens

Transaction

OMT3Z

Settings or Recommended Settings

If you want to display the screen as an Additional screen, don't enter a function code in the FCode field (see Figure 5.58). The system then automatically assigns the next available number to the additional screen.

If you want the screen to be displayed as a secondary screen, enter the relevant function code in the FCode column (e.g., "PB26" for Basic Data Text or "PB13" for Material Forecast). Here, you can choose from a total of 29 function codes.

When you use additional data or function codes to call screens, you must assign a suitable PROCESSING ROUTINE. Here, you can choose from a total of 38 predefined processing routines.

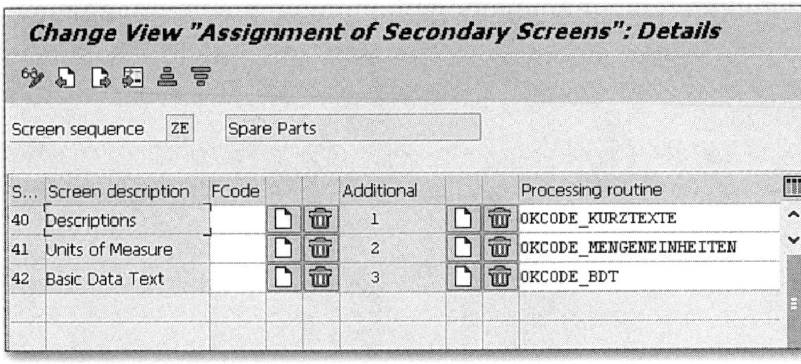

Figure 5.58 Screen Sequence: Secondary Screens

Maintain Order of Main and Additional Screens

You can use this Customizing function to maintain the order in which the *main screens* are displayed when you maintain material master records in the VIEW SELECTION dialog box as well as the order in which the main screens and *additional screens* are displayed as tab pages.

Prerequisites

You've used the Customizing function DEFINE STRUCTURE OF DATA SCREENS FOR EACH SCREEN SEQUENCE to define the basic layout of the screen sequence, and you've used the Customizing function ASSIGN SECONDARY SCREENS to define the layout of the secondary screens.

Customizing Path

LOGISTICS - GENERAL • MATERIAL MASTER • CONFIGURING THE MATERIAL MASTER • MAINTAIN ORDER OF MAIN AND ADDITIONAL SCREENS

Transaction

OMT3R

Settings or Recommended Settings

You define the order in which the main screens and secondary screens are displayed by assigning a status number to each main screen and additional screen (see Figure 5.59). Here, it's best to use intervals of 10 so that you have the option to insert additional screens later. The screens with the lowest number are displayed first.

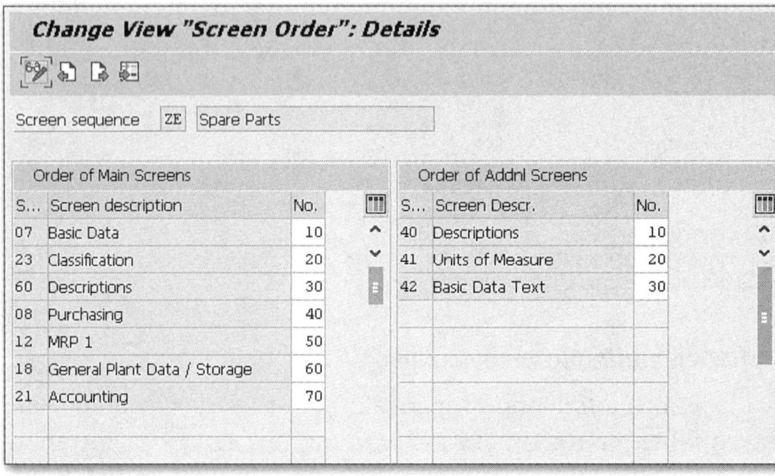

Figure 5.59 Screen Sequence: Order

Assign Screen Sequences to User/Material Type/Transaction/Industry Sector

You use this Customizing function to assign a screen sequence to a combination of one or more transactions, users, material types, or industry sectors. The system automatically displays this screen sequence if the relevant users choose this transaction, material type, or industry sector.

Prerequisites

You've used the Customizing function Define Structure of Data Screens for Each Screen Sequence to define the basic layout of the screen sequence, and you've used the Customizing function Assign Secondary Screens to define the layout of the secondary screens.

Customizing Path

Logistics - General • Material Master • Configuring the Material Master • Assign Screen Sequences to User/Material Type/Transaction/Industry Sector

Transaction

OMT3E

Settings or Recommended Settings

First, you use the SCREEN REFERENCE subfunctions to form groups. For example, you can form groups of transactions, users, material types, and industry sectors, and then use the main function SCREEN SEQUENCE CONTROL to assign the most suitable screen layout to the material master.

For this example, the subfunction MATERIAL TYPE SCREEN REFERENCE is used to define a separate material type reference ERSA for the material type ERSA. As you can see in Figure 5.60, you can also group several material types together to form a material type screen reference. You can also combine or individualize users, transactions, and industry sectors in the same way.

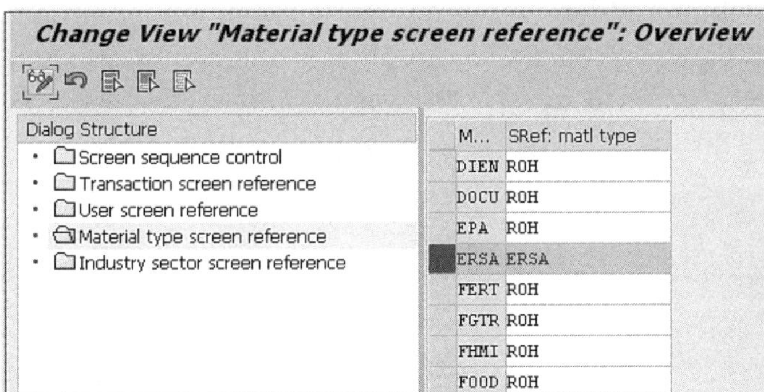

Figure 5.60 Screen Sequence: Reference Material Type

You use the main function SCREEN SEQUENCE CONTROL to assign a screen sequence on the basis of the screen references. It isn't necessary to have created a screen reference for each object. Rather, you can replace each object with a wild card (*). Then, the combination doesn't refer to a particular screen reference. However, SAP has defined some unique priorities here, as shown in Table 5.1. X typically denotes that an entry exists for this value (e.g., material type = ERSA), while * denotes that no entry exists for this value.

Priority	Transaction	User	Material Type	Industry Sector
1	X	X	X	X
2	X	X	*	*
3	X	*	X	X
4	X	*	X	*
5	X	*	*	*

Table 5.1 Priorities for Controlling Screen Sequences

If we apply this to our example, this means the following:

- Because my user isn't assigned to a user group, the TRANSACTION/MATERIAL TYPE combination is used so that the screen sequence ZE (Spare Parts) is displayed when you maintain a material master (see Figure 5.61).
- If, however, a user from the user group BO needs to maintain a material master of the material type ERSA, screen sequence 01 is assigned because the TRANSACTION/USER entry is given greater priority than the TRANSACTION/MATERIAL TYPE entry.

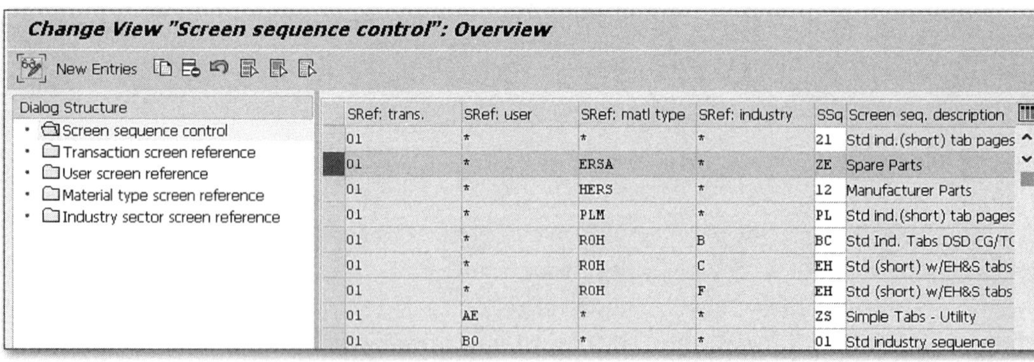

Figure 5.61 Screen Sequence: Assignment

5.7.2 Field Selection

The field selection technique, introduced in Chapter 4, Section 4.9, can be used in conjunction with equipment or orders, for example. You can also use the field selection option with the material master. However, the field selection technique deployed here is completely different from the technique described previously. Therefore, the next section discusses how you can configure the field selection for use with the material master.

In the case of the material master, the field selection depends on three factors (see Figure 5.62):

- Material type
- Industry sector
- Plant

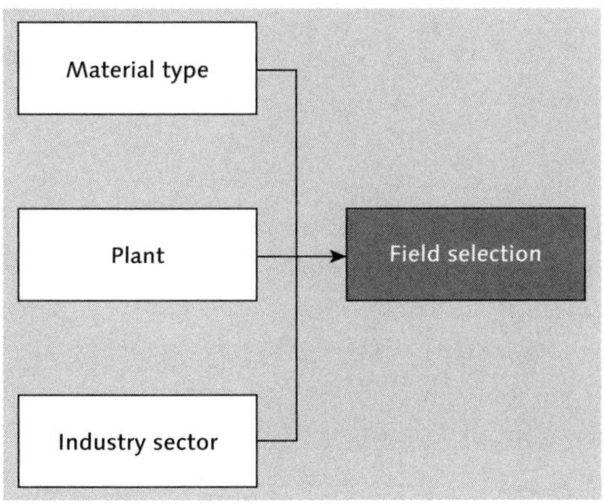

Figure 5.62 Field Selection: Influencing Factors

You've used the Customizing function Define Attributes of Material Types to define a Field reference for the material type (see Figure 5.48 and the Customizing function Define Attributes of Material Types).

Assign Fields to Field Selection Groups

You use this Customizing function to assign the individual fields of the material master to a field selection group (see Figure 5.63). You can assign each field to only one field selection group.

Prerequisites
There are no specific prerequisites.

Customizing Path

Logistics - General • Material Master • Field Selection • Assign Fields to Field Selection Groups

Transaction

OMSR

Settings

Because SAP has reserved certain areas, groups 111 – 120 and 211 – 240 are available to you for your own field selections. However, it's not absolutely necessary for you to form your own groups. You can simply retain the standard SAP assignment and base your field selections on this assignment. Figure 5.63 shows the various field selection groups available.

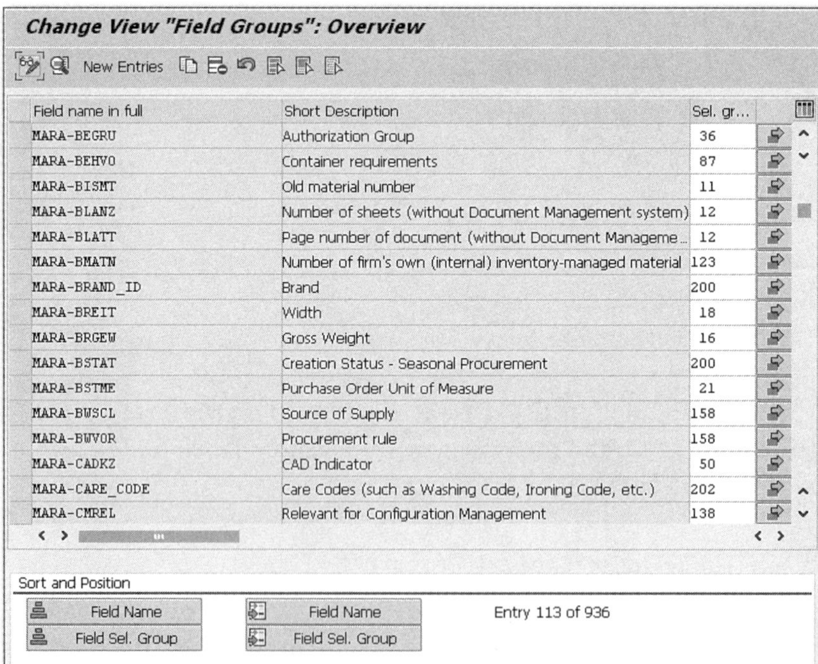

Figure 5.63 Field Selection: Field Selection Groups

Define Industry Sectors and Industry-Sector-Specific Field Selection

You use this Customizing function to define your industry sectors and to determine the field reference that your industry sectors will use to influence the field selection.

Prerequisites

There are no special prerequisites.

Customizing Path

LOGISTICS - GENERAL • MATERIAL MASTER • FIELD SELECTION • DEFINE INDUSTRY SECTORS AND INDUSTRY-SECTOR-SPECIFIC FIELD SELECTION

Transaction

OMS3

Settings

You assign a field reference to the industry sectors that are to influence the field selection. Here, you can assign a custom field reference to each industry sector, or you can combine several industry sectors (see Figure 5.64).

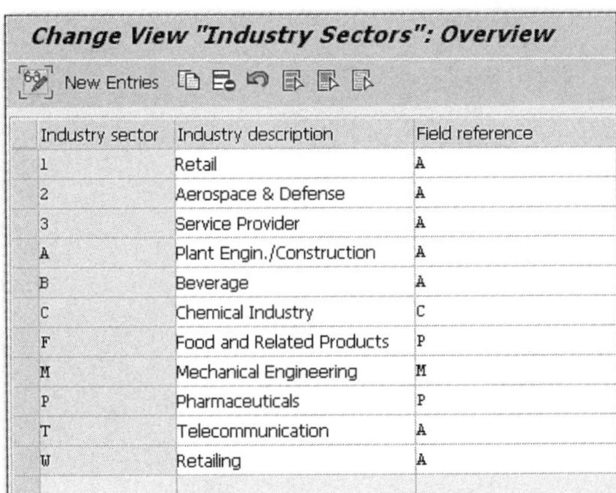

Change View "Industry Sectors": Overview

New Entries

Industry sector	Industry description	Field reference
1	Retail	A
2	Aerospace & Defense	A
3	Service Provider	A
A	Plant Engin./Construction	A
B	Beverage	A
C	Chemical Industry	C
F	Food and Related Products	P
M	Mechanical Engineering	M
P	Pharmaceuticals	P
T	Telecommunication	A
W	Retailing	A

Figure 5.64 Field Selection: Industry Sectors

Define Plant-Specific Field Selection and Plant-Specific Screen Selection

You use this Customizing function to define the field reference that your plants will use to influence the field selection.

Prerequisites

You must define the plants beforehand.

Customizing Path

LOGISTICS - GENERAL • MATERIAL MASTER • FIELD SELECTION • DEFINE PLANT-SPECIFIC FIELD SELECTION AND PLANT-SPECIFIC SCREEN SELECTION

Transaction

OMSA

Settings

You assign a field reference to the plants that are to influence the field selection. Here, you can assign a custom field reference to each plant, or you can combine several plants (see Figure 5.65).

Change View "Field Selection at Plant Level": Overview

New Entries

Plant	Name 1	Maintenance status	Field reference
3000	New York	KDEVALBPQSZXCFG	0001
3050	UK	KDEVALBPQSZXCFG	0001
3100	Chicago	KDEVALBPQSZXCFG	0001
3105	Chicago	KDEVALBPQSZXCFG	0001
3110	Auto Supplier US	KDEVALBPQSZXCFG	0001
3111	Auto OEM US	KDEVALBPQSZXCFG	0001
3112	Auto Wholesale US	KDEVALBPQSZXCFG	0001
3150	Columbus	KDEVALBPQSZXCFG	0001
3200	Atlanta	KDEVALBPQSZXCFG	0001
3300	Los Angeles	KDEVALBPQSZXCFG	0001
3350	San Francisco	KDEVALBPQSZXCFG	0001
3400	Seattle	KDEVALBPQSZXCFG	0001
3500	Dallas	KDEVALBPQSZXCFG	0001
3800	Denver Distribution center	KDEVALBPQSZXCFG	0001
3850	LTL Consolidation Hub	KDEVALBPQSZXCFG	0001
4000	Toronto	KDEVALBPQSZXCFG	0001

Figure 5.65 Field Selection: Plants

Maintain Field Selection for Data Screens

You use this Customizing function to define a particular field reference and its influence on the field selection:

- Display field
- Required entry field
- Normal entry field
- Hide

Prerequisites

You must define the plants beforehand.

Customizing Path

LOGISTICS - GENERAL • MATERIAL MASTER • FIELD SELECTION • MAINTAIN FIELD SELECTION FOR DATA SCREENS

Transaction

OMS9

Settings

Certainly, several influencing factors can now affect the field selection (see Figure 5.66):

- The field selection of the material type ERSA can define a field as a required entry field.
- On the other hand, the field selection of plant 1000 can define the field as a normal entry field.
- At the same time, the field selection of industry sector M wants to hide this field.

If several different influencing factors affect the field selection, the priority rules described earlier also apply here (see Chapter 4, Section 4.9, and Figure 5.31):

- **Priority 1: Hide**
  If one of the influencing factors defines a field as hidden, it's hidden irrespective of the field selection control for other influencing fields.

- **Priority 2: Display**
  If it's not hidden and an influencing factor defines it as a display field, it becomes a display field, even if other influencing factors define it as a required entry or normal entry field.
- **Priority 3: Required entry**
  If it's not a hidden field or display field, and an influencing factor defines it as a required entry field, it becomes a required entry field, even if other influencing factors define it as a normal entry field.
- **Priority 4: Normal entry field**
  If no influencing factor defines it elsewhere, this field remains a normal entry field.

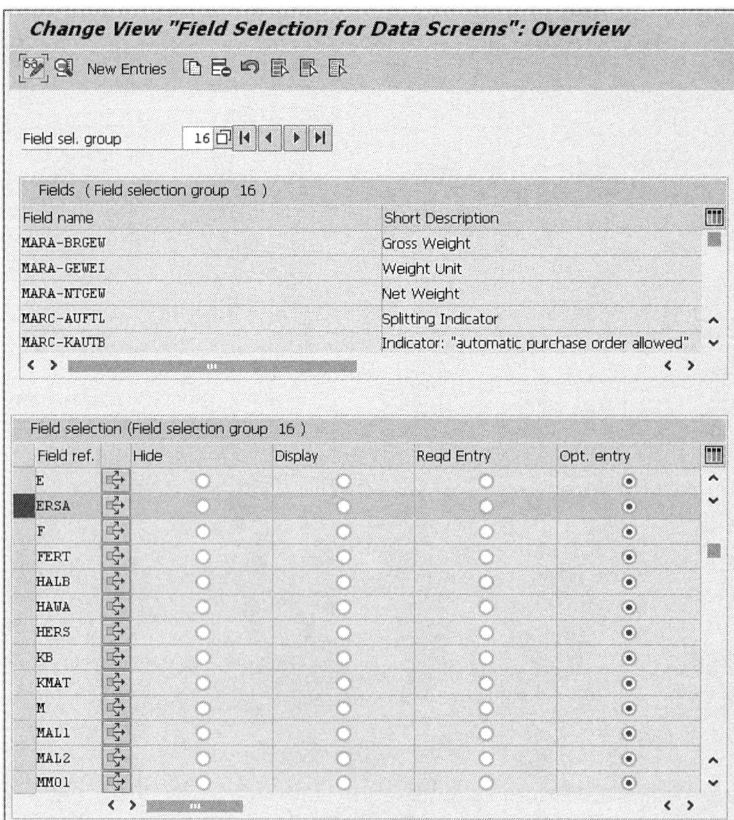

Figure 5.66 Field Selection: Field Selection Control

5.8 Serial Numbers

You create serial numbers for a material number. Here, you can specify as many serial numbers as you want for one material number. A serial number is an individual item and corresponds to a piece of equipment. The serial number function enables you to place the equipment in storage. In terms of usage, a material serial number corresponds to a piece of equipment.

All of the Customizing settings described in Section 5.3 also apply to serial numbers. However, there are also some settings that apply only to serial numbers (e.g., the serial number profile or the serialization attributes of movement types).

If you create serial numbers for materials and you want to maintain these as pieces of equipment, you must first configure certain settings in two of the Customizing functions described previously:

- In the Customizing function CONFIGURING THE MATERIAL MASTER with the SUBSCREENS subfunction, you assign the subscreen SAPLMGD1 5801 (General Plant Data) to one of your data screens. To do this, you select a data screen at the plant level (e.g., the PURCHASING or PLANT/STORAGE DATA data screen) because the serial number profile is always assigned to a plant. The system then displays a subscreen in the material master (as shown in Figure 5.67). Here, you can define serial data such as the serial number profile and the serial number level.

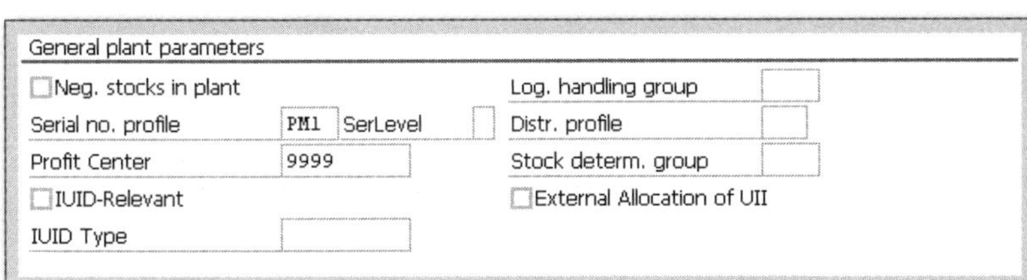

Figure 5.67 Serial Numbers: Subscreen in Material Master

- In the Customizing function DEFINE ADDITIONAL BUSINESS VIEWS FOR EQUIPMENT CATEGORIES, you must set the SD (serial data; refer to Figure 5.16) indicator for the equipment category to which you want to assign the serial number. Then, the system displays the serial data subscreens in the equipment master (see Figure 5.68). Here, you can assign the relevant serial data (e.g., the SERIAL NUMBER, PLANT, STORAGELOCATION, and STOCK BATCH) to the piece of equipment.

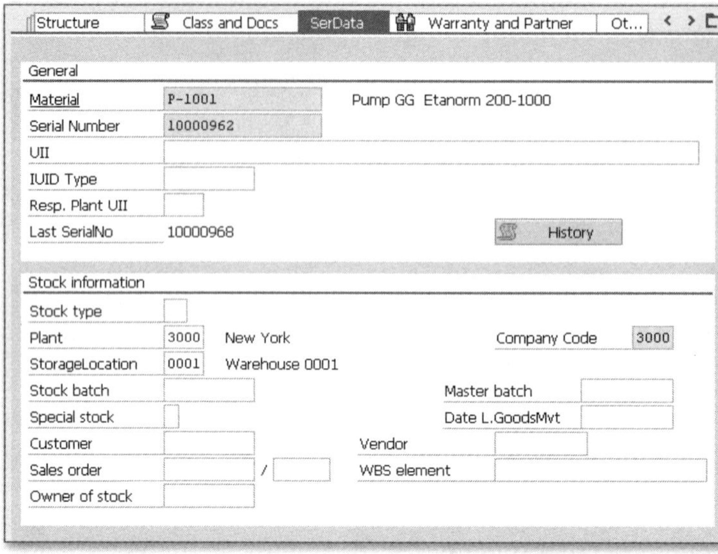

Figure 5.68 Serial Numbers: Subscreens in Equipment Master

Figure 5.69 provides an overview of the Customizing functions for serial numbers.

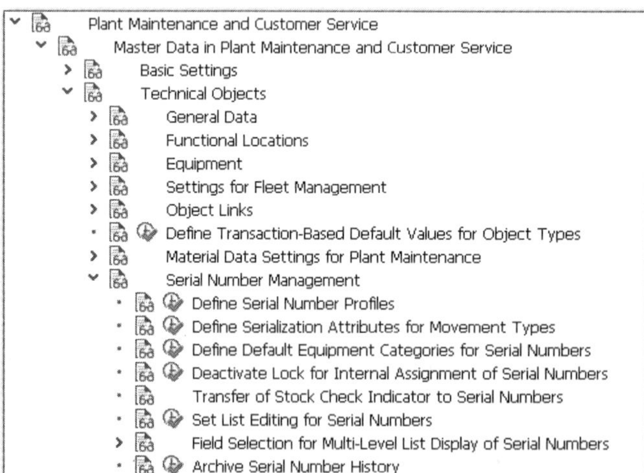

Figure 5.69 Serial Numbers: Customizing Functions

Next, we'll explain the most important of these Customizing functions to you.

Define Serial Number Profiles

You use this Customizing function to define serial number profiles. In the material master, you must assign a serial number profile at the plant level to each *material* that you want to serialize. This means that for each plant, you can assign a custom serial number profile to a material. A material may require serial numbers in one plant, but not in another.

Prerequisites

There are no special prerequisites.

Customizing Path

Plant Maintenance and Customer Service • Master Data in Plant Maintenance and Customer Service • Technical Objects • Serial Number Management • Define Serial Number Profiles

Transaction

OIS2

Settings

On the Overview screen for serial number profiles, you configure the following settings in addition to the profile name and profile description (see Figure 5.70):

- ExistReq.
  Set this indicator if you want the serial number entered for a business process (e.g., for a goods receipt) to have been created as a master record beforehand. Even though this is a good cautionary step, it prevents the system from being able to automatically create serial numbers for a goods receipt, for example.
- Category
  Enter the equipment category that you want the system to assign when automatically creating serial numbers for a business process, such as goods receipt.
- StkCk
  Use this stock check indicator to define whether you want the system to perform a stock check when assigning serial numbers. If so, you also define the type of message (warning message or error message) that you want the system to issue if any stock inconsistencies arise in relation to inventory management.

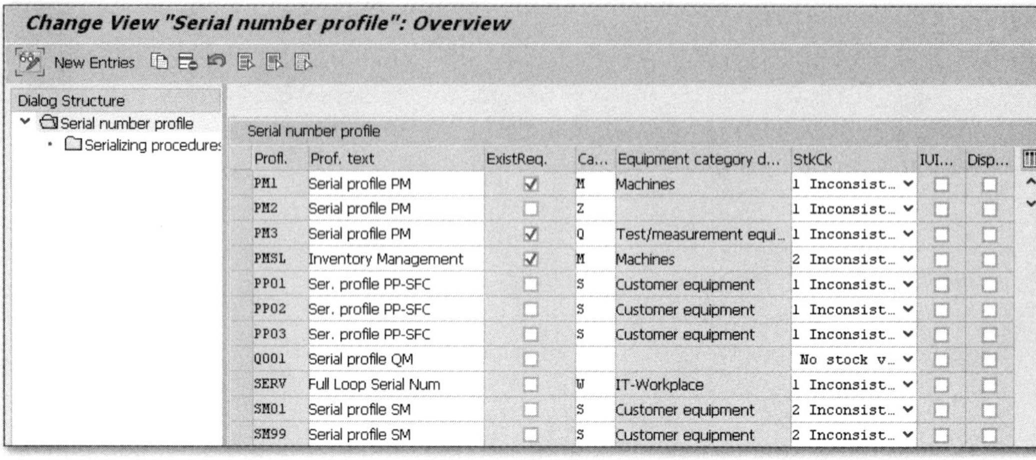

Figure 5.70 Serial Numbers – Serial Number Profile: Overview

Then, on the detail screen, you define the SERIALIZING PROCEDURES that are permitted for this serial number profile (see Figure 5.71). Table 5.2 lists the serializing procedures available.

Serializing Procedure	Description
HUSL	Handling Unit
MMSL	Goods Receipt and Goods Issue
PIAU	Serial Numbers in Process Industry Order
PIRL	Release Process Industry Order
POSL	Serial Numbers in Purchase Orders
PPAU	Serial Numbers in Production Planning (PP) Order
PPRL	Release PP Order
PPSF	Serial Numbers in Repetitive Manufacturing
PRSL	Serial Numbers in Purchase Requisitions
QMSL	Inspection Lot
SDAU	Serial Numbers in Sales Order
SDCC	Completeness Check for Delivery

Table 5.2 Serializing Procedures

Serializing Procedure	Description
SDCR	Completeness Check for Returns Delivery
SDLS	Delivery
SDRE	Returns Delivery
SDVC	Change in Sales Order Version

Table 5.2 Serializing Procedures (Cont.)

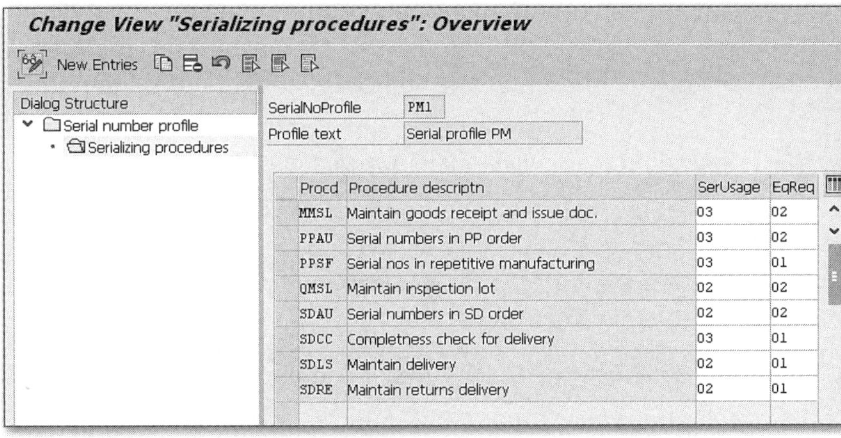

Figure 5.71 Serial Numbers: Serial Number Profile for Procedures

From a maintenance processing perspective, the serializing procedure MMSL is, without doubt, the most important serializing procedure because it enables you to permit goods movements (putaways, stock removals, and stock transfers) for serial numbers.

For each serializing procedure, you also define how you want the system to deal with serial numbers in each case (SerUsage column):

- 01
  No serial numbers are assigned.
- 02
  Serial numbers can be assigned.
- 03
  Serial numbers must be assigned.

- 04
  Serial numbers are automatically assigned.

If serial numbers are automatically assigned to a procedure, the dialog box in which you enter serial numbers isn't displayed. Instead, the system automatically assigns the next available numbers in numeric order.

You use the EqReq (equipment mandatory) column to control whether a piece of equipment must be created for the relevant serializing procedure. If you choose indicator 01 (default: without equipment), you can still decide at a later stage to create an equipment master record when assigning serial numbers in the relevant dialog box. If you choose indicator 02 (always with equipment), this indicator is binding. In other words, you can't undo this indicator.

Define Serialization Attributes for Movement Types

You can use this Customizing function to define serialization attributes for individual *movement types* or *movement type groups*. You only need to maintain this setting if the assignment of serialization attributes at the *serializing procedure level* is too broad, and you want to differentiate it further.

At present, only the serializing procedure MMSL is supported.

Prerequisites

You must maintain the serial number profiles beforehand.

Customizing Path

Plant Maintenance and Customer Service • Master Data in Plant Maintenance and Customer Service • Technical Objects • Serial Number Management • Define Serialization Attributes for Movement Types

Settings or Recommended Settings

For example, in the serial number profile PM1 for the serializing procedure MMSL, you've set SerUsage (serial number usage) to 02. In other words, you can specify a serial number for goods receipts, goods issues, and stock transfers, but it's not mandatory for you to do so.

If, however, you now want to make it mandatory to enter serial numbers for a goods issue from a purchase order or a goods issue from a production order (and only there), you can achieve this by configuring the following settings:

- Use the subfunction Define Movement Type Groups (see Figure 5.72) to define a movement type group (here: PM01) for the serial number profile PM1 and the serializing procedure MMSL. Here, set SerUsage (serial number usage) to 03; that is, serial numbers must be assigned.

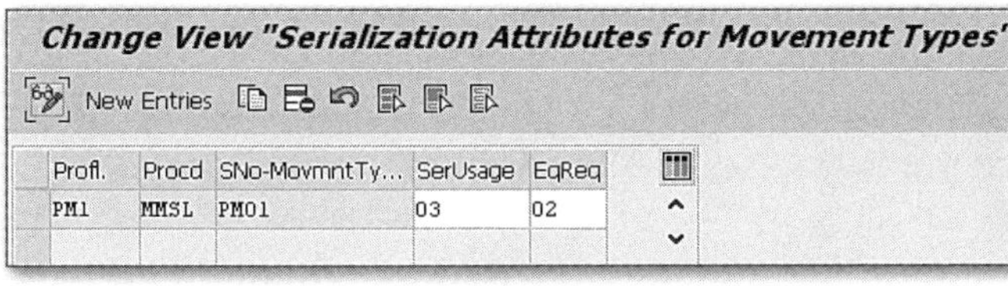

Figure 5.72 Serial Numbers: Movement Type Groups

- You use the subfunction Assign Movement Type Groups to Movement Types (see Figure 5.73) to assign your movement type group PM01 to movement types 101 (Goods Issue from Purchase Order) and 101 (Goods Issue from Production Order), both without consumption.

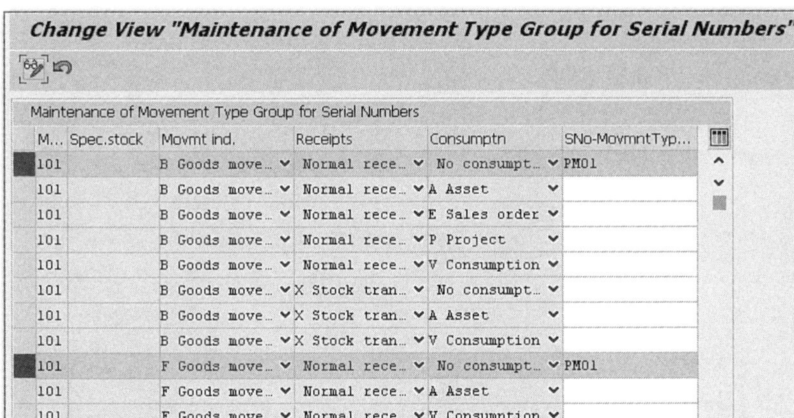

Figure 5.73 Serial Numbers: Movement Types

5.9 Bill of Material

A bill of material (BOM) is essentially a complete, formally structured list of all components that belong to a product or assembly. The components can be parts or assemblies that can or cannot be kept in storage. The assemblies, in turn, can have BOMs that describe them in greater detail. This gives rise to a multilevel BOM structure.

In plant maintenance, you can primarily use BOMs for the following two purposes:

- **Structure description**
  Describes the structure of a technical object or material. You can use the BOM to pinpoint the location of the damage or the location where maintenance tasks are performed for a technical object.
- **Spare parts assignment**
  Describes the assignment of spare parts for a technical object or material.

From a maintenance perspective, we can distinguish between three different types of BOMs:

- Equipment BOM for a special piece of equipment
- Functional location BOM for a special functional location
- Material BOM for a construction type and therefore for a group of functional locations or pieces of equipment

The SAP reference IMG has many Customizing functions for managing BOMs (see Figure 5.74) because they are not only used in plant maintenance but also in PP, Costing, QM, and other areas. However, many of the Customizing functions that relate to BOMs aren't required for plant maintenance. Consequently, we'll only introduce those Customizing functions that are of interest from a maintenance perspective, and therefore should be examined.

Define BOM Status

You use this Customizing function to define the BOM status that you require. The BOM status in the BOM header enables you to control whether the BOM is still locked or released for use. If it's released, you should specify the purpose for which it has been released.

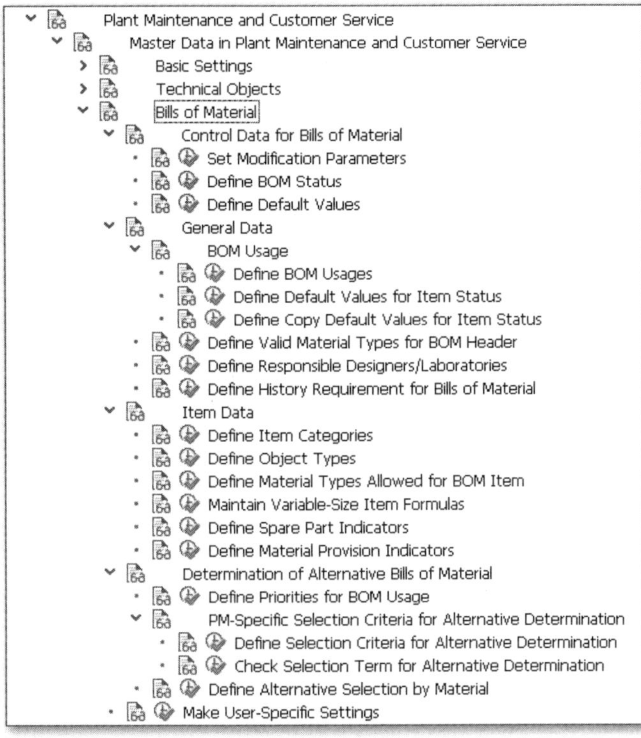

Figure 5.74 Bills of Material: Customizing Functions

Prerequisites

There are no special prerequisites.

Customizing Path

Plant Maintenance and Customer Service • Master Data in Plant Maintenance and Customer Service • Bills of Material • Control Data for Bills of Material • Define BOM Status

Transaction

OICH

Settings

Figure 5.75 shows the standard SAP delivery. If you don't want to release a BOM yet, set status 2 (Inactive) in the BOM header. Otherwise, set status 1 (Active).

If you define your own statuses, take note of the functions that play a role in plant maintenance:

- RELWKSCH
  Released for the task list.
- REL ORDS
  Released for the order.

The remaining functions aren't used in plant maintenance.

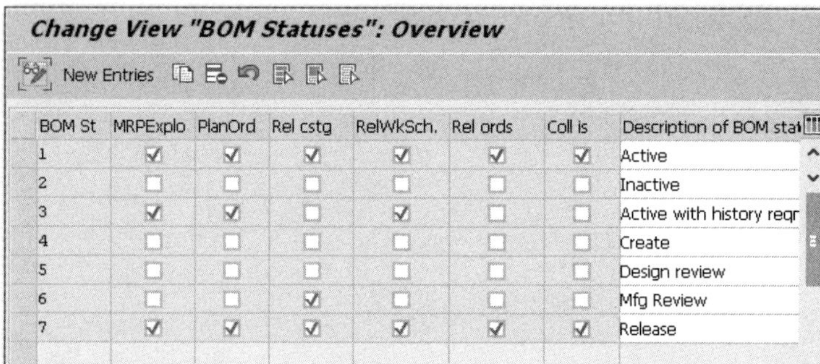

Figure 5.75 BOM Status

Define BOM Usages

You use this Customizing function to define which user departments (e.g., production, plant maintenance, and construction) use BOMs of a certain *usage*.

Prerequisites
There are no special prerequisites.

Customizing Path
PLANT MAINTENANCE AND CUSTOMER SERVICE • MASTER DATA IN PLANT MAINTENANCE AND CUSTOMER SERVICE • BILLS OF MATERIAL • GENERAL DATA • BOM USAGE • DEFINE BOM USAGES

Transaction
OICD

Settings or Recommended Settings

Figure 5.76 shows the standard SAP delivery. SAP delivers usage 4 (PLANT MAINTENANCE) for plant maintenance purposes. Generally, this usage is sufficient. We've only ever encountered one customer who, in addition to usage 4, set up usage 1 for information BOMs, which, for this particular company, contained alternative spare parts for the material number in the BOM header. In other words, the BOM items list all materials that can be used as an alternative to the spare part in the BOM header.

If you define your own BOM usages, make sure that you select the PM (relevant for plant maintenance) column.

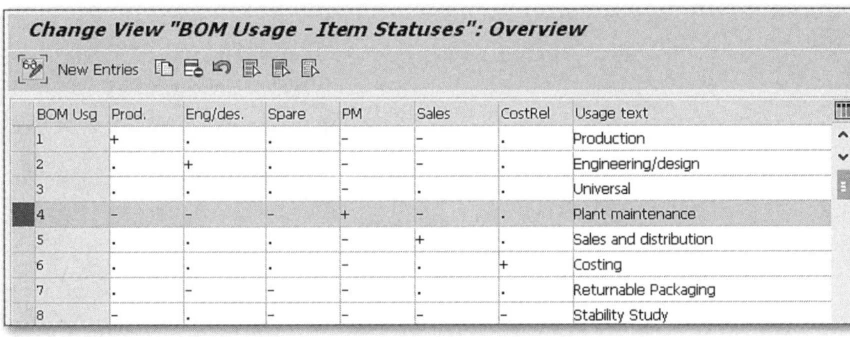

Change View "BOM Usage - Item Statuses": Overview

New Entries

BOM Usg	Prod.	Eng/des.	Spare	PM	Sales	CostRel	Usage text
1	+	.	.	-	-	.	Production
2	.	+	.	-	-	.	Engineering/design
3	.	.	.	-	.	.	Universal
4	-	-	-	+	-	.	Plant maintenance
5	.	.	.	-	+	.	Sales and distribution
6	.	.	.	-	.	+	Costing
7	.	-	-	-	.	.	Returnable Packaging
8	-	.	-	-	-	-	Stability Study

Figure 5.76 BOM Usages

Define Valid Material Types for BOM Header

You require this Customizing function only if you want to create BOMs for construction types. If you want to work solely with equipment BOMs or with BOMs for a functional location, this Customizing function doesn't play any role.

You use this Customizing function to define the material types that you want to create for the BOMs for construction types.

Prerequisites

You must define the necessary material types beforehand (see the Customizing function DEFINE ATTRIBUTES OF MATERIAL TYPES).

Customizing Path

Plant Maintenance and Customer Service • Master Data in Plant Maintenance and Customer Service • Bills of Material • General Data • Define Valid Material Types for BOM Header

Transaction

OICG

Settings

If you don't want to make any restrictions whatsoever, a fully masked entry (BOM usage = * and material type in BOM header = * in Figure 5.77) suffices. In other words, all material types can be used for all BOM usages.

If you want to restrict all BOMs apart from maintenance BOMs, which you don't want to restrict, mask only usage 4 (BOM usage = 4 and material type in BOM header = *). In other words, all material types can be used for maintenance BOMs.

Otherwise, list exactly those material types that are permitted to be used for BOMs for construction types in plant maintenance (e.g., ERSA, FERT, or IBAU in Figure 5.77).

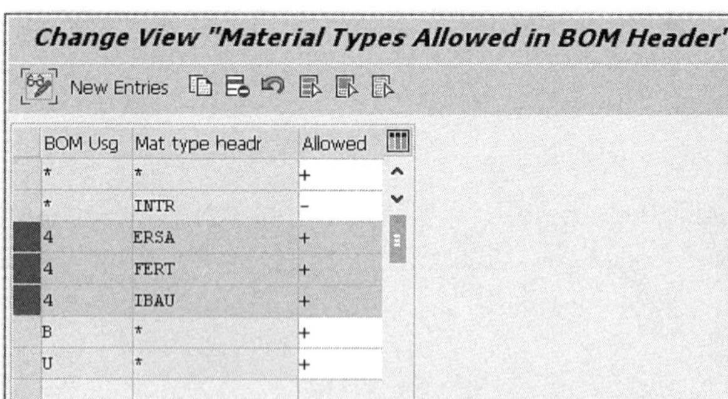

Figure 5.77 Material Types in BOM Header

Define Item Categories

In the BOM, you must assign an item category to each item. You use this Customizing function to define the item categories and their properties, which you require in your BOMs.

Prerequisites
There are no special prerequisites.

Customizing Path
Plant Maintenance and Customer Service • Master Data in Plant Maintenance and Customer Service • Bills of Material • Item Data • Define Item Categories

Transaction
OICK

Settings
Figure 5.78 shows the standard SAP delivery, which generally more than suffices here. From a maintenance perspective, the following item categories play a role:

- L = Stock item
  If you later assign a material with item category L to an order, this triggers a reservation.
- N = Non-stock item
  If you later assign a material with item category N to an order, this triggers a purchase requisition.
- I = PM structure element
  You use these items solely to define a substructure. Neither a reservation nor a purchase requisition is triggered.
- D = Document item
  You use these items as an alternative to the direct document assignment to an object or to assign additional documents to an object. Neither a reservation nor a purchase requisition is triggered.
- T = Text item
  You can use text items to set "headings" above certain sections of the BOM (e.g., the text item "The electrical spare parts start here"). Neither a reservation nor a purchase requisition is triggered.

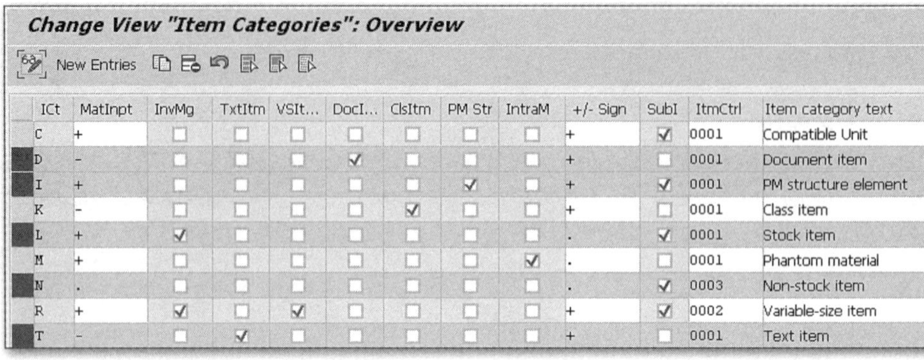

Change View "Item Categories": Overview

New Entries

ICt	MatInpt	InvMg	TxtItm	VSIt...	DocI...	ClsItm	PM Str	IntraM	+/- Sign	SubI	ItmCtrl	Item category text
C	+	[ ]	[ ]	[ ]	[ ]	[ ]	[ ]	[ ]	+	[x]	0001	Compatible Unit
D	-	[ ]	[ ]	[ ]	[x]	[ ]	[ ]	[ ]	+	[ ]	0001	Document item
I	+	[ ]	[ ]	[ ]	[ ]	[ ]	[x]	[ ]	+	[x]	0001	PM structure element
K	-	[ ]	[ ]	[ ]	[ ]	[x]	[ ]	[ ]	+	[ ]	0001	Class item
L	+	[x]	[ ]	[ ]	[ ]	[ ]	[ ]	[ ]	.	[x]	0001	Stock item
M	+	[ ]	[ ]	[ ]	[ ]	[ ]	[ ]	[x]	.	[ ]	0001	Phantom material
N	.	[ ]	[ ]	[ ]	[ ]	[ ]	[ ]	[ ]	.	[x]	0003	Non-stock item
R	+	[x]	[ ]	[x]	[ ]	[ ]	[ ]	[ ]	+	[x]	0002	Variable-size item
T	-	[ ]	[x]	[ ]	[ ]	[ ]	[ ]	[ ]	+	[ ]	0001	Text item

Figure 5.78 BOM: Item Categories

Define Material Types Allowed for BOM Item

You use this Customizing function to define which *material types* are provided in conjunction with the *BOM usage* and the material type in the BOM header for creating BOM items.

Prerequisites

You must maintain the material types in the BOM item, the material types permitted in the BOM header, and the BOM usage beforehand.

Customizing Path

Plant Maintenance and Customer Service • Master Data in Plant Maintenance and Customer Service • Bills of Material • Item Data • Define Material Types Allowed for BOM Item

Transaction

OICP

Settings

Similar to the material types for the BOM header, you also have various options here (see Figure 5.79):

- If you don't want to make any restrictions whatsoever, create a fully masked entry (i.e., BOM usage = *, material type in BOM header = *, and material type in BOM item = *). Specific entries are necessary only if you want to exclude certain material types.

- If you don't want to make any restriction for plant maintenance, mask only the material types (i.e., BOM usage = 4, material type in BOM header = *, and material type in BOM item = *). Specific entries are necessary only if you want to exclude certain material types.
- If you want to make a restriction for only one particular material type for plant maintenance, mask only the material type at the item level (i.e., BOM usage = 4, material type in BOM header = ERSA, and material type in BOM item = *). Specific entries are necessary only if you want to exclude certain material types.
- If you want to define precisely which material types at the item level are to be permitted for which item types at the header level in plant maintenance, create specific entries accordingly (e.g., BOM usage = 4, material type in BOM header = IBAU, material type in BOM item = ERSA).

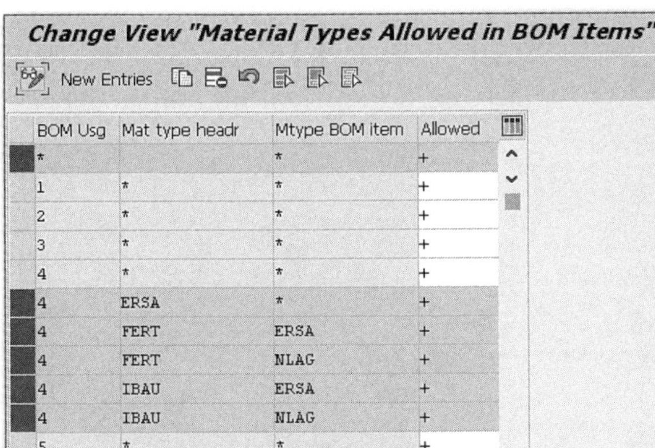
Change View "Material Types Allowed in BOM Items"

New Entries

BOM Usg	Mat type headr	Mtype BOM item	Allowed
*		*	+
1	*	*	+
2	*	*	+
3	*	*	+
4	*	*	+
4	ERSA	*	+
4	FERT	ERSA	+
4	FERT	NLAG	+
4	IBAU	ERSA	+
4	IBAU	NLAG	+
5	*	*	+

Figure 5.79 Material Types for BOM Items

Define Priorities for BOM Usage

You use this Customizing function to define your priorities for the automatic selection of a *BOM usage*.

Prerequisites

You must maintain the BOM usages beforehand.

Customizing Path

Plant Maintenance and Customer Service • Master Data in Plant Maintenance and Customer Service • Bills of Material • Determination of Alternative Bills of Material • Define Priorities for BOM Usage

Transaction

OICI

Settings

In SAP EAM, you require only one priority for BOM usage 4, which you can then process further using the Customizing function Define Selection Criteria for Alternative Determination (see Figure 5.80).

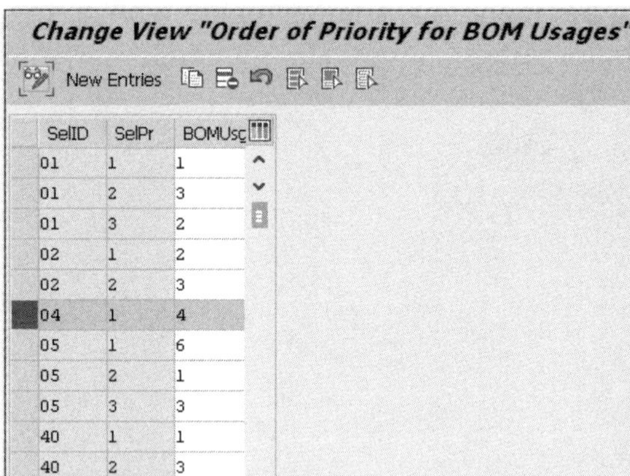

SelID	SelPr	BOMUsg
01	1	1
01	2	3
01	3	2
02	1	2
02	2	3
04	1	4
05	1	6
05	2	1
05	3	3
40	1	1
40	2	3

Figure 5.80 Order of Priority

Define Selection Criteria for Alternative Determination

You use this Customizing function to define, in the case of maintenance BOMs, an application (e.g., INST) that accesses the selection ID for determining *alternative BOMs* according to priorities (e.g., 04). This application is used in the following transactions for selecting BOMs:

- Transaction IH01 (Structural Display for Functional Locations)
- Transaction IH03 (Structural Display for Equipment)
- Transaction IH04 (Structural Display for Equipment Bill of Material)

- Transaction IH05 (Structural Display for Material BOM)
- Transaction IH12 (Structural Display for Bill of Material for Functional Location)

Prerequisites
You must maintain the selection ID for determining *alternative BOMs* according to priorities beforehand.

Customizing Path
PLANT MAINTENANCE AND CUSTOMER SERVICE • MASTER DATA IN PLANT MAINTENANCE AND CUSTOMER SERVICE • BILLS OF MATERIAL • DETERMINATION OF ALTERNATIVE BILLS OF MATERIAL • PM-SPECIFIC SELECTION CRITERIA FOR ALTERNATIVE DETERMINATION • DEFINE SELECTION CRITERIA FOR ALTERNATIVE DETERMINATION

Transaction
OICQ

Settings
Define an application (e.g., INST) in which you assign the selection ID for determining *alternative BOMs* according to priorities (e.g., 04) (see Figure 5.81).

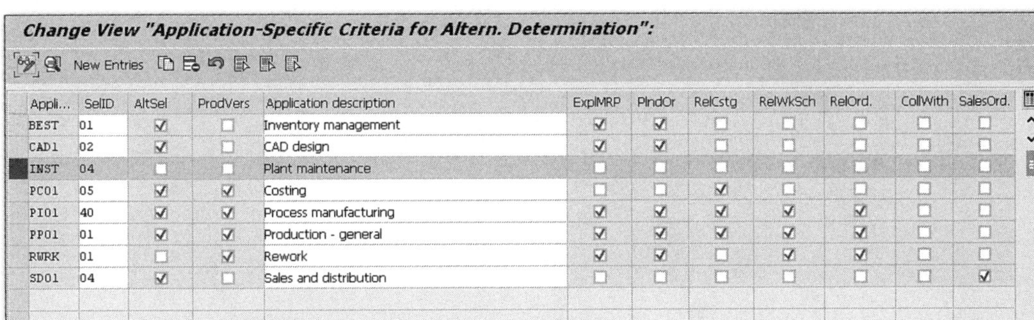

Change View "Application-Specific Criteria for Altern. Determination":

New Entries

Appli...	SelID	AltSel	ProdVers	Application description	ExplMRP	PlndOr	RelCstg	RelWkSch	RelOrd.	CollWith	SalesOrd.
BEST	01	☑	☐	Inventory management	☑	☑	☐	☐	☐	☐	☐
CAD1	02	☑	☐	CAD design	☑	☑	☐	☐	☐	☐	☐
INST	04	☐	☐	Plant maintenance	☐	☐	☐	☐	☐	☐	☐
PC01	05	☑	☑	Costing	☐	☐	☑	☐	☐	☐	☐
PI01	40	☑	☑	Process manufacturing	☑	☑	☑	☑	☑	☐	☐
PP01	01	☑	☑	Production - general	☑	☑	☑	☑	☑	☐	☐
RWRK	01	☐	☑	Rework	☑	☑	☐	☑	☑	☐	☐
SD01	04	☑	☐	Sales and distribution	☐	☐	☐	☐	☐	☐	☑

Figure 5.81 BOM: Alternative Determination

This chapter introduces you to the Customizing functions that will enable you to adjust maintenance processing to your requirements. You'll learn how to configure your tools as well as obtain numerous tips related to configuring Customizing settings for these functions.

6 Maintenance Processing

This chapter explains the core business of plant maintenance, namely business processes. After seeing many companies from the inside, it's clear that each company has its own idea about what business processes in plant maintenance should be like and how they must be mapped in the SAP system. Your first thoughts must be how you can map your day-to-day activities in SAP EAM, and how SAP EAM will support you in accomplishing your tasks. No book on earth can do this work for you. However, this chapter will actively support you in your work.

To give you as complete a picture as possible of the Customizing functions available, this chapter focuses on the *Scheduled Repair*[1] business process because this very complex maintenance process contains most of the Customizing functions you need. Figure 6.1 shows the individual steps within this business process, as described here:

❶ **Notification**
First, you enter the notification in relation to specific damage or any other request (e.g., a request for a modification). The following information is important here: the technical object, a description of what has happened and what repairs are necessary, and the date by which the work is to be completed.

❷ **Planning**
The order is created and planned from the notification. Typical planning tasks include creating operations, reserving spare parts, appointing external companies, and planning operating times. The account assignment rules are also defined in this step.

1 See K. Liebstückel, *Plant Maintenance with SAP—Practical Guide* (3rd edition, SAP PRESS, 2014, pp. 158 – 232).

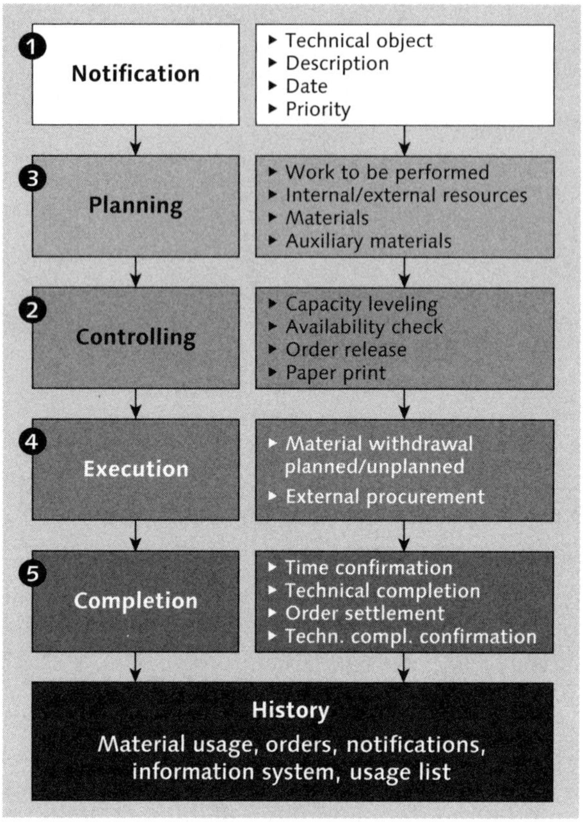

Figure 6.1 Business Process: Scheduled Repair

❸ **Controlling**
The order is transferred to Controlling, where you check the corresponding availability (in particular, material availability), provide the required capacities, and print out the shop papers.

❹ **Execution**
This processing phase involves the withdrawal of the spare parts from the warehouse and the actual processing of the order.

❺ **Completion**
After you complete the tasks, the required actual times are confirmed in completion. Technical completion confirmations are entered here regarding how the damage was handled and the condition of the asset. The order is now complete from both a technical and business perspective. In a final step, the order is settled by Controlling. The information is updated in the history.

This process requires three business objects in the SAP system—notification, order, and completion confirmation—which are described separately in the following sections.

6.1 Notifications

In maintenance processing, you use *notifications* in exceptional operational situations to perform the following activities:

- Describe the technical emergency situation associated with an object.
- Request a required task in the maintenance department.
- Document the work performed.

Consequently, maintenance tasks are documented in notifications so that they can be analyzed in the long term. Figure 6.2 shows the structure and content of a maintenance notification.

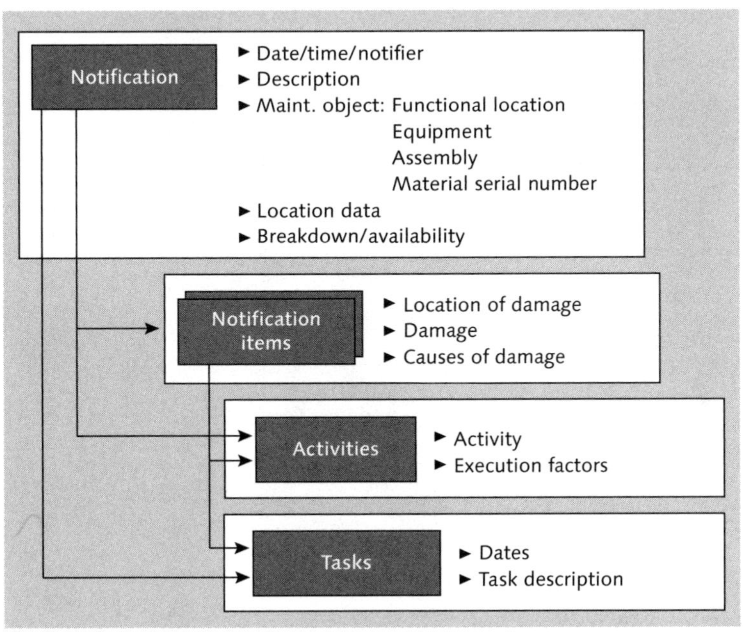

Figure 6.2 Structure and Content of a Notification

The Customizing functions that you can use to adjust notifications to your requirements are described in this subsection. A complete overview of the Customizing functions for notifications is provided in Figure 6.3. The following Customizing functions and their subfunctions also apply to notifications, but we won't discuss them here because they have already been described in Chapter 4:

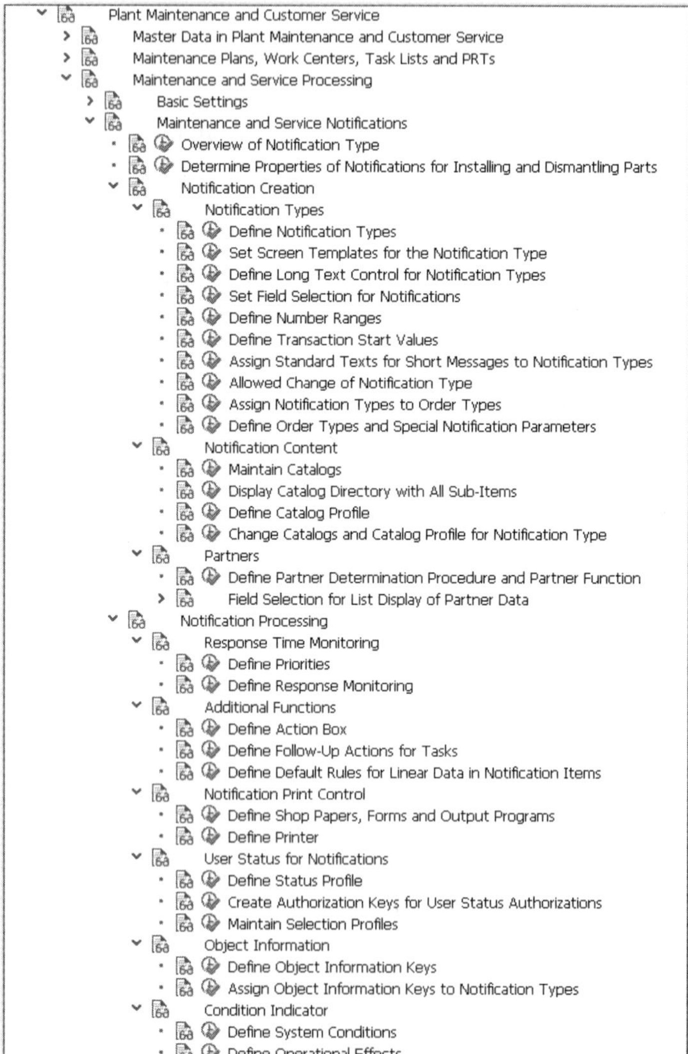

Figure 6.3 Overview of the Customizing Functions for Notifications

- Define Numbers Ranges
- Set Field Selection for Notifications
- Define Partner Determination Procedure and Partner Function
- Define Object Information Keys
- User Status for Notifications
- Define List Variants
- Define Field Selection for Multilevel List Displays for Notifications

The print procedure for notifications isn't described here because, in general, shop papers (not notifications), are printed. If you want to print notification papers, you can refer to the information provided in relation to printing orders because the shop paper printing settings are the same as the notification printing settings.

6.1.1 Notification Types

In Customizing, the notification type is *the* main control element for notification processing. In Customizing, you define the following for the notification type (see Figure 6.4).

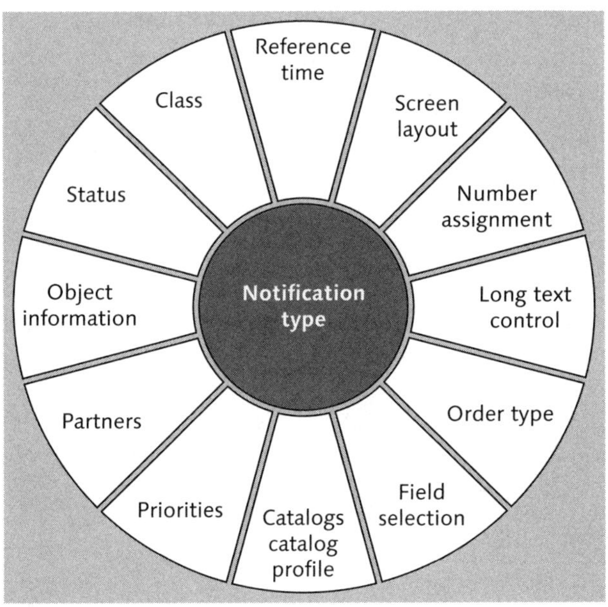

Figure 6.4 Notification Type: Overview

- Screen layout design
- Number to be assigned to the notification
- Long text control
- Order type to which the notification is to be assigned
- Fields to enter
- Catalogs to use
- Priorities to use with the notification
- Partners to assign
- Information to appear as object information
- User status to assign to a notification
- Whether to use a class to include additional information in the notification
- Date to use as the reference time in the history

The first Customizing function you need to execute is DEFINE NOTIFICATION TYPES.

Define Notification Types

You use this Customizing function to define the notification types that you require. The functions described in Figure 6.4 should be taken into consideration when defining the number of notification types you need as well as which particular notification types you want to use. If, for example, you require different data entry layouts, or you want to assign different number ranges for internal reasons, you require different notification types for these purposes.

[+] **Tip**

In practice, it has proven beneficial to use between one and five notification types.

Prerequisites

There are no technical prerequisites. However, you must consult with your colleagues from Quality Management, Service Management, and Claim Management on the naming convention for notification types because they also use notifications.

Customizing Path

Plant Maintenance and Customer Service • Maintenance and Service Processing • Maintenance and Service Notifications • Notification Creation • Notification Types • Define Notification Types

Settings or Recommended Settings

Specifically, you can or should configure the following settings (see Figure 6.5):

- Notification type
  The notification type has a two-character ID. Because notifications are used not only in SAP EAM but also in SAP QM, Customer Service Processing, and SAP Claims Management, and they can be created as *generic notifications*, you must consult with your colleagues on the naming convention for maintenance notification types (e.g., always use M*x* or I*x*.)
- Notif.cat.
  The notification category determines the application in which this notification type is used. Always assign "01 = Plant maintenance" to those notification types that you want to use in SAP EAM. Other notification categories include 02 = Quality notification, 03 = Service notification, 04 = Claim, and 05 = General notification.
- Notification origin
  This setting determines the nature of the notification. Therefore, specify whether it concerns a General maintenance notification (01), a Malfunction report (02), or an Activity report (03). Other origin types (e.g., Q3 = Problem message or 05 = Activity report) don't play any role in SAP EAM. The notification origin is used in the Logistics Information System (LIS) for statistical analyses. Otherwise, it doesn't have any other effect.

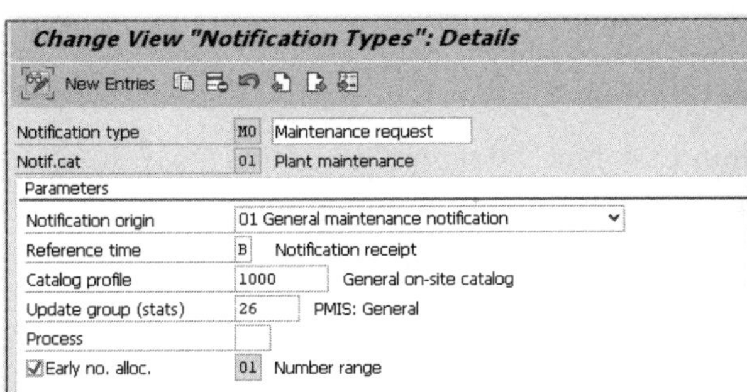

Figure 6.5 Notifications: Define Notification Types

- REFERENCE TIME
  This setting specifies the proposed date and time by which the notification is to be completed:
  - A: Start of breakdown
  - B: Notification receipt
  - C: End of breakdown
  - D: Completion of notification

  This date then controls the chronological order of the notifications in the history.
- CATALOG PROFILE
  The catalog profile entered in the notification type is used if you haven't assigned a catalog profile to the piece of equipment or functional location. For more information, refer to the Customizing functions for catalogs and catalog profiles.
- UPDATE GROUP (STATS)
  This is a control element within the update definition, which different business processes in different applications can use to influence the statistics update in the LIS in a uniform manner. SAP delivers update group 26 for plant maintenance. Therefore, if you haven't set up your own information structures in the LIS, or if you don't require your own update rules, always enter "26" here.
- EARLY NO. ALLOC.
  Setting this early number assignment indicator specifies that the system assigns the notification number immediately after you call this transaction and not when you first save the notification. This is always recommended if, for example, you receive notifications via a central hotline. Then, the hotline representative can immediately quote the notification number to the person who has reported the issue.

Set Screen Templates for the Notification Type

You use this Customizing function to define the screen template layout. In other words, you define the following areas:

- Number of tab pages
- Title to assign to each tab page
- Whether to identify the tab pages by a special icon

- Field groups (subscreens) to appear on each tab page
- Layout changes based on the type of activity (creating, changing, and displaying)

Prerequisites

You must define the notification types beforehand.

Customizing Path

Plant Maintenance and Customer Service • Maintenance and Service Processing • Maintenance and Service Notifications • Notification Creation • Notification Types • Set Screen Templates for the Notification Type

Settings or Recommended Settings

As an example, the following two tab pages have been defined for the notification type M0: Notification 1 and Additional data 1 (see Figure 6.6).

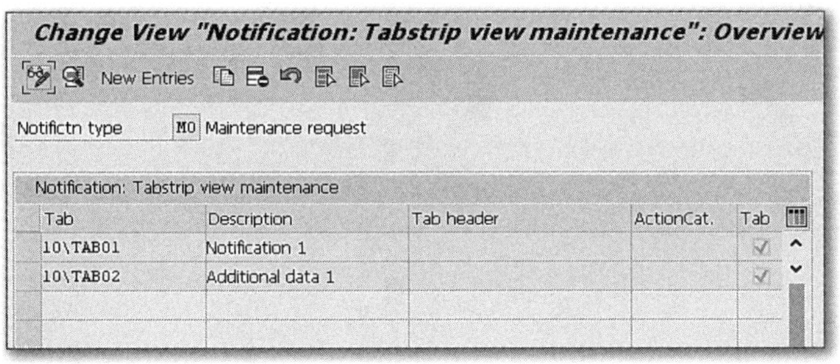

Figure 6.6 Notifications – Define Screen Templates: Overview

On the detail screen, configure the exact settings (see Figure 6.7).

The most important setting here is the definition of the tab page. The following predefined tab pages are available:

- 10\TAB01* – Notification 1
- 10\TAB02* – Additional data 1
- 10\TAB03* – Additional data 2
- 10\TAB04 – System availability
- 10\TAB05 – Breakdown data
- 10\TAB06 – Location
- 10\TAB07 – Dates

- 10\TAB08 – Maintenance plan
- 10\TAB09* – Additional data 3
- 10\TAB10 – Item
- 10\TAB11 – Tasks
- 10\TAB12 – Activities
- 10\TAB13 – Item 2
- 10\TAB14 – All tasks
- 10\TAB15 – All activities
- 10\TAB16 – All causes
- 10\TAB17* – Additional data 4
- 10\TAB18 – Form
- 10\TAB19* – Customer data 1
- 10\TAB20* – Customer data 2
- 10\TAB21* – Customer data 3

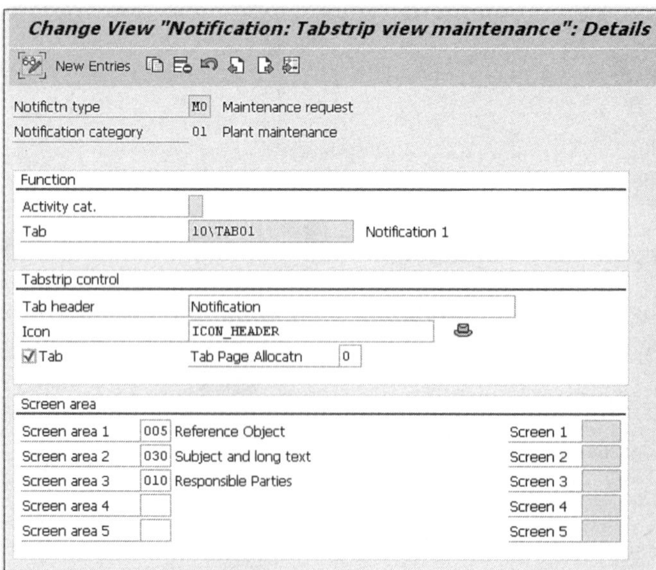

Figure 6.7 Notifications – Define Screen Templates: Details Screen

You can then define the following subtab page for the tab page 10\TAB10 (ITEM):

- 20\TAB01 – ITEM OVERVIEW
- 20\TAB02 – ITEM CAUSES
- 20\TAB03 – ITEM TASKS
- 20\TAB04 – ITEM ACTIVITIES

Because the tab pages not marked with * are predefined by SAP, there are no further settings for these tab pages.

For those tab pages marked with *, you can define the screen areas yourself (as shown in Figure 6.8, which is the second tab page for notification type M0). The following screen areas are available:

- 010 – RESPONSIBLE PARTIES
- 015 – ITEM + CAUSE
- 020 – SYSTEM AVAILABILITY
- 030 – SUBJECT AND LONG TEXT
- 032 – SUBJECT AND LONG TEXT (WINDOWS, 32-BIT)
- 033 – SUBJECT AND LONG TEXT (WEB GUI)
- 035 – BREAKDOWN DATA
- 039 – REFERENCE DOCUMENTS/INSPECTION LOT
- 040 – START/END DATES
- 045 – ACTIVITIES
- 050 – TASKS
- 055 – CAUSES
- 060 – NOTIFICATION AND OBJECT ADDRESS
- 065 – PARTNER OVERVIEW (AS A TABLE CONTROL)
- 070 – INSTALLATION HIERARCHY
- 080 – WARRANTY
- 085 – NOTIFICATION OVERVIEW
- 090 – CUSTOMER SUBSCREEN (one screen)

- 091 – Customer subscreen (several subscreens)
- 092 – DMS links

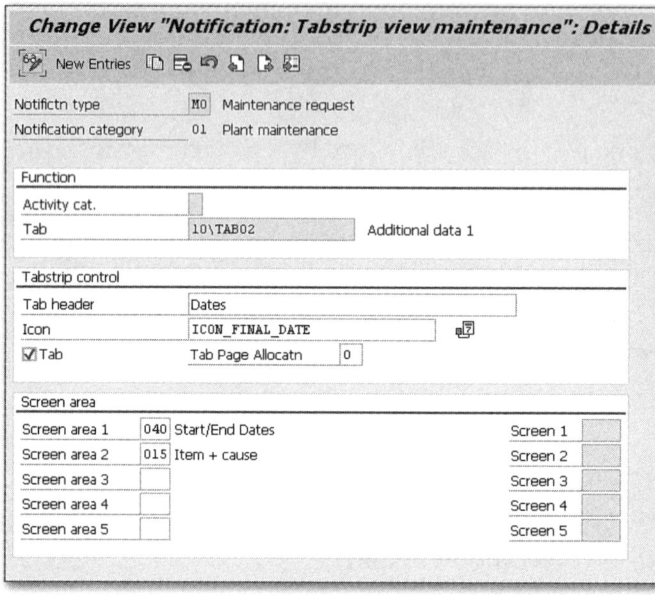

Figure 6.8 Notifications – Define Screen Templates: Details Screen

You still have the following additional settings for each tab page:

- You can define the layout on the basis of the activity category (Activity cat.). You can therefore configure the screen layout in such a way that a different layout appears when you change data compared to when you add data. You might need this option, for example, when you want to provide a production employee with a screen that is as basic as possible for entering a notification. However, if the maintenance employee subsequently calls the same notification at a later stage, he should be able to update the notification with the additional required information. Consequently, if the same notification is called in change mode, it may then contain additional tab pages or other tab pages and subscreens.
- You can define the Tab header yourself; you don't need to adhere to the header defined by SAP.
- If you want a tab page to receive particular attention, you can add an Icon to the header. More than 1,000 predefined icons are available for selection.

[+]

One to Three Tab Pages

In practice, the following principle should always apply: "As much as necessary, but as little as possible." In this spirit, one to three tab pages usually suffices.

In addition, use the option to have different layouts for adding and changing data.

Important: If you want the system to display a particular tab page when you process a notification, you must set the TAB indicator. Otherwise, the tab page is defined in the layout, but it's not shown.

Overview of Notification Type

Within this Customizing function, you use the SCREEN AREAS IN NOTIFICATION HEADER subfunction to define the notification header and reference objects you want to use for each notification type.

Prerequisites

You must define the notification types themselves and the screen layout of these notification types beforehand. In particular, one of the tab pages must contain the entry SCREEN AREA 1 005 = REFERENCE OBJECT as shown in Figure 6.7.

Customizing Path

PLANT MAINTENANCE AND CUSTOMER SERVICE • MAINTENANCE AND SERVICE PROCESSING • MAINTENANCE AND SERVICE NOTIFICATIONS • OVERVIEW OF NOTIFICATION TYPE • SUBFUNCTION SCREEN AREAS IN NOTIFICATION HEADER

Settings or Recommended Settings

Figure 6.9 shows the settings for this subfunction. In SCREEN TYPE HDR (screen type header), always enter "H100" (HEADER MAINTENANCE NOTIFICATION) unless you've defined your own headers. The HEADER MAINTENANCE NOTIFICATION contains information such as the notification type, notification number, short text, status, and assigned order number.

For SCRNTYPE OBJECT (screen type object), the following settings are available:

- 0100 – FUNCTIONAL LOCATION + PIECE OF EQUIPMENT + ASSEMBLY
- 0110 – EQUIPMENT + ASSEMBLY
- 0120 – FUNCTIONAL LOCATION (30) + EQUIPMENT + ASSEMBLY

- 0130 – SERIAL NUMBER + MATERIAL NUMBER + DEVICE DATA
- 0140 – WITHOUT A REFERENCE OBJECT
- 0150 – EQUIPMENT ONLY
- 0160 – FUNCTIONAL LOCATION ONLY
- 0170 – EQUIPMENT + SERIAL NO. + MATERIAL NO.
- 0180 – FUNCTIONAL LOCATION 1:1 + EQUIPMENT + ASSEMBLY

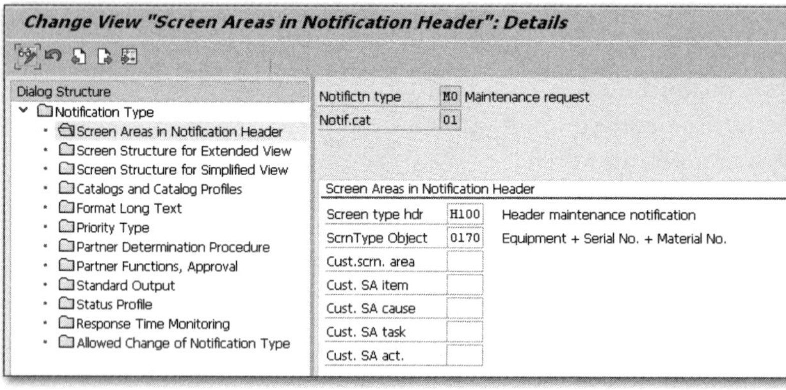

Figure 6.9 Notifications: Screen Areas in Notification Header

[+]

Screen Types H100 and O100

The screen type header (SCREEN TYPE HDR) is always H100. For the screen type object (SCRNTYPE OBJECT), you should enter the screen type that you use most often for this notification type. Most users use O100.

If, however, you require a different screen type in a notification, you can use the menu function EXTRAS • SETTINGS • REFERENCE OBJECT VIEW within the notification to change the screen type.

Define Long Text Control for Notification Types

You use this Customizing function to define how you want to handle long texts within a notification.

Prerequisites

You must define the notification types beforehand.

Customizing Path

Plant Maintenance and Customer Service • Maintenance and Service Processing • Maintenance and Service Notifications • Notification Creation • Notification Types • Define Long Text Control for Notification Types

Settings or Recommended Settings

The following settings are available to you (see Figure 6.10):

- If you set the Log line indicator, the system automatically inserts a line that contains information about the name of the person who entered the long text and the date and time when this entry was made, above the long text entered.
- If you set the No text change indicator, you won't be able to change long texts retroactively.

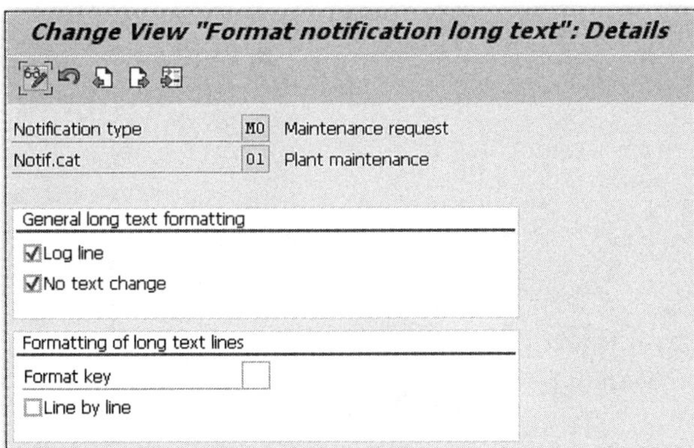

Figure 6.10 Notifications: Long Text Formatting

Log Line and No Text Change Functions [+]

You should activate the Log line function so that you know who entered which texts and when. To prevent any manipulation of the texts, you should also prohibit retroactive changes to the texts by using the No text change checkbox. Unfortunately, these options aren't available elsewhere in the SAP system.

Define Transaction Start Values

You use this Customizing function to define whether default values for calling the entry and maintenance transactions for the notification (in particular, Transactions IW21, IW24, IW25, and IW26) are to be defined.

Prerequisites

You must define the notification types beforehand.

Customizing Path

Plant Maintenance and Customer Service • Maintenance and Service Processing • Maintenance and Service Notifications • Notification Creation • Notification Types • Define Transaction Start Values

Settings or Recommended Settings

In the Notification Type field, enter the notification type that you want the system to propose when you call the transaction (see Figure 6.11).

If you have only one notification type, you don't need to set the Proc.initial screen (process initial screen) indicator, which signifies that you don't access an initial screen. Instead, you immediately access the entry screens for the notification, and the notification type is already set there.

If you have several notification types, you should set the Proc.initial screen indicator so that you access the initial screen for the notification. Although the notification type is also already set there as a default value, you can change it.

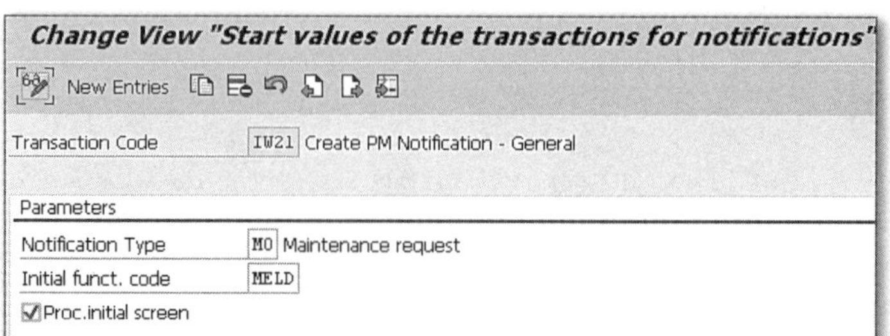

Figure 6.11 Notification Types: Transaction Start Values

Process Initial Screen [+]

We recommend that you set the indicator for processing the initial screen. Otherwise, the system will always jump immediately back to the SAP Easy Access menu.

Allowed Change of Notification Type

You use this Customizing function to define which changes to the notification type are permitted.

Limited Change Options [+]

You can change the notification type in *simple transactions* (Transactions IQS21 and IQS22), but not in Transactions IW21 through IW26.

When you create notifications (Transaction IQS21), you can change the notification type under the following conditions:

- The change was expressly permitted in this table.
- The notification still has the status OPEN.

When you change notifications (Transaction IQS22), you can change the notification type under the following conditions:

- The change was expressly permitted in this table.
- The notification still has the status OPEN.
- The original notification type contains the origin GENERAL NOTIFICATION.

Prerequisites

You must define the notification types beforehand.

Customizing Path

PLANT MAINTENANCE AND CUSTOMER SERVICE • MAINTENANCE AND SERVICE PROCESSING • MAINTENANCE AND SERVICE NOTIFICATIONS • NOTIFICATION CREATION • NOTIFICATION TYPES • ALLOWED CHANGE OF NOTIFICATION TYPE

Settings or Recommended Settings

Enter all permitted changes for Orig. notif. type and Target notif. type (see Figure 6.12).

> **Note**
>
> The original notification type must be of the type General notification. You can't change the type from Maintenance notification to Maintenance notification.

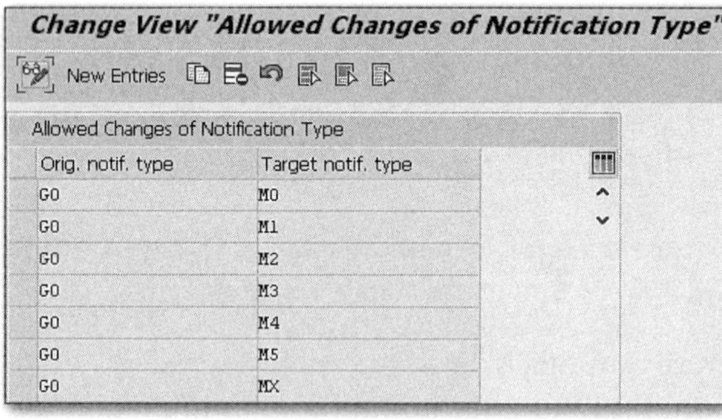

Figure 6.12 Notifications: Allowed Change of Notification Type

Assign Notification Types to Order Types

You use this Customizing function to define which order type is to be proposed for which notification type, if an order is generated from a notification.

Prerequisites

You must define the notification types and order types beforehand.

Customizing Path

Plant Maintenance and Customer Service • Maintenance and Service Processing • Maintenance and Service Notifications • Notification Creation • Notification Types • Assign Notification Types to Order Types

Settings or Recommended Settings

Enter which order type is to be proposed for each notification type (see Figure 6.13).

Change View "Order Type by Notification Type": Overview

Typ	Notification type	Type	Order type text
M0	Maintenance request	PM01	Maintenance Order
M1	Maintenance request	PM01	Maintenance Order
M2	Malfunction report	PM01	Maintenance Order
M3	Activity report	PM01	Maintenance Order
M4	RF & RFID Maint Req	PM01	Maintenance Order
M5	Malfunction report E	PM01	Maintenance Order
M6	Maintenance request	PM01	Maintenance Order
MF	Pool Request	PAM	Maintenance Order
MQ	Error msg. from QM	PM06	Calibration Order
MX		PM01	Maintenance Order

Figure 6.13 Notifications: Assign Notification Types to Order Types

6.1.2 Catalogs and Catalog Profiles

In addition to organizational information (such as dates, responsibilities, or cost centers), you can also store technical information about problems, breakdowns, damage, causes, and solutions or damage repair in a notification. This information is part of the notification and is included in the history. A special feature of this information (when compared with all other information) is the fact that you can formalize this information in catalogs and therefore make it available for analyses.

Damage or *cause of damage* catalogs are generally used in SAP EAM (see Figure 6.14). Each catalog has the following three-tier structure: catalog → code groups → codes.

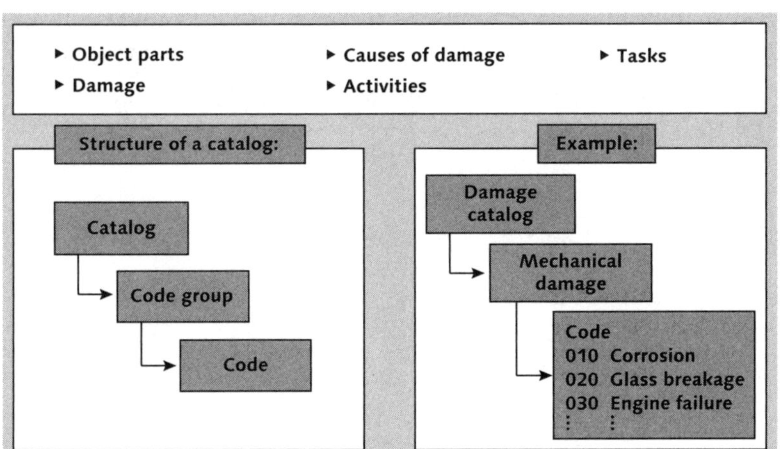

Figure 6.14 Notifications: Catalogs

A code exists for each set of findings. Furthermore, the codes, in turn, are grouped together to form code groups according to particular characteristics.

Next, let's discuss the Customizing functions for creating catalogs and assigning notification types.

Maintain Catalogs

You use this Customizing function to maintain the catalogs, code groups, and codes that you want to use.

Prerequisites

There are no special prerequisites. However, you must consult with your colleagues from QM and SAP Customer Service (SAP CS) on the naming convention for catalogs because they also use catalogs.

Customizing Path

Plant Maintenance and Customer Service • Maintenance and Service Processing • Maintenance and Service Notifications • Notification Creation • Notification Content • Maintain Catalogs

Transaction

QS41

Settings or Recommended Settings

Use the Define Catalogs subfunction to define the first hierarchy level (i.e., the number and description of catalog types that you want to use in SAP EAM). Figure 6.15 shows the type and naming convention for the catalog types that are generally used in SAP EAM:

- 2 = Tasks
- 5 = Causes
- 8 = Activities
- B = Object parts
- C = Overview of damage

Of course, your own naming conventions may differ from these.

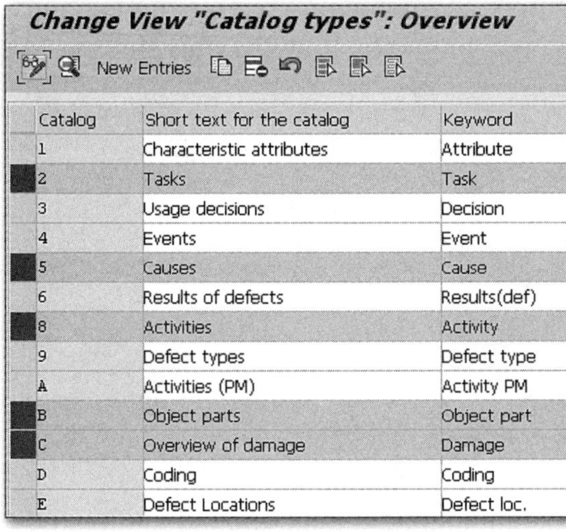

Change View "Catalog types": Overview

New Entries

Catalog	Short text for the catalog	Keyword
1	Characteristic attributes	Attribute
2	Tasks	Task
3	Usage decisions	Decision
4	Events	Event
5	Causes	Cause
6	Results of defects	Results(def)
8	Activities	Activity
9	Defect types	Defect type
A	Activities (PM)	Activity PM
B	Object parts	Object part
C	Overview of damage	Damage
D	Coding	Coding
E	Defect Locations	Defect loc.

Figure 6.15 Notifications: Catalog Types

Use the EDIT CATALOGS subfunction to define the second hierarchy level. Here, define the categorization into code groups for each catalog type. This then raises the key question of how to categorize catalog types.

Categorization Criteria for Catalogs [+]

User companies frequently use the following criteria for categorizing code groups in catalogs:

- **Functional criteria**
  Mechanical damage, electrical causes of damage, hydraulic object parts, and so on.
- **Object-related criteria**
  Damage to engines, causes of damage to pumps, object parts for forklift trucks, and so on.

Figure 6.16 uses the example of CATALOG C = OVERVIEW OF DAMAGE to show one possible categorization. If you want to use the code group in notifications, you must set the STATUS OF CODE GROUP to RELEASED.

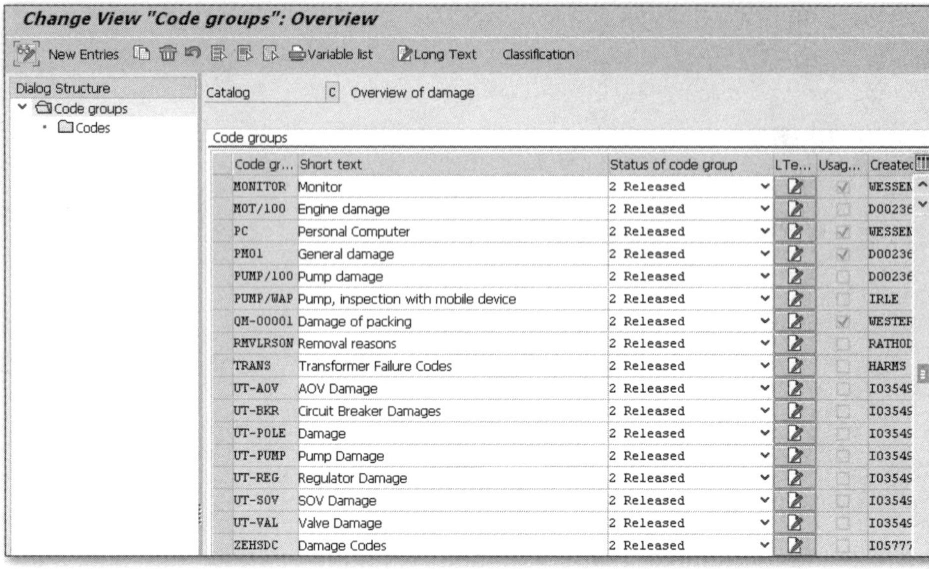

Figure 6.16 Notifications: Code Groups for Catalog Type

Then, in the third hierarchy level, define the actual codes for each code group. Figure 6.17 shows, for example, the code groups defined for CATALOG C = OVERVIEW OF DAMAGE and CODE GROUP PUMP/100 = PUMP DAMAGE.

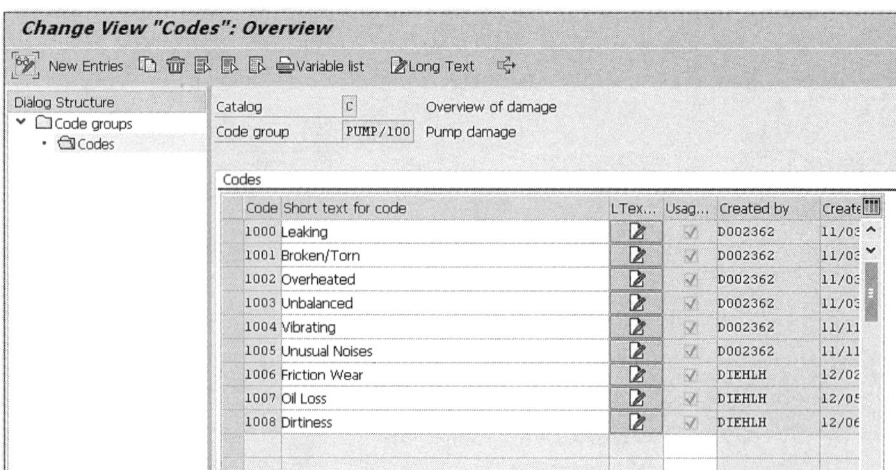

Figure 6.17 Notifications: Codes for Code Group

[+]

Maximum of 25 Entries per Code Group

Define clear damage, cause, and object part codes. A maximum of 25 codes should be available to the user. Otherwise, it will be too time-consuming for users to find the code they need, and data quality and system acceptance will suffer as a result. Once again, the "as much as is necessary, but as little as possible" principle applies here.

Define Catalog Profile

In this work step, you define the *catalog profiles* that you can assign to the technical objects (functional location and equipment) and the notification type. Here, you define the following:

- The catalogs and code groups from the individual catalogs that are to be grouped together. You can then choose one of the code groups contained in the catalog by specifying a generic key for the code groups. You can also make multiple assignments for each catalog type (see Figure 6.18).
- The class in class type 015 to be proposed when classifying notification items. You can also include additional nonstandard fields in the notification.
- The type of message the system will issue if the catalog profile isn't complied with.

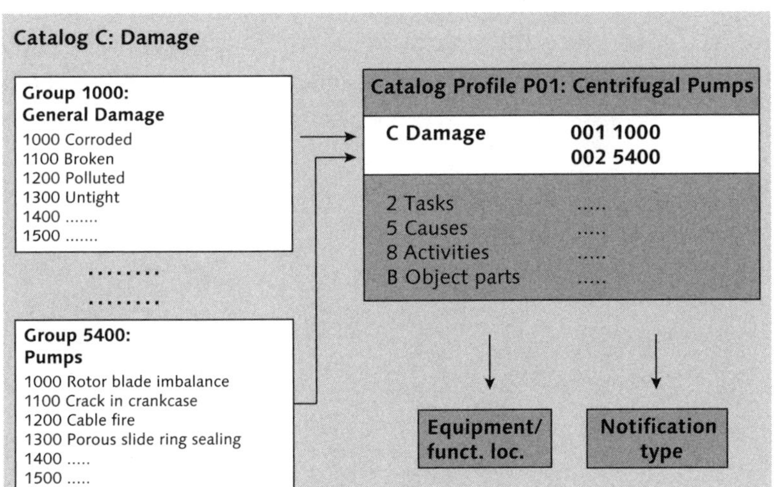

Figure 6.18 Notifications: Catalog Profile

Prerequisites

You must define the catalog types, code groups, and codes beforehand. If you also want to assign a class, you must use Transaction CT04 to define the necessary characteristics and Transaction CL02 to create a class of the class type 015.

Customizing Path

Plant Maintenance and Customer Service • Maintenance and Service Processing • Maintenance and Service Notifications • Notification Creation • Notification Content • Define Catalog Profile

Transaction

OQN6

Settings or Recommended Settings

Figure 6.19 shows, for example, how you can use the subfunction Catalogs/code groups to assign code groups to a catalog profile:

- You can fully qualify the assignment, for example, Catalog C – Code group PUMP/100.
- You can also partly qualify the assignment. For example, the entry Catalog type 5 – Code group P* ensures that, in the notification, all groups whose group name starts with P* are proposed for the damage codes.
- You can assign several groups from the same catalog type. For example, PUMP/100 assigns object-specific entries, while P* assigns generic entries such as pump damage.

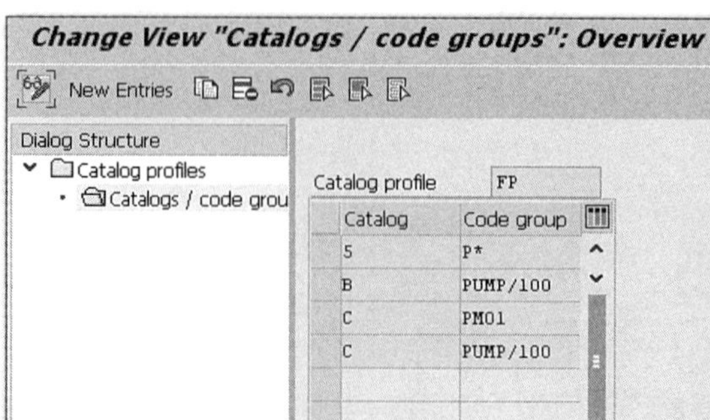

Figure 6.19 Notifications: Catalog Profile with Code Groups

Figure 6.20 shows the CATALOG PROFILES subfunction. Here, you can define the following for each catalog profile:

- Whether you want to assign a class to the profile and, if so, which class you want to assign.
- Whether a message is to be output when you change the catalog profile in a notification, and, if so, what type of message (I = information, W = warning, or E = error; that is, the catalog profile can't be changed later).

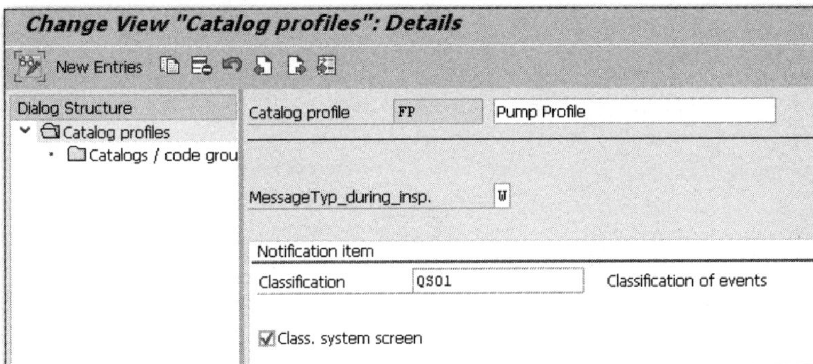

Figure 6.20 Notifications: Catalog Profiles

You can now enter the catalog profile in the following:

- Equipment master
- Master record for the functional location
- Notification type (see the Customizing function CHANGE CATALOGS AND CATALOG PROFILE FOR NOTIFICATION TYPE)

Priority Rule [+]

The following "priority rule" applies to the catalog profile: equipment → functional location → notification type.

In other words, a catalog profile is determined in accordance with the following rules:

- When you determine the catalog profile in the notification, the system checks whether a catalog profile is already entered in the equipment master record. If so, the profile is copied to the notification.

- If not, the system checks whether a catalog profile is already entered in the master record for the functional location. If so, the profile is copied to the notification.
- If not, the system checks whether a catalog profile is already entered in the notification type. If so, the profile is copied to the notification.
- If not, the system doesn't propose any catalog profile in the notification. However, you can use the function EXTRAS • SETTINGS • CATALOG PROFILE • SELECTION at any time to assign a catalog profile to a notification.

Change Catalogs and Catalog Profile for Notification Type

You use this Customizing function to assign the relevant catalogs to the notification types.

Prerequisites

You must define the catalog types, catalog profiles, and notification types beforehand.

Customizing Path

PLANT MAINTENANCE AND CUSTOMER SERVICE • MAINTENANCE AND SERVICE PROCESSING • MAINTENANCE AND SERVICE NOTIFICATIONS • NOTIFICATION CREATION • NOTIFICATION CONTENT • CHANGE CATALOGS AND CATALOG PROFILE FOR NOTIFICATION TYPE

Settings or Recommended Settings

For PROBLEMS (damage), CAUSES, TASKS, ACTIVITIES, and OBJECTPARTS, assign the catalog types that you want to use with this notification type (see Figure 6.21). The CODING entry isn't used in SAP EAM. You can therefore access different catalog types for different notification types.

[!] **Assign Catalog Types to Notification Types**

You must enter the relevant catalogs for the notification types so that the system can display the corresponding fields in the notification. For example, if you don't assign a catalog type to a notification type in the case of catalogs for PROBLEMS, the notification doesn't contain any fields in relation to damage.

You can also assign a CATALOG PROFILE to the notification type. The system then always proposes this catalog profile in the notification if neither the piece of equipment nor the functional location has a catalog profile.

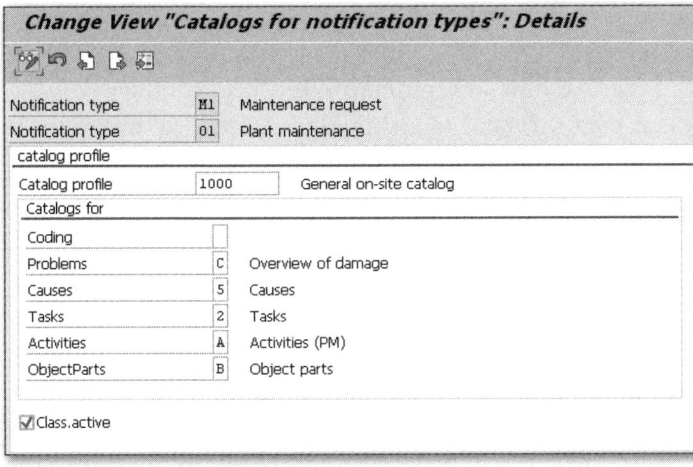

Figure 6.21 Notifications: Assignment of Catalog Types to Notification Types

6.1.3 Other Functions

Of the many other Customizing functions available for the notification, we'll now cover the following:

- Priorities
- Response time monitoring
- Action boxes

Define Priorities

You use this Customizing function to define the *priority types* and associated *priorities* and to assign a notification type.

Prerequisites

You must define the notification types beforehand. You must consult with your colleagues from QM, SM, and SAP Claims Management on the naming convention for priority types because they also use priorities.

Customizing Path

Plant Maintenance and Customer Service • Maintenance and Service Processing • Maintenance and Service Notifications • Notification Processing • Response Time Monitoring • Define Priorities

Settings or Recommended Settings

Use the Define Priority Types subfunction to determine a priority type. A priority type contains the actual priorities and is assigned to a notification type (see Figure 6.22).

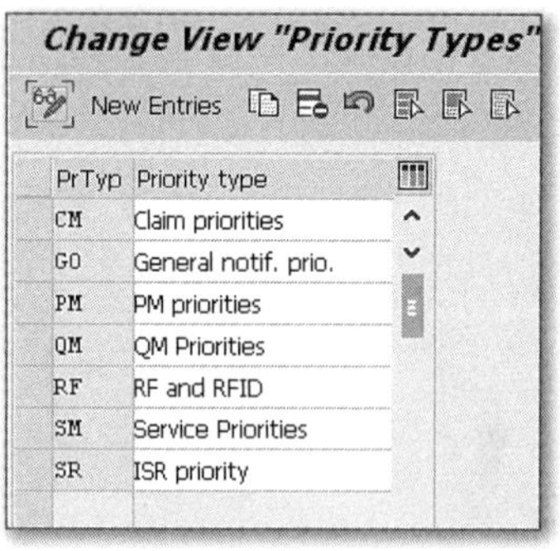

Figure 6.22 Notifications: Priority Types

[+]

One Priority Type for All Notification Types

In most cases, one priority type should suffice for all notification types.

If, however, you want to handle the notification types differently, you'll need several notification types. This opens up the possibility of defining multitiered priorities for your different notification types (e.g., a three-tier priority type for a malfunction report type and a five-tier priority type for normal maintenance requests).

Use the Define Priorities for Each Priority Type subfunction to define the relevant priorities for each priority type. In addition to the priority code and associated text, define the relative start and relative end (in their respective units of measurement) for each priority (see Figure 6.23):

- The system uses the relative start (Rel.start) and its unit of measurement to calculate the *required start* for the notification, starting from the time when the notification was entered. In the example shown, the required start for the notification always corresponds to the date on which the notification was entered because none of the priorities has entered a relative start.

- The system uses the relative end (REL.END) and its unit of measurement to calculate the *required end* for the notification, starting from the time when the notification was entered. In the example shown, priority 3 generates the following required end for the notification: entry date + 14 days.
- The system always bases its calculation on normal calendar days, and not on the factory calendar.

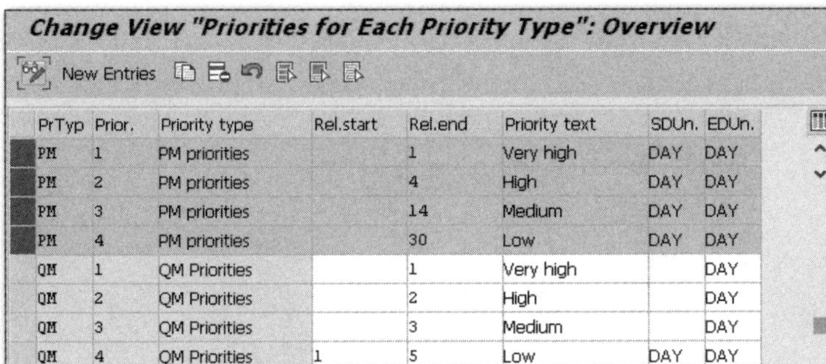

PrTyp	Prior.	Priority type	Rel.start	Rel.end	Priority text	SDUn.	EDUn.
PM	1	PM priorities		1	Very high	DAY	DAY
PM	2	PM priorities		4	High	DAY	DAY
PM	3	PM priorities		14	Medium	DAY	DAY
PM	4	PM priorities		30	Low	DAY	DAY
QM	1	QM Priorities		1	Very high		DAY
QM	2	QM Priorities		2	High		DAY
QM	3	QM Priorities		3	Medium		DAY
QM	4	QM Priorities	1	5	Low	DAY	DAY

Figure 6.23 Notifications: Priorities

Use the ASSIGN PRIORITY TYPES TO NOTIFICATION TYPES subfunction to define which priority type you want to use with which notification type (see Figure 6.24). Generally, you have just one priority type, and this priority type is assigned to all notification types.

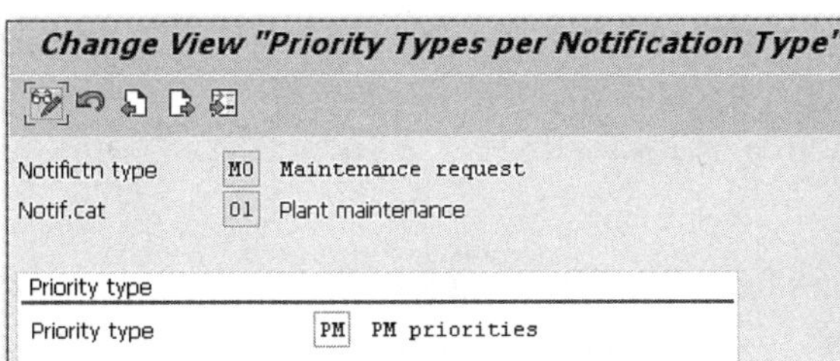

Figure 6.24 Notifications: Assignment of Priority Type to Notification Type

[+]

Always Assign a Priority Type (Even a Blank Priority Type)

If you don't assign a priority type to a notification type and then attempt to enter a notification, the system issues a warning message in a dialog box, indicating that the notification type doesn't have a priority type. To prevent this, you should assign a priority type to each notification type.

If, in general, you work without priorities, or you don't want to have priorities for certain notification types, you should define a blank priority type (i.e., a priority type that doesn't contain any priorities) to circumvent this irritating warning message.

Define Response Monitoring

You use this Customizing function to define response monitoring for notifications. To do this, you determine *response profiles* and *service profiles* and then assign these two profiles to the notification types (see Figure 6.25).

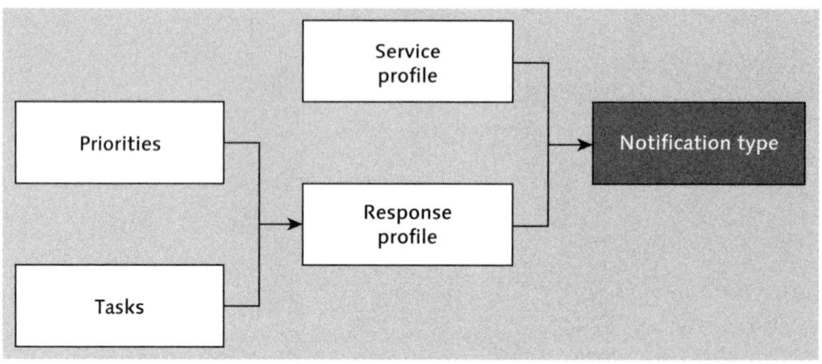

Figure 6.25 Notifications: Overview of Service Profile and Response Profile

Prerequisites

You must define the priority types, and the *Tasks* catalog type and its corresponding code groups and codes beforehand.

You must consult with your colleagues from QM, SM, and SAP Claims Management on the naming convention for service profiles and response profiles because they also use response monitoring.

Customizing Path

PLANT MAINTENANCE AND CUSTOMER SERVICE • MAINTENANCE AND SERVICE PROCESSING • MAINTENANCE AND SERVICE NOTIFICATIONS • NOTIFICATION PROCESSING • RESPONSE TIME MONITORING • DEFINE RESPONSE MONITORING

Transaction

OIMF for service profiles

Settings or Recommended Settings

Use the RESPONSE PROFILE subfunction to define a name and description for response profiles.

Use the RESPONSE TIMES subfunction to define the following:

- Priority (e.g., 1)
- Task (generally, CATALOG TYPE 2 = TASKS)
- Time frame (calculated from the required start for the notification)
- Responsible person (e.g., the PERSON RESPONSIBLE partner function)

Figure 6.26 shows the following examples:

- The NOTIFY PLANT ENGINEER task (catalog type 2, code group PM01, and code 3000) in the first line is to be completed by Mr. Moore (partner function VW = person responsible with personnel number 1208), no later than one hour after the required start for the notification.
- The task CREATE ORDER (catalog type 2, code group PM01, and code 3002) in the third line is to be completed by Mr. Henry (partner function VW = person responsible with personnel number 1000), no later than two hours after the required start for the notification.

Generate Tasks Automatically [+]

If you've defined a response profile with tasks, you can use the function EDIT • TASKS • DETERMINE AUTOMATICALLY to generate the corresponding tasks in the notification.

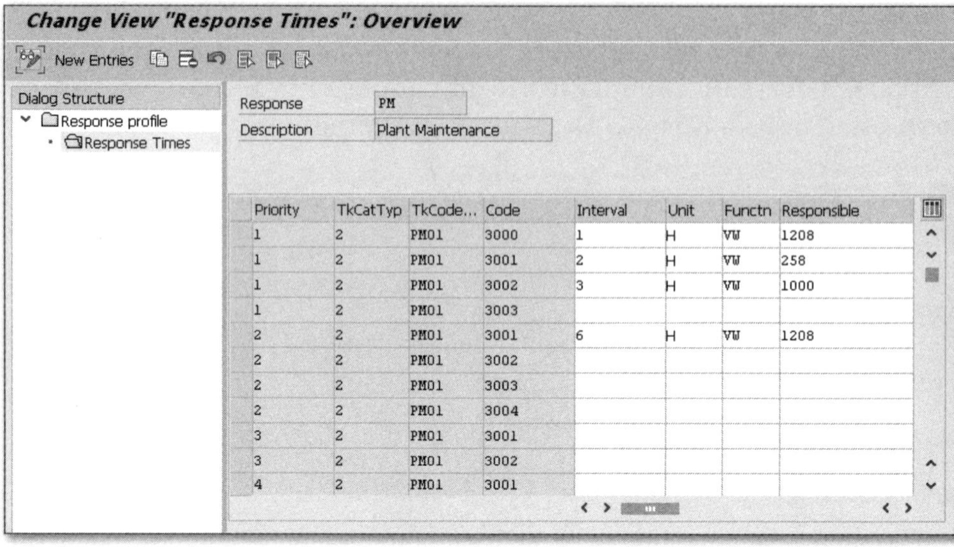

Priority	TkCatTyp	TkCode...	Code	Interval	Unit	Functn	Responsible
1	2	PM01	3000	1	H	VW	1208
1	2	PM01	3001	2	H	VW	258
1	2	PM01	3002	3	H	VW	1000
1	2	PM01	3003				
2	2	PM01	3001	6	H	VW	1208
2	2	PM01	3002				
2	2	PM01	3003				
2	2	PM01	3004				
3	2	PM01	3001				
3	2	PM01	3002				
4	2	PM01	3001				

Figure 6.26 Notifications: Response Profile

Use the DEFINE SERVICE PROFILE subfunction to define the times (within individual weekdays) during which a notification can be processed (see Figure 6.27). You can also schedule several service time intervals for each weekday and assign a separate response profile to each interval.

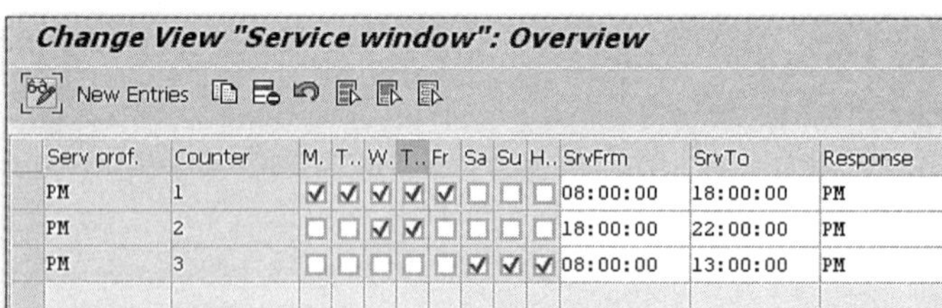

Serv prof.	Counter	M.	T..	W.	T..	Fr	Sa	Su	H..	SrvFrm	SrvTo	Response
PM	1	☑	☑	☑	☑	☑	☐	☐	☐	08:00:00	18:00:00	PM
PM	2	☐	☐	☑	☑	☐	☐	☐	☐	18:00:00	22:00:00	PM
PM	3	☐	☐	☐	☐	☐	☑	☑	☑	08:00:00	13:00:00	PM

Figure 6.27 Notifications: Response Profile

Use the RESPONSE TIME MONITORING FOR NOTIFICATION TYPE subfunction to assign a service profile and response profile to a notification type (see Figure 6.28).

If you've already assigned a response profile to a service profile, this overrides the response profile for the notification type.

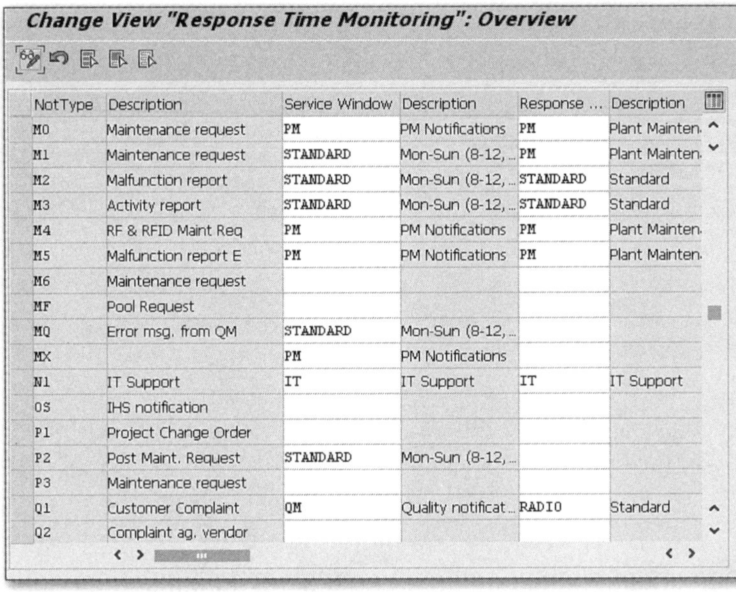

Change View "Response Time Monitoring": Overview

NotType	Description	Service Window	Description	Response ...	Description
M0	Maintenance request	PM	PM Notifications	PM	Plant Mainten.
M1	Maintenance request	STANDARD	Mon-Sun (8-12, ...	PM	Plant Mainten.
M2	Malfunction report	STANDARD	Mon-Sun (8-12, ...	STANDARD	Standard
M3	Activity report	STANDARD	Mon-Sun (8-12, ...	STANDARD	Standard
M4	RF & RFID Maint Req	PM	PM Notifications	PM	Plant Mainten.
M5	Malfunction report E	PM	PM Notifications	PM	Plant Mainten.
M6	Maintenance request				
MF	Pool Request				
MQ	Error msg. from QM	STANDARD	Mon-Sun (8-12, ...		
MX		PM	PM Notifications		
N1	IT Support	IT	IT Support	IT	IT Support
OS	IHS notification				
P1	Project Change Order				
P2	Post Maint. Request	STANDARD	Mon-Sun (8-12, ...		
P3	Maintenance request				
Q1	Customer Complaint	QM	Quality notificat...	RADIO	Standard
Q2	Complaint ag. vendor				

Figure 6.28 Notifications: Assignment of Response and Service Profiles to a Notification Type

Define Action Box

You use this Customizing function to define functions that you can use to generate *tasks* or *activities* for a notification (known as *follow-up activities*). They are defined directly for a notification type and are available when you process notifications at the header level. When you call the function, the system copies the defined task or activity to the notification.

Prerequisites

You must define the notification types and requisite tasks and activities beforehand.

Customizing Path

Plant Maintenance and Customer Service • Maintenance and Service Processing • Maintenance and Service Notifications • Notification Processing • Additional Functions • Define Action Box

Settings or Recommended Settings

Use the Define Follow-Up Activities (Generated Tasks/Activities) subfunction and then Activities to define which follow-up activities you want to assign to the notification type (see Figure 6.29).

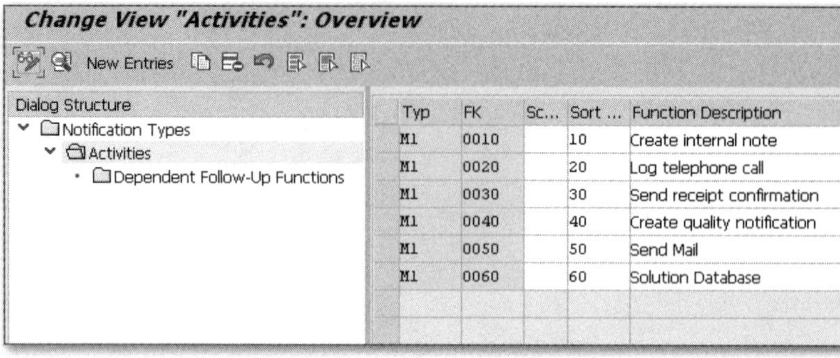

Figure 6.29 Notifications – Action Box: Overview

Use the New Entries button or the button to access the detail screen for new or existing activities. Here, you can define detailed information about each activity (see Figure 6.30).

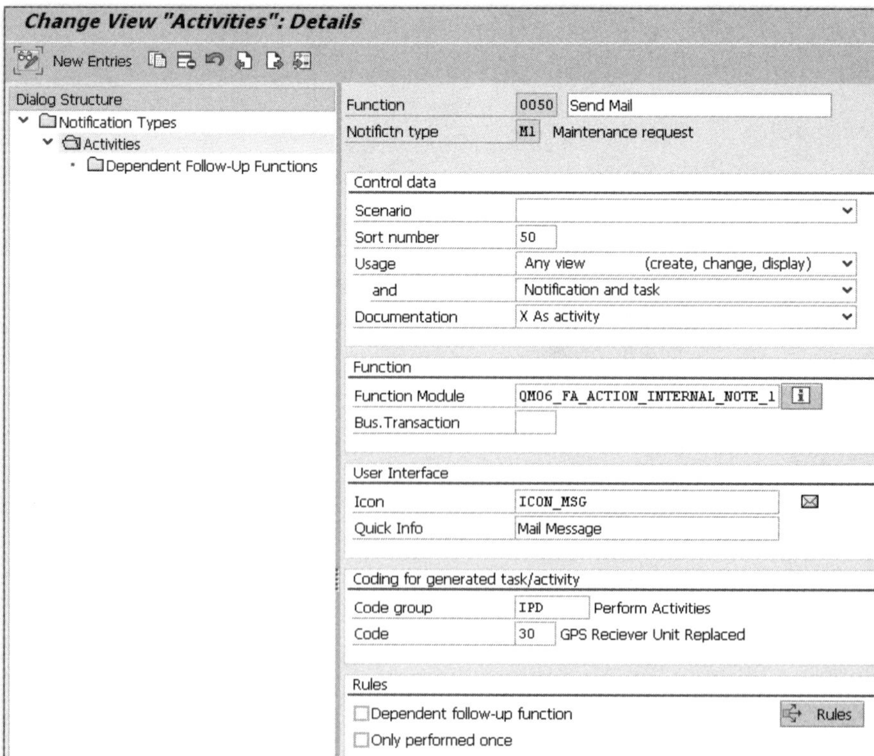

Figure 6.30 Notifications – Activities: Details Screen

The most important piece of information here is, without doubt, the FUNCTION MODULE selected. You can use either custom-programmed function modules or function modules delivered by SAP. Table 6.1 lists the most important function modules.

[+]

The Action Box Contains Certain Function Modules

The function modules QM06_FA* document an activity, while the function modules QM06_FM* document a task.

Function Module	Description
QM06_FA_ACTION_INTERNAL_NOTE_1	Documents a note in the form of a notification activity as a table control
QM06_FA_ACTION_INTERNAL_NOTE_2	Documents a note in the form of a notification activity as an OCX window
QM06_FA_ACTION_TELEPHONCALL_1	Documents a telephone call in the form of a notification activity as a table control
QM06_FA_ACTION_TELEPHONCALL_2	Documents a telephone call in the form of a notification activity as an OCX window
QM06_FA_ACTION_TELEPHONCALL_3	Documents a notification activity when the SAPphone function is used to commence a telephone call
QM06_FM_PARTNER_ACTION	Starts a query to an internal or external notification partner
QM06_FM_TASK_CREATE_8DREPORT	Creates an 8D report from a notification
QM06_FM_TASK_CREATE_QM_NOTIFIC	Creates a quality notification from a notification
QM06_FM_TASK_CRT_8DREPORT_PDF	Creates an 8D report as a PDF from a notification
QM06_FM_TASK_GOODS_MOVEMENT	Branches from a notification to the function for entering goods movements

Table 6.1 Function Modules

Function Module	Description
QM06_FM_TASK_GOODS_ MOVEMENT_2	Defines, from a notification, the quantities that are to be repaired, scrapped, and returned
QM06_FM_TASK_REQ_8DREPORT_PDF	Creates, from a notification, an 8D report as a PDF that can be sent to the person responsible for further processing
QM06_FM_TASK_REQUEST_8DREPORT	Creates, from a notification, an 8D report with print formatting, which can be sent to the person responsible for further processing
QM06_FM_TASK_RMA_ORDER_CREATE	Creates a repair order from a notification
QM06_FM_TASK_SALES_ORDER_STOCK	Transfers, from a notification, the usage decision for reclaimed goods to the corresponding repair order
QM06_FM_TASK_SEND_PAPER	Creates and sends different documents from a notification
QM06_FM_TASK_SEND_PAPER_ATTACH	Creates and sends different documents from a notification
QM06_FM_TASK_SEND_PAPER_PDF	Creates and sends different documents from a notification
QISD_CREATE_DOCUMENT_FLOW	Updates the main document flow
QISD_DISPLAY_ATTACHED_LINKS	Performs follow-up activity for displaying assigned symptoms/solutions
QISD_DISPLAY_SOLUTION	Displays details for a solution
QISD_SOLUTION_DATABASE	Performs follow-up activity for calling the solution database

Table 6.1 Function Modules (Cont.)

In addition, you can specify the following:

- Define the FUNCTION description.
- Use the SORT NUMBER to define the order in which the activities are to appear in the action box.

- Use USAGE to define whether the activity is to appear for every view or only certain views (add, change, or display). Furthermore, define whether the activity is to be available in notification processing, task processing, or both.
- For DOCUMENTATION, define whether the activities performed are to be documented and, if so, whether they are to be defined in the notification as tasks or activities.
- You can assign an ICON to the activity. Here, you can choose from more than 1,000 icons.
- The QUICK INFO is displayed if you position the cursor on the activity.
- Use CODE GROUP and CODE to define the activity or task that is to be used to document the activity in the notification. If you select a function module QM06_FA*, this implies an activity; if you select a function module QM06_FM*, this implies a task.

6.2 Orders

Notifications are primarily used for documentation purposes and to request maintenance tasks. In contrast, *orders* are used by maintenance planners and technicians as a tool for planning and carrying out maintenance work:

- An order is used to plan and carry out a maintenance task.
- Orders primarily contain processing information.
- The order is a highly integrative object with a large number of connections to applications such as Warehouse Management, Purchasing, and Controlling.

Figure 6.31 shows the structure and content of an order, as described here:

- **Order header**
  The data in the order header is information that is used to identify and manage the order. It applies to the entire order.
- **Object list**
  If the order concerns several objects (e.g., in the case of an inspection), you can enter the objects in an object list. The object list contains all objects for which the order is executed (functional locations, equipment, assemblies, and notifications).

- **Operation**
  You can use operations to describe the work that is to be performed when implementing an order. Operations are either carried out by your own employees or external companies.

- **Material list**
  The material list contains spare parts that are required and consumed when an order is being executed. This is either stock material for which a reservation is subsequently generated, or nonstock material for which a purchase requisition is created.

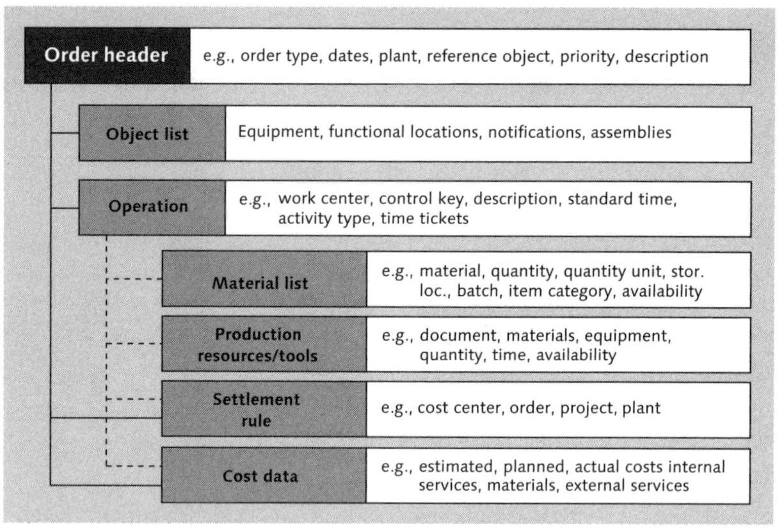

Figure 6.31 Order: Structure and Content

- **Production resources/tools**
  To execute an order, you require production resources/tools (e.g., tools, protective clothing, and hand pallet trucks). However, unlike a material, they aren't consumed.

- **Settlement rule**
  In the settlement rule, you specify the cost object (e.g., cost center) to which the costs must be charged. The settlement rule applies to the entire order. Alternatively, you can assign different account assignments to the operations.

- **Cost data**
  Cost data informs you about how high the estimated, planned, and actual costs are in the value categories for this order and which cost elements are relevant for the order. Furthermore, it specifies which key figures in the Plant Maintenance Information System (PMIS) are updated using the value categories, and how these key figures are updated with the actual costs of the order. You obtain cost information for individual operations and as the total for the complete order.

There are six ways in which you can create an order (see Figure 6.32):

- **Automatically generating an order from a maintenance plan**
  You define a maintenance plan, which automatically creates an order at regular intervals based on the information stored there (e.g., the reference object, order type, and task list).

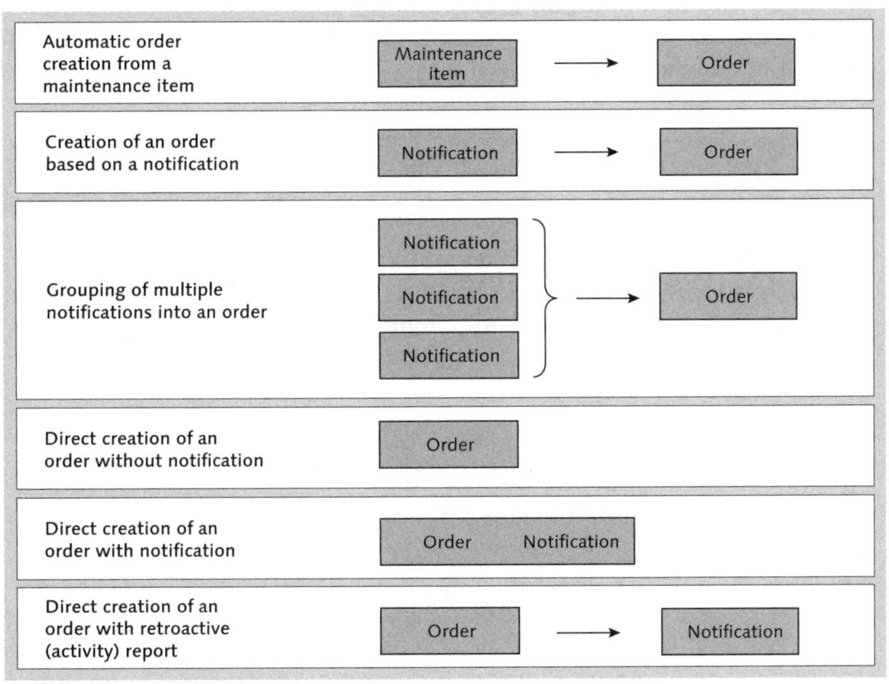

Figure 6.32 Creating an Order

- **Creating an order from a notification**
  You receive an individual notification from a requester (e.g., from Production), and you now create an order from this notification.
- **Grouping several notifications into one order**
  You receive several notifications that are to be processed within a single order (e.g., several malfunction reports that refer to the same asset), and you have the option to create an order from the notification list.
- **Creating an order directly without a notification**
  You want to create an order directly (i.e., without a notification).
- **Creating an order directly (including a notification)**
  Similar to the previous option, you want to create an order directly, but you want to supplement the order with information such as damage and cause of damage or breakdown.
- **Creating an order directly with a retroactive notification**
  You created an order without a notification but, when completing the order, you decided to enter technical confirmation data in a notification, in addition to time confirmation data.

Next, we'll describe the Customizing functions that you can use to adjust orders to your requirements. A complete overview of the Customizing functions for orders is provided in Figure 6.33.

The following Customizing functions and their subfunctions are also relevant for orders, but they've already been described in Chapter 4, so they are simply listed here:

- Number ranges
- Field selection
- Partner determination procedure and partner function
- Object information key
- User status
- List variants
- Field selection for multilevel list displays

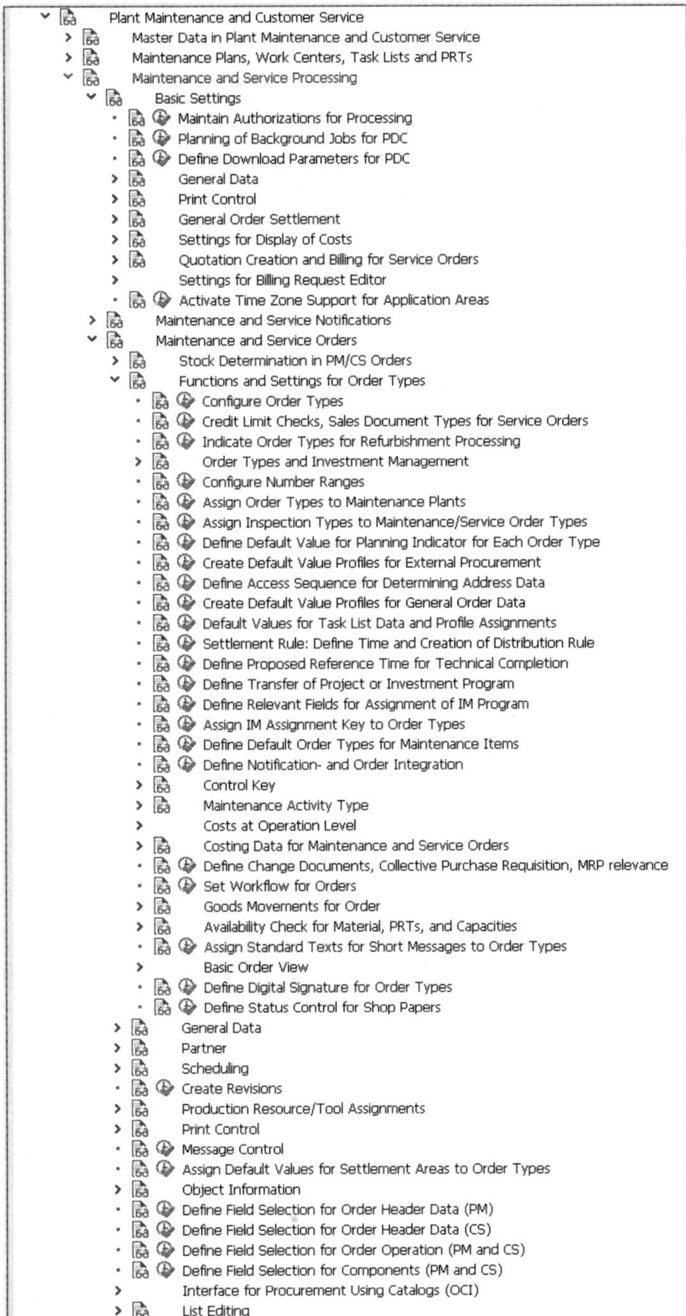

Figure 6.33 Overview of the Customizing Functions for Orders

To provide a clearer overview of the many Customizing functions for orders, this section is subdivided into the following subsections:

- Section 6.2.1, Order Types, contains all of the basic functions that you need to configure an order type.
- Section 6.2.2, Material Planning, shows you all of the Customizing functions for materials within order processing (ranging from the availability check to electronic parts catalogs and material documents).
- Section 6.2.3, Scheduling, and Section 6.2.4, Capacity Planning, contain all of the Customizing functions for scheduling and capacity planning.
- Section 6.2.5, Printing, shows you all of the Customizing functions for print control in relation to shop papers and messages.
- Section 6.2.6, Estimated Costs and Costing, and Section 6.2.7, Order Settlement, contain all of the Customizing functions for cost estimates, costing, and order settlement.
- Section 6.2.8, Project Management and Investment Management, contains all Customizing functions regarding project-based maintenance.
- All other Customizing functions that don't relate to the aforementioned subsections, but are nevertheless important, will be introduced in Section 6.2.9, Other Functions.

6.2.1 Order Types

In Customizing, the order type is *the* main control element for order processing. Unlike the notification type, where the plant plays almost no role whatsoever, it's necessary, in the case of the Customizing functions for orders, to distinguish between the following:

- Whether the Customizing function for the order type is generally applicable (i.e., across all plants)
- Whether the Customizing function for the order type refers to only one particular plant

For each order type without specific reference to a plant (i.e., the settings apply to all plants), you define, among other things, the following in Customizing (see Figure 6.34):

- Which maintenance activity types are permitted for the order type
- Which number is to be assigned to an order
- How you want to settle an order
- How the order is to be integrated with the notification
- Which fields you want to enter
- Which shop papers are to be printed
- Which priorities you want to use in the order
- Which partners you want to assign to an order
- Which object information key is to be used
- How long you want to keep an order in the system
- What effect budgets are to have
- Which status profile is to be used

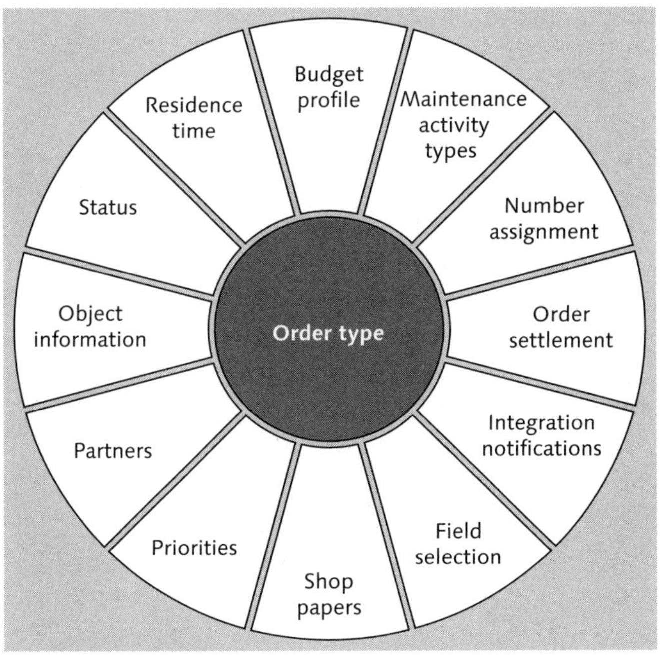

Figure 6.34 Customizing Functions for Each Order Type

For each order type that is to apply to only one particular plant (i.e., with a plant reference), you define, among other things, the following in Customizing (see Figure 6.35):

- The screen layout
- How the address data for the delivery address is to be accessed
- Which values are to be proposed for procurement transactions
- How you want to access task lists
- How you want to schedule orders
- How availability checks are to be performed
- How you want to conduct preliminary and final costings for orders
- Whether you want to show the costs at the header level or operation level
- For which changes the system is to create change documents
- Whether you want the order to contain linear data
- Which goods movements are to be documented
- Which electronic parts catalogs you want to integrate

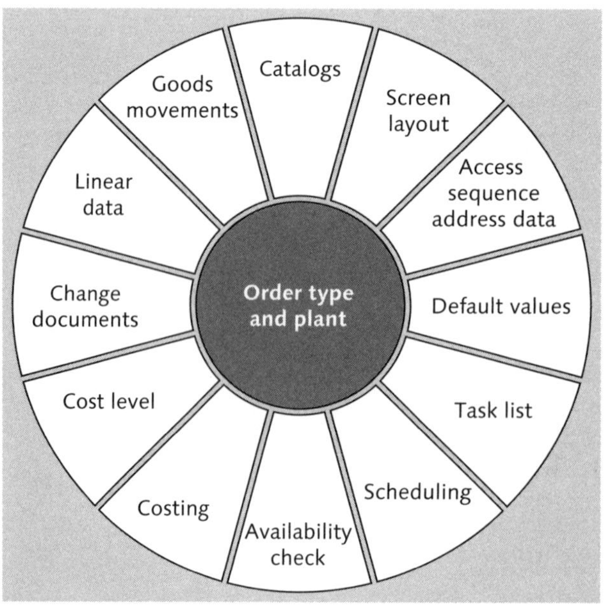

Figure 6.35 Customizing Functions for Each Order Type and Plant

Consultation with All Plants [+]

With regard to the Customizing functions for order types without a plant reference, you must consult with your colleagues in the plants involved.

If this is unsuccessful (e.g., if you don't reach agreement on the number ranges for orders), the only option is to define plant-specific order types.

Next, we'll explain the basic Customizing functions for configuring an order type.

Configure Order Types

You use this Customizing function to define the order type and some basic parameters.

Prerequisites

Because order types are used not only in SAP EAM but also in other user departments and their corresponding SAP applications (see Figure 6.36), you must consult with your colleagues from the other user departments on the naming convention and number ranges for order types.

These different applications result in the use of *order categories*. Each order type must be assigned to an order category, which, in turn, is predefined by SAP. The following order categories exist:

- 01 – Internal order (Controlling)
- 02 – Accrual calculation order (Controlling)
- 03 – Model order (Controlling)
- 04 – CO production order
- 05 – Product cost collector
- 06 – QM order
- 10 – PPS production order
- 20 – Network
- 30 – Service order and maintenance order
- 40 – Process order
- 50 – Inspection lot

- 60 – Personnel order
- 70 – Shipping deadline

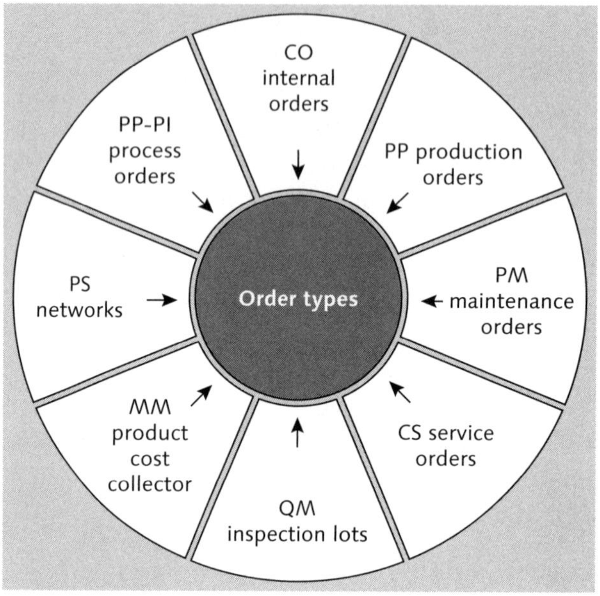

Figure 6.36 Orders: Order Categories and Order Types

Customizing Path

Plant Maintenance and Customer Service • Maintenance and Service Processing • Maintenance and Service Orders • Functions and Settings for Order Types • Configure Order Types

Transaction

OIDA

Settings or Recommended Settings

When defining which order types you want to configure, you should be guided by the customizing functions and customizing options that we explain next.

[+]

Three to Seven Order Types Should Suffice

You'll generally require only the following order types:

- Repair
- Maintenance and inspection
- Investment
- Refurbishment (if used)
- Calibration of test equipment (if used)
- Pool Asset Management (PAM) (if used)
- External processing (optional)

You can use maintenance activity types for further refinements and subdivisions.

You can use this Customizing function to configure the following settings (see Figure 6.37):

- For the ORDER TYPE, specify the naming convention and description.
- In this Customizing function, the system automatically assigns ORDER CATEGORY 30 (MAINTENANCE ORDER).
- Use the OPEN ITEM MANAGEMENT option to define whether you want to use the planned costs associated with the order as a basis for creating *order commitments*.
- The SETTLEMENT PROFILE specifies how the order is to be settled. (For more information, see Section 6.2.6.)
- The BUDGET PROFILE specifies whether a budget availability check is to be performed and, if so, how. (For more information, see Section 6.2.6.)
- RESIDENCE TIME1 specifies, in months, the time gap between the user manually setting the deletion flag and archiving automatically setting the deletion indicator.
- RESIDENCE TIME2 specifies, in months, the time gap between setting the deletion indicator and actually deleting the order.
- The RELEASE IMMEDIATELY indicator only takes effect if the order is automatically retrieved from a maintenance plan or if it's generated from a notification. The effect of this setting is that, in these two scenarios, the system immediately sets the status REL (Released), and not CRTD (Created), in the order.

- Use the SCREEN REFOBJECT field to control which reference object can be entered in the order header (e.g., 0100 = functional location, piece of equipment, and assembly).

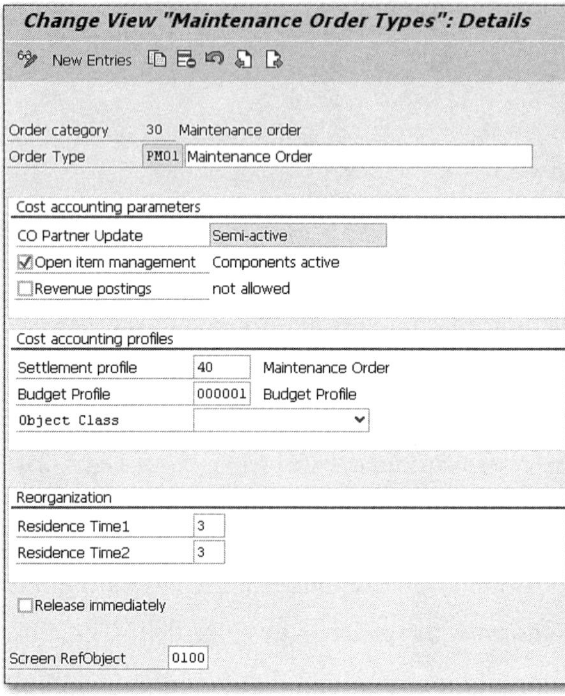

Figure 6.37 Order: Order Type

Assign Order Types to Maintenance Plants

If you want to use an order type in a planning plant, you must use this Customizing function to make the order type available for the plant.

Prerequisites

You must define the order type and plant beforehand.

Customizing Path

PLANT MAINTENANCE AND CUSTOMER SERVICE • MAINTENANCE AND SERVICE PROCESSING • MAINTENANCE AND SERVICE ORDERS • FUNCTIONS AND SETTINGS FOR ORDER TYPES • ASSIGN ORDER TYPES TO MAINTENANCE PLANTS

Transaction

OIOD

Settings or Recommended Settings

Generate the entries that define which order type is to be used in which plant (see Figure 6.38).

Ensure a Complete Assignment [+]

This listing must be complete. Otherwise, you won't be able to use the order type in the plant.

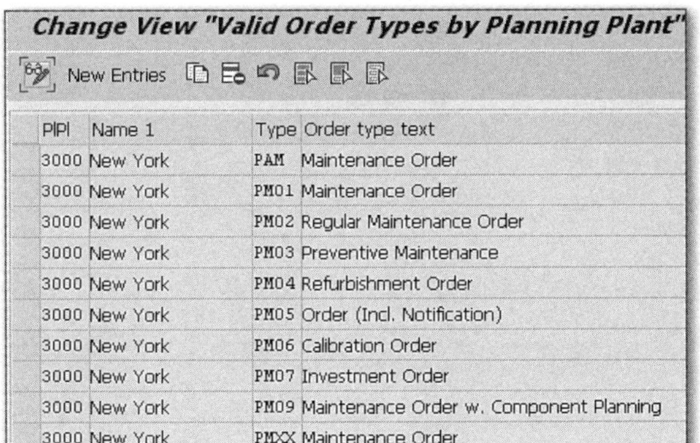

Change View "Valid Order Types by Planning Plant"

New Entries

PlPl	Name 1	Type	Order type text
3000	New York	PAM	Maintenance Order
3000	New York	PM01	Maintenance Order
3000	New York	PM02	Regular Maintenance Order
3000	New York	PM03	Preventive Maintenance
3000	New York	PM04	Refurbishment Order
3000	New York	PM05	Order (Incl. Notification)
3000	New York	PM06	Calibration Order
3000	New York	PM07	Investment Order
3000	New York	PM09	Maintenance Order w. Component Planning
3000	New York	PMXX	Maintenance Order

Figure 6.38 Order: Assign Order Type to Maintenance Plants

Define Default Value for Planning Indicator for Each Order Type

You use this Customizing function to define, for each order type, a default value for the order planning indicator (see Figure 6.39). The order planning indicator is solely used to select and perform statistical analyses in the LIS. It doesn't have any controlling effect whatsoever.

Prerequisites

You must define the order types beforehand.

Customizing Path

Plant Maintenance and Customer Service • Maintenance and Service Processing • Maintenance and Service Orders • Functions and Settings for Order Types • Define Default Value for Planning Indicator for Each Order Type

Transaction

OIDS

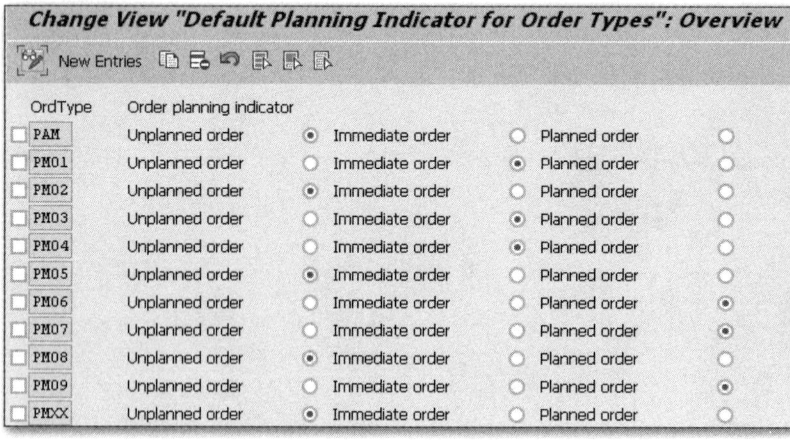

Figure 6.39 Order: Planning Indicator

Settings or Recommended Settings

The LIS contains key figures for the number of *immediate orders*, number of *planned orders*, and number of *unplanned orders*.

[+]

No Fixed Rule for the Planning Indicator

You're completely free to choose which order type you want to assign to which planning indicator. For example, the following assignments are possible:

- Assign the Planned order planning indicator to all order types associated with maintenance and inspection.
- Assign the Immediate order planning indicator to all order types associated with repairing a breakdown.
- Assign the Unplanned order planning indicator to all other order types.

Define Notification and Order Integration

You use this Customizing function to define the following:

- Whether you want to maintain notification and order data on one screen
- Whether you want the system to automatically create a notification when you enter an order
- Whether the system is to automatically copy the long text of the notification header to the order header if an order is generated from a notification
- Whether, and to what extent, you'll permit the assignment of operations to object list entries in an order

Prerequisites

You must define the order types and notification types beforehand.

Customizing Path

PLANT MAINTENANCE AND CUSTOMER SERVICE • MAINTENANCE AND SERVICE PROCESSING • MAINTENANCE AND SERVICE ORDERS • FUNCTIONS AND SETTINGS FOR ORDER TYPES • DEFINE NOTIFICATION- AND ORDER INTEGRATION

Settings or Recommended Settings

The following settings are available for each order type (see Figure 6.40):

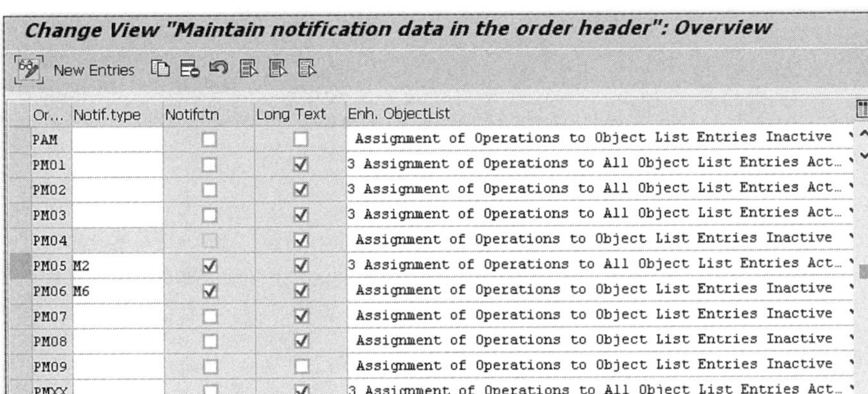

Change View "Maintain notification data in the order header": Overview

New Entries

Or...	Notif.type	Notifctn	Long Text	Enh. ObjectList
PAM		☐	☐	Assignment of Operations to Object List Entries Inactive
PM01		☐	☑	3 Assignment of Operations to All Object List Entries Act...
PM02		☐	☑	3 Assignment of Operations to All Object List Entries Act...
PM03		☐	☑	3 Assignment of Operations to All Object List Entries Act...
PM04		☐	☑	Assignment of Operations to Object List Entries Inactive
PM05	M2	☑	☑	3 Assignment of Operations to All Object List Entries Act...
PM06	M6	☑	☑	Assignment of Operations to Object List Entries Inactive
PM07		☐	☑	Assignment of Operations to Object List Entries Inactive
PM08		☐	☑	Assignment of Operations to Object List Entries Inactive
PM09		☐	☐	Assignment of Operations to Object List Entries Inactive
PMXX		☐	☑	3 Assignment of Operations to All Object List Entries Act...

Figure 6.40 Orders: Order/Notification Integration

- If you set the NOTIFCTN (notification) indicator, you can maintain order and notification data on one screen. Furthermore, if you create an order directly, the system automatically creates the notification in the background.
- In the NOTIF.TYPE column, specify the notification type for the automatically generated notification.
- If you set the LONG TEXT indicator, then when the system generates an order, it copies the long text of the notification header to the long text of the order. However, subsequent changes to the notification long text are no longer copied.

Use the entry in the ENH. OBJECTLIST (enhanced object list) column to control the automatic assignment of operations to object list entries (e.g., a notification, piece of equipment, or functional location) if a notification is assigned to an order. The following settings are available:

- blank = ASSIGNMENT OF OPERATIONS TO OBJECT LIST ENTRIES INACTIVE
  The enhanced functions for the object list are inactive. In other words, the operations of the order can't be assigned to entries in the object list.
- 1 = ASSIGNMENT OF OPERATIONS TO MAINTENANCE PLAN NOTIFICATIONS ACTIVE
  When you assign a notification to an order, the system can automatically assign the operations solely to those notifications in the object list that were generated from maintenance plans. You can display the operations assigned to a notification.
- 2 = ASSIGNMENT OF OPERATIONS TO ALL NOTIFICATIONS ACTIVE
  When you assign a notification to an order, the system can automatically assign the operations to all notifications in the object list (i.e., to notifications created manually and to maintenance plan notifications). You can display the operations assigned to a notification.
- 3 = ASSIGNMENT OF OPERATIONS TO ALL OBJECT LIST ENTRIES ACTIVE
  The system can automatically assign the operations to any object list entry (e.g., to notifications, equipment, or functional locations).

Maintain Default Values for Control Keys for Order Types

In Chapter 3, Section 3.4, we explained both the functionality and the function of the control key. You use this Customizing function to define which control key is to be proposed for which order type if a control key isn't defined in the work center. If a control key is defined in the work center, this has priority.

Prerequisites

You must define the plants, order types, and control keys beforehand.

Customizing Path

PLANT MAINTENANCE AND CUSTOMER SERVICE • MAINTENANCE AND SERVICE PROCESSING • MAINTENANCE AND SERVICE ORDERS • FUNCTIONS AND SETTINGS FOR ORDER TYPES • CONTROL KEY • MAINTAIN DEFAULT VALUES FOR CONTROL KEYS FOR ORDER TYPES

Transaction

OIO6

Settings or Recommended Settings

Define which control key is to be proposed for which order type and in which plant (see Figure 6.41).

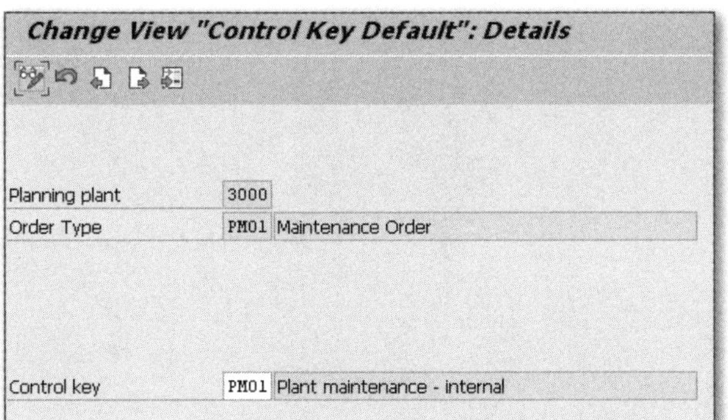

Figure 6.41 Order: Default Values for Control Keys

Customizing Functions for Maintenance Activity Type

You use this Customizing function to define the maintenance activity types and to assign them to the order types. The maintenance activity types are used to refine the order types further and are important selection criteria for order lists. You can do the following with the maintenance activity type:

- Categorize the costs in the LIS. In other words, you can use maintenance activity types to summarize the maintenance costs.
- Indicate whether it concerns a warranty case or an insurance case.

Prerequisites

You must define the order types beforehand.

Customizing Path

Plant Maintenance and Customer Service • Maintenance and Service Processing • Maintenance and Service Orders • Functions and Settings for Order Types • Maintenance Activity Type • Define Maintenance Activity Types or Assign Valid Maintenance Activity Types to Maintenance Order Types or Default Values for Maintenance Activity Type for Each Order Type

Transactions

OIO5 for order types permitted for each maintenance activity type

OIO4 for the default value for each order type

Settings or Recommended Settings

Figure 6.42 lists the maintenance activity types that are commonly set up in projects.

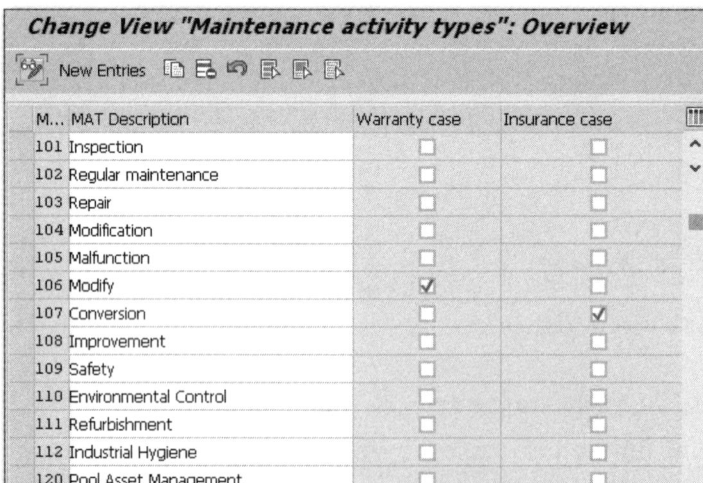

Change View "Maintenance activity types": Overview

New Entries

M...	MAT Description	Warranty case	Insurance case
101	Inspection	☐	☐
102	Regular maintenance	☐	☐
103	Repair	☐	☐
104	Modification	☐	☐
105	Malfunction	☐	☐
106	Modify	☑	☐
107	Conversion	☐	☑
108	Improvement	☐	☐
109	Safety	☐	☐
110	Environmental Control	☐	☐
111	Refurbishment	☐	☐
112	Industrial Hygiene	☐	☐
120	Pool Asset Management	☐	☐

Figure 6.42 Order: Maintenance Activity Types

Four to Ten Maintenance Activity Types Should Suffice [+]

You should always map the following four maintenance types in accordance with the German standard DIN 31051 and the international standard ISO 55000:

- Repair
- Inspection
- Maintenance
- Improvement

If you use these processes, you also require:

- Refurbishment
- Pool Asset Management
- Calibration

Use additional maintenance activity types if necessary.

Then, use the Customizing function Assign Valid Maintenance Activity Types to Maintenance Order Types to define which maintenance activity types are valid for which order types (see Figure 6.43). You're then unable to select any other maintenance activity types for the respective order type.

Change View "Valid Maintenance Activity Types for Each Order Type"

New Entries

Type	Order type text	MAT	MAT Description
PM01	Maintenance Order	101	Inspection
PM01	Maintenance Order	102	Regular maintenance
PM01	Maintenance Order	103	Repair
PM01	Maintenance Order	105	Malfunction
PM01	Maintenance Order	106	Modify
PM01	Maintenance Order	107	Conversion
PM01	Maintenance Order	108	Improvement
PM01	Maintenance Order	109	Safety
PM01	Maintenance Order	110	Environmental Control
PM01	Maintenance Order	112	Industrial Hygiene

Figure 6.43 Order: Maintenance Activity Types for Each Order Type

Finally, use the Customizing function Default Values for Maintenance Activity Type for Each Order Type to define which maintenance activity type is to be proposed for which order type (see Figure 6.44). Here, you can propose only one maintenance activity type, which you must have also previously assigned to the order type.

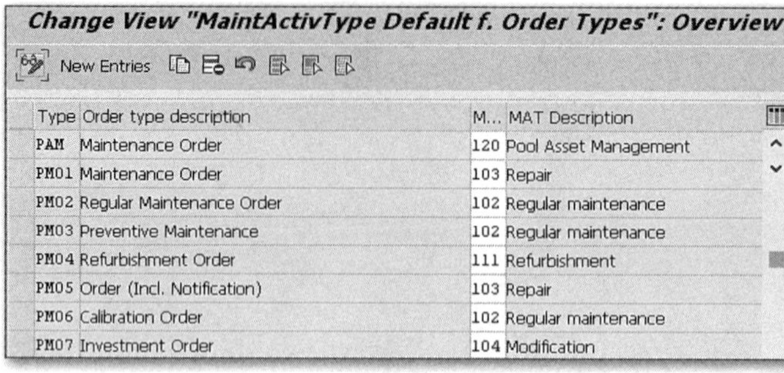

Figure 6.44 Order: Default Values for Maintenance Activity Types for Each Order Type

Customizing Functions for a Basic Order View

You use the Customizing functions for a basic order view to define, in conjunction with *view profiles*, the screen layout for orders and to assign it to the order types for each plant.

Prerequisites

You must define the plants and order types beforehand.

Customizing Path

Plant Maintenance and Customer Service • Maintenance and Service Processing • Maintenance and Service Orders • Functions and Settings for Order Types • Basic Order View • Define View Profiles or Assign View Profiles to Order Types

Settings or Recommended Settings

First, use the Customizing function Define View Profiles to define a name and description for the view profile (see Figure 6.45).

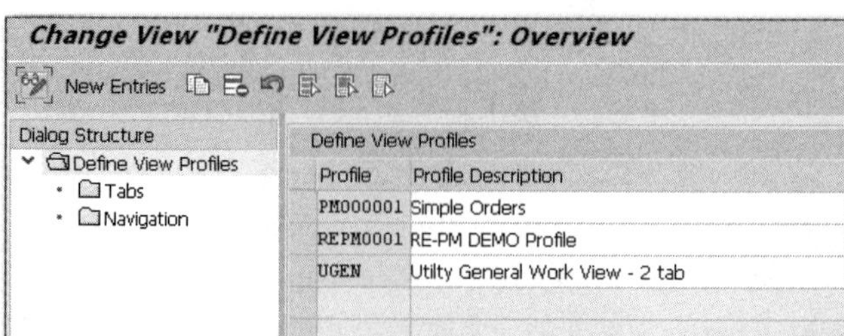

Figure 6.45 Order – View Profiles: Overview

Use the TABS subfunction to determine the tab pages that the view profile is to contain (see Figure 6.46). When defining tab pages, note the following:

- The tab page number is used as a sort key for the order in which the tab pages are displayed.
- You can assign a maximum of eight screen areas to a tab page. The order in which the screen areas are assigned also determines the order in which they are displayed on the tab page.
- Some screen areas occupy the entire tab page (full screen). You must always enter these screen areas in the field for screen area 1.
- You can also assign an icon to each tab page. Here, you can choose from more than 1,000 icons.

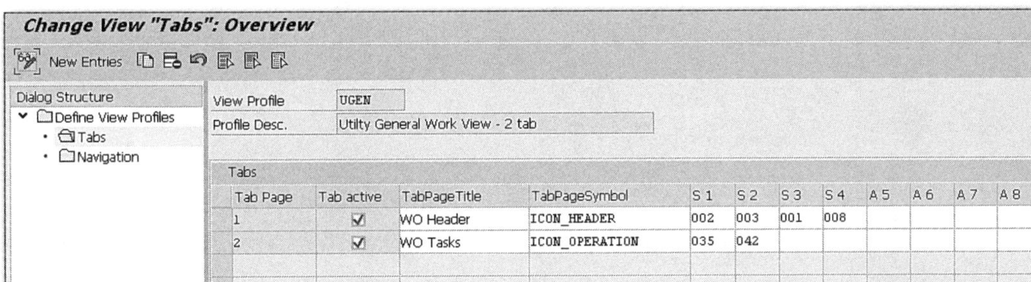

Figure 6.46 Order – View Profile: Detail Screen

Table 6.2 lists the screen areas available.

ID	Screen Area
001	Responsibilities
002	Reference object
003	Dates (overview)
004	Dates (enhanced)
005	Header long text as Control (23 lines)
006	Header long text as Control (10 lines)
007	Operation details (with tab pages)

Table 6.2 Screen Areas

ID	Screen Area
008	Operation details (selected fields)
010	Partner overview
011	Partner overview (small)
012	Addresses
014	Notification data
015	Linear data
021	Cost overview
023	System condition and planning indicator
024	Organization
025	Parameters, administrative data, reservation/PReq
026	Location data
028	Account assignment
033	Operations (full screen)
034	Operations (as a subscreen, 15 lines)
035	Operations (as a subscreen, 6 lines)
040	Components (full screen)
041	Components (as a subscreen, 15 lines)
042	Components (as a subscreen, 6 lines)
050	Object list
090	Customer enhancement
091	Customer enhancement (full screen)

Table 6.2 Screen Areas (Cont.)

[+]

Activate the Tab Page

Don't forget to activate the tab page by setting the TAB ACTIVE indicator. Otherwise, the tab page is defined, but it's not displayed.

Use the NAVIGATION subfunction to define which button will bring you to which of your defined tab pages (see Figure 6.47).

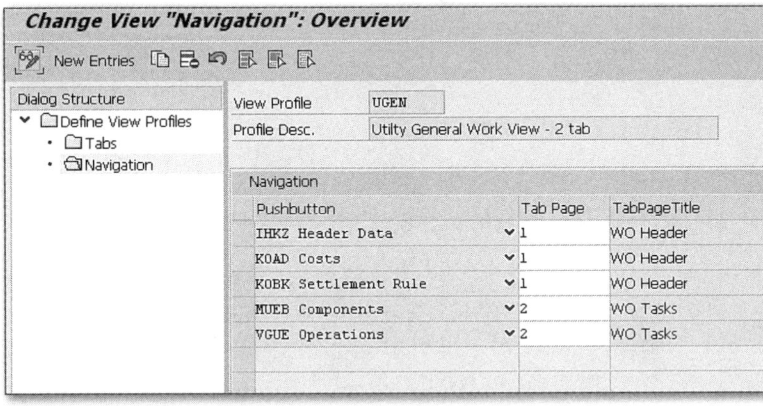

Figure 6.47 Order: View Profile Navigation

Finally, use the Customizing function ASSIGN VIEW PROFILES TO ORDER TYPES to define which view profile is to be used for which order type and in which plant (see Figure 6.48).

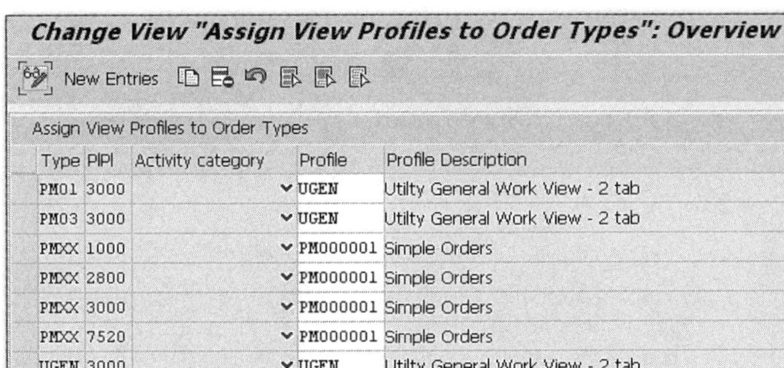

Figure 6.48 Order: Assign View Profile to Order Types

View Profiles for Each Order Type and Plant [+]

View profiles allow different plants to use their own screen layout for the same order type.

6.2.2 Material Planning

To provide a clearer overview of the Customizing functions for materials, we've combined them into the following three groups:

- Availability check
- Catalogs
- Other material functions

Customizing Functions for the Availability Check

The system always performs a dynamic availability check, which checks whether a sufficient quantity of the required material is available in the plant on the requirements date (i.e., when the order is being executed). Figure 6.49 illustrates how the current warehouse stock, based on the current date and taking into account safety stocks, can change up to the requirements date as a result of stock movements (planned goods receipts, planned goods issues, and planned stock transfers). The *ATP quantity* (available to promise) is determined on the requirements date.

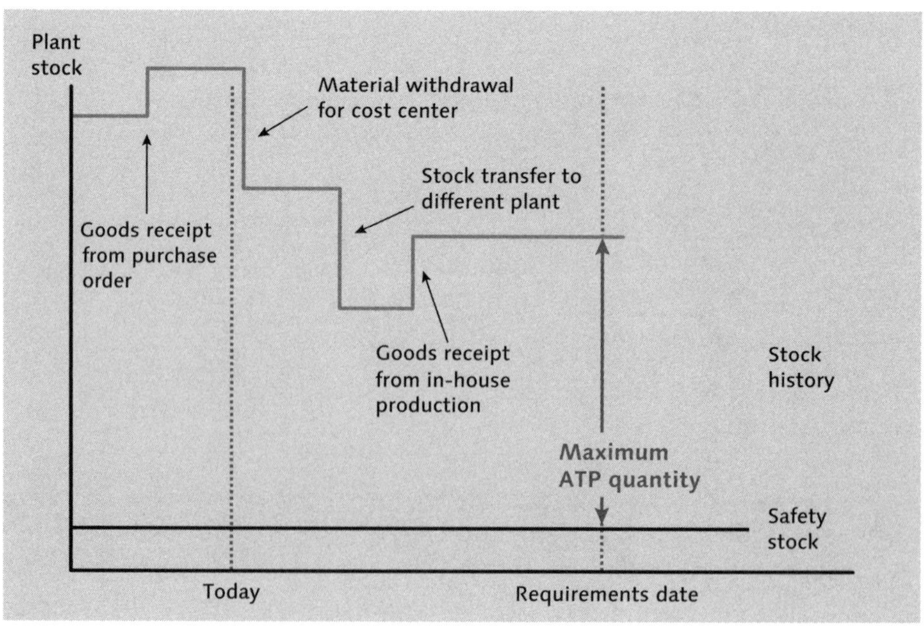

Figure 6.49 Orders: Availability Check

Figure 6.50 shows the relationship between the Customizing functions and the availability check as described here: You define the *checking group* in Customizing and then assign it to the material master in the *MRP* (material requirements planning) view. Next, you define the *checking rule* in Customizing and assign it to an order type for each plant. Finally, on the basis of the checking group and checking rule, you define the *scope of check* in Customizing.

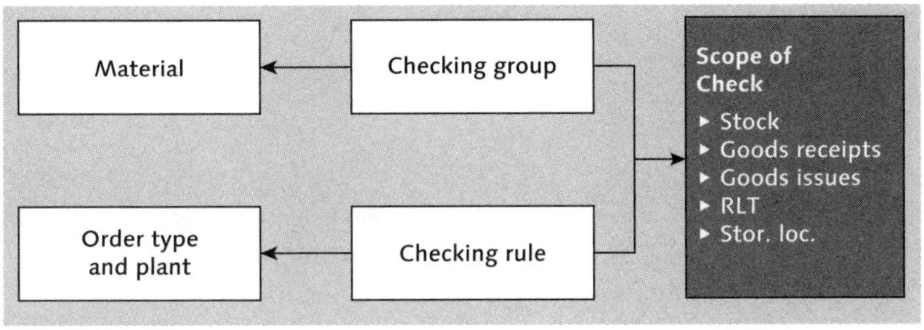

Figure 6.50 Availability Check: Overview

Define Checking Rules

You use this Customizing function to define checking rules.

Prerequisites

There are no specific prerequisites.

Customizing Path

PLANT MAINTENANCE AND CUSTOMER SERVICE • MAINTENANCE AND SERVICE PROCESSING • MAINTENANCE AND SERVICE ORDERS • FUNCTIONS AND SETTINGS FOR ORDER TYPES • AVAILABILITY CHECK FOR MATERIAL, PRTS, AND CAPACITIES • DEFINE CHECKING RULES

Settings or Recommended Settings

Figure 6.51 provides an overview of the checking rules. Define a name and description for each checking rule.

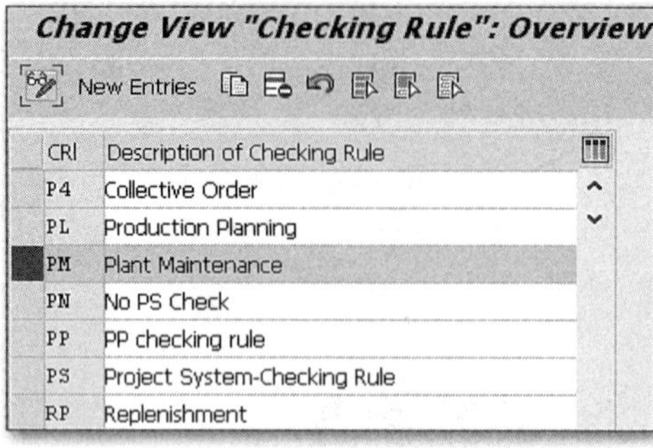

Figure 6.51 Order: Checking Rule

[+] **Separate Checking Rules for Plant Maintenance**

Create a checking rule that you'll use later for maintenance materials only. Initially, the checking rule doesn't have any other function.

Define Inspection Control

You use this Customizing function to define, for each order type, plant, and operation, which checking rule you want to use for the availability check.

Prerequisites

You must define the plants, order types, and checking rules beforehand.

Customizing Path

Plant Maintenance and Customer Service • Maintenance and Service Processing • Maintenance and Service Orders • Functions and Settings for Order Types • Availability Check for Material, PRTs, and Capacities • Define Inspection Control

Transaction

OIOI

Settings or Recommended Settings

As you can see from Figure 6.52, there are two entries for each order type and plant:

- Availability transaction 1
  This refers to created orders. In other words, this material availability check is performed as long as the order has the status CRTD (Created).
- Availability transaction 2
  This applies at the time when an order is released or, if an availability check is performed, after the order has been released.

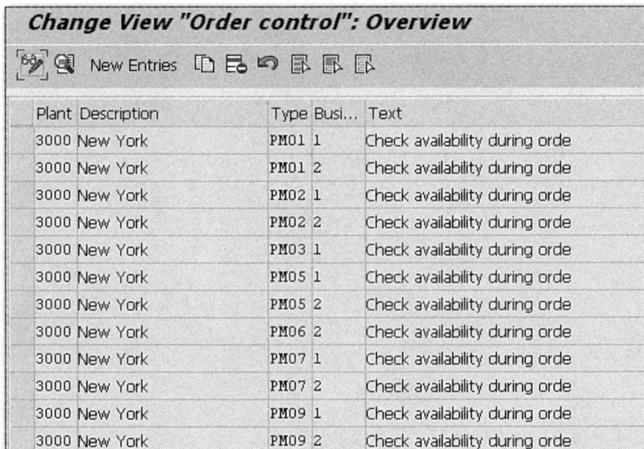

Change View "Order control": Overview

New Entries

Plant	Description	Type	Busi...	Text
3000	New York	PM01	1	Check availability during orde
3000	New York	PM01	2	Check availability during orde
3000	New York	PM02	1	Check availability during orde
3000	New York	PM02	2	Check availability during orde
3000	New York	PM03	1	Check availability during orde
3000	New York	PM05	1	Check availability during orde
3000	New York	PM05	2	Check availability during orde
3000	New York	PM06	2	Check availability during orde
3000	New York	PM07	1	Check availability during orde
3000	New York	PM07	2	Check availability during orde
3000	New York	PM09	1	Check availability during orde
3000	New York	PM09	2	Check availability during orde

Figure 6.52 Orders – Checking Rule for Order Type and Plant: Overview

The following settings are available for each entry (see Figure 6.53):

- You can use the No check indicator to deactivate the material availability check. Then, all other entries have no effect.
- The Status check indicator is available for availability transaction 2 only. You use it to control that the system only ascertains whether the *system status* MSPT (Material shortage) is active when it performs a check before releasing an order. You can use this indicator to prevent the system from triggering a new *availability check* each time you save the order, thus reducing the system load.
- If you set the Check material availability when saving order indicator, the system automatically performs the material availability check when you save the order after having made all of your material-relevant changes to the order (e.g., new components, a changed date, or a new quantity).
- For the Checking Rule, enter the checking rule that you defined for SAP EAM.

- The COMPONENT CHECK TYPE is always ATP CHECK (i.e., a dynamic availability check on the requirements date for the component).
- The following options are available for the RELEASE MATERIAL indicator:
  - 1: USER DECIDES ON RELEASE IF PARTS ARE MISSING
  - 2: RELEASE PERMITTED DESPITE MISSING PARTS
  - 3: NO RELEASE IF PARTS ARE MISSING

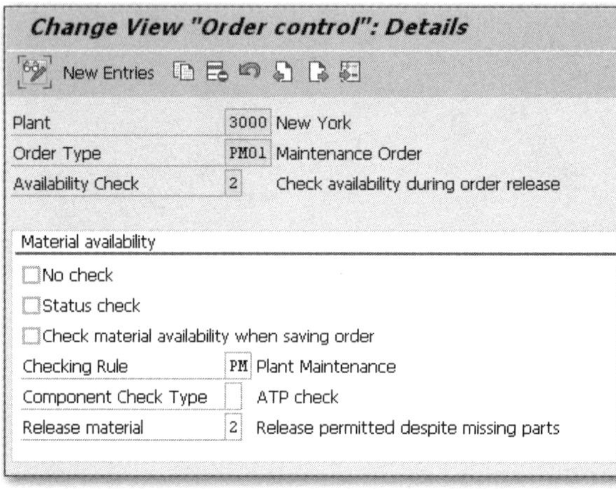

Figure 6.53 Orders – Checking Rule for Order Type and Plant: Details Screen

[+]

Always Let the User Decide

You should always set the RELEASE MATERIAL option to 1 so that the planner is alerted to the fact that there is a material shortfall. Only then can the planner, and not the system, decide how to deal with such a situation.

Define Checking Group

You use this Customizing function to define checking groups (see Figure 6.54).

Prerequisites

There are no specific prerequisites.

Customizing Path

PRODUCTION • SHOP FLOOR CONTROL • OPERATIONS • AVAILABILITY CHECK • DEFINE CHECKING GROUP

Checking Group Not in the Menu for Plant Maintenance [!]

The Customizing function for defining checking groups isn't located in the PLANT MAINTENANCE menu, but is instead in the PRODUCTION menu.

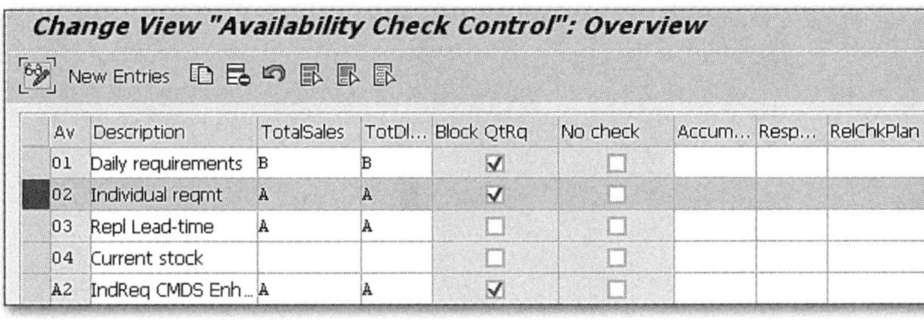

Figure 6.54 Orders: Checking Group

Transaction

OVZ2

Settings or Recommended Settings

For the checking group, you have the following options:

- You can check each individual requirement (entry A in the totals columns).
- You can create totals records for each day or week (entries B – D in the totals columns).

Check Each Individual Requirement [+]

From a maintenance perspective, it doesn't make sense to create MRP total requirements. Therefore, use the total indicator A to check the material requirements individually.

You enter the checking group in the material master (see Figure 6.55). This field belongs to the AVAILABILITY CHECK screen group, which is generally displayed in the MRP view.

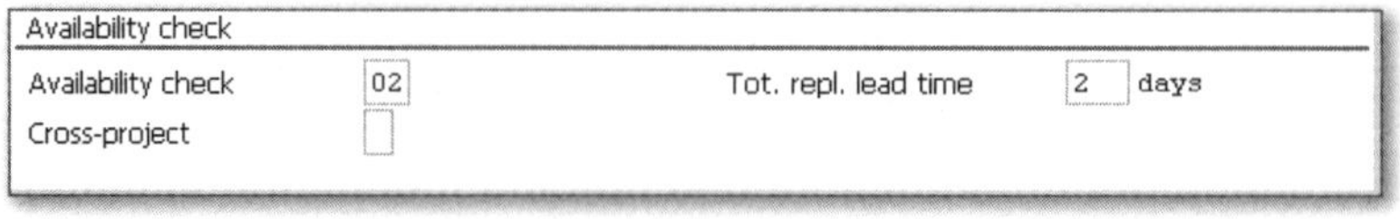

Figure 6.55 Material Master: Availability Check

Define Scope of Check

You use this Customizing function to control, for each checking group and checking rule, how you want to perform the availability check (i.e., which stock types and which goods receipts/goods issues you want the system to check).

Prerequisites
You must define the checking group and checking rule beforehand.

Customizing Path
Plant Maintenance and Customer Service • Maintenance and Service Processing • Maintenance and Service Orders • Functions and Settings for Order Types • Availability Check for Material, PRTs, and capacities • Define Scope of Check

Transaction
OPJJ

Settings or Recommended Settings
You can make the following settings (see Figure 6.56):

- The left-hand side of the screen contains the *stock types* to be included in the calculation. Therefore, if you want the material currently located in the quality inspection stock to be taken into account, for example, set the Incl.quality insp. stock (include quality inspection stock) indicator.
- The right-hand side of the screen contains the movement types that cause a planned inward movement (also known as a planned goods receipt). Therefore, if you want to take into account purchase orders whose goods receipt, as a result of the delivery date, is to be expected up to the requirements date of the material in the order, for example, set the Incl.Purchase Orders indicator.

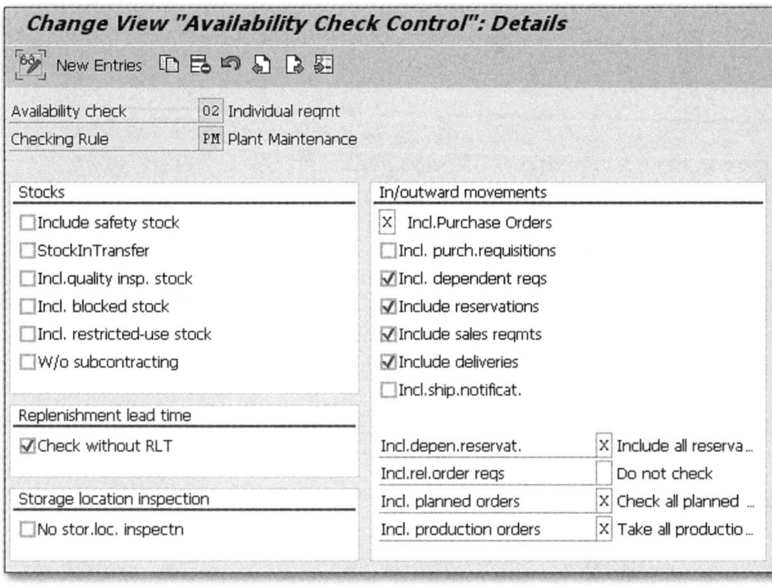

Figure 6.56 Orders: Scope of Availability Check

- The right-hand side of the screen also contains the movement types that cause a planned outward movement (also known as a planned goods issue). Therefore, if, for example, you want to take into account reservations whose goods issue is to be expected up to the requirements date of the material in the order, set the INCLUDE RESERVATIONS indicator (i.e., reservations that were manually created using Transaction MB21) and INCL.DEPEN.RESERVAT. (include dependent reservations, i.e., reservations that were created as a result of another order).
- The CHECK WITHOUT RLT option specifies whether or not the system considers the *replenishment lead time* during the availability check. Materials whose requirements date lies within the replenishment lead time, but for which there is no stock, are confirmed after the replenishment lead time. In the case of materials whose delivery date lies outside the replenishment lead time, the system assumes that a complete confirmation is possible. If you want the system to take the replenishment lead time into account, you must leave this field blank in the standard system.
- You can set the NO STOR.LOC. INSPECTN (no storage location inspection) indicator if, for this scope of check, you want to deactivate the availability check at the storage location level. In other words, the check is performed at the plant level; that is, across all storage locations.

[+]

Settings for the Scope of Check

If you want to perform a reliable material availability check, proceed as follows:

1. Deactivate all stock types.
2. Activate all requirements that result in a planned goods issue (e.g., reservations).
3. Deactivate all requirements that result in a planned goods receipt (e.g., purchase requisitions).
4. Deactivate the CHECK WITHOUT RLT and NO STOR.LOC. INSPECTN indicators.

Customizing Functions for the Electronic Parts Catalogs

You can connect electronic parts catalogs to SAP ERP and thereby provide support for the procurement of spare parts. If you connect more than one catalog, the system always navigates directly to the catalog that you select. There, you select the parts you require. You then use the ADD function to place them in your shopping basket and use the CHECK OUT function to transfer them to your SAP system (see Figure 6.57).

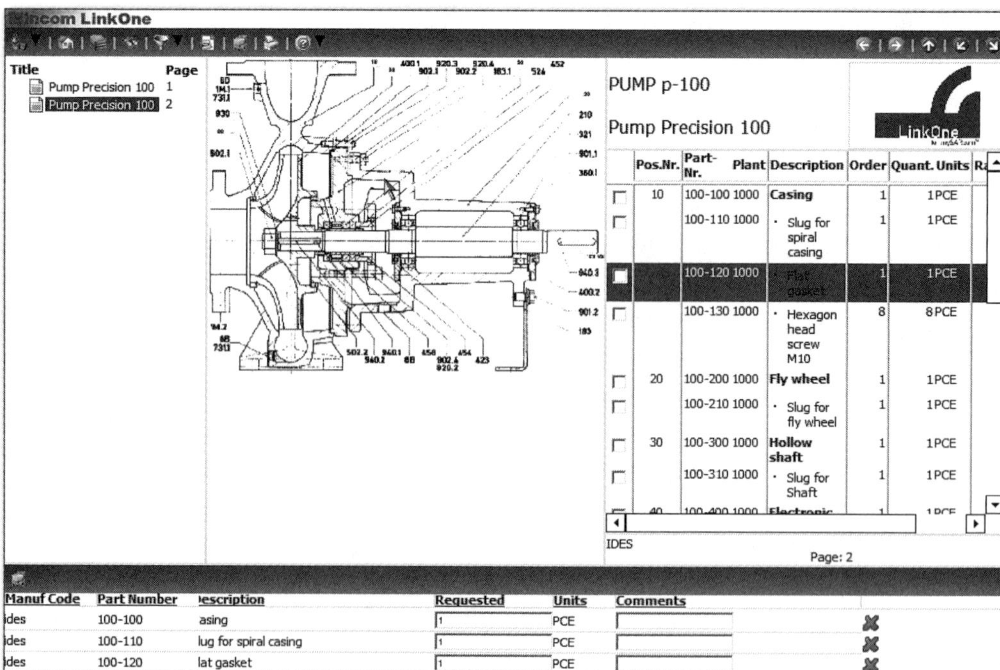

Figure 6.57 Orders: Electronic Parts Catalog

[+]

Activate Enterprise Extension EA-PLM

If you want to use the catalog connection, you must first use the switch framework (Transaction SFW5) to activate *Enterprise Extension EA-PLM*.

The Customizing function INTERFACE FOR PROCUREMENT USING CATALOGS (OCI) won't appear in the menu until you activate Enterprise Extension EA-PLM.

Now we'll describe how to configure the necessary settings for making electronic parts catalogs available for material planning within an order.

Define Catalogs

You use this Customizing function to define all catalogs and their call structure.

Prerequisites

You must activate Enterprise Extension EA-PLM beforehand.

Customizing Path

PLANT MAINTENANCE AND CUSTOMER SERVICE • MAINTENANCE AND SERVICE PROCESSING • MAINTENANCE AND SERVICE ORDERS • INTERFACE FOR PROCUREMENT USING CATALOGS (OCI) • DEFINE CATALOGS

Settings or Recommended Settings

First, define the CATALOG ID and CATALOG NAME (see Figure 6.58).

If you want to call a catalog from material planning within an order, you must define its internal call structure (i.e., the requisite data for addressing the catalog).

Note

Because the internal call structure of a spare parts catalog is always catalog-specific, ask the catalog provider for the catalog's internal call structure.

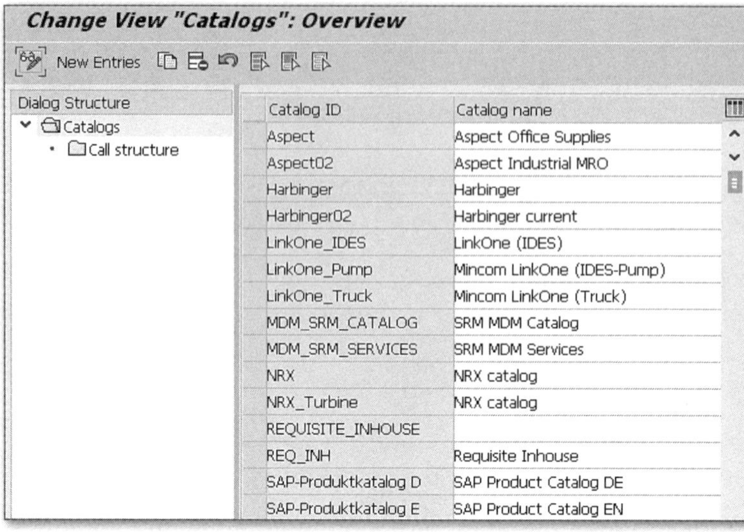

Figure 6.58 Orders – Catalogs: Overview

This generally comprises information such as the following (see Figure 6.59):

- URL
- Internal user name
- Password
- Language

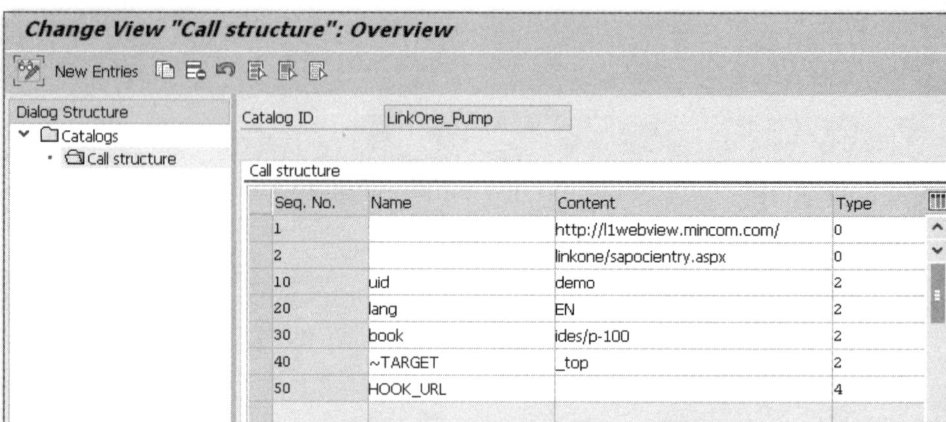

Figure 6.59 Orders – Catalogs: Detail Screen

Assign Catalog to Order Type

You use this Customizing function to define which catalogs are to be used in which plants and for which order types. Several catalogs are permitted for each order type.

Prerequisites

You must activate Enterprise Extension EA-PLM and define the catalogs, order types, and plants beforehand.

Customizing Path

Plant Maintenance and Customer Service • Maintenance and Service Processing • Maintenance and Service Orders • Interface for Procurement Using Catalogs (OCI) • Assign Catalog to Order Type

Settings or Recommended Settings

Define which catalogs are to be used for each order type and plant. As you can see in Figure 6.60, this concerns an *n:m* relationship. In other words, you can assign several catalogs to each order type for each plant, and you can use the same catalog in several order type/plant combinations.

Change View "OCI Interface: Control Parameter PM/CS Order"

New Entries

OCI Interface: Control Parameter PM/CS Order

Type	PlPl	Seq...	Catalog ID	Organization
PM01	2800	0	LinkOne_Pump	
PM01	2800	1	SAPCATALOG	
PM01	2821	0	SAP-Produktkatalog	
PM01	2821	1	SAP_CCM	
PM01	2821	5	SAP_CCM	
PM01	3000	0	LinkOne_Pump	
PM01	3000	1	SRM_MDM_CATALOG	
PM01	3150	0	LinkOne_Pump	

Figure 6.60 Orders: Catalogs, Plant, and Order Type

Convert HTML Fields to SAP Fields

You use this Customizing function to define how the HTML fields from the catalog are assigned to the fields in SAP EAM.

Prerequisites

You must activate Enterprise Extension EA-PLM beforehand.

Customizing Path

Plant Maintenance and Customer Service • Maintenance and Service Processing • Maintenance and Service Orders • Interface for Procurement Using Catalogs (OCI) • Convert HTML Fields to SAP Fields

Settings or Recommended Settings

Table 6.3 shows the HTML fields and an explanation in relation to how they are defined for the Open Catalog Interface (OCI).

HTML Field	Explanation
NEW_ITEM-MATNR	Material number
NEW_ITEM-VENDORMAT	Material number for the vendor
NEW_ITEM-VENDOR	Vendor number
NEW_ITEM-DESCRIPTION	Description of the item to be ordered
NEW_ITEM-MATGROUP	Material group
NEW_ITEM-CONTRACT	Contract number
NEW_ITEM-CONTRACT_ITEM	Contract item
NEW_ITEM-CURRENCY	Currency, which is required when a price is returned, otherwise, optional; must have the standard ISO code for currencies
NEW_ITEM-PRICEUNIT	Number of units that must be ordered at a certain price
NEW_ITEM-PRICE	Price according to the catalog
NEW_ITEM-QUANTITY	Required amount
NEW_ITEM-UNIT	Unit of measure (in standard ISO code)
NEW_ITEM-MANUFACTCODE	Manufacturer number
NEW_ITEM-MANUFACTMAT	Product number of manufacturer
NEW_ITEM-LEADTIME	Delivery time in days

Table 6.3 HTML Fields for OCI

HTML Field	Explanation
NEW_ITEM-SERVICE	Required if services are to also be copied from the catalog
NEW_ITEM-CUST_FIELD1	Customer-specific field
NEW_ITEM-CUST_FIELD2	Customer-specific field
NEW_ITEM-CUST_FIELD3	Customer-specific field
NEW_ITEM-CUST_FIELD4	Customer-specific field
NEW_ITEM-CUST_FIELD5	Customer-specific field

Table 6.3 HTML Fields for OCI (Cont.)

Table 6.4 shows which fields are copied to the target fields of the component in the order.

SAP Target Field	SAP Receiver Structure Field	HTML Field
RESBD-MATNR	RIHFCOM_XL-MATNR	NEW_ITEM-MATNR
RESBD-IDNLF	RIHFCOM_XL-IDNLF	NEW_ITEM-VENDORMAT
RESBD-POSTP	RIHFCOM_XL-POSTP	
RESBD-POTX1	RIHFCOM_XL-KTEXT	NEW_ITEM-DESCRIPTION
RESBD-WERKS	RIHFCOM_XL-WERKS	
RESBD-BDMNG	RIHFCOM_XL-MENGE	NEW_ITEM-QUANTITY
RESBD-MENGE	RIHFCOM_XL-MENGE	NEW_ITEM-QUANTITY
RESBD-ERFMG	RIHFCOM_XL-MENGE	NEW_ITEM-QUANTITY
RESBD-ERFME	RIHFCOM_XL-MEINH	NEW_ITEM-UNIT
RESBD-EINHEIT	RIHFCOM_XL-MEINH	NEW_ITEM-UNIT
RESBD-LIFNR	RIHFCOM_XL-FLIEF	NEW_ITEM-VENDOR
RESBD-INFNR	RIHFCOM_XL-INFNR	
RESBD-KONNR	RIHFCOM_XL-CONTRACT	NEW_ITEM-CONTRACT
RESBD-KTPNR	RIHFCOM_XL-CONTRACT_ITEM	NEW_ITEM-CONTRACT_ITEM

Table 6.4 HTML Fields and SAP Fields

SAP Target Field	SAP Receiver Structure Field	HTML Field
RESBD-MFRPN	RIHFCOM_XL-MFRPN	NEW_ITEM-MANUFACTMAT
RESBD-MFRNR	RIHFCOM_XL-MFRNR	NEW_ITEM-MANUFACTCODE
RESBD-LIFZT	RIHFCOM_XL-LEADTIME	NEW_ITEM-LEADTIME
RESBD-MATKL	RIHFCOM_XL-MATKL	NEW_ITEM-MATGROUP
RESBD-EKGRP	RIHFCOM_XL-EKGRP	
RESBD-EKORG	RIHFCOM_XL-EKORG	
RESBD-GPREIS	RIHFCOM_XL-PREIS	NEW_ITEM-PRICE
RESBD-PEINH	RIHFCOM_XL-PEINH	NEW_ITEM-PRICEUNIT
RESBD-SAKNR	RIHFCOM_XL-SAKTO	
RESBD-CATALOGID	RIHFCOM_XL-CATALOGID	
RESBD-CATALOG_PROD_ID	RIHFCOM_XL-CATALOG_PROD_ID	

Table 6.4 HTML Fields and SAP Fields (Cont.)

Figure 6.61 shows an example of such an assignment.

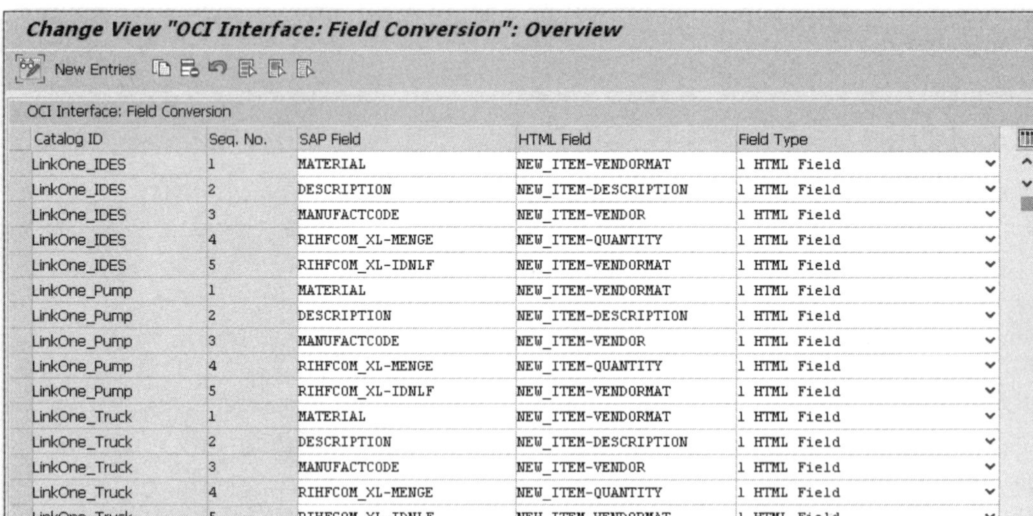

Change View "OCI Interface: Field Conversion": Overview

New Entries

OCI Interface: Field Conversion

Catalog ID	Seq. No.	SAP Field	HTML Field	Field Type
LinkOne_IDES	1	MATERIAL	NEW_ITEM-VENDORMAT	1 HTML Field
LinkOne_IDES	2	DESCRIPTION	NEW_ITEM-DESCRIPTION	1 HTML Field
LinkOne_IDES	3	MANUFACTCODE	NEW_ITEM-VENDOR	1 HTML Field
LinkOne_IDES	4	RIHFCOM_XL-MENGE	NEW_ITEM-QUANTITY	1 HTML Field
LinkOne_IDES	5	RIHFCOM_XL-IDNLF	NEW_ITEM-VENDORMAT	1 HTML Field
LinkOne_Pump	1	MATERIAL	NEW_ITEM-VENDORMAT	1 HTML Field
LinkOne_Pump	2	DESCRIPTION	NEW_ITEM-DESCRIPTION	1 HTML Field
LinkOne_Pump	3	MANUFACTCODE	NEW_ITEM-VENDOR	1 HTML Field
LinkOne_Pump	4	RIHFCOM_XL-MENGE	NEW_ITEM-QUANTITY	1 HTML Field
LinkOne_Pump	5	RIHFCOM_XL-IDNLF	NEW_ITEM-VENDORMAT	1 HTML Field
LinkOne_Truck	1	MATERIAL	NEW_ITEM-VENDORMAT	1 HTML Field
LinkOne_Truck	2	DESCRIPTION	NEW_ITEM-DESCRIPTION	1 HTML Field
LinkOne_Truck	3	MANUFACTCODE	NEW_ITEM-VENDOR	1 HTML Field
LinkOne_Truck	4	RIHFCOM_XL-MENGE	NEW_ITEM-QUANTITY	1 HTML Field
LinkOne_Truck	5	RIHFCOM_XL-IDNLF	NEW_ITEM-VENDORMAT	1 HTML Field

Figure 6.61 Orders: Catalogs for HTML Fields

Convert HTML Field Values

You use this Customizing function to define which standard SAP values are used to overwrite the values from the catalog.

Prerequisites

You must activate Enterprise Extension EA-PLM and define the units of measurement beforehand.

Customizing Path

PLANT MAINTENANCE AND CUSTOMER SERVICE • MAINTENANCE AND SERVICE PROCESSING • MAINTENANCE AND SERVICE ORDERS • INTERFACE FOR PROCUREMENT USING CATALOGS (OCI) • CONVERT HTML FIELD VALUES

Settings or Recommended Settings

Make corresponding entries in terms of which catalog values are to be overwritten with fixed SAP values. If, for example, this concerns an American catalog, the unit of measure EA or the currency $ is possibly used in the catalog. However, you can overwrite these with the SAP values ST and USD (see Figure 6.62).

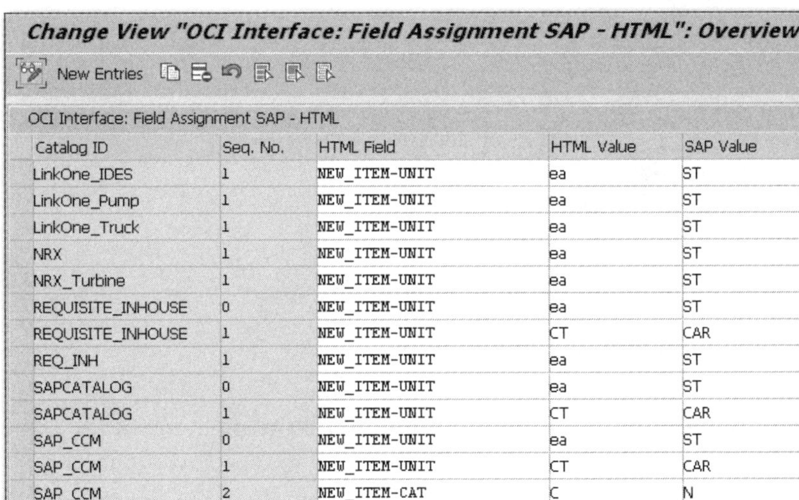

Change View "OCI Interface: Field Assignment SAP - HTML": Overview

New Entries

OCI Interface: Field Assignment SAP - HTML

Catalog ID	Seq. No.	HTML Field	HTML Value	SAP Value
LinkOne_IDES	1	NEW_ITEM-UNIT	ea	ST
LinkOne_Pump	1	NEW_ITEM-UNIT	ea	ST
LinkOne_Truck	1	NEW_ITEM-UNIT	ea	ST
NRX	1	NEW_ITEM-UNIT	ea	ST
NRX_Turbine	1	NEW_ITEM-UNIT	ea	ST
REQUISITE_INHOUSE	0	NEW_ITEM-UNIT	ea	ST
REQUISITE_INHOUSE	1	NEW_ITEM-UNIT	CT	CAR
REQ_INH	1	NEW_ITEM-UNIT	ea	ST
SAPCATALOG	0	NEW_ITEM-UNIT	ea	ST
SAPCATALOG	1	NEW_ITEM-UNIT	CT	CAR
SAP_CCM	0	NEW_ITEM-UNIT	ea	ST
SAP_CCM	1	NEW_ITEM-UNIT	CT	CAR
SAP_CCM	2	NEW_ITEM-CAT	C	N

Figure 6.62 Orders: HTML Field Values

Define Conversion Modules

You can use this Customizing function to define conversion modules for converting the data from the catalog into valid SAP values.

Prerequisites

You must activate Enterprise Extension EA-PLM and create and program the conversion modules beforehand.

Customizing Path

Plant Maintenance and Customer Service • Maintenance and Service Processing • Maintenance and Service Orders • Interface for Procurement Using Catalogs (OCI) • Define Conversion Modules

Settings or Recommended Settings

Enter which function module is to be processed for which catalog and in which order (see Figure 6.63).

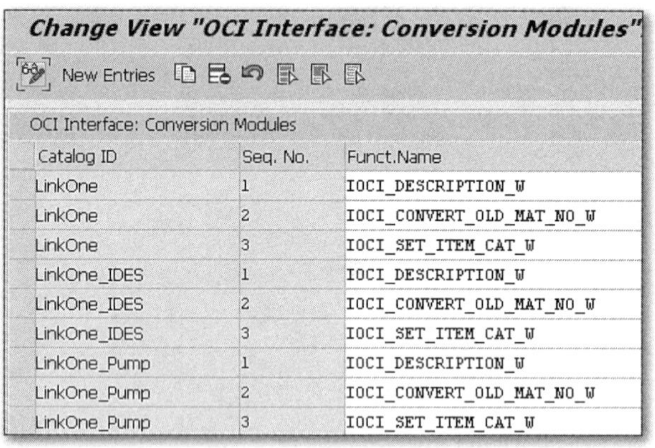

Catalog ID	Seq. No.	Funct.Name
LinkOne	1	IOCI_DESCRIPTION_W
LinkOne	2	IOCI_CONVERT_OLD_MAT_NO_W
LinkOne	3	IOCI_SET_ITEM_CAT_W
LinkOne_IDES	1	IOCI_DESCRIPTION_W
LinkOne_IDES	2	IOCI_CONVERT_OLD_MAT_NO_W
LinkOne_IDES	3	IOCI_SET_ITEM_CAT_W
LinkOne_Pump	1	IOCI_DESCRIPTION_W
LinkOne_Pump	2	IOCI_CONVERT_OLD_MAT_NO_W
LinkOne_Pump	3	IOCI_SET_ITEM_CAT_W

Figure 6.63 Orders: Catalogs of Conversion Modules

Table 6.5 shows the conversion modules delivered with the interface in the standard SAP system.

Name	Explanation
IOCI_CONVERT_INFO_REC_W	Converts the catalog material number to an SAP material number in accordance with the purchasing info record
IOCI_CONVERT_MPM_W	Converts the catalog material number to an SAP material number in accordance with the manufacturer number
IOCI_CONVERT_OLD_MAT_NO_W	Converts to an existing material number using the OLD MATERIAL NUMBER field
IOCI_DESCRIPTION_W	Copies the catalog material number and description to the SAP field for the short text
IOCI_SET_ITEM_CAT_W	Searches for the item category in accordance with the Customizing settings for the material type
IOCI_LONGTEXT_W	Copies the catalog long text to the material long text or the service long text
IOCI_EXAMPLE_W	Copies the template for customer-specific conversion modules

Table 6.5 OCI Conversion Modules

However, you can program your own conversion modules at any time. You can use the function module IOCI_EXAMPLE_W as a template for your own conversion modules.

To use the supplied conversion modules properly, proceed as follows:

1. Call the Customizing function CONVERT HTML FIELDS TO SAP FIELDS.
2. In the SAP FIELD column, specify the fields listed in Table 6.6.
3. Fill the columns with the catalog fields you require. Table 6.6 shows which field is used in which conversion module.

Conversion Module	SAP Field 1	SAP Field 2
IOCI_CONVERT_INFO_REC_W	VENDOR	VENDORMAT
IOCI_CONVERT_MPM_W	VENDOR	VENDORMAT
IOCI_CONVERT_OLD_MAT_NO_W	MATERIAL	
IOCI_DESCRIPTION_W	MATERIAL	DESCRIPTION
IOCI_LONGTEXT_W	LONGTEXT	

Table 6.6 Fields in the Conversion Modules

[+]

Use Function Modules Purposefully

To receive the right data at the right location, use the function modules in a purposeful manner. For example, you should always process the IOCI_SET_ITEM_CAT_W function module because it determines whether or not a reservation or purchase requisition is triggered.

Other Customizing Functions for Material Planning

In addition to the preceding groups (availability check and electronic parts catalogs), there are still other Customizing functions that you can use to influence or support material planning within an order, as described next.

Define Default Values for Component Item Categories

If you plan a component in an order, you must assign an item category. The most frequently used item categories in plant maintenance are L (stock material) and N (nonstock material).

You can use this Customizing function to define which item category is to be proposed for a component for each plant and material type. If you don't assign a default value, you must enter the item category manually.

Prerequisites

You must define the plants, material types, and item categories beforehand.

Customizing Path

PLANT MAINTENANCE AND CUSTOMER SERVICE • MAINTENANCE AND SERVICE PROCESSING • MAINTENANCE AND SERVICE ORDERS • GENERAL DATA • DEFINE DEFAULT VALUES FOR COMPONENT ITEM CATEGORIES

Settings or Recommended Settings

If, for example, you want to trigger a reservation for spare parts (material type ERSA), assign item category L (see Figure 6.64). If you want to generate a purchase requisition for all nonstock materials (material type NLAG), assign item category N.

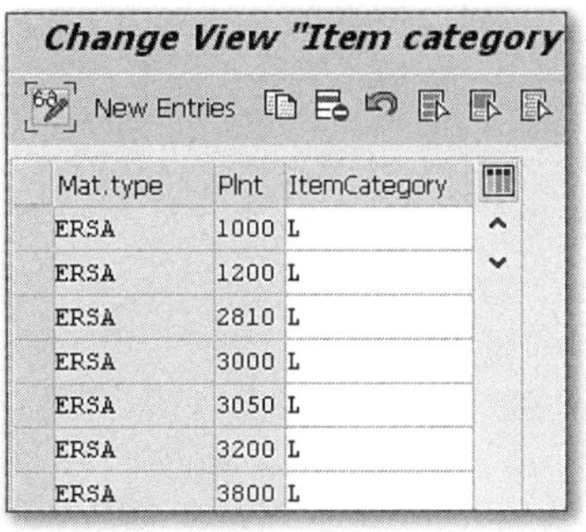

Change View "Item category

New Entries

Mat.type	Plnt	ItemCategory
ERSA	1000	L
ERSA	1200	L
ERSA	2810	L
ERSA	3000	L
ERSA	3050	L
ERSA	3200	L
ERSA	3800	L

Figure 6.64 Orders: Default Values for Item Category

"Blank" Entry for Components without a Material Number [+]

You should always include an entry without a material type ("blank" value for the material type). This entry is always used if you want to procure a component without a material number within the order. You should assign the item category for nonstock components to this entry. Usually, this is the value N.

Determine Movement Types for Material Reservations

During material planning, reservations are created for components that can be kept in stock. Such *reservations* receive a movement type for the material withdrawal. You use this Customizing function to define which movement types are to be entered.

Prerequisites

You must define the movement types beforehand.

Customizing Path

Plant Maintenance and Customer Service • Maintenance and Service Processing • Maintenance and Service Orders • General Data • Define Movement Types for Material Reservations

Settings or Recommended Settings

In this table, the entry IW01 (Appl. Development R/3 Plant Maintenance Order Processing; see Figure 6.65) is crucial here.

For normal maintenance orders, only the entries Movement Type Goods Issue and Mvt Type Reversal of Goods Issue are used with standard movement types 261 and 262, respectively.

For refurbishment orders, you also require the entries Movement Type Goods Receipt and Mvt Type Reversal of Goods Receipt with standard movement types 101 and 102, respectively.

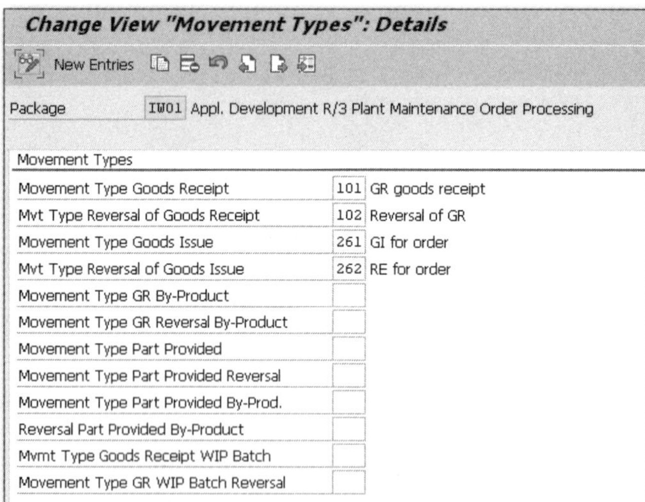

Figure 6.65 Orders: Movement Types

[+]

No Change to Movement Types Necessary

Generally, you can use the standard entries delivered by SAP. You only need to adjust them if you've defined your own movement types.

Note: The entry applies to all orders. Unfortunately, it's not possible to differentiate according to order type or plant.

Define Account Assignment Category and Document Type for Purchase Requisitions

You use this Customizing function to define the *account assignment category* and the document type for purchase requisitions created from orders.

Prerequisites

You must define the document type and account assignment category beforehand.

Customizing Path

PLANT MAINTENANCE AND CUSTOMER SERVICE • MAINTENANCE AND SERVICE PROCESSING • MAINTENANCE AND SERVICE ORDERS • GENERAL DATA • DEFINE ACCOUNT ASSIGNMENT CATEGORY AND DOCUMENT TYPE FOR PURCHASE REQUISITIONS

Settings or Recommended Settings

For maintenance orders, only the entries DOCUMENT TYPE and ACCT.ASSGMT.CAT. GEN. (general account assignment category) are used with the standard entries NB (PURCH.REQUIS. STAND.) and F (ORDER). See Figure 6.66.

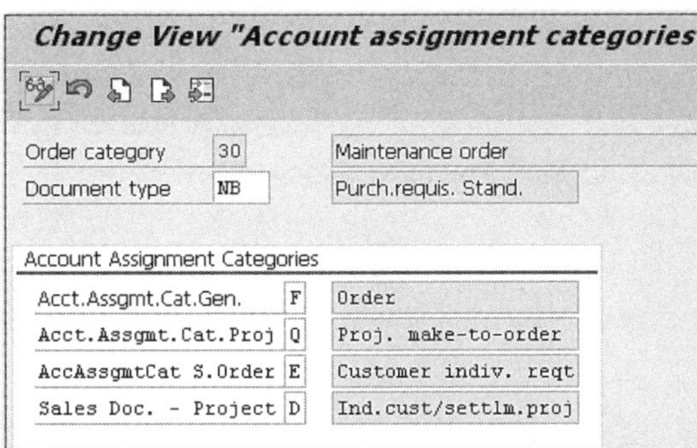

Figure 6.66 Orders: Account Assignment Category and Document Type for Purchase Requisitions

[+]

No Change to Document Type and Account Assignment Category Necessary

Generally, you can use the standard entries delivered by SAP. You only need to adjust them if you've defined your own document types and account assignment categories.

Note: The entry applies to all orders. Unfortunately, it's not possible to differentiate according to order type or plant.

Create Default Value Profiles for External Procurement

You use this Customizing function to define profiles for the external procurement of services and materials. In the profile, you can define default values for generating purchase requisitions from orders. In each profile, you can define default values for the following:

- Cost element
- Purchasing organization
- Purchasing group
- Material group

For external procurement, it's always necessary to differentiate between the external procurement of materials and the external procurement of services.

Prerequisites

You must define the requisite values for the cost element, purchasing organization, purchasing group, and material group beforehand.

Customizing Path

PLANT MAINTENANCE AND CUSTOMER SERVICE • MAINTENANCE AND SERVICE PROCESSING • MAINTENANCE AND SERVICE ORDERS • FUNCTIONS AND SETTINGS FOR ORDER TYPES • CREATE DEFAULT VALUE PROFILES FOR EXTERNAL PROCUREMENT

Settings or Recommended Settings

Make all of the entries that are to be copied to purchase requisitions later. Figure 6.67 shows some examples.

Then, use the Customizing function DEFAULT VALUES FOR TASK LIST DATA AND PROFILE ASSIGNMENTS (see Section 6.2.9) to make the relevant assignment for each order type and plant.

Define Access Sequence for Determining Address Data

You can assign an address other than the plant address to the purchase requisitions generated from the orders. You can assign an address to maintenance objects (i.e., a piece of equipment, functional location, notification, or order).

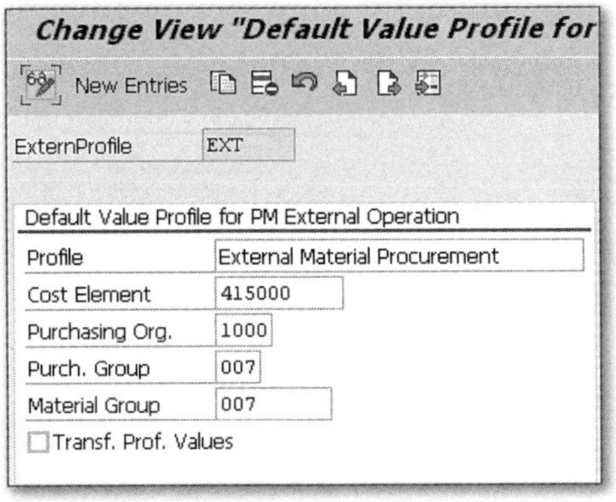

Figure 6.67 Orders: Default Value Profiles for External Procurement

You use this Customizing function to define, for each plant and order type, the sequence in which the delivery address of nonstock components is to be determined from the addresses of maintenance objects (see Figure 6.68).

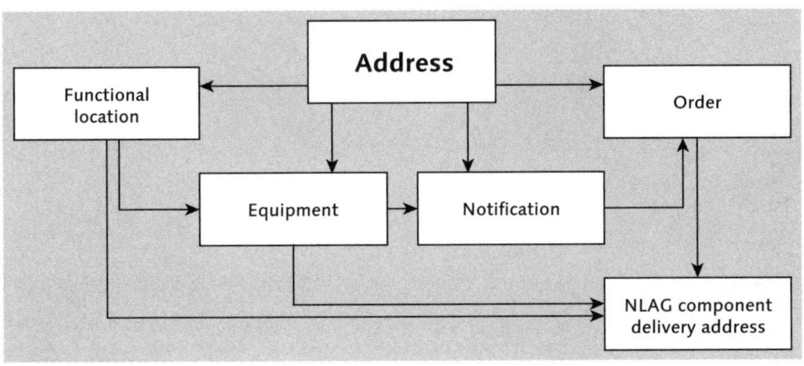

Figure 6.68 Addresses

Prerequisites

You must define the plant and order type beforehand.

Customizing Path

Plant Maintenance and Customer Service • Maintenance and Service Processing • Maintenance and Service Orders • Functions and Settings for Order Types • Define Access Sequence for Determining Address Data

Settings or Recommended Settings

For each order type and plant, define an access sequence with up to five priorities. The following objects are available to you:

1. Order
2. Functional location
3. Equipment
4. Sold-to party (order partner)
5. Ship-to party (order partner)

In fields A1 to A5, define the sequence in which the system is to access the relevant addresses (see Figure 6.69).

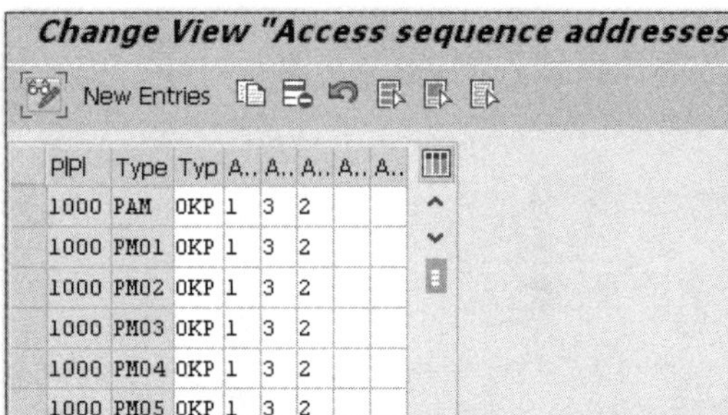

PlPl	Type	Typ	A..	A..	A..	A..	A..
1000	PAM	OKP	1	3	2		
1000	PM01	OKP	1	3	2		
1000	PM02	OKP	1	3	2		
1000	PM03	OKP	1	3	2		
1000	PM04	OKP	1	3	2		
1000	PM05	OKP	1	3	2		

Figure 6.69 Orders: Define Access Sequence for Address Data

[+]

Access Sequences Make Sense if the Delivery Address Differs from the Plant Address

An access sequence always makes sense if the delivery address for the components differs from the plant address (e.g., if objects such as the power supply or the infrastructure measures lie outside the plant premises).

If you operate typical on-site maintenance, you generally don't require access sequences.

Define Change Documents, Collective Purchase Requisition, MRP Relevance

You use this Customizing function to define, for each order type and plant, various order parameters (including parameters that you can use to control material planning).

Prerequisites

You must define the plants and order types beforehand.

Customizing Path

PLANT MAINTENANCE AND CUSTOMER SERVICE • MAINTENANCE AND SERVICE PROCESSING • MAINTENANCE AND SERVICE ORDERS • FUNCTIONS AND SETTINGS FOR ORDER TYPES • DEFINE CHANGE DOCUMENTS, COLLECTIVE PURCHASE REQUISITION, MRP RELEVANCE

Settings or Recommended Settings

Figure 6.70 shows the fields within this table, which you can use to influence material planning. (A description of the other parameters is provided in Section 6.2.6.)

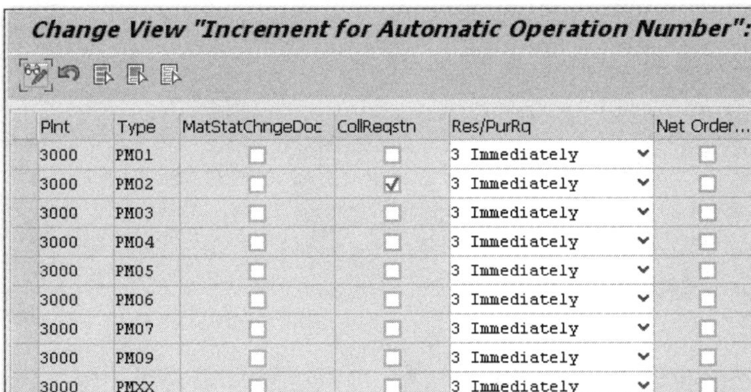
Change View "Increment for Automatic Operation Number":

Plnt	Type	MatStatChngeDoc	CollReqstn	Res/PurRq	Net Order...
3000	PM01	☐	☐	3 Immediately	☐
3000	PM02	☐	☑	3 Immediately	☐
3000	PM03	☐	☐	3 Immediately	☐
3000	PM04	☐	☐	3 Immediately	☐
3000	PM05	☐	☐	3 Immediately	☐
3000	PM06	☐	☐	3 Immediately	☐
3000	PM07	☐	☐	3 Immediately	☐
3000	PM09	☐	☐	3 Immediately	☐
3000	PMXX	☐	☐	3 Immediately	☐

Figure 6.70 Orders: Define Change Documents, Collective Purchase Requisition, MRP Relevance

- Use the MATSTATCHNGEDOC indicator to control whether change documents are to be generated for changes to the material status (e.g., MSPT = Material shortage and MACM = Material committed).
- If you set the COLLREQSTN (collective purchase requisition) indicator, all of the purchase requisition items that result from an order number are grouped together under a single purchase requisition number. If you don't set this indicator, the system generates a separate purchase requisition number for each purchase requisition item.

- Use the Res/PurRq (reservation/purchase requisition) indicator to define the point in time when purchase requisitions or reservations are to become MRP-relevant. The following options are available:
  - 1: Never
  - 2: From Release
  - 3: Immediately
- Use the Net Order Price indicator to specify that the *net price* from the *purchase requisition* must be copied unchanged to the *purchase order* and can't be changed there.

[+]

Note on the "Res/PurRq" Indicator

- Unfortunately, it's not possible to differentiate between reservations and purchase requisitions. Instead, these are always activated simultaneously. In other words, you can't control that purchase requisitions are generated immediately but that reservations don't become MRP-relevant until they are released, for example.
- You can't prevent the system from generating reservations. If you plan a stock material, the system always generates a reservation document even if the indicator is set to 0 (Never) or 2 (From Release). The sole purpose of these indicators is to ensure that the reservation never becomes MRP-relevant or only becomes MRP-relevant after orders have been released.

Define Documentation for Goods Movements for the Order

You use this Customizing function to define which material movements are to be recorded for each order type and plant. You can then view these in the order (either in the document flow or in the documented goods movements).

Prerequisites

You must define the order type and plant beforehand.

Customizing Path

Plant Maintenance and Customer Service • Maintenance and Service Processing • Maintenance and Service Orders • Functions and Settings for Order Types • Goods Movements for Order • Define Documentation for Goods Movements for the Order

Settings or Recommended Settings

You can document the following goods movements for the maintenance orders separately using these indicators (see Figure 6.71):

- PURCHORDER
  Goods receipts for the purchase order.
- PLGOODSISS
  Planned goods issues, that is, goods issues for a reservation.
- UNPLGDSISS
  Unplanned goods issues, that is, goods issues without a reservation.
- GR REFRBSHMNT
  Goods receipts for refurbishment orders.

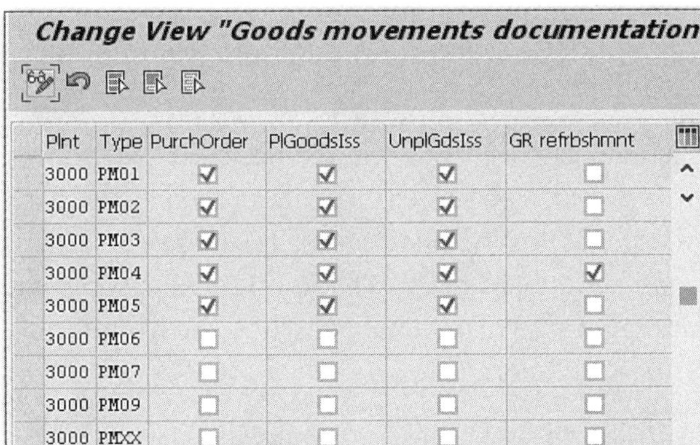

Figure 6.71 Orders: Documentation for Goods Movements

[+]

Notes on Documenting the Goods Movements

- You may be inclined to document all types of material movements. However, this increases the number of documents used.
- Irrespective of these settings, the material movements are always documented in Materials Management (MM), and you can use the MM transactions (e.g., MB51) to display these documents.
- Activate only the material movements that are also to be documented in the order.
- If you change the settings retroactively, you must start the Report RIAUFM00, which recreates the documents for the material movements in maintenance orders.

6.2.3 Scheduling

Scheduling involves calculating the scheduled dates at the operation level and header level, based on the basic dates manually defined in the order and taking into account the durations at the operation level.

[+]

Scheduling Doesn't Always Make Sense

Scheduling in the SAP system only makes sense if you've saved standard times. If you don't have these, you shouldn't activate scheduling initially. You can ensure this by using a control key for which the SCHEDULE option isn't activated.

In Chapter 3, Section 3.4, you learned how to define control keys (see also the Customizing function MAINTAIN CONTROL KEYS and Figure 6.26).

There are many different ways to perform scheduling:

- Lead-time scheduling is executed either as forward scheduling or backward scheduling.
- Both forward and backward scheduling are executed for network scheduling, while taking relationships into account. The basic start date is used as the basis for calculating the earliest scheduled dates, while the basic finish date is used as the basis for calculating the latest scheduled dates. The difference between the earliest date and the latest date gives rise to a *buffer*, which is shown not only for each operation, but also at header level.

To perform scheduling, you must fulfill the following prerequisites:

- The relevant duration must be entered in all operations.
- A control key must be assigned to the operation, and the SCHEDULE option must be activated for this control key.
- A formula must be assigned for the *duration of internal processing* to the work center. This must point to the DAUNO field, that is, the duration from the operation. In the standard system, this is the formula SAP004.
- If you want to perform lead-time scheduling, the Customizing function SET SCHEDULING PARAMETERS must first be used to assign the scheduling type (forwards or backwards) to the order type in the plant.

Figure 6.72 shows you the relationships between scheduling, capacity requirements, and costing, and the procedures associated with each, as described here:

- In Customizing, you define *formula parameters* for the fields DAUNO and ARBEI. (In the standard delivery, these are the formula parameters SAP_10 and SAP_07.)
- In Customizing, you create *formula definitions* for the formula parameters. (In the standard delivery, these are the formulas SAP004 and SAP008.)
- In the work center, you enter the formula for duration in the DURATION OF INTERNAL PROCESSING field on the SCHEDULING tab page.
- In the work center, you enter the formula for work in the FORMULA CONDITION FOR INTERNAL PROCESSING field on the CAPACITIES tab page and in the FORMULA KEY FOR COSTING field on the COSTING tab page.
- You can then schedule the order on the basis of the duration. Furthermore, you can generate capacity requirements on the basis of the work and perform a costing.

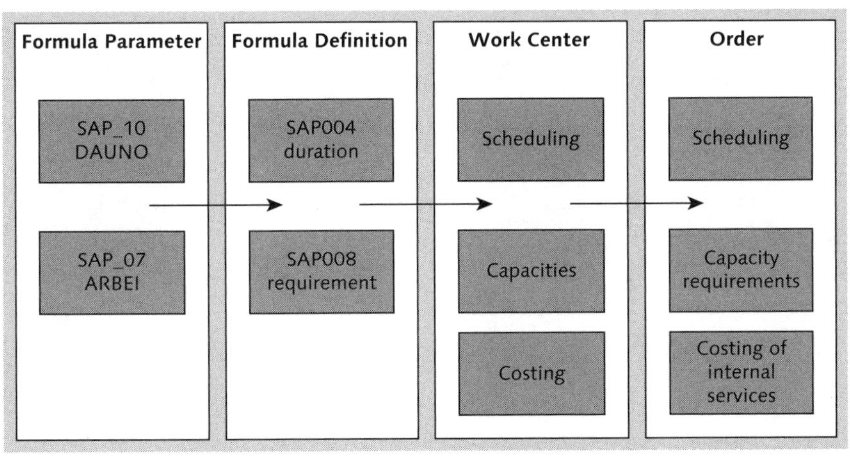

Figure 6.72 Order: Relationships among Scheduling, Capacity Requirements, and Costing

Scheduling Only Partly Available in the Plant Maintenance Menu [!]

For the most part, the Customizing functions for configuring scheduling aren't available in the PLANT MAINTENANCE menu. Therefore, use the Customizing menu for PRODUCTION or simply call the relevant transactions directly.

Define Formula Parameters

You use this Customizing function to define the parameters that you'll later use as a basis for scheduling, costing, and capacity requirements determination. From a maintenance perspective, you require the fields DAUNO (Duration) and ARBEI (Work) for this purpose.

Prerequisites
There are no special prerequisites.

Customizing Path
PRODUCTION • SHOP FLOOR CONTROL • OPERATIONS • SCHEDULING • DEFINE FORMULA PARAMETERS

Transaction
OP7B

Settings or Recommended Settings
In the standard delivery, the formula parameters SAP_07 for ARBEI and SAP_10 for DAUNO are available in Customizing for this purpose (see Figure 6.73).

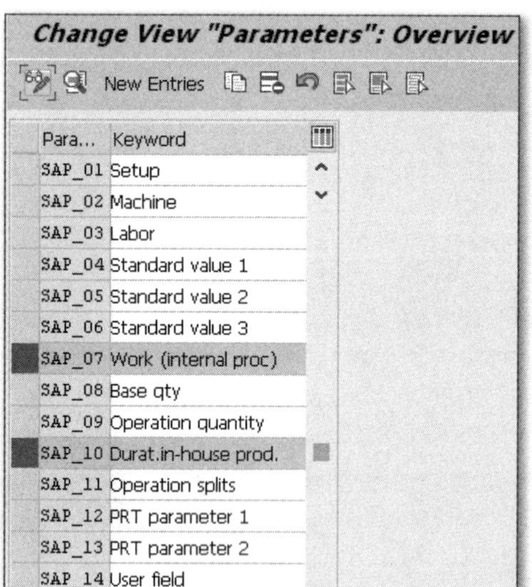

Figure 6.73 Orders: Define Formula Parameters for Scheduling

[+]

Use the Formula Parameters in the Standard SAP System

Generally, you can use the formula parameters delivered in the standard SAP system without having to make any adjustments to them. You only require other formula parameters if you want to define your own calculations.

Configure Formula Definition

You use this Customizing function to define the formulas used to calculate the results on the basis of formula parameters that contain mathematical functions, namely, for the following:

- Capacity planning
- Scheduling
- Costing

Prerequisites
You must define the formula parameters beforehand.

Customizing Path
PRODUCTION • SHOP FLOOR CONTROL • OPERATIONS • SCHEDULING • DEFINE FORMULAS

Transaction
OPK3

Settings or Recommended Settings
In the standard delivery, the formula definition SAP004 for the duration and the formula definition SAP008 for the requirement are defined for plant maintenance (see Figure 6.74).

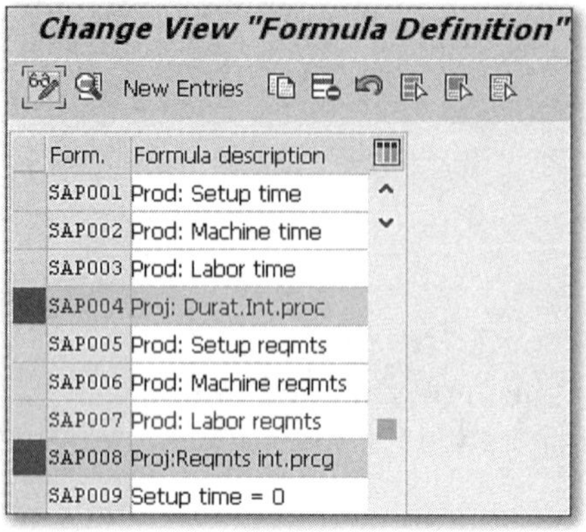

Figure 6.74 Orders: Define Scheduling Formulas

The formula key SAP004 (see Figure 6.75), which calculates the resulting lead time, is based on the formula parameter SAP_10 and therefore on the field DAUNO (Duration). It's important that you set the Allowed for Scheduling indicator here.

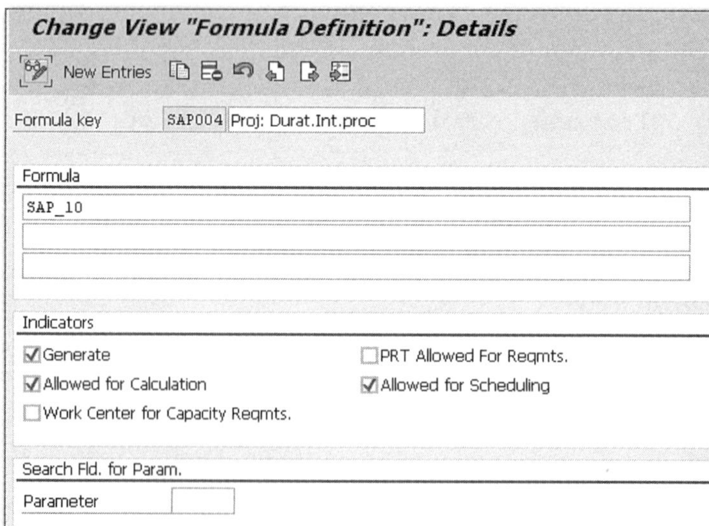

Figure 6.75 Orders: Formula Definition for Duration

The formula key SAP008 (see Figure 6.76) is based on the formula parameter SAP_07 and therefore on the field ARBEI (Work). It can calculate capacity requirements and costs for internal services. It's important that you set the Allowed for Calculation and Work Center for Capacity Reqmts. indicators here.

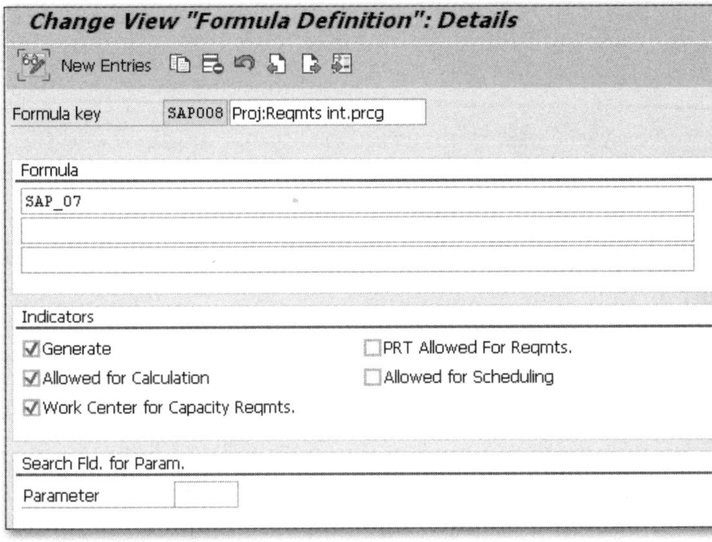

Figure 6.76 Orders: Formula Definition for Requirements

Use the Formula Definitions in the Standard SAP System [+]

Generally, you can use the standard SAP definitions for formula definitions to schedule orders, calculate costs, and generate capacity requirements. You only need to change the standard SAP definitions if you want to perform your own calculations.

Specify Scheduling Type

You use this Customizing function to define the scheduling types that you want to use. In the standard delivery, SAP supports the following scheduling types (see Figure 6.77):

- FORWARDS
  The system uses the basic start date and adds the operation durations to calculate the *earliest scheduled start and end dates* at both header level and operation level.
- FORWARDS WITH TIME ENTRY
  This is the same as FORWARDS, except that you specify the required start time in addition to the basic start date.
- BACKWARDS
  The system uses the basic finish date and subtracts the operation durations to

calculate the *latest scheduled start and end dates* at both header level and operation level.

- Backwards in time
  This is the same as Backwards, except that you specify the required end time in addition to the basic finish date.
- Current date
  This is the same as Forwards with time entry, except that the basic start date and time are the current date and current time.
- Only capacity requirements
  The scheduled dates are equated with the basic dates, and the work is distributed evenly. Example: You have an order with the basic start date 11.01., and the basic finish date 11.10. For the operation, you've scheduled 40 hours of work. Result: From 11.01. to 11.10., a capacity requirement of four hours is allocated to each day.
- Do not schedule

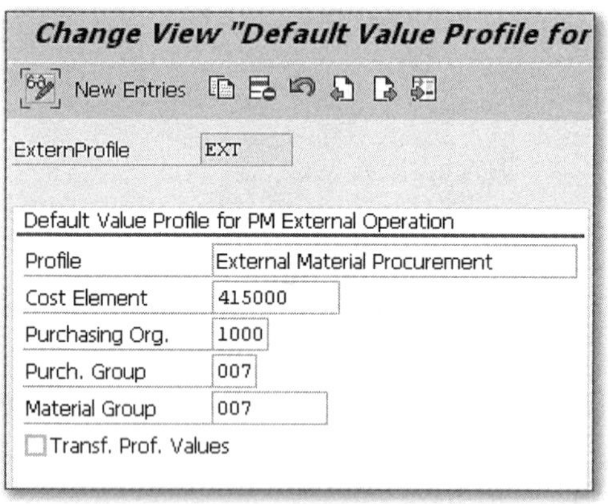

Figure 6.77 Orders: Scheduling Types

Prerequisites

There are no special prerequisites.

Customizing Path

Production • Shop Floor Control • Operations • Scheduling • Specify Scheduling Type

Transaction
OPJN

Settings or Recommended Settings
Figure 6.78 shows the table of scheduling types delivered by SAP.

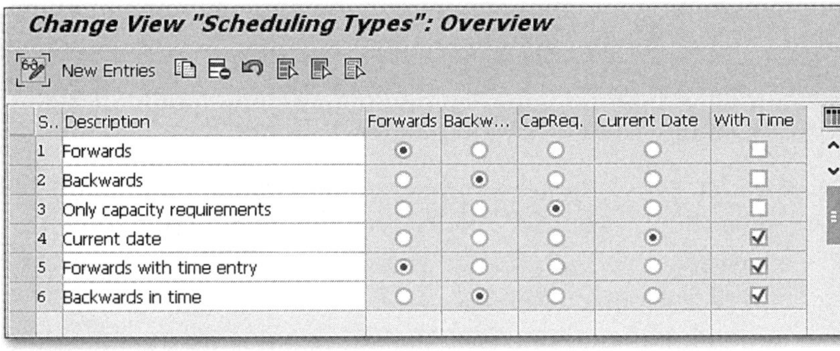
Change View "Scheduling Types": Overview

New Entries

S..	Description	Forwards	Backw...	CapReq.	Current Date	With Time
1	Forwards	◉	○	○	○	☐
2	Backwards	○	◉	○	○	☐
3	Only capacity requirements	○	○	◉	○	☐
4	Current date	○	○	○	◉	☑
5	Forwards with time entry	◉	○	○	○	☑
6	Backwards in time	○	◉	○	○	☑

Figure 6.78 Orders: Specify Scheduling Types

All Scheduling Types Available [+]

Generally, you can use the standard SAP definitions for scheduling types. You only need to change the standard SAP definitions if you want to exclude scheduling types or change their descriptions.

Set Scheduling Parameters

You use this Customizing function to define, for each order type and plant, which scheduling procedure is to be used.

Production Scheduler = * [+]

When you assign the scheduling parameters to the order type and plant, the production scheduler must always be * because plant maintenance doesn't use the production scheduler. Only production uses the production scheduler.

Prerequisites
You must define the scheduling type, order type, and plant beforehand.

Customizing Path

Plant Maintenance and Customer Service • Maintenance and Service Processing • Maintenance and Service Orders • Scheduling • Set Scheduling Parameters

Transaction

OPU7

Settings or Recommended Settings

This Customizing function contains the following settings (see Figure 6.79):

- Use the Adjust Dates option to specify whether, as a result of scheduling, the basic dates are to be automatically adjusted to the scheduled dates and whether the requirements dates for the material components are to be based on the basic start date or the respective operation dates.
- For the Scheduling Type, specify the preferred scheduling type for the plant and order type. This scheduling type is then copied to the order as a default value. However, it can be changed there.
- Automatic Scheduling ensures that an order is automatically scheduled when you choose Save.

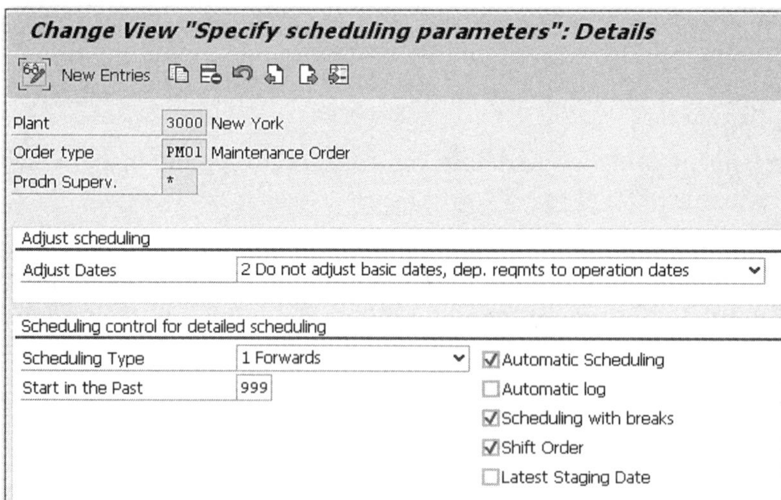

Figure 6.79 Orders: Maintain Scheduling Types

- AUTOMATIC LOG ensures that a scheduling log is automatically displayed after scheduling.
- Use the SCHEDULING WITH BREAKS indicator to determine that break times are to be taken into careful consideration. If you set this indicator, a date calculated by the system can no longer occur during a break time.
- Use the SHIFT ORDER indicator to specify that, in the case of partially confirmed operations, existing actual dates aren't taken into account when scheduling is repeated.
- The LATEST STAGING DATE indicator only plays a role in network scheduling. Use it to specify whether the requirements date for the material components is to be based on the earliest or latest dates.

Scheduling Recommendations [+]

- Set the ADJUST DATES indicator to DO NOT ADJUST BASIC DATES. Otherwise, during scheduling, the basic dates that you preassigned manually are overwritten by the scheduled dates. In other words, they are lost and can't be restored.
- In plant maintenance, FORWARD SCHEDULING is the normal scenario for lead-time scheduling.
- If you haven't deactivated scheduling, you should set the AUTOMATIC SCHEDULING indicator.
- To ensure that the system doesn't display the scheduling results in irritating pop-up logs each time you save the order, deactivate the AUTOMATIC LOG indicator.
- To obtain as up-to-date scheduling data as possible, set the SCHEDULING WITH BREAKS indicator.
- To ensure that scheduling is adjusted to the actual situation, set the SHIFT ORDER indicator.
- Because the earliest material requirements dates should always be used in the order, don't set the LATEST STAGING DATE indicator.

Activate Default Value for Current Date as Basic Date

You use this Customizing function to specify, for each plant, whether the current date is to be proposed as the basic date.

Prerequisites

You must define the plants beforehand.

Customizing Path

Plant Maintenance and Customer Service • Maintenance and Service Processing • Maintenance and Service Orders • General Data • Activate Default Value for Current Date as Basic Date

Settings or Recommended Settings

Select the plants for which the current date is to be proposed as the basic date (see Figure 6.80).

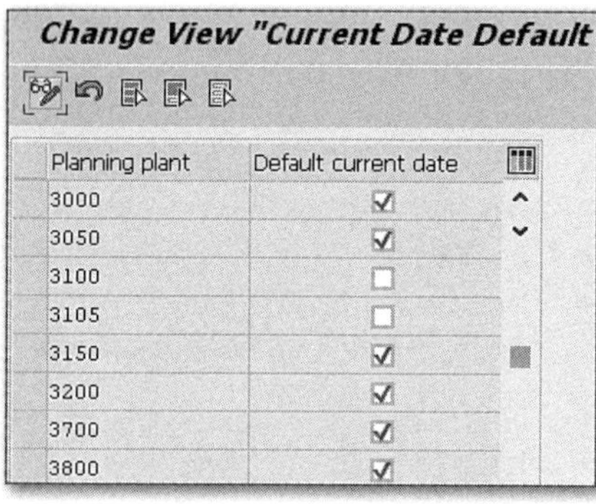

Figure 6.80 Orders: Default Current Date

Define Default Values for Units for Operation

You use this Customizing function to define, for each plant, which units of measurement are to be proposed for the Work and Duration fields.

Prerequisites

You must define the plants and units of measurement beforehand.

Customizing Path

Plant Maintenance and Customer Service • Maintenance and Service Processing • Maintenance and Service Orders • General Data • Define Default Values for Units for Operation

Transaction

OIO9

Settings or Recommended Settings

Define which units of measurement are to be proposed for the WORK and DURATION fields for each plant (see Figure 6.81). If, however, the work center for the operation contains different units of measurement as default values, the default values for the plant are overwritten. If necessary, you can always overwrite these default values manually.

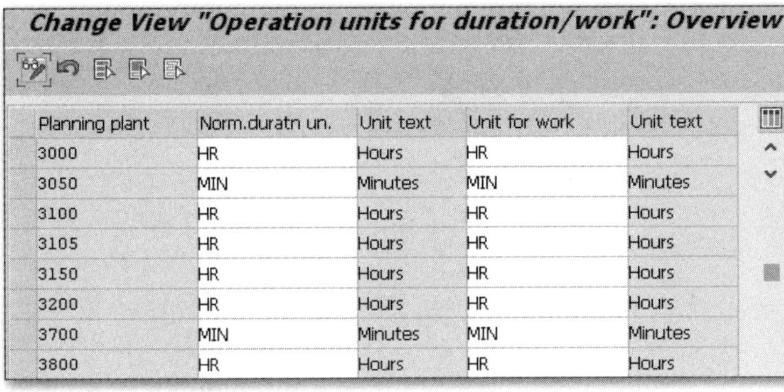

Change View "Operation units for duration/work": Overview

Planning plant	Norm.duratn un.	Unit text	Unit for work	Unit text
3000	HR	Hours	HR	Hours
3050	MIN	Minutes	MIN	Minutes
3100	HR	Hours	HR	Hours
3105	HR	Hours	HR	Hours
3150	HR	Hours	HR	Hours
3200	HR	Hours	HR	Hours
3700	MIN	Minutes	MIN	Minutes
3800	HR	Hours	HR	Hours

Figure 6.81 Orders: Default Values for Units of Measurement

Bar Chart—Define Graphic Profiles

You use this Customizing function to define a graphic profile that you can use as a basis for displaying a bar chart in Transaction I_GRAPH_MONITOR. This bar chart shows you which technician processed which orders and on which dates. You call the graphic profile immediately after you start the transaction.

Prerequisites

There are no special prerequisites.

Customizing Path

SAP NETWEAVER • APPLICATION SERVER • FRONTEND SERVICES • BAR CHART • DEFINE GRAPHIC PROFILES

Settings or Recommended Settings

The graphic profile defines the complete graphical layout. It comprises the following parts (see Figure 6.82):

- The chart types made available by the application are grouped together in the Chart group, where you can define the structure and content of the graphic (i.e., the title, areas, headings, columns, etc.), the background color for the areas, and similar.
- The graphical element types made available by the application are grouped together in the Graph. element group, where you can define, for example, the type, size, and color of the symbols and lines used.
- The color categories made available by the application are grouped together in the Color group, where you can define the color categories that are subsequently used in the graphical element group.
- The form types made available by the application are grouped together in the Form group, where you can define the form types that are subsequently used in the graphical element group.
- The Options profile specifies data in relation to the window and the processing modes, where you can define, for example, the window sizes, time interval, scroll bar, options available, and other properties of the graphic.

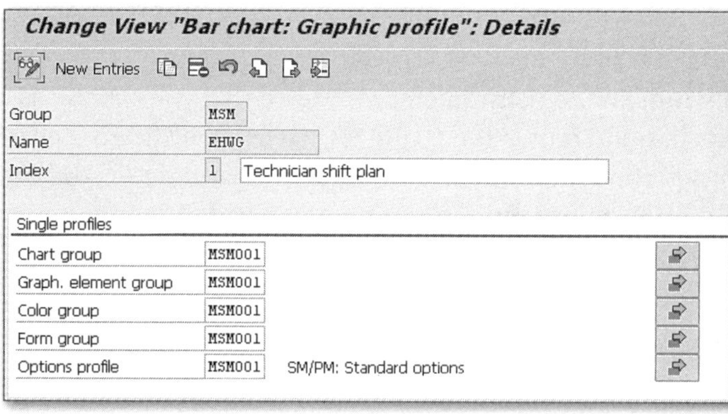

Figure 6.82 Orders – Bar Chart: Graphic Profile

[+]

Use Bar Chart Graphic Profiles from the Standard SAP System

You don't have to define bar chart graphic profiles in the first phase of the implementation project. At first, use one of the profiles delivered in the standard SAP system (e.g., MSM). Then, only create your own profiles if the standard SAP profiles prove to be unsuitable. The best way to do this is to copy a standard SAP profile and adjust it accordingly.

Network—Define Graphic Profiles

You use this Customizing function to define a graphic profile that you can use as a basis for displaying the network in Transaction IW32 (under GOTO • GRAPHIC • NETWORK STRUCTURE) or a bar chart of the order in Transaction IW32 (under GOTO • GRAPHIC • GANTT CHART).

Prerequisites

There are no special prerequisites.

Customizing Path

SAP NETWEAVER • APPLICATION SERVER • FRONTEND SERVICES • NETWORK/HIERARCHY • DEFINE GRAPHIC PROFILES

Settings or Recommended Settings

The graphic profile defines the complete graphical layout. A graphic profile comprises the following parts (see Figure 6.83):

- The node types made available by the application are grouped together in the NODE GROUP, where you can define colors, shapes, lines, and sizes of nodes (operations) in the network.
- The link types made available by the application are grouped together in the LINK GROUP, where you can define colors, shapes, lines, and thickness of the links in the network.
- The frame types made available by the application are grouped together in the FRAME GROUP, where you can define colors and shapes as well as the width and thickness of the frames in the network.
- The OPTIONS PROFILE defines data in relation to the window, processing modes, and colors, where you can define, for example, the window sizes, time interval, area sequence, options available, and other properties of the graphic.

Don't call the graphic profile for the network directly within an order. Instead, use the Customizing functions CREATE DEFAULT VALUE PROFILES FOR GENERAL ORDER DATA and DEFAULT VALUES FOR TASK LIST DATA AND PROFILE ASSIGNMENTS to assign it to each plant and order type.

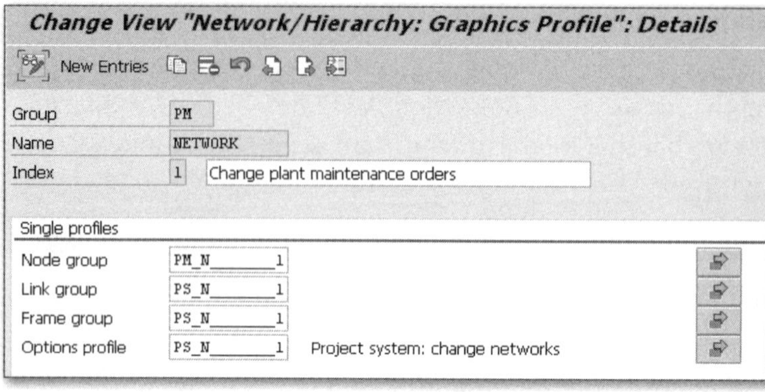

Figure 6.83 Orders – Network: Graphic Profile

[+]

Use Network Graphic Profiles from the Standard SAP System

You don't have to define network graphic profiles in the first phase of the implementation project. At first, use one of the profiles delivered in the standard SAP system (e.g., PM NETWORK).

Then, only create your own profiles if the standard SAP profiles prove to be unsuitable. The best way to do this is to copy a standard SAP profile and adjust it accordingly.

6.2.4 Capacity Planning

In capacity planning, you compare the available capacity of the capacity requirement. You maintain the available capacity in the work center, and the capacity requirement is determined by the standard times associated with the operations in the order (WORK field). However, capacity planning doesn't always make sense.

[+]

First Check Whether Capacity Planning Makes Sense

Essentially, capacity planning only makes sense if you have a sufficiently large order volume and reasonably accurate standard times. Unlike in production, these prerequisites are frequently not fulfilled in plant maintenance. There are either no exact specifications in relation to the standard times, and/or you have a large number of unplanned orders.

If these prerequisites aren't fulfilled, you shouldn't activate capacity planning. You can ensure this by using a control key for which the DETERMINE CAPACITY REQUIREMENTS option isn't activated.

In Section 3.4, you learned how to define control keys (see also the Customizing function MAINTAIN CONTROL KEYS and refer to Figure 6.26).

Performing Capacity Planning

If you perform capacity planning, this always comprises three stages (see Figure 6.84).

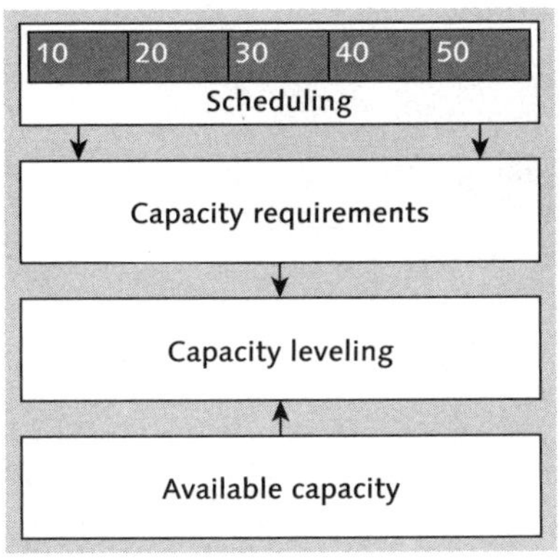

Figure 6.84 Orders: Capacity Planning

You define the *available capacity* on the CAPACITIES tab page within the work center (either as the standard available capacity or with shift models).

Capacity requirements are created on the basis of orders (more specifically, on specifications in the operations in the WORK field). To generate capacity requirements, you must fulfill the following prerequisites:

- In the work center, you've entered a formula for calculating the capacity requirement for internal processing. This formula must point to the field ARBEI; that is, the work from the operation. In the standard system, this is the formula SAP008 (see also the Customizing functions DEFINE FORMULA PARAMETERS and DEFINE FORMULAS).
- In the case of operations, you use a control key for which the CAPACITY REQUIREMENTS function is activated.
- You've scheduled the order (see Section 6.2.3).

For *capacity leveling*, you have the following options:

- You can perform a capacity evaluation (Transaction CM01); that is, a clear comparison of the available capacity and the capacity requirement, and ensure capacity leveling (e.g., by shifting the orders).
- Alternatively, you can use the planning table (Transaction CM33 or Transaction CM34) to perform order scheduling on a capacity basis.

Next, we'll describe the most important Customizing settings for capacity planning in SAP EAM.[2]

[!] **Capacity Planning Not in the Plant Maintenance Menu**

The Customizing functions for configuring capacity planning aren't available in the PLANT MAINTENANCE menu. Therefore, use the Customizing menu for PRODUCTION, or simply call the relevant transactions directly.

Define Capacity Category

This Customizing function is the starting point for capacity planning. Here, you define the capacity types that you want to assign to a work center and use as a basis for capacity planning.

Prerequisites

There are no special prerequisites.

Customizing Path

PRODUCTION • CAPACITY REQUIREMENTS PLANNING • MASTER DATA • CAPACITY DATA • DEFINE CAPACITY CATEGORY

Settings or Recommended Settings

From a maintenance perspective, labor capacity is the most important capacity category. Therefore, it's important to use a capacity category for which the CAPCAT. PER indicator is set. In the standard delivery, this is capacity category 002 (LABOR) (see Figure 6.85).

2 For a complete overview of capacity planning, we recommend the following: F. Gulyassy and B. Vithayathil, *SAP Capacity Planning* (Bonn: SAP PRESS, 2014). German language only.

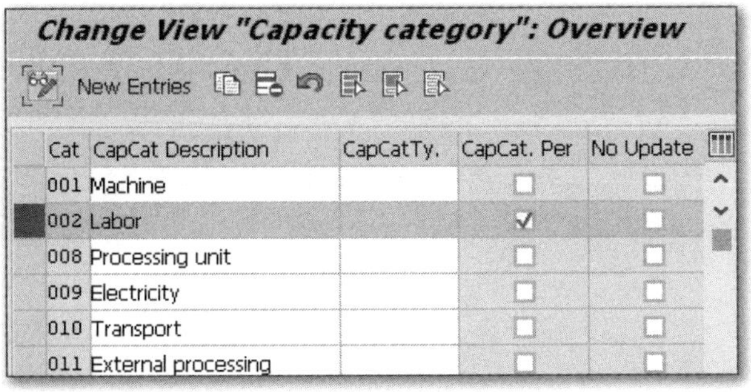

Figure 6.85 Orders: Capacity Category

[+]

Use Capacity Category 002

You can also use the standard SAP capacity category for plant maintenance (002). You only need to define your own capacity category if you want to differentiate yourself from production. Ensure that you set the CapCat. Per indicator.

Define Shift Sequences

If you want to define only the standard available capacity for the work center, you don't have to configure any further Customizing settings for the available capacity.

If, however, you want to define different available capacities and shifts, you must use this Customizing function to define the shift models that you'll use in SAP EAM.

Prerequisites

There are no special prerequisites.

Customizing Path

Production • Capacity Planning • Operations • Available Capacity • Define Shift Sequences

Transaction

OP4A

Settings or Recommended Settings

Define shift sequences in the three consecutive, interrelated subfunctions of WORK BREAK SCHEDULES, SHIFT DEFINITIONS, and SHIFT SEQUENCES.

Use the WORK BREAK SCHEDULES subfunction to define parameters that describe the individual break schedules in more detail (see Figure 6.86):

- ▸ Start
- ▸ End
- ▸ Break duration

You can enter these either as fixed times (e.g., GLZ) or relative to the start of the shift (e.g., M3).

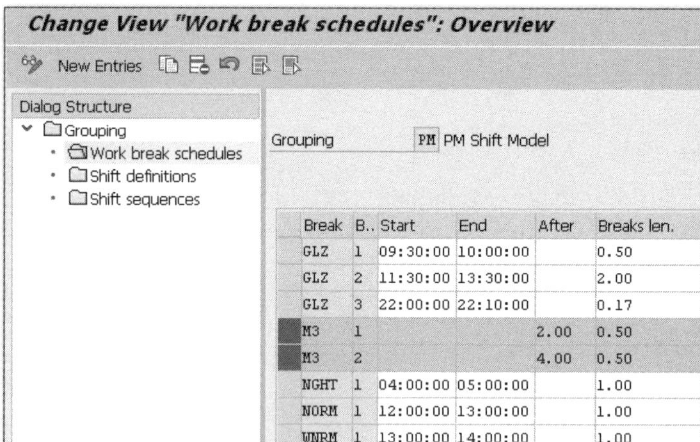

Figure 6.86 Orders: Work Break Schedules

Use the SHIFT DEFINITIONS subfunction to define cross-work center parameters that describe the individual shifts in more detail (see Figure 6.87). Examples include the following:

- ▸ Start
- ▸ End
- ▸ Break times
- ▸ Validity period of a shift

Here, incorporate one of the work break schedules that were defined in the work break schedules (BREAK field).

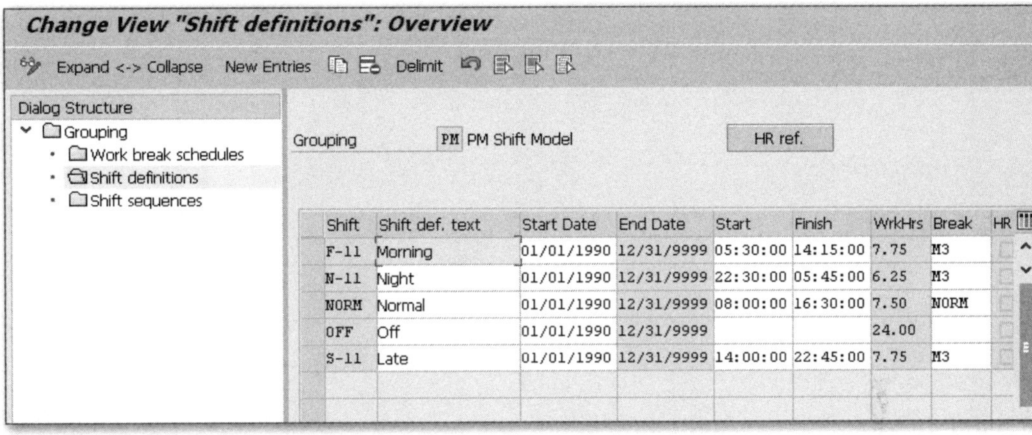

Figure 6.87 Orders: Shift Definitions

Use the SHIFT SEQUENCE subfunction to define a series of shifts per weekday (see Figure 6.88). The shift sequence is therefore a grouping of *shift definitions*.

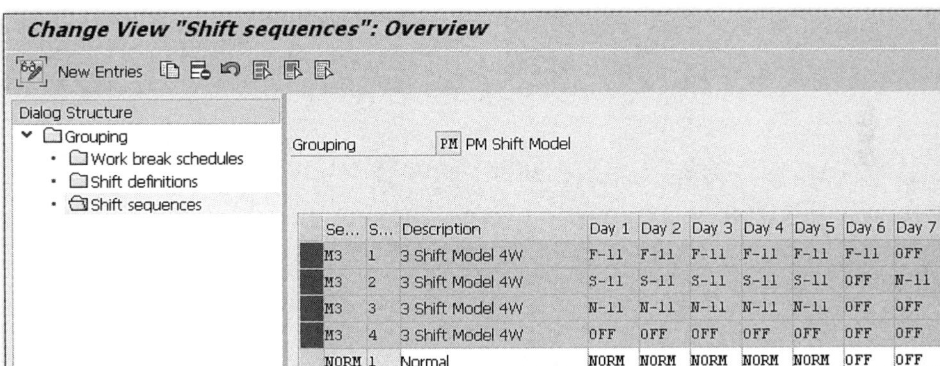

Figure 6.88 Orders: Shift Sequences

In the work center, you can assign the *shift sequence* to a *capacity category* as *available capacity*. In conjunction with the number of persons assigned to the capacity category, the SAP system uses the predefined values from the shift definition to determine the available capacity for a work center.

Define Overall Profiles in Capacity Evaluation

For capacity evaluation (Transaction CM01), you use this Customizing function to define the selection and layout.

Prerequisites

There are no special prerequisites.

Customizing Path

Production • Capacity Requirements Planning • Evaluation • Profiles • Define Overall Profiles

Transaction

OPA6

Settings or Recommended Settings

An overall profile comprises the following single profiles (see Figure 6.89):

- In the Sel. profile subprofile, define the units (work center types, order categories, and capacity categories) and periods (start date, end date, and unit) for which you want to perform capacity planning.
- In the Option profile subprofile, define, among other things, whether you want to cumulate the capacity evaluation, whether you want a hierarchical analysis, or which period type is to be the underlying basis here.

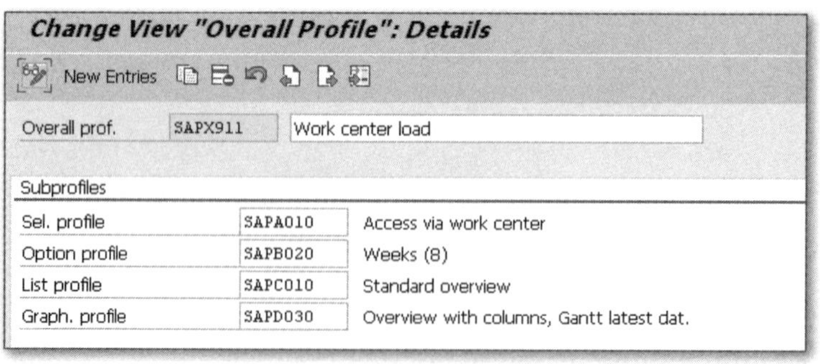

Figure 6.89 Orders: Overall Profile in Capacity Evaluation

- In the LIST PROFILE subprofile, define the level of detail for the capacity evaluation and the interface to Microsoft Excel.
- In the GRAPH. PROFILE subprofile, define the analysis display (columns, bars, and lines) and the sort order for the capacity details.

If you start Transaction CM01, you can use PLANNING • PROFILES to select an overall profile and each of the subprofiles separately.

[+]

Capacity Evaluation Profiles from the Standard SAP System

You don't have to define capacity evaluation profiles in the first phase of the implementation project. At first, use one of the profiles delivered in the standard SAP system (e.g., SAPX911).

Then, only create your own profiles if the standard SAP profiles prove to be unsuitable. The best way to do this is to copy a standard SAP profile and adjust it accordingly.

Define Overall Profile in Capacity Leveling

For capacity leveling (Transaction CM33 – Graphical; Transaction CM34 – Tabular), you use this Customizing function to define the selection, control, and layout for the planning table in which you'll perform capacity leveling.

Prerequisites

There are no special prerequisites.

Customizing Path

PRODUCTION • CAPACITY PLANNING • CAPACITY LEVELING AND EXTENDED EVALUATION • DEFINE OVERALL PROFILE

Transaction

OPDO

Settings or Recommended Settings

An overall profile contains the following subprofiles (see Figure 6.90):

- In the SELECTION PROFILE subprofile, define the primary selection criteria (generally, the plant and work center) for accessing the transaction. You can also specify additional filtering criteria here (e.g., maintenance planner group, functional location, and plant section).

- In the CONTROL PROFILE subprofile, define whether you want to use the graphical or tabular planning table.
- In the TIME PROFILE subprofile, define which past or future periods are to appear in the planning table.
- In the EVALUATION PROFILE subprofile, define the unit of measurement and whether you want to perform capacity cumulation over this period.
- In the STRATEGY PROF. (strategy profile) subprofile, define which functions are to be available when scheduling orders (e.g., operation date check) as well as the sequence in which scheduling is to take place (e.g., latest start date).
- In the PERIOD PROFILE subprofile, define whether capacities are to be planned on a daily, weekly, or monthly basis.
- In the PLANNING TAB. PROFILE subprofile (profile of graphical planning table), define the layout of the graphical planning table (e.g., screen areas, time axis, and graphical elements).
- In the PROF. PLN.TAB. (TAB) subprofile (profile of tabular planning table), define the layout of the tabular planning table (e.g., columns, rows, and totaling).
- In the LIST PROFILE subprofile, define which information is to be available in the detail list and how this list is to be sorted.

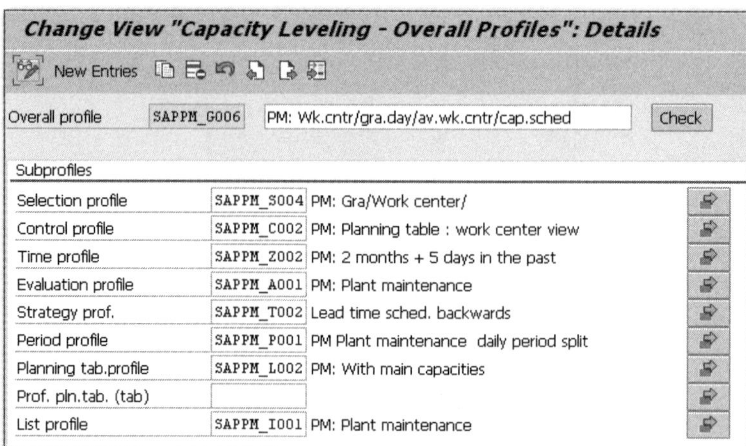

Figure 6.90 Orders: Overall Profile in Capacity Leveling

Use Capacity Leveling Profiles from the Standard SAP System [+]

You don't have to define capacity leveling profiles in the first phase of the implementation project. At first, use one of the profiles delivered in the standard SAP system (e.g., SAPPM_G006). Then, only create your own profiles if the standard SAP profiles prove to be unsuitable. The best way to do this is to copy a standard SAP profile and adjust it accordingly.

6.2.5 Printing

The SAP system has many design options in relation to printing orders and notifications. Generally, only the order is printed when notifications and orders are used, so we'll just concentrate on order printing here.

Freedom of Choice with Shop Paper Printing [+]

You have considerable freedom of choice in relation to the following:

- How many shop papers to print
- Which shop papers to print
- The titles to assign to shop papers
- The layout of these shop papers
- Which shop paper is to be output and on which output medium

For example, you could print the following documents as shop papers (see Figure 6.91):

- Operation control ticket for the maintenance supervisor
- Job ticket for the part to be repaired
- Time tickets and completion confirmation slips for the technician
- Material pick list for the warehouse
- Material issue slips for the technician
- Object list for inspections

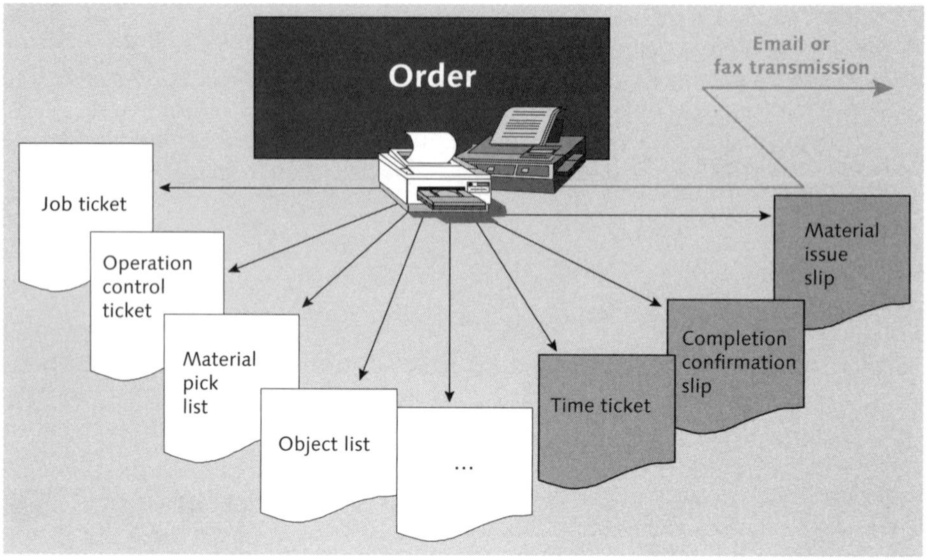

Figure 6.91 Orders: Print

The following output media are available:

- Local printers
- Network printers
- Fax machines
- Email
- PC downloads

The following Customizing functions are available for controlling the output of order documents.

Define Shop Papers, Forms, and Output Programs

You can use this Customizing function to define your shop papers and their print layout. You can also determine which shop paper is to be printed for which order type.

Prerequisites

You must program the print report and define the layout in an SAPscript form beforehand. You must also define the order type.

Customizing Path

PLANT MAINTENANCE AND CUSTOMER SERVICE • MAINTENANCE AND SERVICE PROCESSING • MAINTENANCE AND SERVICE ORDERS • PRINT CONTROL • DEFINE SHOP PAPERS, FORMS, AND OUTPUT PROGRAMS

Transactions

- OIDF for defining shop papers
- OIDG for assigning shop papers to order types

Settings or Recommended Settings

This Customizing function has two subfunctions, namely DEFINE SHOP PAPERS and DEFINE SHOP PAPERS FOR ORDER TYPE. Use the DEFINE SHOP PAPERS subfunction (see Figure 6.92) to define the following:

- A four-digit number (max.) for the SHOP PAPER.
- The report name (PROGRAM NAME) and the form routine (ABAP FORM) used for the print control.
- The SAPscript form (FORM) used as a basis for designing the pages. The form contains different elements that are used to control the layout of the individual pages and to show design information for the texts to be output on the individual pages.

You also have the following control options:

- Use the DELTA PRINT indicator to define whether delta print is to be the default print setting for this shop paper. Delta print means that only those operations and components that don't yet have the status PRINTED are printed.
- Set the OPERATION STAT. (operation status) indicator for operation-based shop papers (e.g., time tickets and completion confirmation slips). This causes the system to set the status PRINTED when a shop paper is printed. In other words, the system sets the status PRINTED as soon as you print a shop paper for which the OPERATION STAT. indicator has been set. If you don't set this indicator for any of your shop papers, the operation can never obtain the status PRINTED. Therefore, in the case of a delta print, the system will print the shop papers for all operations.

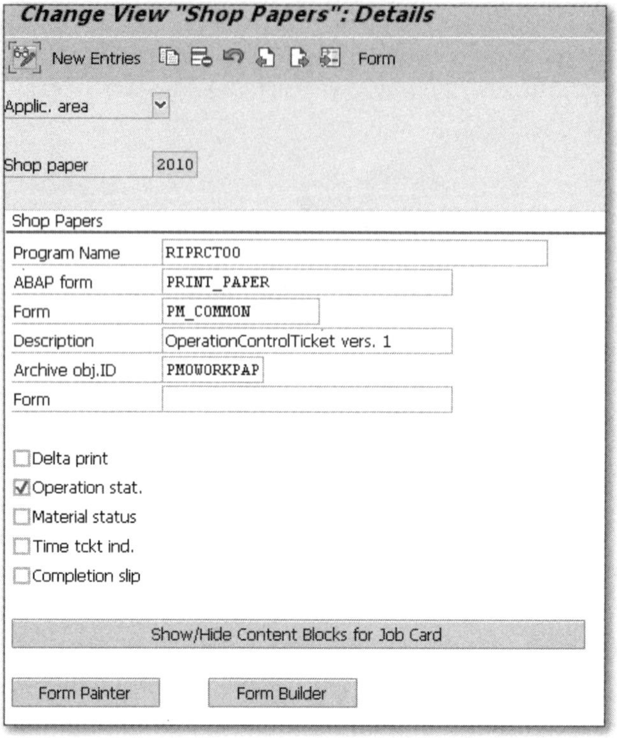

Figure 6.92 Orders: Define Shop Papers

- Set the Material status indicator for component-based shop papers (material issue slips). This causes the system to set the status Printed when a shop paper is printed. In other words, the system sets the status Printed as soon as you print a shop paper for which the Material status indicator has been set. If you don't set this indicator for any of your shop papers, the component can never obtain the status Printed. Therefore, in the case of a delta print, the system will print the shop papers for all components.
- The Time tckt ind. (time ticket indicator) indicator identifies the shop paper as a time ticket. In other words, it's printed only if, in the case of an operation, you use a control key for which the Print time tickets indicator has been set.
- The Completion slip indicator identifies the shop paper as a completion confirmation slip. In other words, it's printed only if, in the case of an operation, you use a control key for which the Print completion confirmation indicator has been set.

Use the subfunction DEFINE SHOP PAPERS FOR ORDER TYPE to define which shop paper is to be printed for which order type (DO... column for document type) (see Figure 6.93). You can use masking here whereby the masked entries apply to all order types that aren't explicitly specified.

Use the SELECTION column to specify which assigned shop papers are to be proposed for printing. If you suppress the print dialog box, the system prints exactly these shop papers.

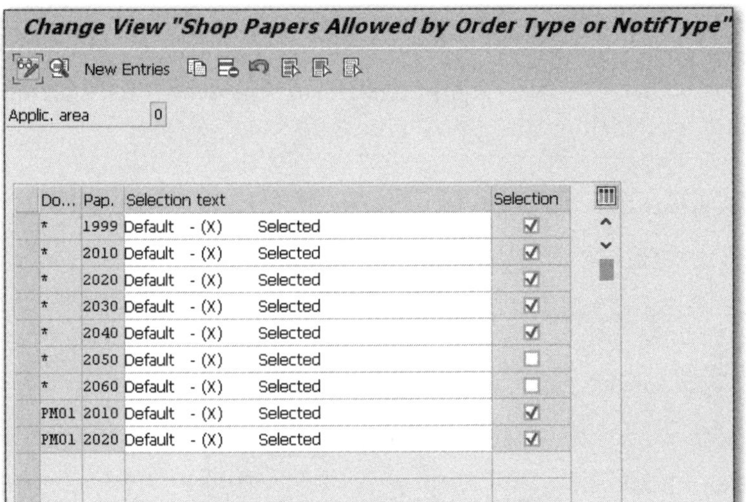

Figure 6.93 Orders: Assign Shop Papers to Order Types

Define Printer

You use this Customizing function to control which shop paper is to be printed on which printer. User-specific print control and print diversion functions are available here.

Prerequisites

You must define the user master records, shop papers, plants, maintenance planner groups, and printers beforehand.

Customizing Path

Plant Maintenance and Customer Service • Maintenance and Service Processing • Maintenance and Service Orders • Print Control • Define Printer

Transaction

- OIDH for user-specific print control
- OIDI for defining fields for the print diversion
- OIDJ for defining the field contents for the print diversion

Settings or Recommended Settings

Use the User-Specific Print Control subfunction to define which user prints which shop paper and on which printer (see Figure 6.94).

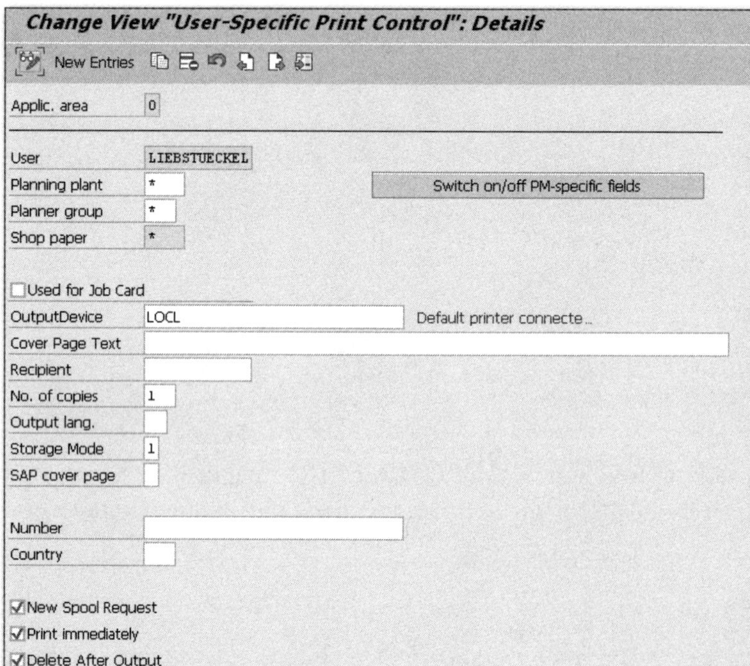

Figure 6.94 Orders: User-Specific Print Control

You can define the plant and maintenance planner group as additional influencing factors. However, you can also use masking, whereby the respective shop paper is always printed on the same printer, irrespective of the plant and/or maintenance planner group. You can choose one of the following printers:

- The user's local Windows printer (printer LOCL).
- A network printer.
- Email. (A PDF attachment is generated instead of a printout. For more information, see the section entitled "Send Shop Papers by Email.")

You can use the DEFINE PRINT DIVERSION subfunction to activate a print diversion. A print diversion means that the shop papers aren't printed using the standard printer defined in the USER-SPECIFIC PRINT CONTROL subfunction. Instead, the print output is diverted to another printer. There are two types of print diversion:

- **Print diversion of shop papers to the work center printer**
  The work center printer is used for completion confirmation tickets and time tickets. In this case, you define the printer in the work center (Transaction IR02, DEFAULT VALUES tab page, SHOP PAPER PRINTER field).
- **Print diversion via Customizing**
  When determining a printer, print diversion (if activated) always has the highest priority. In other words, it has priority over the user-specific selection and the work center printer.

Depending on which shop paper it concerns, the print diversion via Customizing occurs at the header, operation, or component level. Table 6.7 provides an overview of which shop paper is diverted via which level (H = header, O = operation, M = material components, and W = work center printer).

Shop Paper	Description	H	O	M	W
2010	Operation control ticket	X			
2020	Material pick list	X			
2030	Job ticket	X			
2040	Material issue slip	X	X	X	
2060	Time ticket	X	X		X
2062	Time ticket for split records	X	X		X
2070	Completion confirmation slip	X	X		X

Table 6.7 Orders: Assignment of Print Diversion

Shop Paper	Description	H	O	M	W
2072	Completion confirmation slip for split records	X	X		X

Table 6.7 Orders: Assignment of Print Diversion (Cont.)

Accordingly, the fields from the following tables are available for activating the print diversion (see Figure 6.95):

- CAUFVD: Orders (header)
- AFVGD: Order operations
- RESBD: Material components
- RIPW0: Technical objects
- ILOA: Technical objects

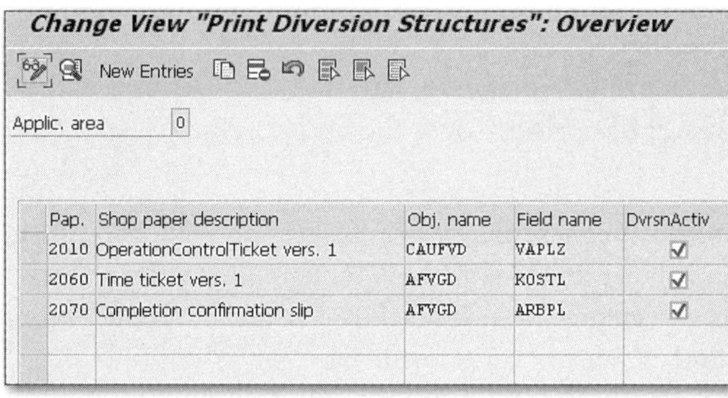

Figure 6.95 Orders: Activate Print Diversion

If you've activated print diversion, you can use the subfunction PRINT DIVERSION ACCORDING TO FIELD CONTENTS to define special printers and print parameters for a specific shop paper on the basis of particular field contents (see Figure 6.96). At any given time, only one table entry can be active for each shop paper. Otherwise, the uniqueness of the printer assignment can't be ensured.

For each table entry, you can define any number of records that differ in terms of their field content or field name (e.g., different printers for the mechanical engineering work center and the electronics work center).

You can define further information for the following scenarios:

- To output the shop paper to a fax number (with a country key) instead of a printer
- To output the shop paper several times (number of printouts)
- To output the shop paper immediately
- To use a spool to control shop paper printing

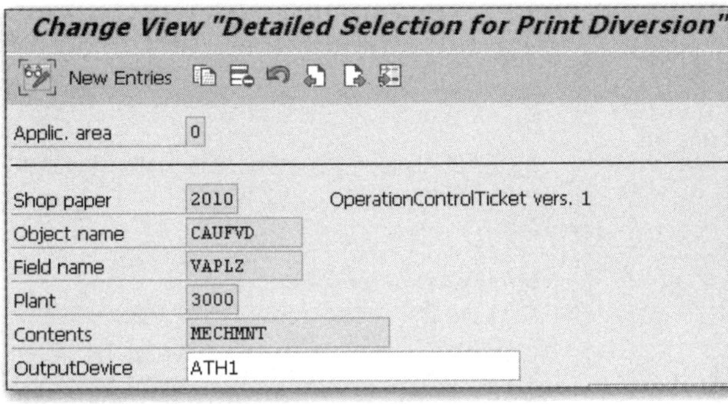

Figure 6.96 Orders: Control Print Diversion

Send Shop Papers by Email

A special feature here is that you can also send your shop papers by email (e.g., to a smartphone) rather than printing them out. Because this is becoming more and more common, the instructions are provided here.

[+]

Instructions for Sending Shop Papers by Email

To send shop papers by email, proceed as follows:

- Set up automatic forwarding in the workplace (Transaction SBWP) (Settings • Office Settings • Automatic Forwarding). Here, enter the email address to which you want to send emails.
- Use Transaction SCOT to set up a send job for INT (Business Communication Administration • Settings • Outbound Message • Send Jobs • Schedule a Send Job for INT). Here, define, in particular, the frequency with which emails are to be sent from the SAP system to the external email address (e.g., every five minutes).
- Use Transaction SPAD to set up a printer called EMAIL. Assign the device type PDF1 to this printer. You must also set the access method for the host spool to M (email to recipient/owner).
- Use the User-Specific Print Control or Define Print Diversion subfunction to assign the printer EMAIL as an output device.

If you then print an order, the SAP system sends emails with a PDF attachment (see Figure 6.97) instead of outputting the document to a physical printer. You can then decide whether you want to output the PDF as a shop paper.

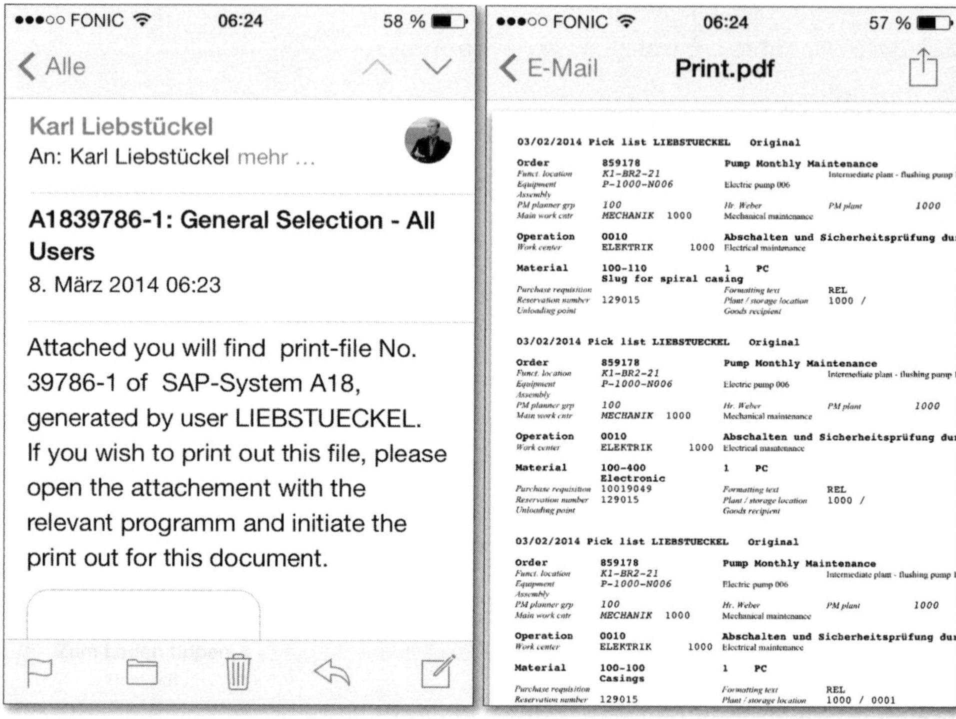

Figure 6.97 Orders: Shop Papers by Email on iPhone

Define Status Control for Shop Papers

You can use this Customizing function to specify, in the case of maintenance orders, whether the timeliness of printed shop papers should be tracked.

If an order or operation is changed after printing the shop papers, the system displays the corresponding status in the order. Furthermore, this option means that, in this case, the status of the order can't be technically completed before the shop papers are reprinted.

Prerequisites

You must define the plants and order types beforehand.

Customizing Path

PLANT MAINTENANCE AND CUSTOMER SERVICE • MAINTENANCE AND SERVICE PROCESSING • MAINTENANCE AND SERVICE ORDERS • FUNCTIONS AND SETTINGS FOR ORDER TYPES • DEFINE STATUS CONTROL FOR SHOP PAPERS

Settings or Recommended Settings

Activate status control for each plant and order type (see Figure 6.98).

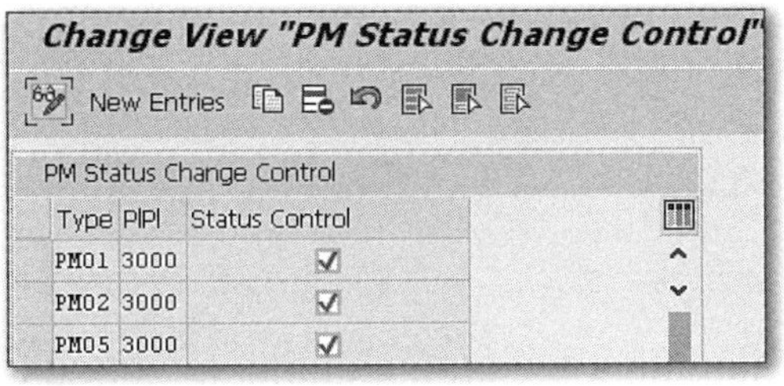

Figure 6.98 Orders: Status Control for Shop Papers

6.2.6 Estimated Costs and Costing

In *costing*, costs are automatically calculated on the basis of the cost rates defined in the system and on the amounts specified in resource planning. Consequently, the planned costs (and also the actual costs subsequently accrued) aren't planned manually.

You can also enter *estimated costs*. However, the estimated costs aren't calculated. Instead, they are specified manually because they are solely based on a maintenance planner's empirical values and not on the planned resources. There are three different ways in which to enter estimated costs:

- At the operation level, differentiated for each value category. The total for all operations is then shown at header level for each value category and as a total.
- At the header level, differentiated for each value category. The total is then created automatically.
- As the total for the entire order.

The planning of resources leads to the accrual of planned costs on the order. Here, there are two different ways in which to display the costs:

- **According to cost elements**
  The cost display according to cost elements shows all cost elements included in the costing. This is more like a controlling-orientated view because it displays a comparison of planned and actual costs.
- **According to value categories**
  The cost display according to value categories groups several cost elements into *value categories*. In general, this is a clearer layout for maintenance purposes because it displays a comparison of estimated, planned, and actual costs.

The costs of an order are calculated at two *summarization levels*. You can also display the costs at these two levels:

- **Operation level**
  The cost display at the operation level shows you all costs included in the relevant operation, either on the basis of the cost elements as a planned/actual comparison or on the basis of the value categories as a comparison of the estimated/planned and actual costs. Orders whose costs are displayed at the operation level obtain the status ACAS. In other words, you can use this status to select the orders in question.
- **Header level**
  The cost display at the header level summarizes all costs at the operation level and shows you all costs included in the order, either on the basis of the cost elements as a planned/actual comparison or on the basis of the value categories as a comparison of the estimated/planned and actual costs. The costs at the header level are always displayed irrespective of whether you've activated costs at the operation level.

To be able to assign estimated costs within an order and to display costs at the value category level, you must configure the following Customizing functions.

Maintain Value Categories

You use this Customizing function to define, for each controlling area, all of the value categories you want to manage within plant maintenance (COMPONENT entry "PM").

Prerequisites

Because value categories are always defined for each controlling area, you must define the controlling areas beforehand.

Customizing Path

PLANT MAINTENANCE AND CUSTOMER SERVICE • MAINTENANCE AND SERVICE PROCESSING • BASIC SETTINGS • SETTINGS FOR DISPLAY OF COSTS • MAINTAIN VALUE CATEGORIES

Settings or Recommended Settings

In addition to a number and description for the value category, define whether the category concerns costs or revenues (see Figure 6.99).

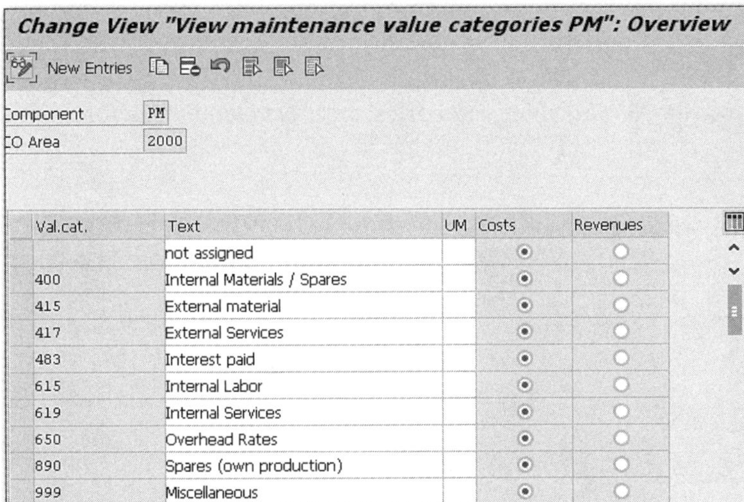

Figure 6.99 Orders: Maintain Value Categories

Internal maintenance orders always fall into the COSTS category. The REVENUES category only comes into play if you render services for a third party and want to use maintenance orders to process these services.

Furthermore, specify whether you want to maintain other values (e.g., hours) in addition to costs (UM column).

Generally Five Value Categories Are Required [+]

In plant maintenance, you generally require the following value categories:

- Internal service
- External processing
- Stock material

- Nonstock material
- Other

If necessary, you can add other value categories.

Assign Cost Elements to Value Categories

You use this Customizing function to control, for each controlling area, which cost elements are to be displayed in which value categories.

Prerequisites

You must define controlling areas, value categories, and cost elements beforehand.

Customizing Path

Plant Maintenance and Customer Service • Maintenance and Service Processing • Basic Settings • Settings for Display of Costs • Assign Cost Elements to Value Categories

Transaction

OIK2

Settings or Recommended Settings

When assigning cost elements to value categories, you have the following options (see Figure 6.100):

- One or more individual cost elements
- One or more intervals
- One or more cost element groups

[!]

Take Care When Making Retroactive Changes to Value Categories

If you subsequently change the assignment of cost elements to value categories, you must use Program RIPMCO00 to recreate the files PMCO and PMCOQT. You must also use Report RIPMS001 to rebuild the information structures in PMIS.

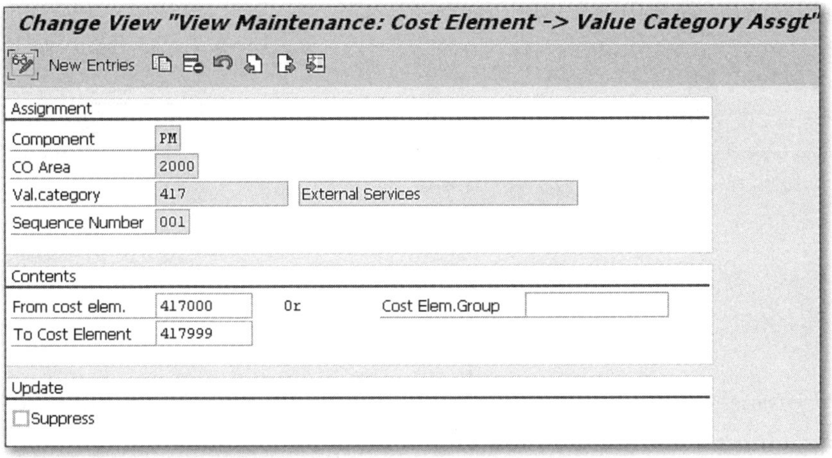

Figure 6.100 Orders: Assign Cost Elements to Value Categories

Define Default Values for Value Categories

You use this Customizing function to select those value categories that are defined using the Customizing function Maintain Value Categories and are to appear in the order for entering estimated costs as default values (see Figure 6.101).

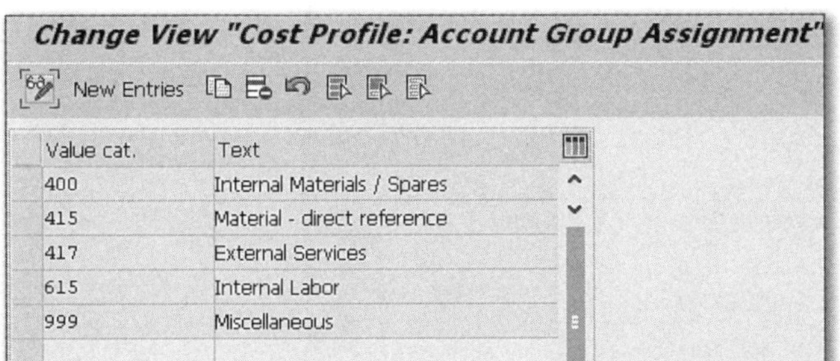

Figure 6.101 Orders: Default Values for Value Categories

Prerequisites

You must define the value categories beforehand.

Customizing Path

Plant Maintenance and Customer Service • Maintenance and Service Processing • Basic Settings • Settings for Display of Costs • Define Default Values for Value Categories

Settings or Recommended Settings

[+]

Check Consistency

You can use the Customizing function CHECK CONSISTENCY OF VALUE CATEGORY ASSIGNMENT to check whether errors occur when you assign cost elements to value categories.

The system displays the following results (see Figure 6.102):

- 
  Everything is okay.
- 
  Warning messages are issued; for example, if cost elements haven't been assigned.
- 
  Errors have occurred; for example, if cost elements have been assigned several times.

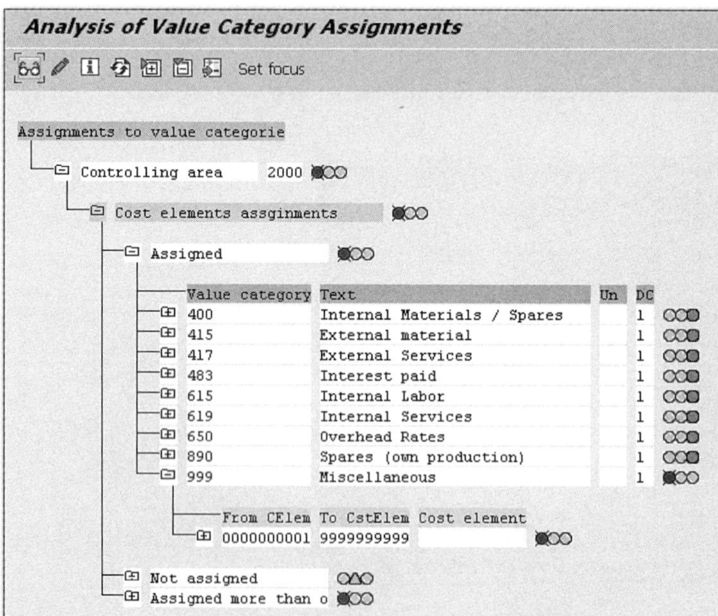

Figure 6.102 Orders: Analysis of Value Category Assignments

We'll now calculate the costs associated with the orders. Figure 6.103 shows the relationship between the Customizing functions so that you can calculate the costs associated with an order (i.e., the planned costs and actual costs):

- In the COSTING SHEET, define the cost elements that the costing is to take into consideration. Furthermore, determine which charge lines are to be used to calculate the costs (e.g., overhead rates).
- Use the VALUATION VARIANT to control which cost rates are to be used to calculate the costs (e.g., material price control).
- The COSTING TYPE isn't important in plant maintenance. However, it's significant in applications such as PP or CO. Together with the valuation variant, its sole purpose is to define the costing type of a costing variant.
- Finally, use the COSTING PARAMETER to assign the costing variant for both the preliminary costing and the actual costing for each order type and plant.

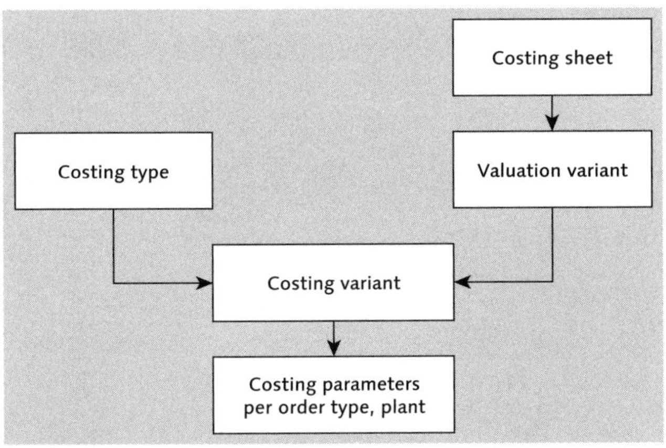

Figure 6.103 Orders: Costing Overview

Separate Costing [+]

Because costing is used not only in plant maintenance, but also in other applications such as Project Processing, PP, SM, and CO, you should always define a separate costing for maintenance processing to make it more flexible and to differentiate it from other types of costing.

Maintain Costing Sheet

You use this Customizing function to define the costing sheet. You then use the costing sheet to determine the costing structure and content.

Prerequisites

You must define the cost elements beforehand. If you work with overhead rates, you must also define the objects to be credited (e.g., the warehouse cost center) beforehand.

Customizing Path

PLANT MAINTENANCE AND CUSTOMER SERVICE • MAINTENANCE AND SERVICE PROCESSING • MAINTENANCE AND SERVICE ORDERS • FUNCTIONS AND SETTINGS FOR ORDER TYPES • COSTING DATA FOR MAINTENANCE AND SERVICE ORDERS • MAINTAIN COSTING SHEET

Settings or Recommended Settings

Use the COSTING SHEETS subfunction to define a name and description for your costing sheet (see Figure 6.104).

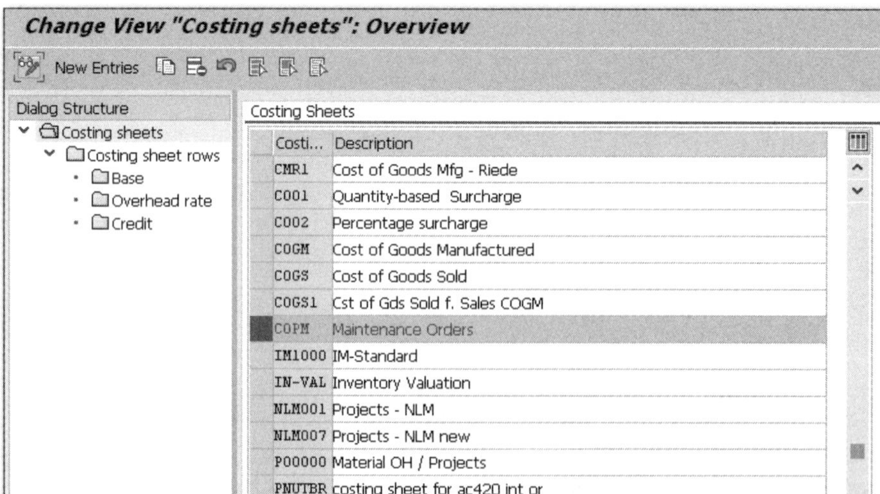

Figure 6.104 Orders – Costing Sheets: Overview

Use the COSTING SHEET ROWS subfunction to define the costing elements and structure. The following elements should be contained here (see Figure 6.105):

- Internal service
- Overhead rate for internal service
- External processing
- Overhead rate for external processing
- Stock material
- Overhead rate for stock material

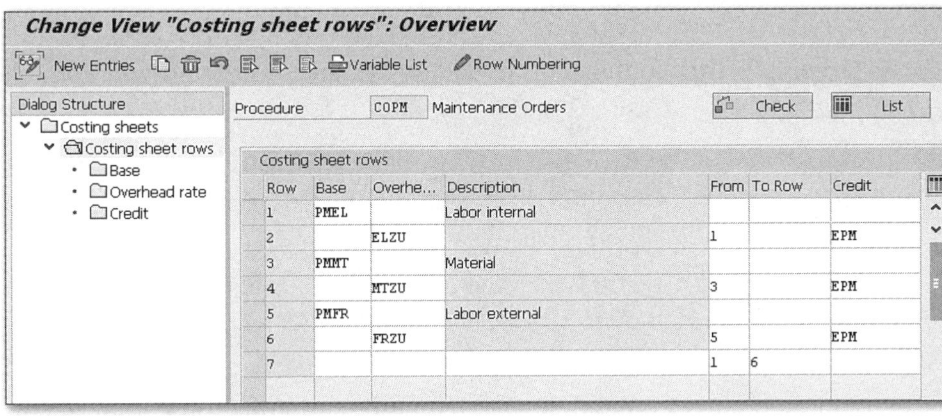

Figure 6.105 Orders: Costing Sheet Rows

Take Care When Assigning Rows [+]

In the FROM ROW and TO ROW columns, ensure the following:

- The subtotals and totals are calculated correctly.
- The overhead rows refer to the correct base rows.

Use the BASE subfunction to define the assignment of cost elements as a basis for calculating overheads (see Figure 6.106). The following options are associated with assigning cost elements:

- Individual cost elements
- Cost element intervals
- Cost element groups

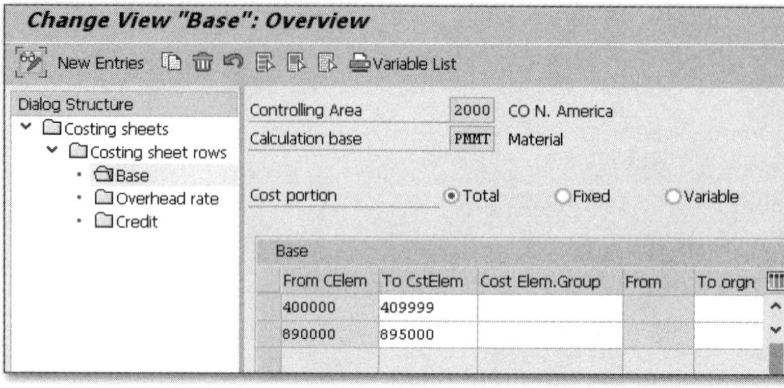

Figure 6.106 Orders – Costing Sheet: Base

Use the Overhead rate subfunction to define your overhead rates with the following information (see Figure 6.107):

- Valid from/To
  This can be used to manage different overhead rates in different validity periods.
- Ovrhd type
  This is the overhead type for which the overhead rate is to apply (1 = standard cost estimate, 2 = actual costing, 3 = commitment costing).
- Percentage/Unit
  This is the amount and unit of measurement for the overhead rate (e.g., 20% or USD 5)

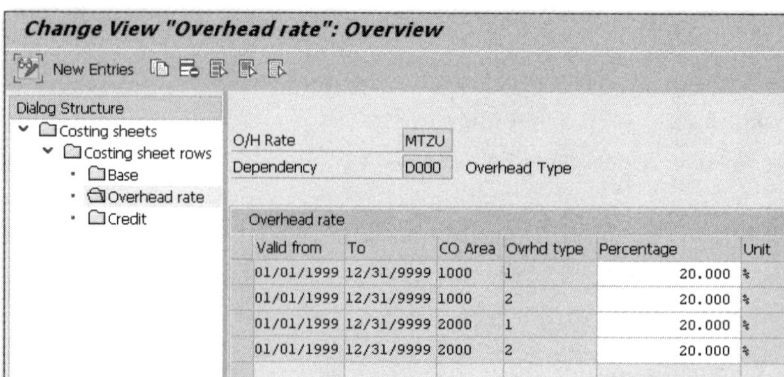

Figure 6.107 Orders – Costing Sheet: Overhead Rates

If you work with overhead rates, you *must* assign an account assignment object that is to be credited with the values from the overhead calculation. Use the CREDIT subfunction to define this account assignment object (see Figure 6.108). For example, you've defined an overhead rate for material postings. In other words, you award percentage or absolute overhead costs to each goods receipt or goods issue. You should then specify the warehouse cost center as a credit object to which the corresponding amounts are credited.

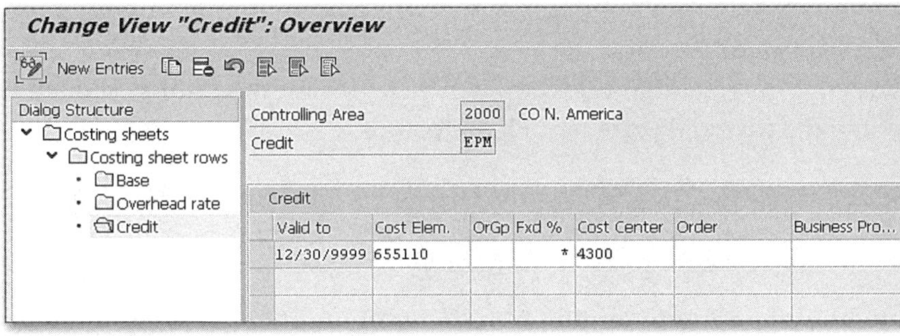

Figure 6.108 Orders – Costing Sheet: Credit

Recommendations for Overhead Costs [+]

- You should always use the same overhead rates for all overhead types, especially planned costing and actual costing. At some point, all other overhead rates will cease to make sense.
- You can specify percentage or absolute overhead rates. For example, you can define a cost of USD 5 for each material withdrawal.
- You must specify an account assignment object that is credited (e.g., the warehouse cost center, which is credited with the material overhead costs).

Define Valuation Variants

You use this Customizing function to control which valuation methods (e.g., which price control for the material) are to be used for the costing.

Prerequisites

You must define the costing sheet(s) beforehand.

Customizing Path

Plant Maintenance and Customer Service • Maintenance and Service Processing • Maintenance and Service Orders • Functions and Settings for Order Types • Costing Data for Maintenance and Service Orders • Define Valuation Variants

Transaction

OKP8

Settings or Recommended Settings

On the Material Val. (material valuation) tab page, define the priority sequence for determining the material price (see Figure 6.109).

Recommendation

7 Valuation Price According to Price Control in Mat. Master

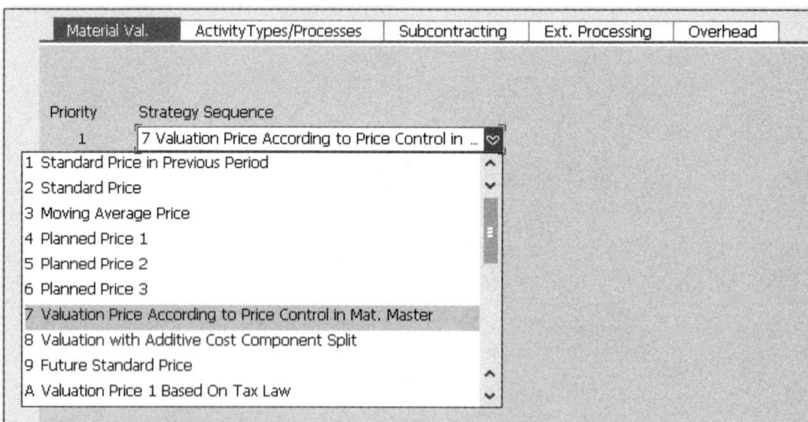

Figure 6.109 Orders – Costing: Material Valuation

On the ActivityType/Processes tab page, define the priority sequence for determining the prices for calculating the cost of the internal service (see Figure 6.110).

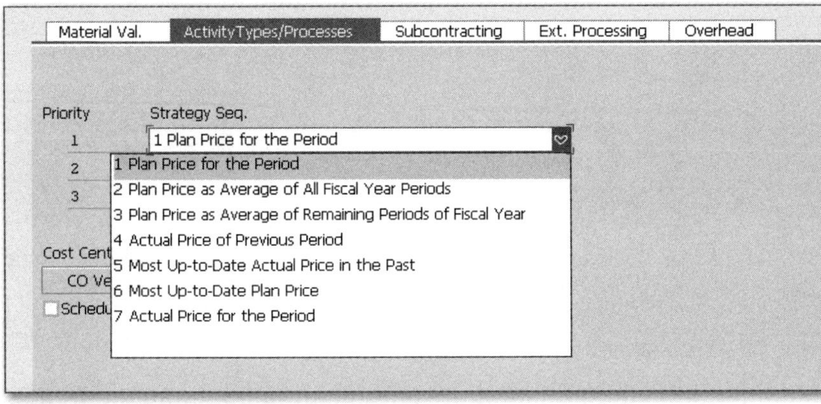

Figure 6.110 Orders – Costing: Internal Service

Recommendation

1 Plan Price for the Period

On the Ext. Processing tab page, define the priority sequence for determining the prices for external processing (see Figure 6.111).

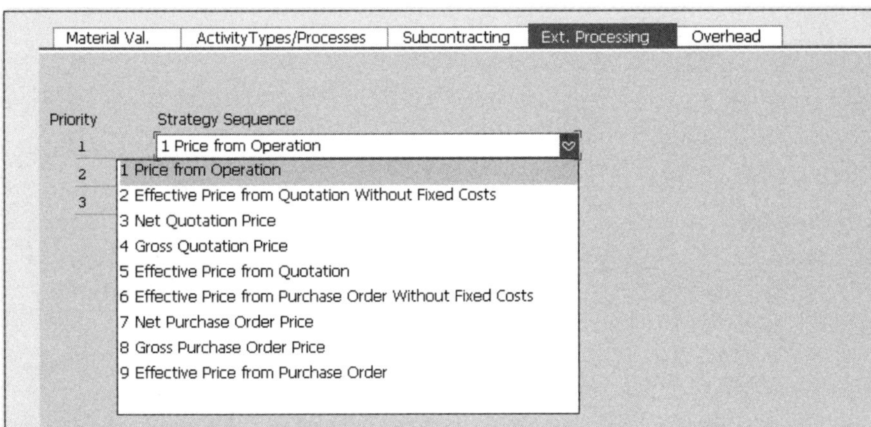

Figure 6.111 Orders – Costing: External Processing

[+]

Recommendation

1 PRICE FROM OPERATION (for the planned costing) and 9 EFFECTIVE PRICE FROM PURCHASE ORDER (for the actual costing). If you require two different valuation approaches, define two different valuation variants.

The information on the SUBCONTRACTING tab page will be required during the *subcontracting* business process (see Chapter 8, Section 8.2) and will therefore be discussed there.

On the OVERHEAD tab page, assign the costing sheet.

Maintain Costing Variants

You use this Customizing function to group the costing type and valuation variant together to form a costing variant.

Prerequisites

You must define the costing type and valuation variant beforehand. The costing type doesn't have any controlling effect in SAP EAM (unlike in the SAP applications PP, CO, and SAP PS) and is therefore not discussed further here. The only thing to ensure here is that a separate costing type (e.g., 05 = MAINTENANCE ORDER) is defined here for SAP EAM.

Customizing Path

PLANT MAINTENANCE AND CUSTOMER SERVICE • MAINTENANCE AND SERVICE PROCESSING • MAINTENANCE AND SERVICE ORDERS • FUNCTIONS AND SETTINGS FOR ORDER TYPES • COSTING DATA FOR MAINTENANCE AND SERVICE ORDERS • MAINTAIN COSTING VARIANTS

Transaction

OKP6

Settings or Recommended Settings

Define a name and description for the costing variant(s). Assign the costing type and valuation variant (see Figure 6.112).

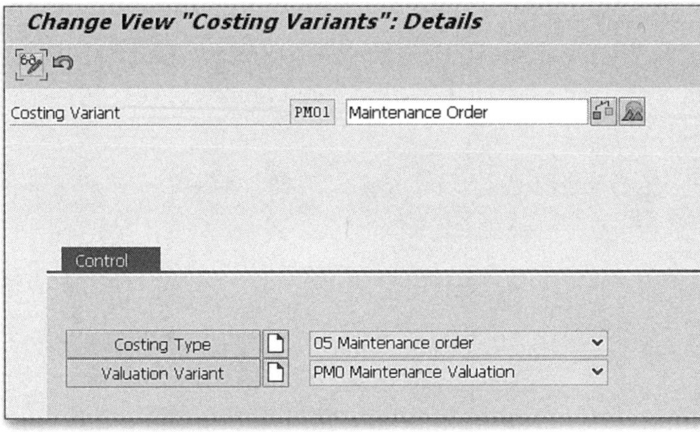

Figure 6.112 Orders: Maintain Costing Variant

[+]

One or Two Costing Variants

If you have two different valuation variants (e.g., one for planned costing and one for actual costing), define two different costing variants here.

Assign Costing Parameters and Results Analysis Keys

You use this Customizing function to assign, for each order type and plant, the costing variant(s) that you want to use for the planned costing and actual costing.

Prerequisites

You must define the plants, order types, and costing variants beforehand.

Customizing Path

Plant Maintenance and Customer Service • Maintenance and Service Processing • Maintenance and Service Orders • Functions and Settings for Order Types • Costing Data for Maintenance and Service Orders • Assign Costing Parameters and Results Analysis Keys

Transaction

OIOF

Settings or Recommended Settings

For PLANNED VAR. CAL. or ACTUAL VAR. CALC., enter the costing variant(s) that you want to use for the planned costing or actual costing (see Figure 6.113).

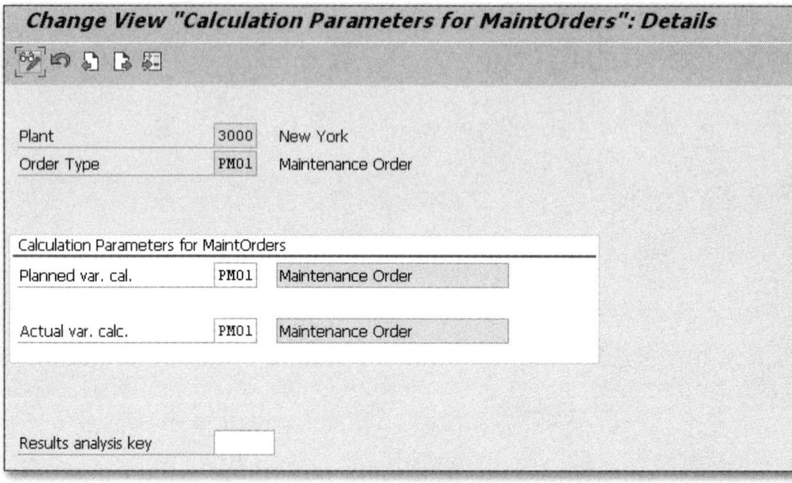

Figure 6.113 Orders: Assign Costing Variant(s)

[+] **Planned Costing and Actual Costing: The Same or Different?**

You have the option of using two different costing variants to structure planned costing and actual costing differently. For example, in the planned costing, you can use the price from the operation to valuate external processing, whereas in the actual costing, you can use the effective price from the purchase order to valuate external processing.

Define Cost Settings

You use this Customizing function to define, for each order type and plant, whether you want to manage costs at the operation level or the header level within the order.

Prerequisites

You must define the plants and order types beforehand.

Customizing Path

Plant Maintenance and Customer Service • Maintenance and Service Processing • Maintenance and Service Orders • Functions and Settings for Order Types • Costs at Operation Level • Define Cost Settings

Settings or Recommended Settings

In the case of those plants and order types for which you want to activate costs at the operation level, set the indicator in column A (see Figure 6.114).

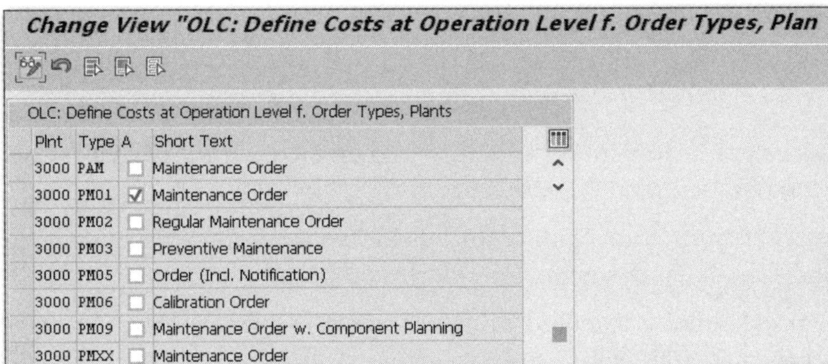

Plnt	Type	A	Short Text
3000	PAM	☐	Maintenance Order
3000	PM01	☑	Maintenance Order
3000	PM02	☐	Regular Maintenance Order
3000	PM03	☐	Preventive Maintenance
3000	PM05	☐	Order (Incl. Notification)
3000	PM06	☐	Calibration Order
3000	PM09	☐	Maintenance Order w. Component Planning
3000	PMXX	☐	Maintenance Order

Figure 6.114 Orders: Costs at the Operation Level

[+]

Effects of Costs at the Operation Level

If you activate the costs at the operation level, this affects the entire cost processing of the order:

- The costs associated with the planned costing and actual costing are calculated at the operation level, and the totals are displayed at the header level.
- The estimated costs are assigned at the operation level and totaled at the header level.
- The order is settled with reference to the operation. In other words, each operation obtains a separate settlement rule.
- All documents (e.g., purchase requisitions, purchase orders, material withdrawals, and invoice receipts) must be assigned to the order number *and* operation number.
- If you activate the costs at the operation level retroactively, you must use Report RIPMCO01 or Transaction OLI5N to rebuild the tables with the order costs PMCO and PMCO_OP.
- The same applies if you undo the activation retroactively.

To enable the system to determine the costs on the basis of your resource planning, you must fulfill the following prerequisites in addition to the preceding Customizing settings:

- Use the COSTING function to assign a control key to the operations that are to be included in the costing.
- Assign WORK to the operations that are to be included in the costing.
- For the cost center of the work center, use Transaction KP26 to define a fixed price (FIXED PRICE option) and/or variable price (VARIABLE PRICE option) for the version you use, the fiscal year, cost center, and activity type.
- In the work center, enter the standard value key "SAP0" on the BASIC DATA tab page.
- In the work center, maintain the following data on the COSTING tab page: COST CENTER, ACTIVITY TYPE, and FORMULA KEY.
- The formula key must point to the ARBEI field (the work from the operation). In the standard system, this is the formula SAP008.
- Transfer prices (either a standard price or a moving average price) must be defined for the materials whose costs are to be calculated.
- Enter a value for nonstock materials.
- Enter a value for external processing.

[+] **Missing Order Costs**

If no costs are displayed in your order (either none at all or only costs for certain value categories), check all of the preceding Customizing settings and other prerequisites. This should rectify the problem.

6.2.7 Order Settlement

Figure 6.115 shows the relationship between the Customizing functions for order settlement:

- In the allocation structure, you define how you want to convert the debit cost elements into credit cost elements.

- In the settlement profile, you define the allocation structure, the permitted account assignment objects, and other settlement parameters.
- You then assign the settlement profile to the order type.

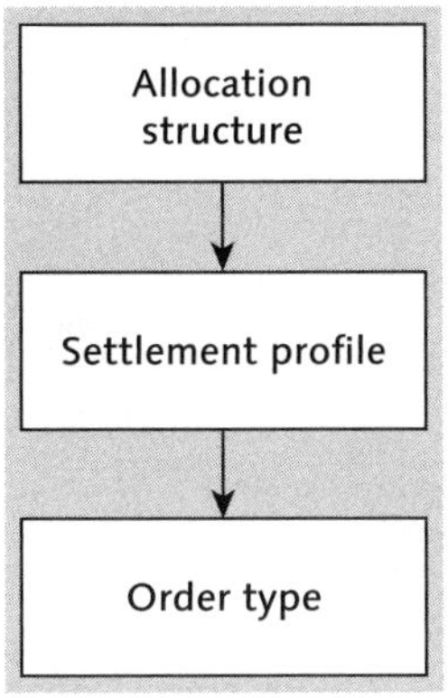

Figure 6.115 Orders – Settlement: Overview

[+]

Separate Order Settlement

Because order settlement is used not only in SAP EAM, but also in other applications such as Project Processing, PP, SM, and CO, you should always define a separate order settlement for maintenance processing to make it more flexible and to differentiate it from other types of order settlement.

Maintain Allocation Structures

You use this Customizing function to define how the cost elements used to debit the orders are to be converted into credit cost elements.

Prerequisites

You must define the cost elements and, if necessary, the cost element groups beforehand.

Customizing Path

Plant Maintenance and Customer Service • Maintenance and Service Processing • Basic Settings • General Order Settlement • Maintain Allocation Structures

Settings or Recommended Settings

Use the Define Allocation Structures subfunction to define a name and description for the allocation structure that you want to use in plant maintenance (see Figure 6.116). You can have more than one allocation structure.

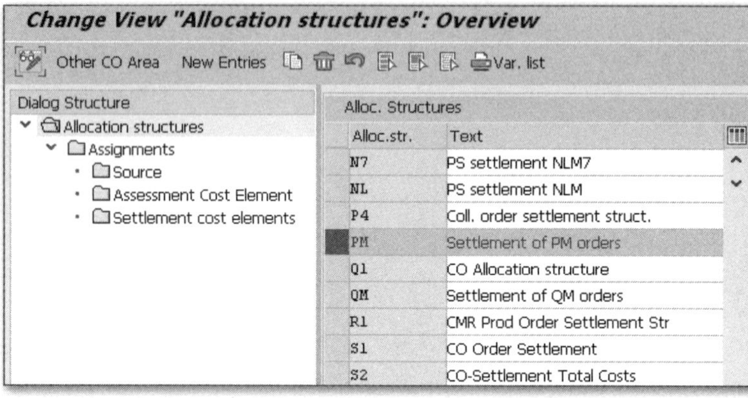

Figure 6.116 Orders: Allocation Structures

Use the Assignments subfunction to define the groupings (not to be confused with cost element groups) and therefore the level of detail for the order settlement (see Figure 6.117). Common assignments include, for example:

- Internal services
- External processing
- Stock material
- External material
- Overhead costs
- Settlement costs (if orders have been debited with order settlements that are to be allocated further)
- Other costs

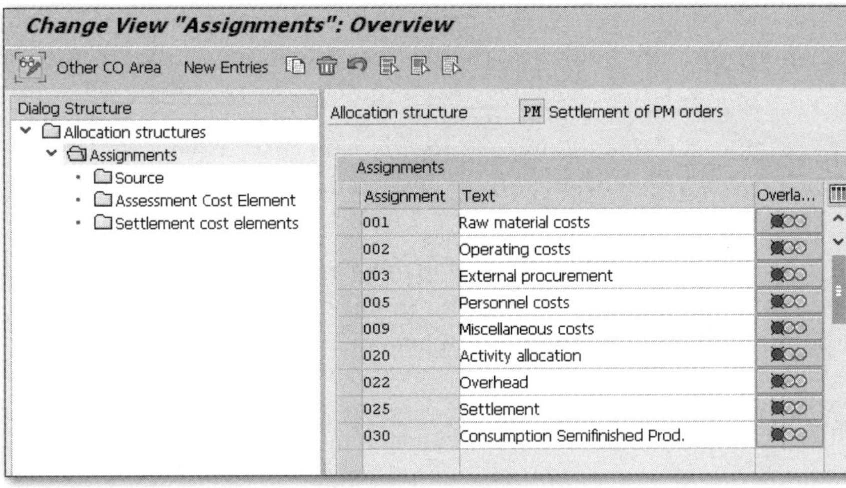

Figure 6.117 Orders – Allocation Structure: Assignments

Use the SOURCE subfunction to define, for each assignment and controlling area, the cost elements or cost element groups that need to be settled (see Figure 6.118).

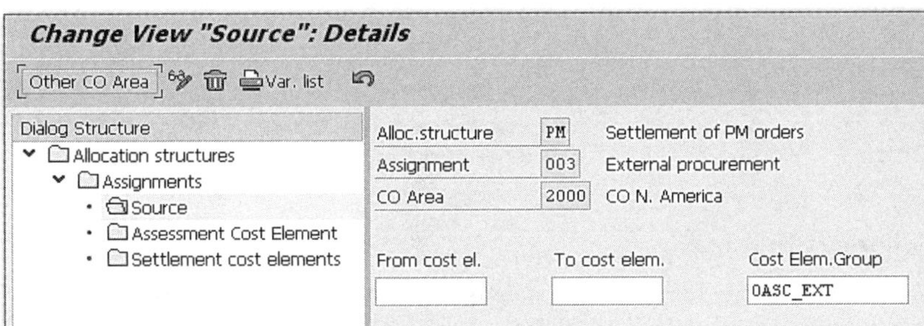

Figure 6.118 Orders – Allocation Structures: Source

The allocation structure must fulfill the following criteria:

- **Completeness**
  All cost elements under which debits accrue must be represented in the relevant allocation structure.
- **Uniqueness**
  Each cost element under which debits accrue must occur only once within an allocation structure. In other words, within an allocation structure, only one settlement cost element is permitted to be assigned to a source.

Use the SETTLEMENT COST ELEMENTS subfunction to define the settlement procedure for each assignment, controlling area, and receiver category (e.g., cost center or order) (see Figure 6.119). Here, you have two options:

- Assign the debit cost element groups to a settlement cost element. In this case, the settlement cost element must have cost element category 21 (INTERNAL SETTLEMENT).
- Perform a cost element-based settlement. In other words, the debit cost element is also used as the settlement cost element. In this case, all cost element categories are permitted as the settlement cost element.

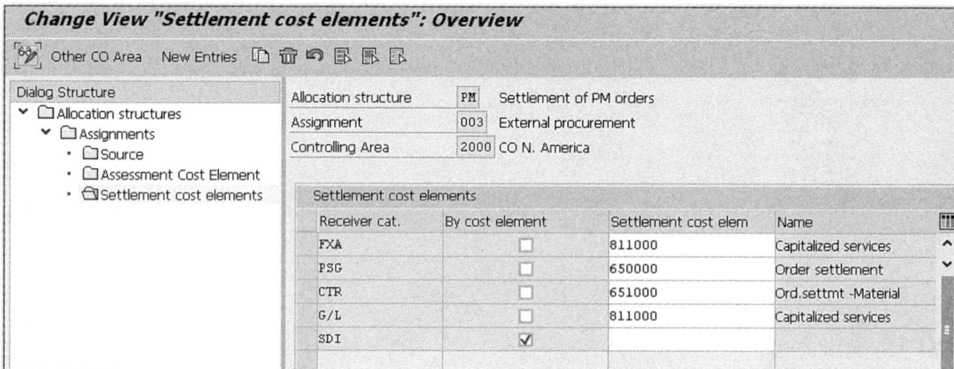

Figure 6.119 Orders – Allocation Structures: Settlement Cost Elements

Maintain Settlement Profiles

You use this Customizing function to define the allocation structure, the permitted account assignment objects, and other settlement parameters.

Prerequisites

You must define the allocation structures and document types for the settlement documents beforehand.

Customizing Path

PLANT MAINTENANCE AND CUSTOMER SERVICE • MAINTENANCE AND SERVICE PROCESSING • BASIC SETTINGS • GENERAL ORDER SETTLEMENT • MAINTAIN SETTLEMENT PROFILES

Settings or Recommended Settings

You must define the following (see Figure 6.120):

- Use the radio buttons in the ACTUAL COSTS/COST OF SALES area to specify whether or not your maintenance orders are to be settled. Generally, you specify that the orders are to be settled in full. Only in exceptional cases (e.g., if you don't use CO) are orders not settled.
- In the VALID RECEIVERS field group, define the account assignment objects that you want to permit as settlement receivers for maintenance orders. The cost center, order, WBS element, and network are the most common receivers for costs from maintenance orders.

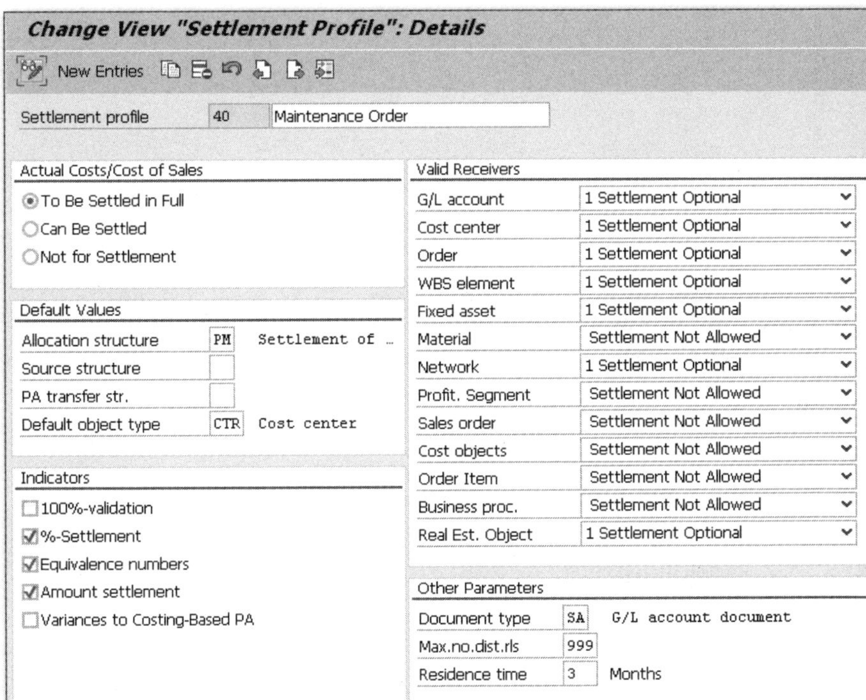

Figure 6.120 Orders: Settlement Profile

- You can then enter one of these valid receivers as a DEFAULT OBJECT TYPE, which is then used to automatically create the settlement rule. For example, if you specify COST CENTER under VALID RECEIVERS, and you specify CTR as the DEFAULT OBJECT TYPE, the cost center from the functional location or piece of equipment is automatically entered in the settlement rule for the order when the order is released or technically completed.

- If you set the %-SETTLEMENT indicator, you can define, within the *settlement rule*, distribution rules governing the settlement of costs according to a percentage settlement.
- If you set the AMOUNT SETTLEMENT indicator, you can define, within the *settlement rule*, distribution rules governing the settlement of costs according to fixed amounts.
- Use MAX.NO.DIST.RLS (maximum number of distribution rules) to determine the upper limit for the maximum number of distributions permitted.

[+]

Recommendations for the Settlement Profile

- Settle the maintenance orders in full.
- Permit at least one cost center and one order as a VALID RECEIVER.
- Set the most common receiver as the DEFAULT OBJECT TYPE. This is generally the cost center.
- Activate %-SETTLEMENT so that you can distribute, across several receivers, the entire amount to be settled.
- Activate AMOUNT SETTLEMENT so that you implement settlement according to fixed prices, which is a very popular form of settlement in plant maintenance. Amount settlement is very much in keeping with the sold-to party/contractor principle.

Settlement Rule: Define Time and Creation of Distribution Rule

You use this Customizing function to control the point in time (release, technical completion) when the settlement rule is to be created and to control how you want to handle an order hierarchy in terms of its settlement rule.

Prerequisites

You must define the order types beforehand.

Customizing Path

PLANT MAINTENANCE AND CUSTOMER SERVICE • MAINTENANCE AND SERVICE PROCESSING • MAINTENANCE AND SERVICE ORDERS • FUNCTIONS AND SETTINGS FOR ORDER TYPES • SETTLEMENT RULE: DEFINE TIME AND CREATION OF DISTRIBUTION RULE

Settings or Recommended Settings

Use the DISTRIB. RULE SUB-ORDER option to define how you want to handle the settlement rule for suborders. You have five options (see Figure 6.121):

- 1
  Copy the distribution rule(s) for the main order.
- 2
  Copy the distribution rule(s) for the superior order.
- 3
  Main order will be settlement receiver.
- 4
  Superior order will be settlement receiver.
- blank
  No special handling of suborders.

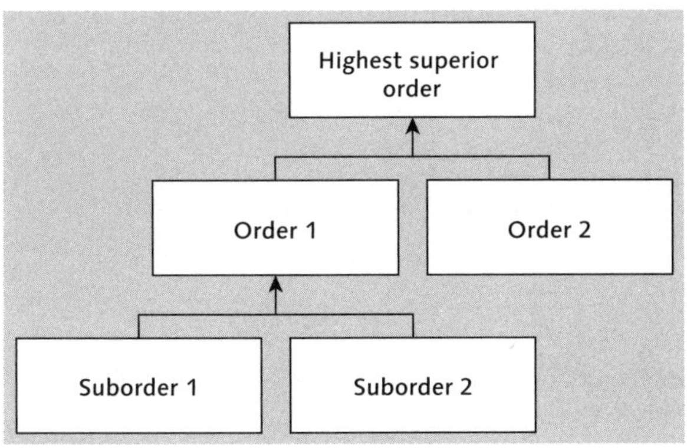

Figure 6.121 Orders: Order Hierarchy

For settings 1–4, the default values from the settlement profile are overridden. If you leave the entry blank, the settings from the settlement profile apply.

Use the SETTLEMENT RULE TIME option to define when and how the settlement rule is to be created. You have three options (see Figure 6.122):

- 1: MANDATORY FOR RELEASE
  On the basis of the settlement profile, the system tries to automatically create the settlement rule when the order is released. If it's unsuccessful, you must maintain the settlement rule manually.

- 2: DISTRIBUTION RULE FOR RELEASE
  On the basis of the settlement profile, the system tries to automatically create the settlement rule when the order is released. If it's unsuccessful, the order is released without a settlement rule.
- blank: MANDATORY FOR TECHNICAL COMPLETION
  On the basis of the settlement profile, the system tries to automatically create the settlement rule when the order is technically completed. If it's unsuccessful, you must maintain the settlement rule manually.

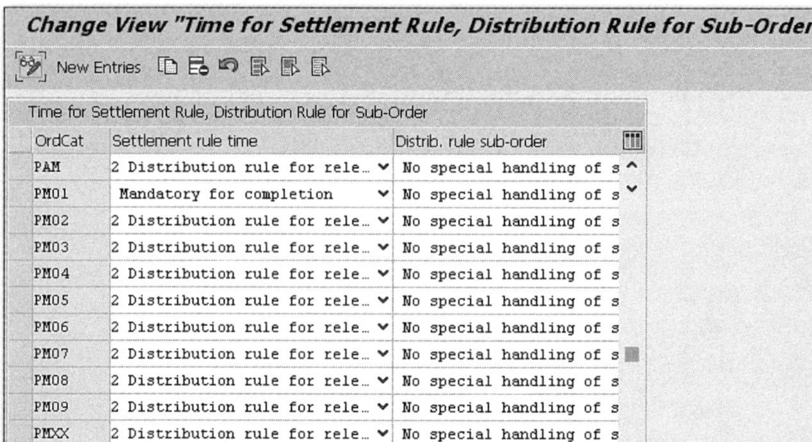

OrdCat	Settlement rule time	Distrib. rule sub-order
PAM	2 Distribution rule for rele…	No special handling of s
PM01	Mandatory for completion	No special handling of s
PM02	2 Distribution rule for rele…	No special handling of s
PM03	2 Distribution rule for rele…	No special handling of s
PM04	2 Distribution rule for rele…	No special handling of s
PM05	2 Distribution rule for rele…	No special handling of s
PM06	2 Distribution rule for rele…	No special handling of s
PM07	2 Distribution rule for rele…	No special handling of s
PM08	2 Distribution rule for rele…	No special handling of s
PM09	2 Distribution rule for rele…	No special handling of s
PMXX	2 Distribution rule for rele…	No special handling of s

Figure 6.122 Orders: Settlement Rule Time

[+] **Settlement Rule When Order Is Released**

Always create the settlement rule when an order is released. Otherwise, you can't settle the order before the order is technically completed.

6.2.8 Project Management and Investment Management

You can assign SAP EAM orders to SAP Project System (SAP PS) objects (project, WBS element, and network) and to SAP Investment Management (SAP IM) objects (investment program and investment program position).[3] For these assignments to work, they require some Customizing settings.

3 This is explained in Chapter 5, Section 5.12.1, and Chapter 7, Section 7.3.3, in *Plant Maintenance with SAP—Practical Guide* (3rd edition, SAP PRESS, 2014).

First, you must decide, for each order type, whether you want to assign SAP PS objects or SAP IM objects to the SAP EAM order (see Figure 6.123). It's not possible to assign both simultaneously.

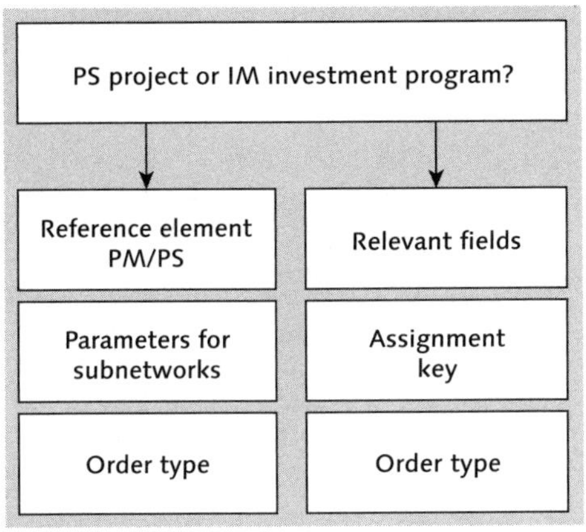

Figure 6.123 Orders: SAP PS or SAP IM

Then, depending on whether you want to assign SAP PS objects or SAP IM objects, additional settings are necessary:

- If you want to assign SAP PS objects, use the *PM/PS reference element* to determine the parameters for an automatic assignment, and use the parameters for subnetworks to determine the permitted SAP PS network type/SAP EAM order type combinations.
- If you want to assign SAP IM objects, determine the relevant field for the assignment and then use an *assignment key* to assign them to an SAP EAM order type.

Define Transfer of Project or Investment Program

You use this Customizing function to control, for each object type, whether project numbers maintained in the master data of maintenance objects *or* investment programs are to be automatically copied to the maintenance order. It isn't possible to copy both to an SAP EAM order simultaneously.

Prerequisites

You must define the order types beforehand.

Customizing Path

Plant Maintenance and Customer Service • Maintenance and Service Processing • Maintenance and Service Orders • Functions and Settings for Order Types • Define Transfer of Project or Investment Program

Settings or Recommended Settings

If you want to copy project numbers from the piece of equipment or functional location to the maintenance order, enter "X" in the Copy assignment column (see Figure 6.124).

If you want to copy investment programs from the piece of equipment or functional location to the maintenance order, enter "1" in the Copy assignment column.

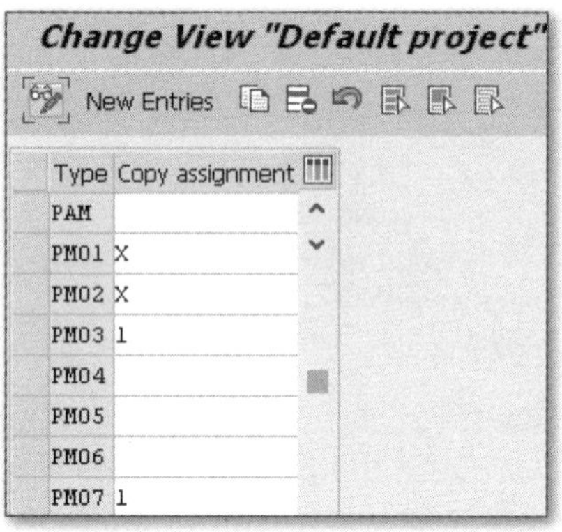

Figure 6.124 Orders: Project Number or Investment Program

Define Relevant Fields for Assignment of IM Program

You can use this Customizing function to define, for each assignment key, the fields that are to be used to automatically determine an *investment program*. This concerns fields in the maintenance order or fields from the master data of the reference object (e.g., the functional location or piece of equipment) assigned to the order.

Prerequisites

There are no specific prerequisites.

Customizing Path

PLANT MAINTENANCE AND CUSTOMER SERVICE • MAINTENANCE AND SERVICE PROCESSING • MAINTENANCE AND SERVICE ORDERS • FUNCTIONS AND SETTINGS FOR ORDER TYPES • DEFINE RELEVANT FIELDS FOR ASSIGNMENT OF IM PROGRAM

Settings or Recommended Settings

You can use the following fields to control the automatic assignment of the investment program (see Figure 6.125):

- FISCAL YR
  Fiscal year.
- COMP. CODE
  Company code.
- BUS. AREA
  Business area.
- CO AREA
  Controlling area.
- PLNT
  Choose from 1 = maintenance plant, 2 = planning plant, or blank = no effect.
- CCTR
  Choose from 1 = cost center of reference object, 2 = cost center of main work center, or blank = no effect.
- PROFIT CTR
  Profit center.
- PLANT SEC.
  Plant section.
- FL
  Choose from 1 = functional location of order, 2 = top functional location of order, or blank = no effect.

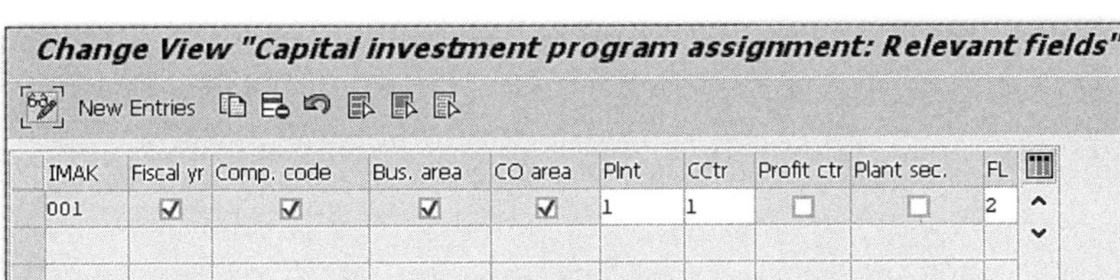

Figure 6.125 Orders: Fields for Assignment of Investment Program

Assign IM Assignment Key to Order Types

You can use this Customizing function to define, for each order type, an assignment key that can be used to automatically find a specific investment program.

Prerequisites

You must define the order types and assignment keys beforehand.

Customizing Path

Plant Maintenance and Customer Service • Maintenance and Service Processing • Maintenance and Service Orders • Functions and Settings for Order Types • Assign IM Assignment Key to Order Types

Settings or Recommended Settings

Enter the corresponding assignment key for the order types for which you want the system to automatically propose investment programs (see Figure 6.126).

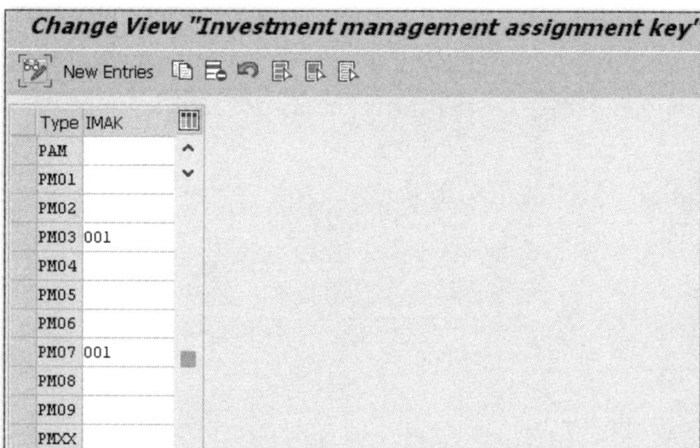

Figure 6.126 Orders: Assign Investment Management to Order Type

Define Field Values for PM/PS Reference Element

You use this Customizing function to define the fields that you want to use for the automatic assignment of maintenance orders to WBS elements. To ensure that this automatic assignment works, you must fulfill the following prerequisites:

- Use the Customizing function DEFINE FIELD VALUES FOR PM/PS REFERENCE ELEMENT to maintain fields that will later establish a direct connection.
- Assign the PM/PS reference element to a task list (see Figure 6.127) or maintenance item.
- Assign the PM/PS reference element to standard objects (standard WBS, standard network) or operational objects (WBS element, network) (see Figure 6.127).

You can then use Transaction ADPMPS to perform an automatic or semi-automatic assignment of maintenance orders to WBS elements or networks.

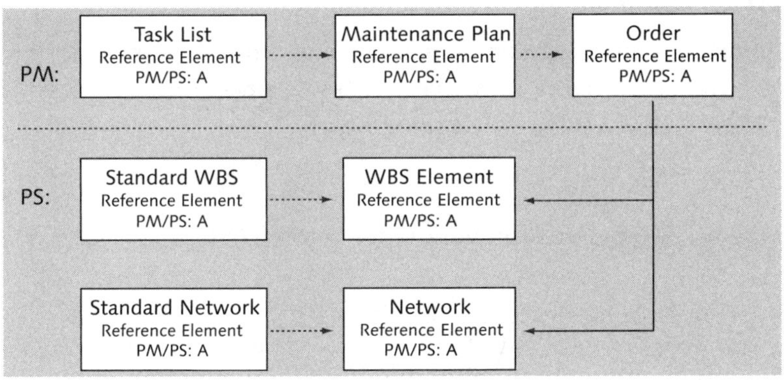

Figure 6.127 Orders – PM/PS Relationships

Prerequisites

There are no specific prerequisites.

Customizing Path

PLANT MAINTENANCE AND CUSTOMER SERVICE • MAINTENANCE AND SERVICE PROCESSING • DEFINE FIELD VALUES FOR PM/PS REFERENCE ELEMENT

Settings or Recommended Settings

Define the PM/PS reference elements as a grouping for subsequent maintenance tasks (see Figure 6.128).

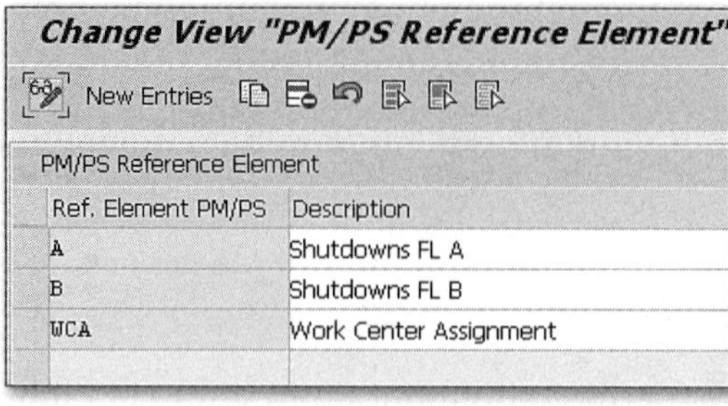

Figure 6.128 Orders: PM/PS Reference Element

Define Parameters for Subnetworks

You use this Customizing function to define possible combinations for the assignment of SAP EAM order types to SAP PS network types. From a project processing perspective, SAP EAM orders are known as *subnetworks*.

Prerequisites

You must define the order types and network types beforehand.

Customizing Path

PROJECT SYSTEM • STRUCTURES • OPERATIVE STRUCTURES • NETWORK • DEFINE PARAMETERS FOR SUBNETWORKS

Transaction

OPTP

Settings or Recommended Settings

In this Customizing function, define each SAP EAM order type that you want to assign to a SAP PS network type. In addition, define which control key and dates are to be copied from the SAP PS network to the SAP EAM order (see Figure 6.129).

Define Investment Profiles

You use this Customizing function to control, in the case of investment measures, the integration into Asset Accounting. In particular, you control whether an asset under construction (AuC) is to be created automatically and, if so, to which asset class this is to belong.

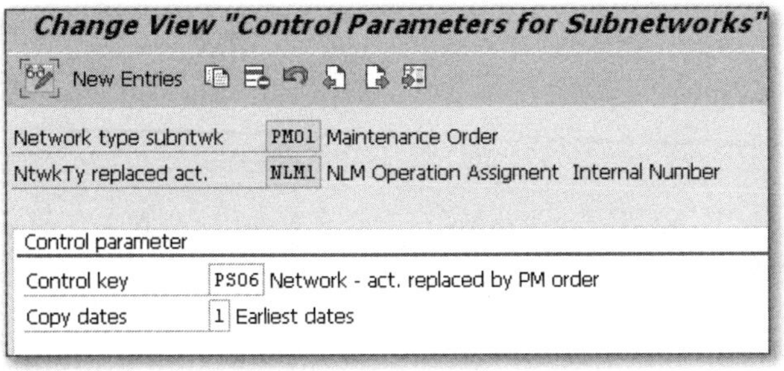

Figure 6.129 Orders: Parameters for Subnetworks

Prerequisites

You must define the asset classes beforehand.

Customizing Path

Plant Maintenance and Customer Service • Maintenance and Service Processing • Maintenance and Service Orders • Functions and Settings for Order Types • Order Types and Investment Management • Define Investment Profiles

Transaction

OITA

Settings or Recommended Settings

Figure 6.130 shows investment profile 000001, which is an investment profile template provided by SAP. You can use this template as a basis for defining your own investment profiles:

- Set the Manage AuC indicator to define the parallel management of an AuC. If you don't set this indicator, parallel management of the associated investment measures isn't possible in Asset Accounting.
- Use the Inv.meas. ast.class (assets under construction in investment measures) option to determine the asset class for your automatically created AuC.
- If you want to manage more than one AuC for each investment measure, you must set the AuC per source structure/assignmt indicator.

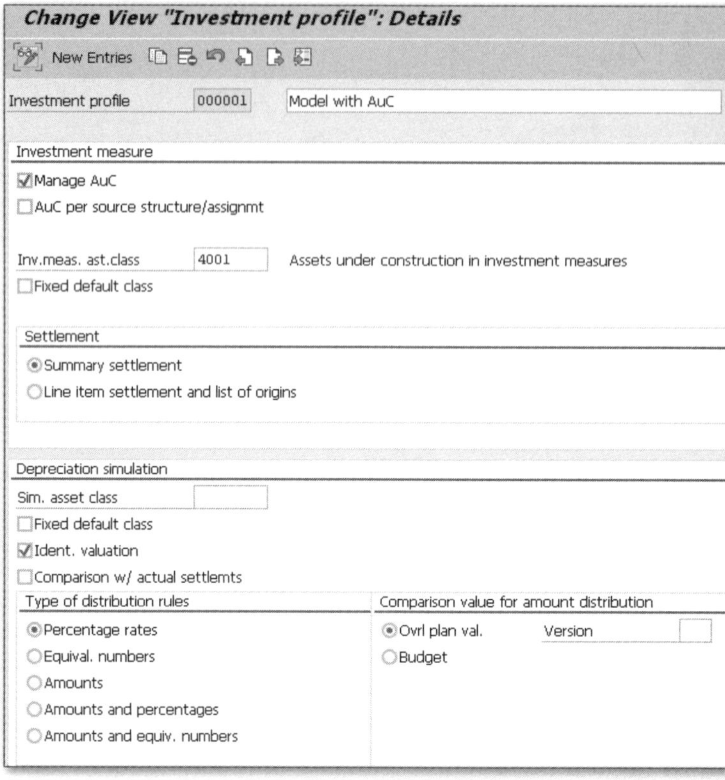

Figure 6.130 Orders: Investment Profile

- Use the radio buttons in the SETTLEMENT screen area to determine whether investment measures are to be managed on a line item or summary basis. SUMMARY means that only flat-rate distribution rules can be entered for the entire debit or for specific cost elements. In contrast with LINE ITEM SETTLEMENT AND LIST OF ORIGINS, the SUMMARY settlement has the advantage of better performance. However, when you use SUMMARY settlement, the system can't display the origins of the line items.
- In the TYPE OF DISTRIBUTION RULES area, define how rules are to be entered for the depreciation forecast (distribution according to percentages, amounts, equivalence numbers, etc.).

Indicate Order Types for Investment Measures

You use this Customizing function to define which of your order types is available for investments and which investment profile is to be used.

Prerequisites

You must define the order types and investment profiles beforehand.

Customizing Path

PLANT MAINTENANCE AND CUSTOMER SERVICE • MAINTENANCE AND SERVICE PROCESSING • MAINTENANCE AND SERVICE ORDERS • FUNCTIONS AND SETTINGS FOR ORDER TYPES • ORDER TYPES AND INVESTMENT MANAGEMENT • INDICATE ORDER TYPES FOR INVESTMENT MEASURES

Settings or Recommended Settings

If you want to manage investment orders, activate an order type by setting the relevant indicator in the INVEST. ORDER column. Then, assign a suitable investment profile (see Figure 6.131).

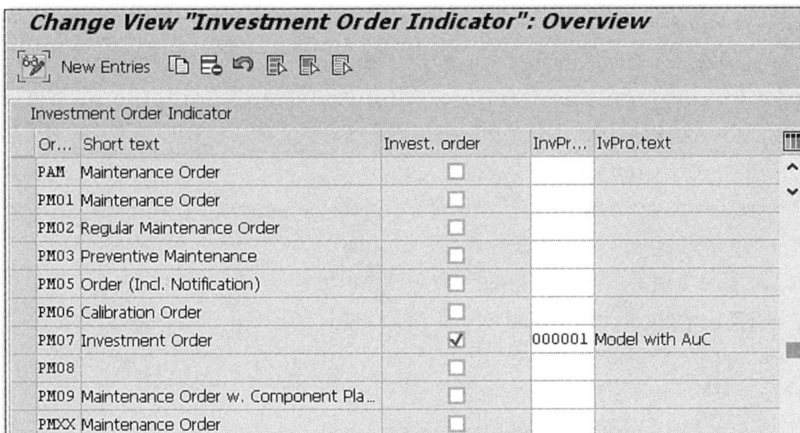

Figure 6.131 Orders: Investment Order Indicator

6.2.9 Other Functions

This section introduces you to some other Customizing functions for orders, which don't relate to the aforementioned subsections but nevertheless need to be configured in the system to complete all of the necessary settings for working with orders.

Create System Conditions or Operating Conditions

You use this Customizing function to define the operating condition that the piece of equipment or functional location is to have during order processing (e.g., IN OPERATION or NOT IN OPERATION). If necessary, you also define the capacity load on the production work center here.

Prerequisites

There are no specific prerequisites.

Customizing Path

PLANT MAINTENANCE AND CUSTOMER SERVICE • MAINTENANCE AND SERVICE PROCESSING • MAINTENANCE AND SERVICE ORDERS • GENERAL DATA • CREATE SYSTEM CONDITIONS OR OPERATING CONDITIONS

Settings or Recommended Settings

The integration between plant maintenance and production is described in *Plant Maintenance with SAP—Practical Guide* (Section 6.2.2, Production Planning and Control). Essentially, you can define a production work center (WORK CENTER field) in the reference objects (functional location and/or a piece of equipment). If your maintenance tasks are to influence the capacities in the production work center, you must define an operating condition by selecting the relevant indicator in the RESERVATION BY PM column (see Figure 6.132). The maintenance order then generates a capacity load for the production work center. However, this doesn't affect the capacity load on the maintenance work center.

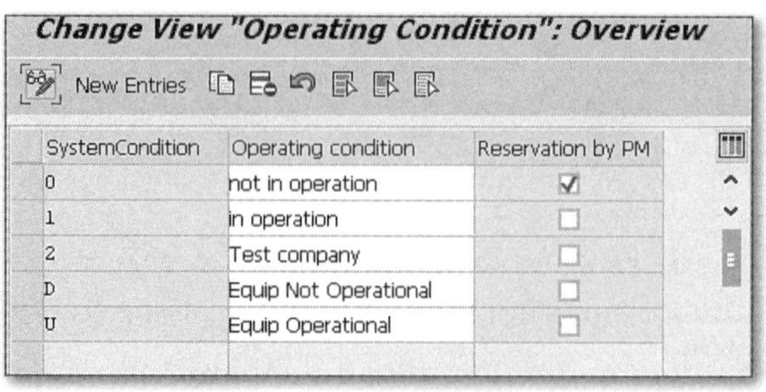

Figure 6.132 Orders: Operating Condition

Define PRT Control Keys

When you process orders, you can assign production resources/tools (PRTs) at the operation level. PRTs such as protective clothing, hand pallet trucks, drawings, or tools are needed to process the order. However, unlike components, they aren't consumed. For each assignment, you must specify a *PRT control key* for the production resource/tool. You use this Customizing function to define PRT control keys.

[+]

PRT Control Keys versus Control Keys

Don't confuse PRT control keys with the control keys that you assign to a work center, to an operation in a task list, or to an operation in an order. PRT control keys have nothing to do with these control keys.

Prerequisites

There are no specific prerequisites.

Customizing Path

PLANT MAINTENANCE AND CUSTOMER SERVICE • MAINTENANCE AND SERVICE PROCESSING • MAINTENANCE AND SERVICE ORDERS • PRODUCTION RESOURCE/TOOL ASSIGNMENTS • DEFINE PRT CONTROL KEYS

Settings or Recommended Settings

You can use PRT control keys to specify the following (see Figure 6.133):

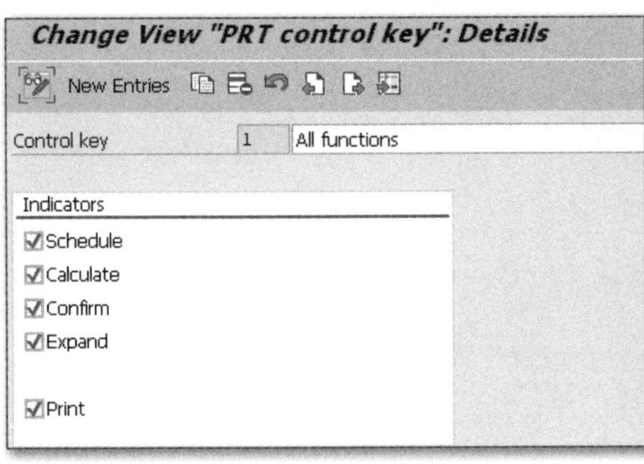

Figure 6.133 Orders: PRT Control Keys

- Schedule
- Calculate
- Confirm
- Expand (additional information when printing)
- Print

[+]

Only the Print Indicator Is Relevant

At present, only the PRINT indicator is relevant for maintenance orders. You also control whether the PRTs are to be output during shop paper printing. As of yet, none of the other functions have been programmed for maintenance orders.

Message Control

You use this Customizing function to define the manner in which a system message concerning orders is output (e.g., error or warning).

Prerequisites

There are no specific prerequisites.

Customizing Path

PLANT MAINTENANCE AND CUSTOMER SERVICE • MAINTENANCE AND SERVICE PROCESSING • MAINTENANCE AND SERVICE ORDERS • MESSAGE CONTROL

Settings or Recommended Settings

Define how each message is to be output. The following options are available (see Figure 6.134):

- Blank: No message
- W: Warning
- E: Error message
- S: Result message
- P: Dialog box
- I: Information
- X: Program termination (a little joke among developers)

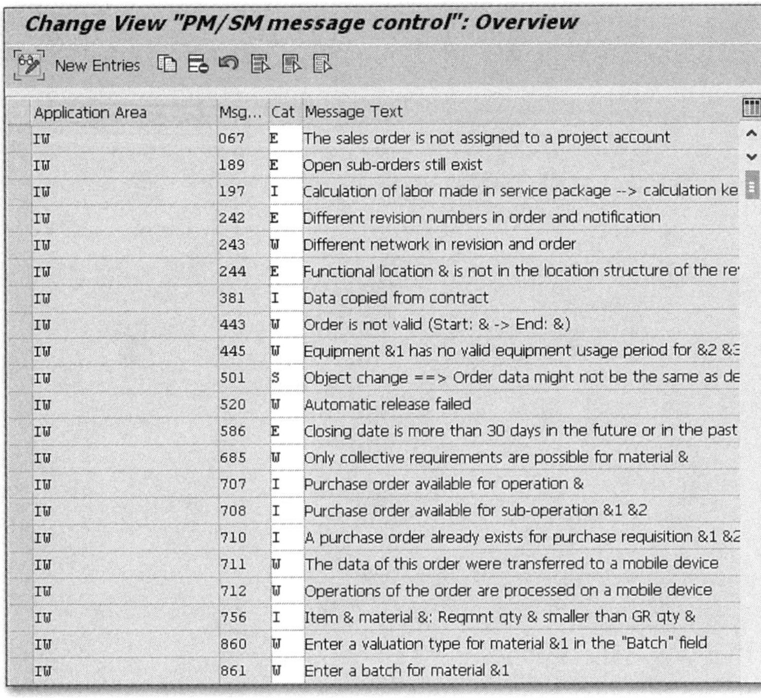

Change View "PM/SM message control": Overview

New Entries

Application Area	Msg...	Cat	Message Text
IW	067	E	The sales order is not assigned to a project account
IW	189	E	Open sub-orders still exist
IW	197	I	Calculation of labor made in service package --> calculation ke
IW	242	E	Different revision numbers in order and notification
IW	243	W	Different network in revision and order
IW	244	E	Functional location & is not in the location structure of the re
IW	381	I	Data copied from contract
IW	443	W	Order is not valid (Start: & -> End: &)
IW	445	W	Equipment &1 has no valid equipment usage period for &2 &3
IW	501	S	Object change ==> Order data might not be the same as de
IW	520	W	Automatic release failed
IW	586	E	Closing date is more than 30 days in the future or in the past
IW	685	W	Only collective requirements are possible for material &
IW	707	I	Purchase order available for operation &
IW	708	I	Purchase order available for sub-operation &1 &2
IW	710	I	A purchase order already exists for purchase requisition &1 &2
IW	711	W	The data of this order were transferred to a mobile device
IW	712	W	Operations of the order are processed on a mobile device
IW	756	I	Item & material &: Reqmnt qty & smaller than GR qty &
IW	860	W	Enter a valuation type for material &1 in the "Batch" field
IW	861	W	Enter a batch for material &1

Figure 6.134 Orders: Message Control

Table 6.8 contains the most important constellations in which system messages occur.

System Message	Description
189 (Open suborders still exist)	You want to technically complete a superior order. However, at least one suborder has yet to be technically completed.
242 (Different revision numbers in order and notification)	You've assigned different revision numbers to a notification and order.
243 (Different network in revision and order)	You've assigned different networks to a revision and order.
244 (Functional location & is not in the location structure of the revision)	You've assigned a functional location number to an order. However, the functional location number isn't in the functional location structure that you've assigned to a revision number.

Table 6.8 System Messages

System Message	Description
381 (Data copied from contract)	You've assigned a contract number to a nonstock component or an external processing operation, and, from there, you've copied information such as the vendor, material group, and price to the order.
445 (Equipment &1 has no valid equipment usage period for &2 &3)	You've assigned a piece of equipment to an order. However, this piece of equipment isn't valid on the order date. Either the date isn't valid yet (i.e., the piece of equipment has a valid-from date that is later than the basic finish date of the order), or the piece of equipment is no longer valid (i.e., it has a valid-to date that is earlier than the basic start date of the order).
501 (Object change → Order data might not be the same as default values)	You've retroactively assigned another piece of equipment and/or another functional location to the order.
520 (Automatic release failed)	You've set the AUTOMATIC RELEASE indicator for the order type because you want the system to automatically release an order from a maintenance plan or notification. However, the release fails (e.g., due to a material shortage).
707 (Purchase order available for operation &)	You want to set a deletion flag in an order but there is still at least one open purchase order (i.e., a purchase order for which there is no goods receipt or invoice receipt).
708 (Purchase order available for suboperation &1 &2)	You want to set a deletion flag in a suborder, but there is still at least one open purchase order (i.e., a purchase order for which there is no goods receipt or invoice receipt).
860 (Enter a valuation type for material &1 in the "Batch" field)	This concerns a refurbishment order (see Chapter 8, Section 8.2). You've assigned a component to an order. The component is subject to a separate valuation, but you haven't specified a valuation type.
861 (Enter a batch for material &1)	You're currently planning a component that is subject to a batch management requirement. However, you haven't entered a batch.

Table 6.8 System Messages (Cont.)

Define Proposed Reference Time for Technical Completion

You use this Customizing function to define which date is to be proposed as the date for the technical completion of an order. This is crucial when sorting orders into chronological order in the history (e.g., the month to which the order is assigned in PMIS).

Prerequisites

You must define the order types beforehand.

Customizing Path

Plant Maintenance and Customer Service • Maintenance and Service Processing • Maintenance and Service Orders • Functions and Settings for Order Types • Define Proposed Reference Time for Technical Completion

Settings or Recommended Settings

Define, for each order type, which date is to be proposed as the date for the technical completion of the order (see Figure 6.135).

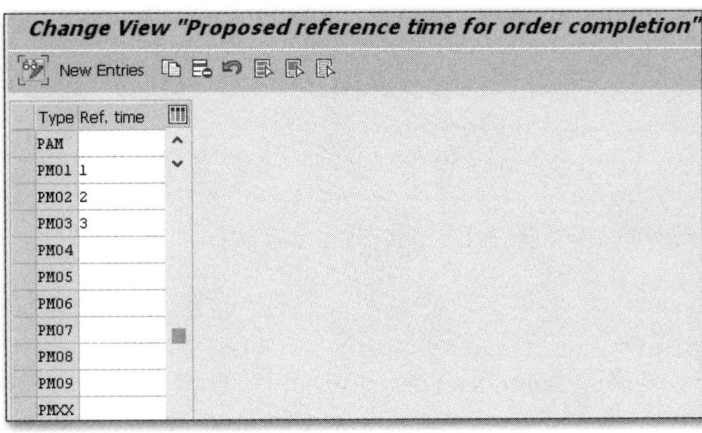

Figure 6.135 Orders: Reference Time for Technical Completion

Here, you have the following options:

- Current date
- 1: Entry date

- 2: Basic start date
- 3: Basic finish date

Irrespective of the date proposed, you can overwrite the technical completion date.

Create Default Value Profiles for General Order Data

You use this Customizing function to define certain default values that you assign, together with other default values in the Customizing function DEFAULT VALUES FOR TASK LIST DATA AND PROFILE ASSIGNMENTS, to the order type and plant. Figure 6.136 shows the relationships between these default values.

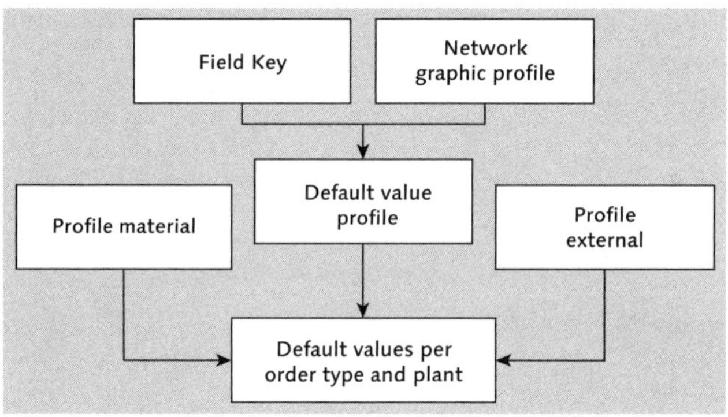

Figure 6.136 Orders: Relationships among Default Values

Prerequisites
You must define the field key and network graphic profiles beforehand.

Customizing Path
PLANT MAINTENANCE AND CUSTOMER SERVICE • MAINTENANCE AND SERVICE PROCESSING • MAINTENANCE AND SERVICE ORDERS • FUNCTIONS AND SETTINGS FOR ORDER TYPES • CREATE DEFAULT VALUE PROFILES FOR GENERAL ORDER DATA

Settings or Recommended Settings
You can define the following default values in a default value profile (see Figure 6.137):

- You require the FIELD KEY if you need additional nonstandard fields in the operation.
- Use the CALCULATION KEY to control the relationship between the WORK and DURATION fields. Here, you have the following options:
  - 0: Maintain manually
  - 1: Calculate duration
  - 2: Calculate work
  - 3: Calculate number of capacities

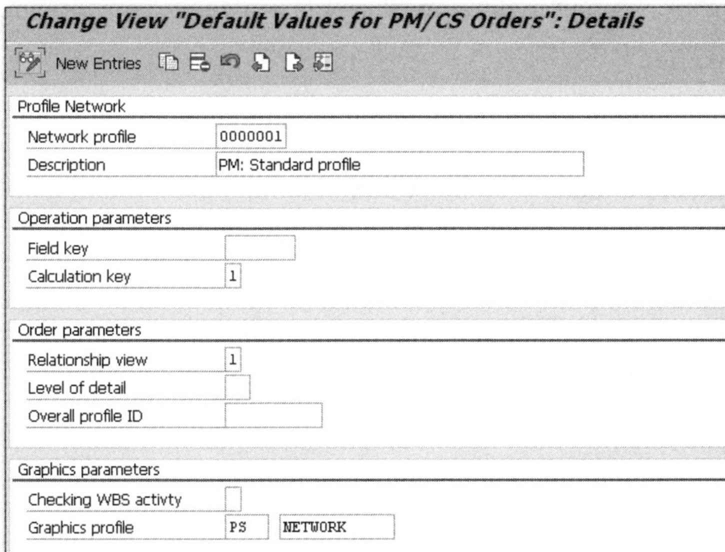

Figure 6.137 Orders: Default Value Profiles

- In the RELATIONSHIP VIEW field, define the view from which the relationships in the operations are to be displayed. Here, you have the following options:
  - 1: Predecessor
  - 2: Successor
  - 3: Mixed
  - 4: As created

- Use the Graphics profile to define the complete layout of the network graphic. Detailed explanations are provided in the subsection "Network—Define Graphic Profiles."

Note

Don't be confused by the use of the term Network Profile in this Customizing function. This is a term from Project Processing. The Default value profile is actually meant here.

Default Values for Task List Data and Profile Assignments

You use this Customizing function to define profiles and other default values for each order type and plant.

Prerequisites

You must define the order types, plants, default value profiles, and the profiles for external material and external processing beforehand.

Customizing Path

Plant Maintenance and Customer Service • Maintenance and Service Processing • Maintenance and Service Orders • Functions and Settings for Order Types • Default Values for Task List Data and Profile Assignments

Settings or Recommended Settings

Make the following assignments for each order type and plant (see Figure 6.138):

- ExternProfile
  Assigns default values for the cost element, purchasing organization, material group, and purchasing group for external processing (see also Chapter 8, Section 8.1, External Processing).
- Mat.Profile
  Assigns default values for the cost element, purchasing organization, material group, and purchasing group for external material (see also Section 6.2.2).
- MaintProfile
  Assigns the default value profiles.

- ACT./OP. UOM
  Defines the default value for the unit of measure for the operation.

In addition to these default values, you can use this Customizing function to define how tasks lists are to be transferred to orders:

- Whether a dialog box for selecting operations is to appear
- Whether a dialog box for selecting work centers is to appear
- Whether the operations selected are to be renumbered consecutively in intervals of 10 operations
- Whether a task list can be fully integrated once only
- Whether the sequence of operations is the same as in the task list

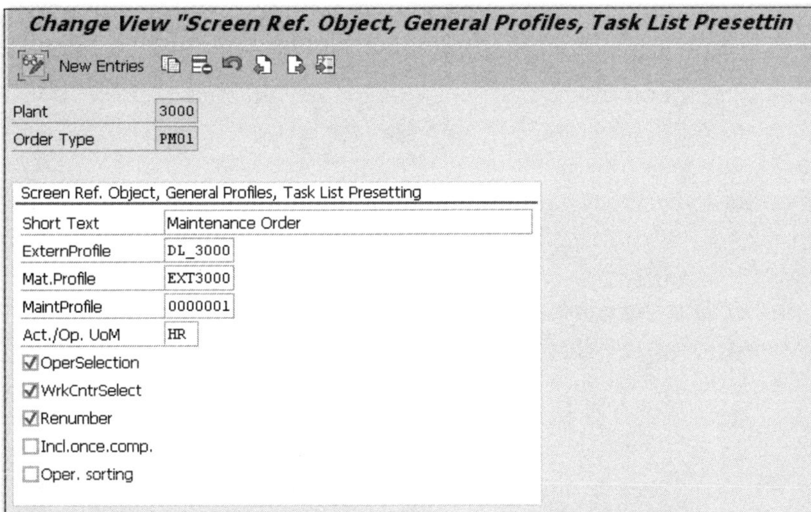

Figure 6.138 Orders: Default Values

Note

These default values and settings are overwritten by user-specific default values, which the user can configure in the order under EXTRAS • SETTINGS • DEFAULT VALUES.

6.3 Completion Confirmations

The Customizing functions available to you for confirmation reporting are the maintenance transactions (Transaction IW41 et seq.) and the Customizing functions for the Cross-Application Time Sheet (CATS).

6.3.1 SAP EAM Completion Confirmations

The following Customizing functions are available to you if you want to use completion confirmations in conjunction with the maintenance Transactions IW41 et seq.

Define Control Parameters for Completion Confirmations

You use this Customizing function to define, for each order type and plant, the control parameters for confirmation reporting in SAP EAM. This is a basic prerequisite for all other functions, irrespective of whether you use maintenance transactions, CATS transactions, or other transactions to execute your completion confirmations.

Prerequisites
You must define the order types and plants beforehand.

Customizing Path
Plant Maintenance and Customer Service • Maintenance and Service Processing • Completion Confirmations • Define Control Parameters for Completion Confirmations

Transaction
OIOR

Settings or Recommended Settings
The following control options are available for each order type and plant (see Figure 6.139):

- Set the Final Confirmation indicator if you want the system to propose the final confirmation indicator for the completion confirmation.
- Set the Post open reservs. indicator if you want the system to propose the Post open reservs. indicator for the completion confirmation. When you set this

indicator, open reservations (i.e., reservations that haven't been withdrawn) are deleted with the final confirmation.

- Set the PROPOSE DATES indicator if you want the system to propose the operation dates (scheduled start date and scheduled finish date) as completion confirmation dates.
- Set the PROPOSE ACTIVITIES indicator if you want the system to propose the work as actual work or, in the case of existing completion confirmations, the remaining work as actual work.
- Set the CALC. PERFORMANCE indicator to determine whether the system is to calculate the actual work if a change is made to the dependent fields (e.g., the FINAL CONFIRMATION or the REMAINING WORK field).

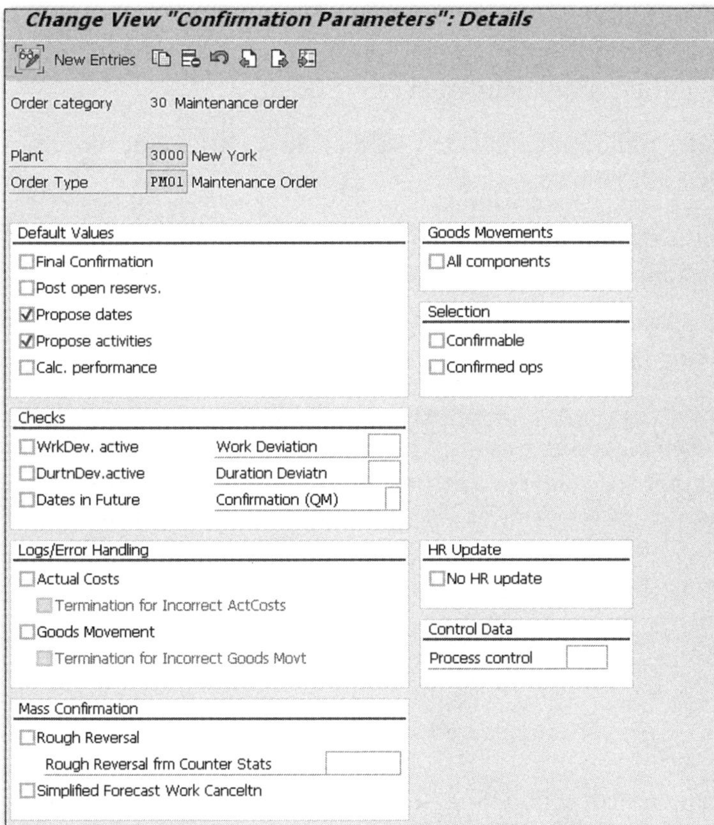

Figure 6.139 Completion Confirmations: Control Parameters

- Set the WrkDev. active indicator if you want to activate the deviation check for work. Then, in the Work Deviation field, specify the percentage that this deviation must exceed before the system issues a warning message.
- Set the DurtnDev.active indicator if you want to activate the deviation check for the duration. Then, in the Duration Deviatn field, specify the percentage that this deviation must exceed before the system issues a warning message.
- Set the Dates in Future indicator if you want the system to permit the use of dates that lie in the future (e.g., the posting date, end of execution time, etc.). For example, this is necessary if you enter a time recording on October 31st, but you want to add this time to the month of November (posting date).
- In the Logs/Error Handling screen area, specify that the system will automatically display a log if the actual cost calculation or material posting is incorrect. Set the Termination for Incorrect ActCosts indicator or the Termination for Incorrect Goods Movt indicator if you want to ensure that the time confirmation isn't posted if the aforementioned errors occur.

Use the All components indicator to decide which components are displayed in the backflushing overview:

- If you set this indicator, the system displays all components irrespective of whether the backflushing indicator is set for the component in the order.
- If you don't set this indicator, the system displays only those components for which the backflushing indicator is set.

[+]

Recommendations for the Control Parameters for Completion Confirmations

- Make sure that you define the control parameters for all plants and order types that are to be used. Otherwise, you may not be able to confirm an order type in the plant in question.
- Set the Propose dates and Propose activities indicators, and possibly the All components indicator.
- Leave all other indicators and default values blank.

Define Causes for Variances

You use this Customizing function to define, for each plant, reasons for variances between the actual times and planned times. You can then use these reasons in your analyses.

Prerequisites

You must define the plants beforehand.

Customizing Path

PLANT MAINTENANCE AND CUSTOMER SERVICE • MAINTENANCE AND SERVICE PROCESSING • COMPLETION CONFIRMATIONS • DEFINE CAUSES FOR VARIANCES

Settings or Recommended Settings

For each plant, define the variance reasons that you want to track later (see Figure 6.140).

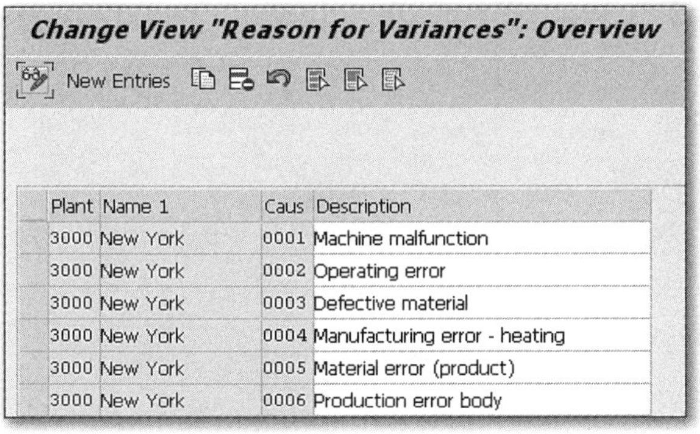
Change View "Reason for Variances": Overview

New Entries

Plant	Name 1	Caus	Description
3000	New York	0001	Machine malfunction
3000	New York	0002	Operating error
3000	New York	0003	Defective material
3000	New York	0004	Manufacturing error - heating
3000	New York	0005	Material error (product)
3000	New York	0006	Production error body

Figure 6.140 Completion Confirmations: Reasons for Variances

Set Screen Templates for Completion Confirmation

You can use this Customizing function to compile the individual screen areas for the overall completion confirmation (Transaction IW42). You can define different profiles here. Each user can later select the most appropriate profile in his user settings for the overall completion confirmation (in Transaction IW42 under EXTRAS • SETTINGS).

Prerequisites

There are no specific prerequisites.

Customizing Path

PLANT MAINTENANCE AND CUSTOMER SERVICE • MAINTENANCE AND SERVICE PROCESSING • COMPLETION CONFIRMATIONS • SET SCREEN TEMPLATES FOR COMPLETION CONFIRMATION

Settings or Recommended Settings

Each profile contains five rows, which, in turn, contain one or two screen areas (see Figure 6.141).

IW42 – Overall Completion Confirmation	
Subscreen 1	Subscreen 2
Subscreen 3	Subscreen 4
Subscreen 5	Subscreen 6
Subscreen 7	Subscreen 8
Subscreen 9	Subscreen 10

Figure 6.141 Completion Confirmations: Profile for the Overall Completion Confirmation

The following screen areas are available (see Figure 6.142):

- TIMES
- SERVICES
- GOODS MOVEMENTS
- NOTIFICATION ITEMS
- CAUSES OF DAMAGE
- MEAS./CNTR READINGS
- TASKS
- ACTIVITIES
- ITEMS

If, in a particular row, you fill only the left screen area, the screen area in the Transaction IW42 entry screen spans the entire width of the screen. If you fill the left and right screen areas, the two screen areas are displayed side by side.

In addition to the layout of the screen areas, you can specify which pushbuttons are to appear (e.g., Times or Goods movements). Then, in the overall completion confirmation, use the pushbuttons to switch to the corresponding full screen. You then no longer have six entry rows but rather an entire overview screen.

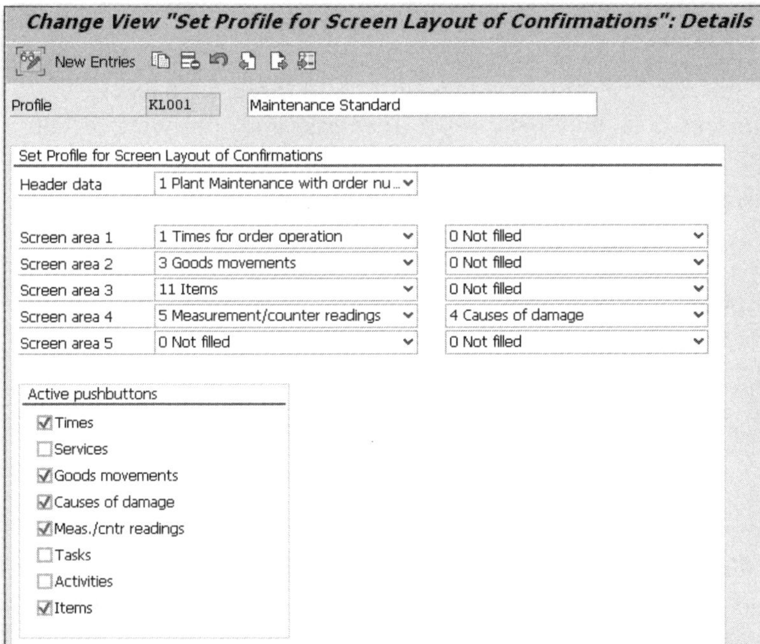

Figure 6.142 Completion Confirmations Profile

Define Separate Layouts for the Overall Completion Confirmation [+]

Seize this opportunity to provide your users with suitable profiles for their work environment. The "as much as necessary, but as little as possible" principle from earlier in the chapter applies here as well.

6.3.2 CATS Confirmations

Cross-Application Time Sheet (CATS) is an application that you can use to enter several person-related actual times. This application is not only available in plant maintenance but also in other application areas such as Human Resources or Production, for example.

Next, we'll discuss the most important Customizing functions for use in plant maintenance.

Set Up Data Entry Profiles

This Customizing function is a prerequisite for using CATS. In the data entry profiles, you define, for example, whether a separate release and approval procedure is required for all actual times entered, the number of periods for which times can be entered simultaneously, and whether you want to work with worklists.

Prerequisites
There are no specific prerequisites.

Customizing Path
Cross-Application Components • Time Sheet • Specific Settings for CATS regular • CATS regular • Record Working Time • Set Up Data Entry Profiles

Settings or Recommended Settings
Because the data entry profile is a long screen with a large number of settings, we've divided it into two parts. Figure 6.143 shows the upper part of the data entry profile. The following settings are the most important settings from an SAP EAM perspective:

- Set the Profile changeable indicator if you want the user to be able to change the profile settings (with restrictions, e.g., display or hide a totals column) while entering the relevant data.
- Set the With totals line indicator if you want the system to display a daily total of all times entered.
- Set the Release on saving indicator if you want the system to release all entered times when the user chooses Save. If you don't set this indicator, the user must release all entered times separately.

- Set the PERIOD TYPE field to control the period view used with the profile (e.g., 2 = week and 4 = month), while the PERIODS field controls the number of periods to be entered.
- If you want to enter this data for several persons, set the ENTER FOR SEVERAL PERSONNEL NOS indicator, and specify how these persons are to be selected (e.g., according to cost center or organizational unit).

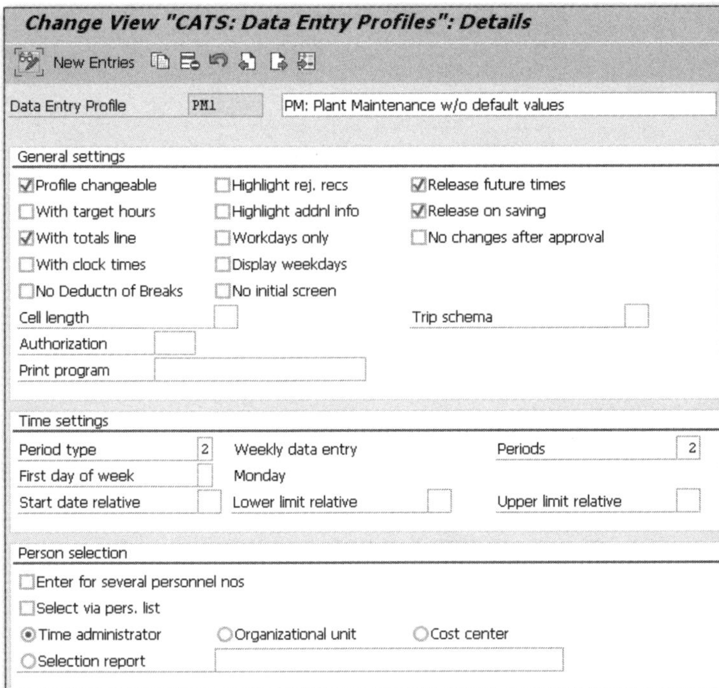

Figure 6.143 Confirmations: CATS Data Entry Profile 1

Figure 6.144 shows the lower part of the data entry profile. Here, you can make the following settings:

- In the APPROVAL screen area, specify whether all entered and released times are to be subject to an approval procedure before they are transferred to SAP EAM. (Because we're covering a standard scenario in which such times aren't subject to an additional approval procedure, the approval procedure isn't discussed here.)

- In the COST ACCOUNTING VARIANT area, specify how personnel costs are to be handled. In SAP EAM, the standard scenario is that the personnel costs are assigned to the master cost center.
- In the DEFAULT VALUES area, define the default values that the system is to automatically preassign when entering times. From an SAP EAM perspective, the following default values make sense: CONTROLLING AREA, ACTIVITY TYPE, and ATT./ABSENCE TYPE.

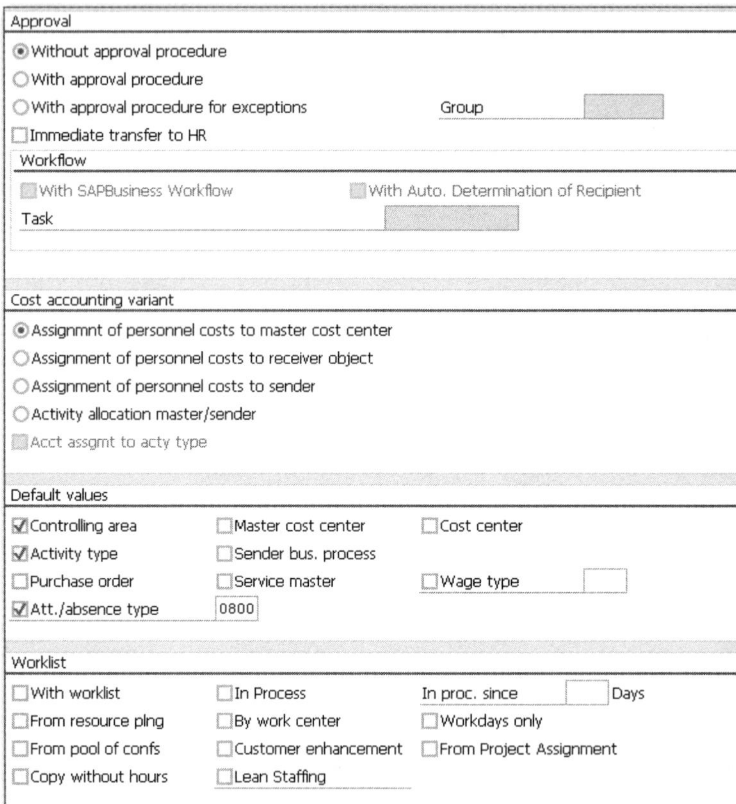

Figure 6.144 Confirmations: CATS Data Entry Profile 2

- In the WORKLIST area, set the WITH WORKLIST indicator if you want to work with a worklist. We recommend that you don't work with a worklist because it makes the entry of actual times very unclear. If, however, you want to work with a worklist, you must define what type of worklist is to be created and how (e.g., on the basis of the operations for the work center).

[+]

Recommendations for the CATS Data Entry Profile

- Allow the user to change the profile while entering the relevant data.
- Allow the system to release all entered times when the user chooses SAVE.
- Display a totals row for each day.
- Enter the times on a weekly basis.
- Don't enter the times for more than one person simultaneously.
- Transfer the times without an approval procedure.
- Set the default values for CONTROLLING AREA and ACTIVITY TYPE.
- Enter the times without a worklist.

Define Field Selection

The Customizing functions for the field selection were explained in Chapter 4, Section 4.9, as generic functions. However, because the field selection is particularly important for CATS, we'll discuss some of its special features here.

Prerequisites

You must set up the data entry profiles beforehand.

Customizing Path

CROSS-APPLICATION COMPONENTS • TIME SHEET • SPECIFIC SETTINGS FOR CATS REGULAR • CATS REGULAR • RECORDWORKING TIME • DEFINE FIELD SELECTION

Transaction

CAC2

Settings or Recommended Settings

Set up the field selection for the DATA ENTRY SECTION screen group, the influencing value DATA ENTRY PROFILE, and the data entry profile that you've set up for plant maintenance (PM1 in our example, see Figure 6.145).

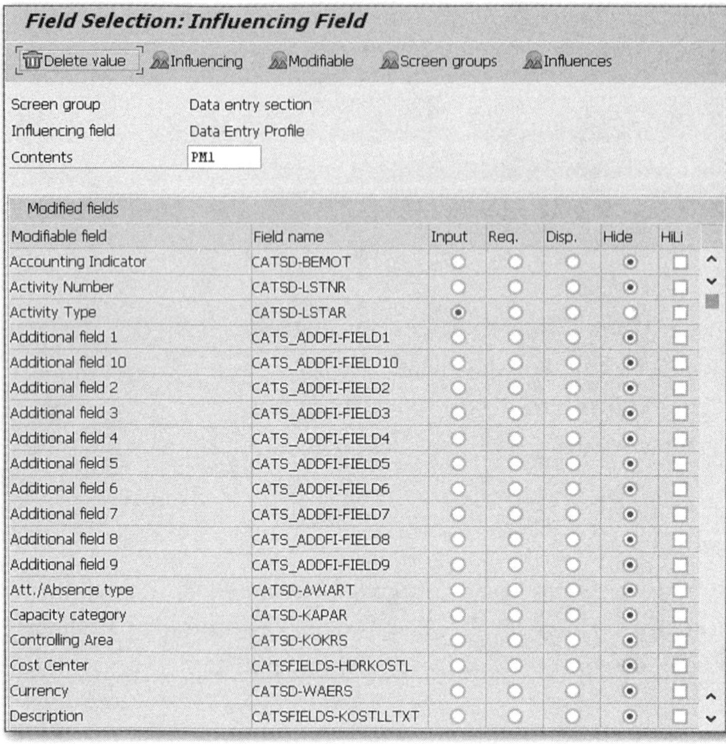

Figure 6.145 Completion Confirmations: CATS Field Selection

[+]

Hide Fields

Because the data entry section has a large number of fields that aren't required from an SAP EAM perspective, it makes sense to hide all of the fields that you don't require and to leave the remaining fields ready for input so that the data entry section remains clear.

From a maintenance perspective, you require only the following fields (all others can and should be hidden):

- Work center: CATSD-ARBPL
- Recipient order: CATSD-RAUFNR
- Final confirmation: CATSDB-AUERU
- Activity Type: CATSD-LSTAR
- Total: CATSD-SUMDAYS
- Partial confirmation: CATSDB-ERUZU
- Operation: CATSD-VORNR
- Plant: CATSD-WERKS

In this chapter, you'll learn about the Customizing functions for preventive maintenance. Here, the focus is on those options available to you within the SAP system for adjusting task lists and maintenance plans to your requirements.

7 Preventive Maintenance

Preventive maintenance is characterized by the fact that you can plan ahead required resources (work centers, materials, external companies, etc.) in terms of content and scheduling. This type of business process is associated with the following scenarios:

- A pump undergoes a visual and functional inspection every 6 months, and the mechanical seal is changed every 12 months.
- Regularly, after 1,000 operation hours, the hydraulic oil in a forklift truck is changed and after 2,000 operation hours, the brake fluid is changed.
- The fire extinguisher in the building is refilled every two years.
- Test equipment is recalibrated every 120 days.

To perform the business processes associated with preventive maintenance, several objects are used in SAP Enterprise Asset Management (SAP EAM), whose significance and interrelationships are briefly explained in the following section (see Figure 7.1):

- **Maintenance strategy**
  A maintenance strategy contains the chronological sequence of maintenance activities (e.g., maintenance packages of 3-, 6-, 12-, and 24 months for time-based maintenance, or maintenance packages of 1,000-, 2,000-, and 5,000 operation hours for performance-based maintenance). The maintenance strategy does not provide any details about the activity, object, or date; it's only required for strategy-based maintenance plans.
- **Task list**
  The task list describes the activities (operations) and contains materials and deadlines (maintenance packages). Here, a distinction is made between object-specific task lists (equipment task lists and functional location task lists), and

neutral task lists (general maintenance task lists). Strategy-based maintenance plans *must* contain a task list, while all other maintenance plans *can* contain a task list (optional).

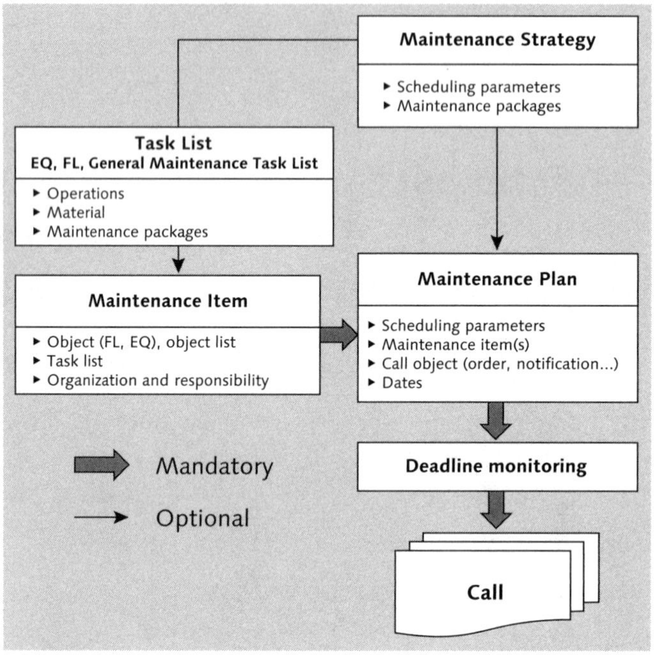

Figure 7.1 Preventive Maintenance: Interrelationships

- **Maintenance item**
  The maintenance item describes the activities to be performed, contains the reference object (or the object list), and has organizational data relevant for subsequent processing.
- **Maintenance plan**
  The maintenance plan contains one or more maintenance items and determines the maintenance dates and maintenance call object (order, notification, etc.).
- **Deadline monitoring**
  Deadline monitoring (Program RISTRA20) runs automatically as a batch job and ensures that the maintenance call objects (e.g., the orders) are automatically generated on the due date.

7.1 Task Lists

A task list essentially describes activities (operations) and contains materials that are required during activity processing. From a maintenance perspective, we can distinguish between three different types of *task list*:

- **Equipment task list**
  You create an equipment task list for exactly one piece of equipment if you want to map its specific features. However, you can only use this task list in connection with this one piece of equipment.
- **Functional location task list**
  You create a functional location task list for exactly one functional location if you want to map its specific features. Similarly, you can only use this task list in connection with this one functional location.
- **General maintenance task list**
  A general maintenance task list isn't object-specific, which means that it isn't assigned to any particular piece of equipment or functional location. However, you can indirectly make a general maintenance task list available for several pieces of equipment and/or functional locations. To do this, use the CONSTRUCTION TYPE field in the STRUCTURING screen group in the master record for the piece of equipment or functional location. All pieces of equipment and functional locations for which a material number is entered in the CONSTRUCTION TYPE field can then access general maintenance task lists for which the same material number is entered in the CONSTRUCTION TYPE field in the general maintenance task list header.

A task list comprises the following elements (see Figure 7.2):

- **Header data**
  Header data is information that is used to identify and manage the task list. This data applies to the entire task list (e.g., number, group counter, plant, responsible work center, etc.).
- **Operations**
  Operations are used to describe the work that is to be performed when implementing the task list.
- **Material list**
  The material list contains spare parts that are required and consumed when implementing the task list.

- **Production resources/tools**
  Production resources/tools (PRTs) (e.g., tools, protective clothing, and hand pallet trucks) are also required for implementing the task list. However, unlike a material, they aren't consumed.
- **Inspection characteristics**
  If inspections are to be conducted within an operation (e.g., inspections of length, weight, and function), you can define these as inspection characteristics.

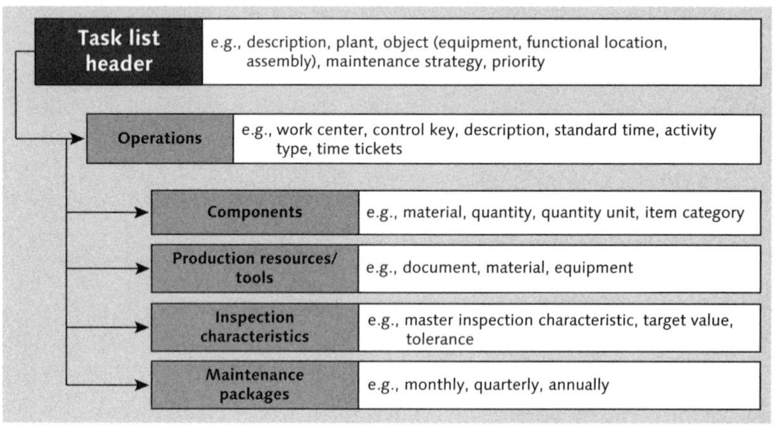

Figure 7.2 Task Lists: Structure and Content

- **Maintenance packages**
  If the task list is used in a strategy maintenance plan, you use maintenance packages to control the frequency with which the maintenance work is performed—either on a time-dependent basis (e.g., once every three months) or on a performance-dependent basis (e.g., once every 1,200 operating hours).

The following Customizing functions are available for adjusting the management and use of task lists to your requirements.

Maintain Task List Status

You use this Customizing function to define which task list status you want to manage. You then use the task list status to control whether the task list is still locked, or to control the functions for which it's released.

Prerequisites

There are no special prerequisites.

Customizing Path

PLANT MAINTENANCE AND CUSTOMER SERVICE • MAINTENANCE PLANS, WORK CENTERS, TASK LISTS AND PRTS • TASK LISTS • GENERAL DATA • MAINTAIN TASK LIST STATUS

Transaction

OIL1

Settings or Recommended Settings

The standard SAP delivery contains the following statuses, among others (see Figure 7.3):

- 1 CREATION PHASE
  This status does not release the task lists for orders (RELIND) or costing (CSTNG).
- 4 RELEASED (GENERAL)
  This status releases the task lists for orders and costing.

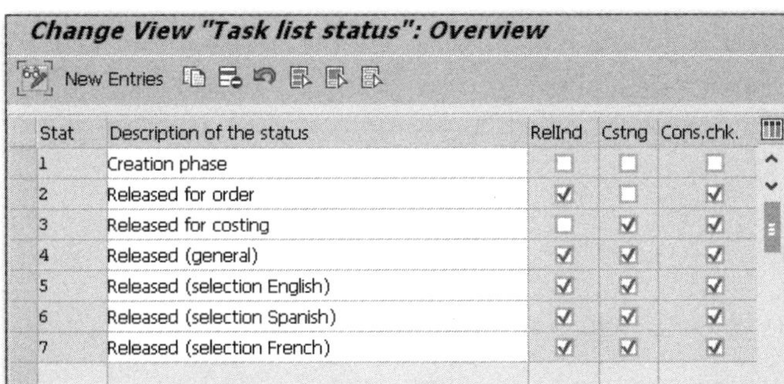

Stat	Description of the status	RelInd	Cstng	Cons.chk.
1	Creation phase	☐	☐	☐
2	Released for order	☑	☐	☑
3	Released for costing	☐	☑	☑
4	Released (general)	☑	☑	☑
5	Released (selection English)	☑	☑	☑
6	Released (selection Spanish)	☑	☑	☑
7	Released (selection French)	☑	☑	☑

Figure 7.3 Task Lists: Task List Statuses

Task List Statuses 1 and 4 Usually Suffice [+]

If you nevertheless want to define your own statuses, make sure that you define at least one status that releases the task lists for both costing and orders.

Define Task List Usage

In the SAP system, task lists are used not only in SAP EAM, but in other applications as well:

- Production Planning, where they are used as normal task lists or standard task lists
- Production Planning for Process Industry, where they are used as master recipes
- SAP Project System (SAP PS), where they are used as standard networks
- SAP Quality Management (QM), where they are used as inspection plans

You use this Customizing function to define task list usages.

Prerequisites

There are no special prerequisites.

Customizing Path

Plant Maintenance and Customer Service • Maintenance Plans, Work Centers, Task Lists and PRTs • Task Lists • General Data • Define Task List Usage

Settings or Recommended Settings

The standard SAP delivery provides task list Usage 4 for Plant maintenance (see Figure 7.4).

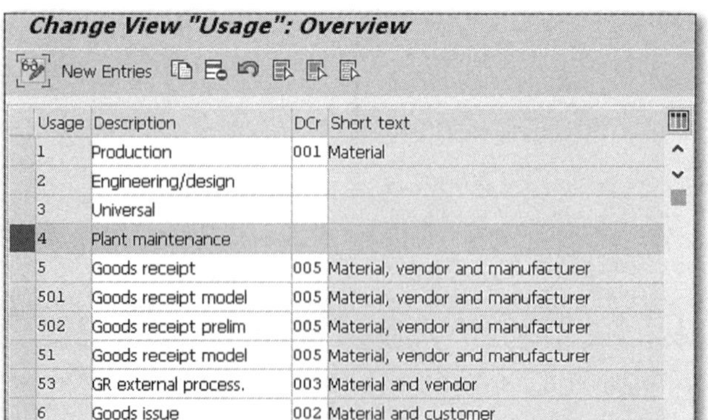

Usage	Description	DCr	Short text
1	Production	001	Material
2	Engineering/design		
3	Universal		
4	Plant maintenance		
5	Goods receipt	005	Material, vendor and manufacturer
501	Goods receipt model	005	Material, vendor and manufacturer
502	Goods receipt prelim	005	Material, vendor and manufacturer
51	Goods receipt model	005	Material, vendor and manufacturer
53	GR external process.	003	Material and vendor
6	Goods issue	002	Material and customer

Figure 7.4 Task Lists: Task List Usage

Task List Usage 4 Usually Suffices [+]

If you nevertheless want to define your own task list usages, make sure to make the right selection in all follow-up actions (e.g., when assigning the task list usage in the work center).

Configure Planner Group

You use this Customizing function to define the groups or persons responsible for maintaining task lists for each plant.

Prerequisites

You must maintain the plants beforehand.

Customizing Path

Plant Maintenance and Customer Service • Maintenance Plans, Work Centers, Task Lists, and PRTs • Task Lists • General Data • Configure Planner Group

Settings or Recommended Settings

At the plant level, you can define descriptive or numeric planner groups that contain a maximum of three numbers or characters (see Figure 7.5).

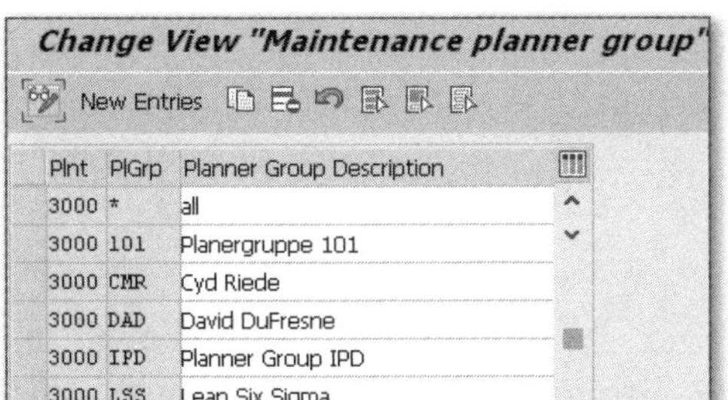

Figure 7.5 Task Lists: Planner Group

Not All Planner Groups Are the Same [+]

Note that the SAP system recognizes two different planner groups:

- **Work scheduler group**
  You assign this group as the responsible planner group to a task list at the header level.

> ▸ **Maintenance planner group**
> You assign this group as the responsible planner group to a piece of equipment, functional location, notification, order, or maintenance item.

Define Profiles with Default Values

You use this Customizing function to maintain default values such as increment or unit. When you create a task list, you can call one of the profiles and transfer its values to the new task list as default values.

Prerequisites

If necessary, you must maintain the following values as default values beforehand:

- Graphic profile (see the Customizing function NETWORK • DEFINE GRAPHIC PROFILES)
- Field key for user fields (see the Customizing function DEFINE USER FIELDS)
- Purchasing organization
- Purchasing group
- Vendor
- Cost element
- Material group
- Task list usage (see the Customizing function DEFINE TASK LIST USAGE)
- Task list status (see the Customizing function MAINTAIN TASK LIST STATUS)

Customizing Path

PLANT MAINTENANCE AND CUSTOMER SERVICE • MAINTENANCE PLANS, WORK CENTERS, TASK LISTS AND PRTS • TASK LISTS • CONTROL DATA • DEFINE PROFILES WITH DEFAULT VALUES

Transaction

OIL6

Settings or Recommended Settings

In particular, you can assign the following default values to a profile (see Figure 7.6):

- Use the Oper./act. increment option to determine the number assignment for operations (generally, 0010).
- Use Group and Name to assign a previously defined graphic profile so that you can create a network within a task list.
- Use Field key to assign additional fields to the task list that are relatively freely definable and don't exist in the standard system.
- Use Calculation key to control the relationship between the Work and Duration fields. Here, you have the following options:
  - 0 – Maintain manually
  - 1 – Calculate duration
  - 2 – Calculate work
  - 3 – Calculate no. of capacities

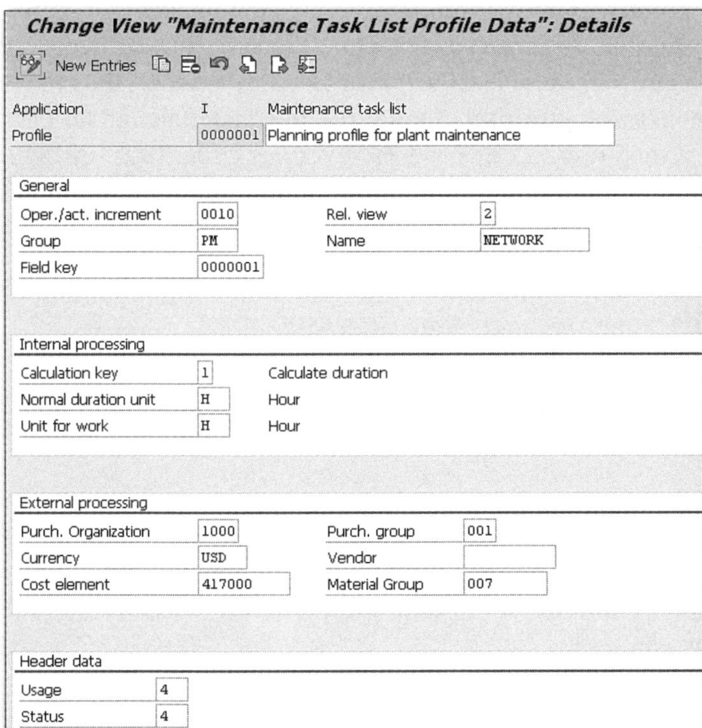

Figure 7.6 Task Lists: Profiles with Default Values

- Use UNIT FOR WORK to determine the units to be defined as default values for the WORK and DURATION fields.
- In the EXTERNAL PROCESSING field group, define default values for the purchasing organization, purchasing group, currency, vendor, cost element, and material group if an operation is processed externally.
- Use USAGE and STATUS to assign default values to the task list for task list usage and the task list status.

[+] **Define and Use Default Value Profiles**

Create different default value profiles so that your users don't have to repeatedly enter the same data manually.

Define Presetting for Free Assignment of Material

In the maintenance task lists, you can use free material assignment. In other words, you don't have to create a bill of material (BOM) beforehand. Instead, you can assign the materials directly to the task list. Purely for technical reasons, the system automatically generates a BOM in the background for these materials and task list. Don't use the normal dialog transactions (e.g., Transactions CS02, IB02, or IB12) to maintain this BOM.

Every BOM requires a *BOM usage*. You therefore use this Customizing function to define the BOM usage, which is assigned to those BOMs that are automatically generated by the system when you use free material assignment.

Prerequisites

You must define the BOM usage beforehand (see the Customizing function DEFINE BOM USAGES).

Customizing Path

PLANT MAINTENANCE AND CUSTOMER SERVICE • MAINTENANCE PLANS, WORK CENTERS, TASK LISTS AND PRTS • TASK LISTS • CONTROL DATA • DEFINE PRESETTING FOR FREE ASSIGNMENT OF MATERIAL

Settings or Recommended Settings

In the standard SAP delivery, the maintenance BOMs are assigned BOM USG 4 (see Figure 7.7).

Figure 7.7 Task Lists: Default Value for BOM Usage

> **BOM Usage 4 Generally Used** [+]
>
> The only time you have to enter a different BOM usage is if you define your own usages for maintenance BOMs.

Define User Fields

You can use this Customizing function to define fields that aren't contained in the standard system but are required in your maintenance task lists. You can then use the menu option OPERATION • USER DATA to call or assign these fields in the task list. Then, you can use the Customizing function DEFINE PROFILE WITH DEFAULT VALUES to propose field keys.

Prerequisites

There are no special prerequisites.

Customizing Path

PLANT MAINTENANCE AND CUSTOMER SERVICE • MAINTENANCE PLANS, WORK CENTERS, TASK LISTS AND PRTS • TASK LISTS • OPERATION DATA • DEFINE USER FIELDS

Transaction

OILJ

Settings or Recommended Settings

In particular, the following field types are available (see Figure 7.8):

- Two 20-digit character fields
- Two 10-digit character fields
- Two quantity fields
- Two value fields

- Two date fields
- Two checkboxes

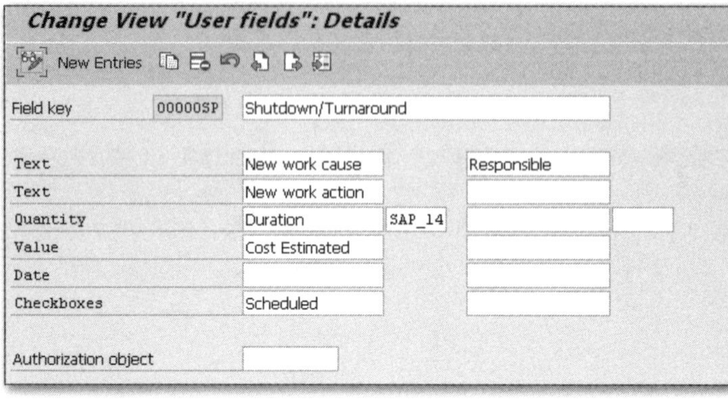

Figure 7.8 Task Lists: User Fields

Maintain Suitabilities

You use this Customizing function to define, for each plant, the requisite *suitability* for performing a particular activity. Then, on the master data screen DEFAULT VALUES IN WORK CENTER, you decide whether you want to define the suitability as a reference or as a default value in the task list. Otherwise, you can assign a suitability for each operation manually.

Prerequisites

You must define the plants beforehand.

Customizing Path

PLANT MAINTENANCE AND CUSTOMER SERVICE • MAINTENANCE PLANS, WORK CENTERS, TASK LISTS AND PRTS • TASK LISTS • OPERATION DATA • MAINTAIN SUITABILITIES

Settings or Recommended Settings

Figure 7.9 shows a sample list of potential suitabilities. In each case, you assign a two-digit SUITABILITY number and a description that contains a maximum of 30 characters.

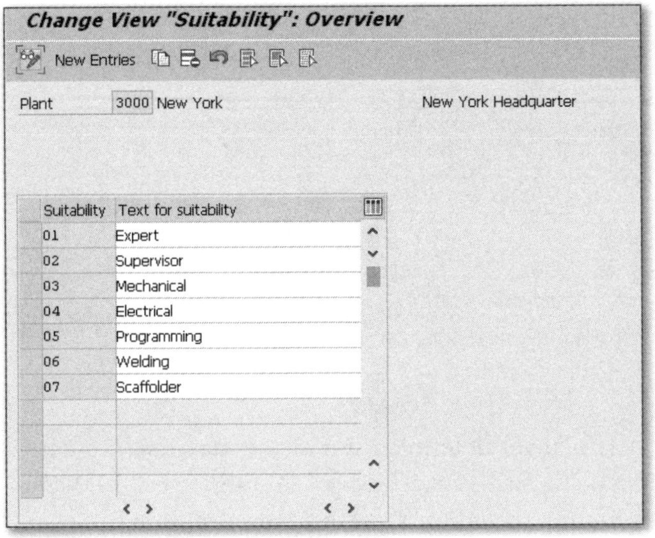

Figure 7.9 Task Lists: Suitabilities

7.2 Maintenance Plans

SAP EAM provides you with maintenance plans that support the business processes associated with preventive maintenance. The following maintenance plan types are available for selection (see Figure 7.10):

- **Single cycle plans**
  You create single cycle plans when you have to perform the same maintenance action tasks in full at regular intervals (time-based or performance-based). In this case, including a task list is optional.
- **Maintenance strategies and strategy plans**
  You create maintenance strategies and strategy plans when you have to perform maintenance action tasks that are interrelated or supersede each other—either as a time-based strategy (e.g., every 3-, 6-, or 12 months) or as a performance-based strategy (e.g., every 10,000-, 20,000-, or 40,000km). In this case, you must include a task list, specifically one that pursues the same strategy as the maintenance plan.

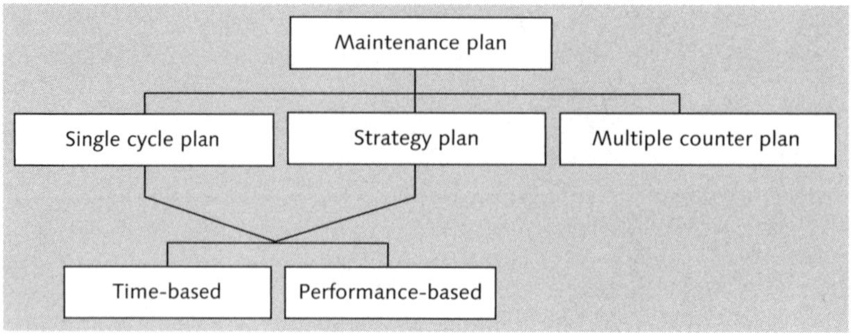

Figure 7.10 Maintenance Plans: Types of Maintenance Plans

- **Multiple counter plans**
  You create multiple counter plans if determination of the maintenance date is dependent on several influencing factors (e.g., every 6 months, every 10,000km, or every 1,000 operation hours). Including a task list here is optional.

The following Customizing functions are available for adapting the management and use of maintenance plans to your needs.

Set Maintenance Plan Categories

You use this Customizing function to define all of the control parameters for the maintenance plans (the maintenance call object or reference object, among others).

Prerequisites

There are no special prerequisites.

Customizing Path

PLANT MAINTENANCE AND CUSTOMER SERVICE • MAINTENANCE PLANS, WORK CENTERS, TASK LISTS AND PRTS • MAINTENANCE PLANS • SET MAINTENANCE PLAN CATEGORIES

Settings or Recommended Settings

Figure 7.11 shows the available settings. The most important setting is the CALL OBJECT. You can use this setting to define which of the following maintenance call objects are to be called from the maintenance plan when they are due:

- MAINTENANCE OR SERVICE ORDER
  Select this if you generate an order and want to execute the maintenance plan specifications unchanged; that is, without any further planning.

- NOTIFICATION
  Select this if you want to perform other detailed planning when it's due (e.g., if you want to combine several tasks into an order on the basis of the current capacity usage). In such cases, several task lists are then copied into a single order as an operation list.
- INSPECTION LOT
  Select this if you conduct stability studies in QM and therefore require corresponding inspection lots at regular intervals.
- SERVICE ENTRY SHEET
  Select this if you have an outline agreement on regular services with an external company and you want the system to automatically generate, at regular intervals, the service entry sheets required for acceptance.

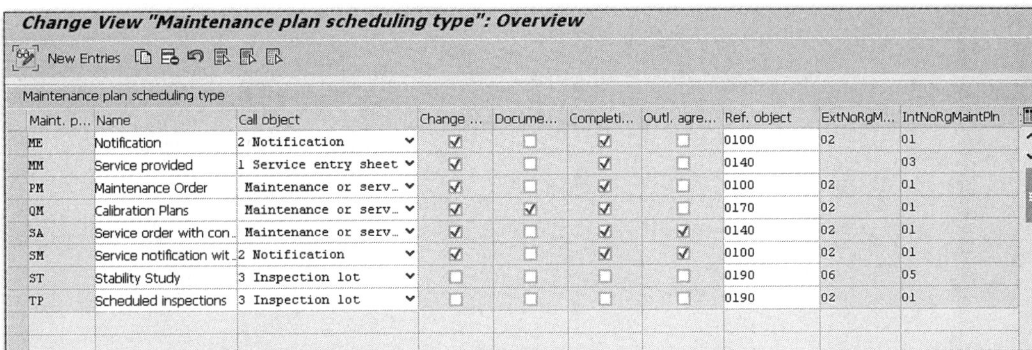

Maint. p...	Name	Call object	Change ...	Docume...	Completi...	Outl. agre...	Ref. object	ExtNoRgM...	IntNoRgMaintPln
ME	Notification	2 Notification	☑	☐	☑	☐	0100	02	01
MM	Service provided	1 Service entry sheet	☑	☐	☑	☐	0140		03
PM	Maintenance Order	Maintenance or serv...	☑	☐	☑	☐	0100	02	01
QM	Calibration Plans	Maintenance or serv...	☑	☑	☑	☐	0170	02	01
SA	Service order with con...	Maintenance or serv...	☑	☐	☑	☑	0140	02	01
SM	Service notification wit...	2 Notification	☑	☐	☑	☑	0100	02	01
ST	Stability Study	3 Inspection lot	☐	☐	☐	☐	0190	06	05
TP	Scheduled inspections	3 Inspection lot	☐	☐	☐	☐	0190	02	01

Figure 7.11 Maintenance Plans: Maintenance Plan Categories

In addition to the maintenance call object, you can use the maintenance plan category to specify the following:

- Set the CHANGE DOCUMENT indicator if you want to activate change documents. In other words, the change documents will later contain information for this maintenance plan category, namely who changed the maintenance plan, what change was made, and when the change was made.
- Set the DOCUMENT RELEASE indicator if you want to activate change documents when a maintenance call is released. In other words, the change documents will later contain information for this maintenance plan category, namely who released the maintenance call from the maintenance plan, which maintenance call was released, and when the maintenance call was released.

- Set the COMPLETION DATA indicator to do the following: When you complete one of the maintenance call objects (notification or order) generated from a maintenance plan, you can enter a *reference time* that applies to the maintenance plan only. When you complete a notification or order, the fields related to the maintenance plan, COMPLETION DATE and COMPLETION TIME, are also displayed, which you can use as a basis for future maintenance plan scheduling. These fields are prefilled with the original planned date and the current time. If, however, the original planned date lies in the future, the system date is proposed.
- Set the OUTL. AGREEMENT indicator to establish a reference to an outline agreement within the maintenance plan (e.g., if you call service entry sheets).
- Set the REF. OBJECT to define the technical object that you want to assign to the maintenance item:
  - 0100: Functional location + piece of equipment + assembly
  - 0110: Equipment + assembly
  - 0120: Functional location (30) + equipment + assembly
  - 0130: Serial number + material number + device data
  - 0140: Without a reference object
  - 0150: Equipment only
  - 0160: Functional location only
  - 0170: Equipment + serial number + material number
  - 0180: Functional location 1:1 + equipment + assembly
  - 0190: Physical sample

[+]

Recommendations Concerning the Maintenance Plan Category

The following settings are used most often:

- MAINTENANCE AND SERVICE ORDER call object
- Change documents (CHANGE DOCUMENT and DOCUMENT RELEASE)
- Completion data
- Reference object (functional location or equipment)

Define Sort Fields for Maintenance Plan

You use this Customizing function to define the potential content of the sort field for the maintenance plan. You assign the sort field to a maintenance plan on the MAINTENANCE PLAN: ADDITIONAL DATA tab page (see the next section). The sort field then becomes an important grouping and selection criterion during maintenance plan scheduling (Transaction IP30, Program RISTRA20).

Prerequisites

There are no special prerequisites.

Customizing Path

PLANT MAINTENANCE AND CUSTOMER SERVICE • MAINTENANCE PLANS, WORK CENTERS, TASK LISTS AND PRTS • MAINTENANCE PLANS • DEFINE SORT FIELDS FOR MAINTENANCE PLAN

Settings or Recommended Settings

You can freely determine the sort field (see Figure 7.12).

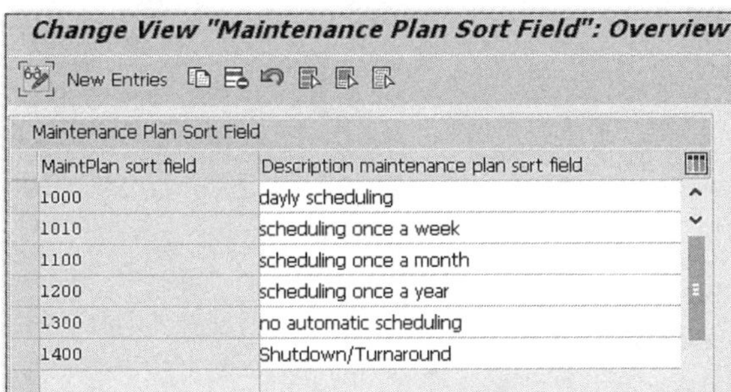

Figure 7.12 Maintenance Plans: Sort Field

Recommendations Concerning the Sort Field [+]

You can use the following categorization criteria:

- Object type (e.g., PC, pump, engine, and fleet object)
- Scheduling frequency (daily, weekly, and monthly)
- Object location (e.g., country, plant, and building)

Because maintenance planning scheduling has very few selection criteria, many users use the sort field to determine the scheduling frequency (e.g., D = daily, W = weekly, and M = monthly).

Configure Special Functions for Maintenance Planning

You use this Customizing function to activate the *enhanced multiple counter plan*, which enables you to define several interrelated cycles, as illustrated in the following example:

- **Cycle set 1**
  Every 1,000 operation hours, every 5,000km, or every 12 months.
- **Cycle set 2**
  Every 3,000 operation hours, every 15,000km, or every 36 months.

If you don't activate this enhancement function, you can define only one cycle set in the multiple counter plan (Transaction IP43).

Prerequisites

There are no special prerequisites.

Customizing Path

PLANT MAINTENANCE AND CUSTOMER SERVICE • MAINTENANCE PLANS, WORK CENTERS, TASK LISTS AND PRTS • MAINTENANCE PLANS • CONFIGURE SPECIAL FUNCTIONS FOR MAINTENANCE PLANNING

Settings or Recommended Settings

If you want to activate the enhancement function, set the ENH. MULTIPLECOUNTERPLAN indicator (see Figure 7.13).

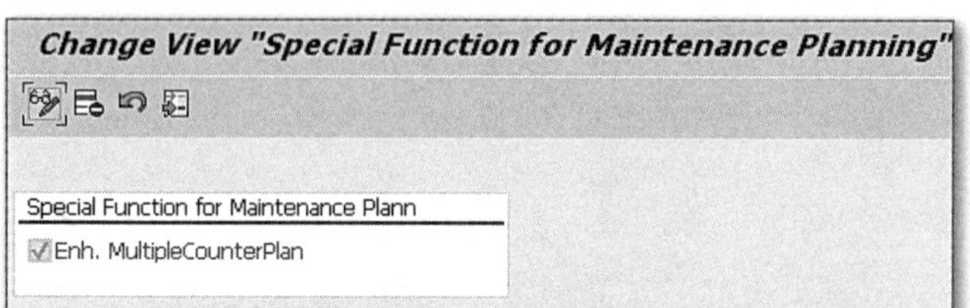

Figure 7.13 Maintenance Plans: Special Functions

[+] **Take Care When Enhancing the Multiple Counter Plan**

After you set the ENH. MULTIPLECOUNTERPLAN indicator, you can no longer deselect it. However, there are no negative consequences if you set the indicator and don't create any enhanced multiple counter plans.

This chapter describes the Customizing functions for those business processes that build on basic maintenance processing and give you enhanced options in relation to your plant maintenance (e.g., refurbishment, subcontracting, and calibration processes).

8 Additional Business Processes

In Chapter 6, you learned about the Customizing functions that can influence basic maintenance processing, specifically notifications, orders, and completion confirmations. This chapter introduces you to the Customizing functions for those business processes that build on basic maintenance processing and provide you with additional business processes for your plant maintenance. These processes include the following:

- External processing of maintenance work, either as an individual purchase order or as a service specification
- Processing of refurbishment tasks
- Subcontracting—external processing with refurbishment
- Calibrating test/measurement equipment
- Managing objects with Pool Asset Management (PAM)
- Using shift notes and shift reports
- Processing maintenance projects using the Maintenance Event Builder (MEB)

Because the following explanations are based on the information provided in Chapter 6, all of the Customizing functions aren't repeated here. Instead, we'll discuss only the following:

- Customizing functions that you now require *in addition* to those described in Chapter 6
- *Special attributes, features, or changes* concerning the Customizing functions discussed earlier and now used in one of the aforementioned processes

8.1 External Processing

External processing plays a major role in plant maintenance, considerably more than in production, for example. A nonrepresentative short survey of SAP user companies in the German-speaking SAP user group showed that, on average, approximately half of their maintenance costs result from external processing. Some companies don't have their own maintenance workshops. Instead, they have coordination points (e.g., work scheduling or planners), which are responsible for planning, monitoring, and approving external processing.

You use a control key to initiate external processing. Furthermore, you use different control keys, depending on what type of external processing is to take place.

The control key attributes control the type of external processing that is to take place (see Figure 8.1):

- **Work center for external company**
  If you've configured a work center for an external company and you want to execute external processing using an internal order in the same way as you do for your own work centers, initiate this type of external processing using a CONTROL KEY (PM01 or similar) for which the EXT. PROCESSING option is set to INTERNALLY PROCESSED OPERATION and the SERVICE indicator isn't set. In Chapter 3, Section 3.4, you learned how to define control keys (see also the Customizing function MAINTAIN CONTROL KEYS).

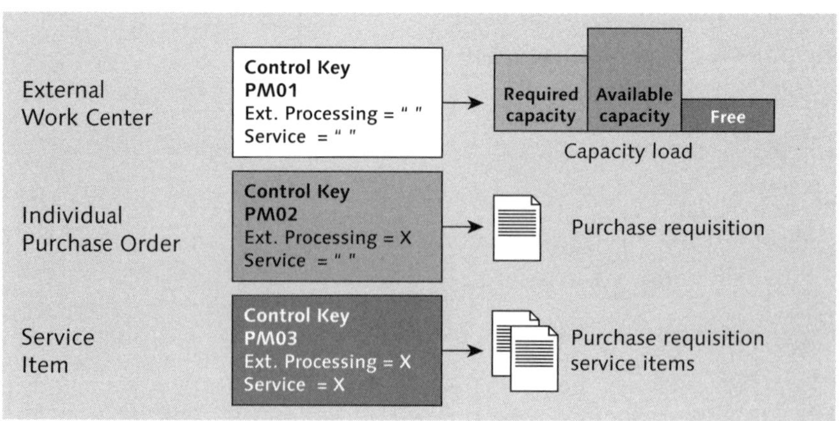

Figure 8.1 External Processing – Control Keys: Overview

- **Individual purchase order**
  If you want to execute external processing using a purchase requisition and an individual standard purchase order, initiate this type of external processing using a control key (PM02 or similar) for which the EXT. PROCESSING option is set to EXTERNALLY PROCESSED OPERATION and the SERVICE indicator isn't set. The additional Customizing functions required for this business process, and the Customizing settings that need to be changed, will be discussed in Section 8.1.1.

- **Service specifications**
  If you want to execute external processing using service items or service specifications and a subsequent service entry sheet, initiate this type of external processing using a CONTROL KEY (PM03 or similar) for which the EXT. PROCESSING option is set to EXTERNALLY PROCESSED OPERATION and the SERVICE indicator is set. The additional Customizing functions required for this business process and the Customizing settings that need to be changed will be discussed in Section 8.1.2.

8.1.1 External Processing as an Individual Purchase Order

If you want to commission external processing as an individual purchase order, the process is similar to the following (see Figure 8.2):

1. When you plan external processing in an order, a purchase requisition is automatically triggered in the background.
2. The purchasing department (or maintenance planner in SAP EAM) converts the purchase requisition into a purchase order.
3. After the external company has rendered the services, you enter them in the system. However, you don't confirm external processing in the same way as normal time confirmations. Instead, you enter a service confirmation as a goods receipt for the purchase order. If the goods receipt is valuated (an option available for the service item in the purchase order), actual costs are posted to the order at this point in time.
4. The invoice receipt concludes this process. If the invoice amount differs from the amount on the purchase order, the costs in the order are automatically corrected, and the order shows the net costs from the invoice.

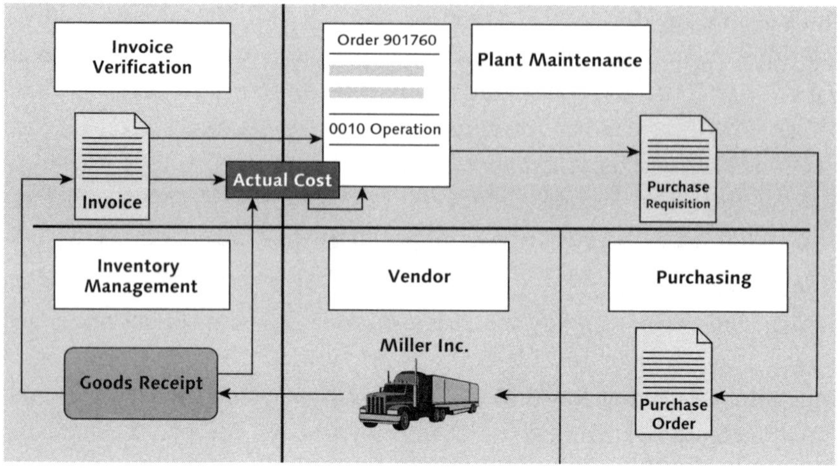

Figure 8.2 External Processing with Individual Purchase Order: Process

The settings required for this business process are described next.

Maintain Control Keys

To initiate external processing as an individual purchase order, you assign a specific control key to an operation associated with a task list (but more importantly here, to an operation associated with an order).

Prerequisites

There are no specific prerequisites.

Customizing Path

PLANT MAINTENANCE AND CUSTOMER SERVICE • MAINTENANCE PLANS, WORK CENTERS, TASK LISTS AND PRTS • WORK CENTERS • TASK LIST DATA • MAINTAIN CONTROL KEYS

Settings or Recommended Settings

The settings for the control key (see Figure 8.3) have already been explained in detail in Chapter 3, Section 3.4.

[+]

Separate Control Key for External Processing

For external processing as an individual purchase order, configure a special control key that has the following properties:

- The Ext. processing option *must* be set to Externally processed operation.
- The Cost indicator *should* be set so that the external processing costs are included in the costing.
- The Confirmation option *should* be set to Confirmation not possible so that none of the internal employees can write their hours to the same operation.
- To prevent the system from creating a service specification, you're *not permitted* to set the Service indicator.
- You *can* set the Sched.ext.op. (schedule external operation) indicator if you order external processing on an hourly basis, and you want to include it in order scheduling.

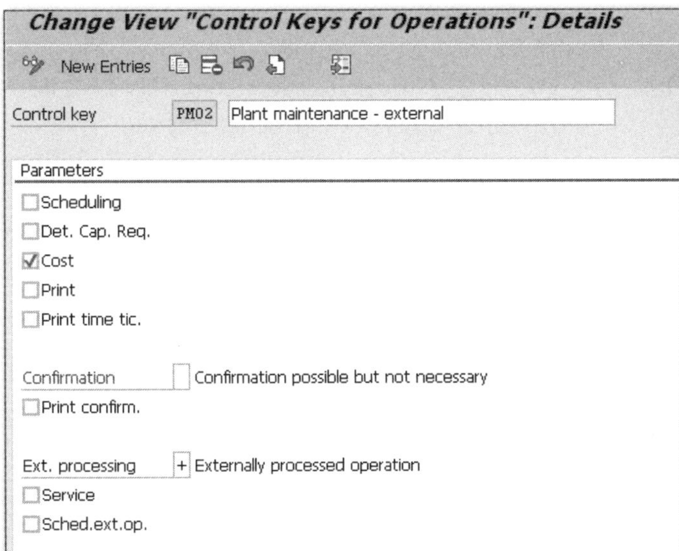

Figure 8.3 External Processing as an Individual Purchase Order: Control Key

Create Default Value Profiles for External Procurement

You use this Customizing function to define profiles for the external procurement of services and materials. In the profile, you can define default values for generating purchase requisitions from orders.

In each profile, you can define default values for the following:

- Cost element
- Purchasing organization
- Purchasing group
- Material group

In the case of external procurement, it's always necessary to differentiate between the external procurement of materials and the external procurement of services.

Prerequisites
You must define the requisite values for the cost element, purchasing organization, purchasing group, and material group beforehand.

Customizing Path
PLANT MAINTENANCE AND CUSTOMER SERVICE • MAINTENANCE AND SERVICE PROCESSING • MAINTENANCE AND SERVICE ORDERS • FUNCTIONS AND SETTINGS FOR ORDER TYPES • CREATE DEFAULT VALUE PROFILES FOR EXTERNAL PROCUREMENT

Settings or Recommended Settings
Make all of the entries that will be copied to the purchase requisitions at a later stage (see Figure 8.4).

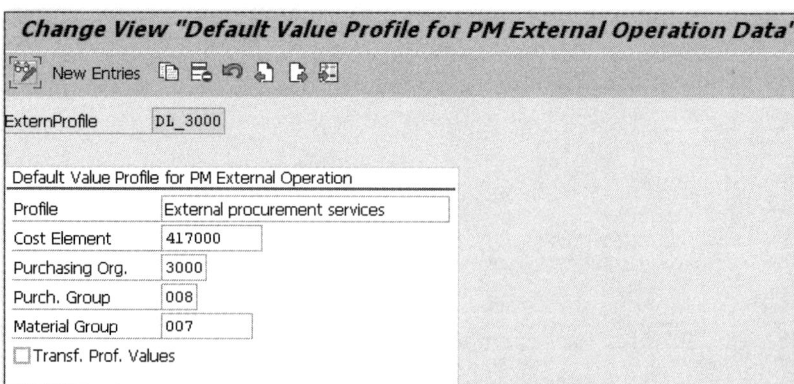

Figure 8.4 External Processing with an Individual Purchase Order—Default Value Profile: Definition

Then, use the Customizing function DEFAULT VALUES FOR TASK LIST DATA AND PROFILE ASSIGNMENTS to make the relevant assignment for each order type and plant.

Default Values for Task List Data and Profile Assignments

You use this Customizing function to define, for each order type and plant, the profiles for external procurement as well as other default values.

Prerequisites
You must define the order types, plants, default value profiles, and external material and external processing profiles beforehand.

Customizing Path
PLANT MAINTENANCE AND CUSTOMER SERVICE • MAINTENANCE AND SERVICE PROCESSING • MAINTENANCE AND SERVICE ORDERS • FUNCTIONS AND SETTINGS FOR ORDER TYPES • DEFAULT VALUES FOR TASK LIST DATA AND PROFILE ASSIGNMENTS

Settings or Recommended Settings
Make the assignment for each order type and plant (see Figure 8.5).

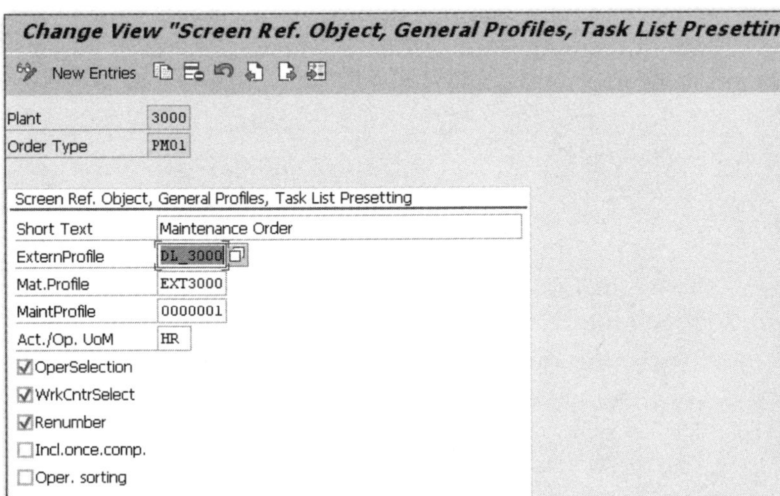

Figure 8.5 External Processing with an Individual Purchase Order—Default Value Profile: Assignment

Use EXTERNPROFILE to assign the profile with default values for the cost element, purchasing organization, material group, and purchasing group for external processing.

[+]

One Profile for All Types of External Processing

The profile for external processing applies not only to external processing with an individual purchase order, but also to external processing with service specifications. The SAP system doesn't make any distinction here.

Define Change Documents, Collective Purchase Requisition, MRP Relevance

You use this Customizing function to define various order parameters (including parameters that you can use to control external processing) for each order type and plant.

Prerequisites

You must define the plants and order types beforehand.

Customizing Path

PLANT MAINTENANCE AND CUSTOMER SERVICE • MAINTENANCE AND SERVICE PROCESSING • MAINTENANCE AND SERVICE ORDERS • FUNCTIONS AND SETTINGS FOR ORDER TYPES • DEFINE CHANGE DOCUMENTS, COLLECTIVE PURCHASE REQUISITION, MRP RELEVANCE

Settings or Recommended Settings

The table shown in Figure 8.6 contains the following parameters for controlling external processing:

- If you set the COLLREQSTN (collective purchase requisition) indicator, all of the purchase requisition items that result from an order number are grouped together under a single purchase requisition number. If you don't set this indicator, the system generates a separate purchase requisition number for each purchase requisition item.
- Use the indicator RES/PURRQ to define the time at which purchase requisitions are to be generated. The following options are available:

 - 1 = NEVER
 - 2 = FROM RELEASE
 - 3 = IMMEDIATELY

- Set the NET ORDER PRICE indicator to specify that the *net price* from the *purchase requisition* must be copied unchanged to the *purchase order* and can't be changed there.
- Set the CHANGE DOCUMENT indicator to activate the creation of change documents if a user changes fields in the external operation (e.g., DATE, VENDOR, or MATERIAL GROUP).
- Set the OPSTATCHANGEDOC indicator to activate the creation of change documents if the status of the operation changes. This is important because the statuses EOPD (external operation – partial delivery) and EODL (external operation – final delivery) are set for the goods receipt. In other words, it should be noted who entered the goods receipt or set the final delivery indicator.

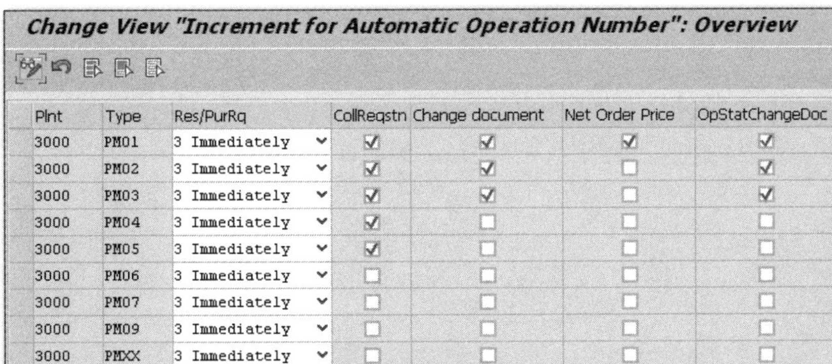

Change View "Increment for Automatic Operation Number": Overview

Plnt	Type	Res/PurRq	CollReqstn	Change document	Net Order Price	OpStatChangeDoc
3000	PM01	3 Immediately	☑	☑	☑	☑
3000	PM02	3 Immediately	☑	☑	☐	☑
3000	PM03	3 Immediately	☑	☑	☐	☑
3000	PM04	3 Immediately	☑	☐	☐	☐
3000	PM05	3 Immediately	☑	☐	☐	☐
3000	PM06	3 Immediately	☐	☐	☐	☐
3000	PM07	3 Immediately	☐	☐	☐	☐
3000	PM09	3 Immediately	☐	☐	☐	☐
3000	PMXX	3 Immediately	☐	☐	☐	☐

Figure 8.6 External Processing with Individual Purchase Order: Define Change Documents, Collective Purchase Requisition, MRP Relevance

One Setting for All Types of External Processing [+]

The settings for the Customizing function DEFINE CHANGE DOCUMENTS, COLLECTIVE PURCHASE REQUISITION, MRP RELEVANCE apply not only to external processing with an individual purchase order, but also to external processing with service specifications. Once again, the SAP system doesn't make any distinction here.

8.1.2 External Processing with Service Specifications

The *external processing with service specifications* business process differs from the *external processing with individual purchase order* business process in that the services to be rendered by the external company aren't broadly outlined in the short and long text of the purchase order item, but rather they are specifically listed individually in a service specification. This gives rise to the following differences in the process flow (see Figure 8.7):

- You use a *service specification* to execute order planning.
- Within order planning, you can define *limits for planned and unplanned services*.
- You execute a *service entry sheet* rather than entering a goods receipt. This can also be carried out by the vendor.

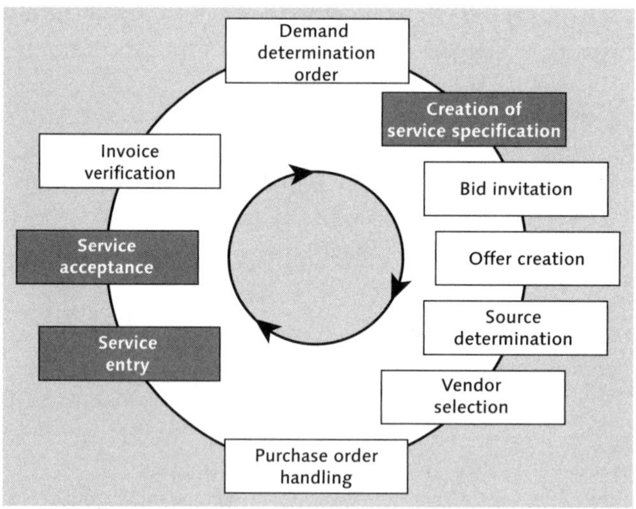

Figure 8.7 External Processing with Service Specifications: Process

- The document that you generate here is called a *service entry sheet*.
- Unlike the goods receipt, the service entry sheet enables you to add *unplanned items*.
- You must use a service acceptance to *release* the services entered (principle of dual control). Only then are they integrated into Materials Management (MM) and an invoice can be issued.

Some of the settings required for this business process are described next.

Maintain Control Keys

To initiate external processing with a service specification, you once again assign a specific control key to an operation associated with a task list or an operation associated with an order.

Prerequisites

There are no specific prerequisites.

Customizing Path

PLANT MAINTENANCE AND CUSTOMER SERVICE • MAINTENANCE PLANS, WORK CENTERS, TASK LISTS AND PRTS • WORK CENTERS • TASK LIST DATA • MAINTAIN CONTROL KEYS

Settings

Figure 8.8 shows the settings for the control key for external processing with service specifications.

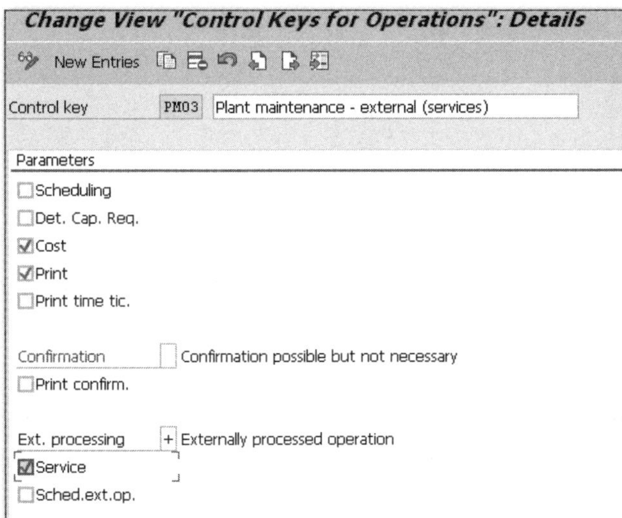

Figure 8.8 External Processing with Service Specifications: Control Key

Separate Control Key for Service Processing [+]

To initiate external processing with a service specification, you should configure a control key that contains the following indicators:

- The EXT. PROCESSING indicator *must* be set to EXTERNALLY PROCESSED OPERATION.

- The SERVICE indicator *must* be set so that a service specification can be created.
- The COST indicator *should* be set so that the external processing costs are included in the costing.
- The PRINT indicator *should* be set so that the service entry sheet can be printed.
- The CONFIRMATION option *should* be set to CONFIRMATION NOT POSSIBLE so that none of the internal employees can write their hours to the same operation.
- You *can* set the SCHED.EXT.OP. (schedule external operation) indicator if you've converted the services to an hourly basis and want to include them in order scheduling.

Define Service Category

The service master is the master record that describes a service (i.e., the counterpart to the material master record). In the same way as a material type must be assigned to each material master record, a service category must be assigned to a service master record. The service category is the most important grouping criterion for service master records, and it affects the screen sequence, field selection, and number assignment, among other things. You use this Customizing function to define your service categories.

Prerequisites

You must define the account category references and the organizational statuses beforehand.

Customizing Path

MATERIALS MANAGEMENT • EXTERNAL SERVICES MANAGEMENT • SERVICE MASTER • DEFINE SERVICE CATEGORY

Settings or Recommended Settings

The following service categories are available (see Figure 8.9):

- ASO1
  Master records that describe internal services.
- ASO3
  Master records that describe services to be procured externally.

- AS04
  Master records that describe services for third parties. Use the Customizing function Define Organizational Status for Service Categories to maintain the organizational status along with the assignment of the relevant functions.

Change View "Service Category": Overview

New Entries

SrvCat	OrgSrvCat	Ex	ARef	Service Cat. Desc.
ALL	AS02	☐	0006	Complete service
BULD	AS03	☑	0006	Construction
GRND	AS01	☐	0006	Basis service
INST	AS03	☐	0006	Plant maintenance
MINI	AS03	☐	0006	
SERV	AS03	☑	0006	Service: purchasing
SRV1	AS01	☐	0006	General Services
SRV2	AS04	☐	0006	Hardware Services
SRV3	AS04	☐	0006	Software Services
SRV4	AS04	☐	0006	Train.&Consult. Services
SRV5	AS04	☐	0006	Technical Services
SRV6	AS04	☐	0006	Contract Services

Figure 8.9 External Processing with Service Specifications: Service Category

Our example uses organizational status AS03 for services to be procured externally. The service category also contains the assignment to an Account Category Reference that influences the automatic account determination for Financial Accounting (FI).

Define Attributes of System Messages

You use this Customizing function to define the manner in which a system message concerning service entry sheets is to be output (e.g., error or warning).

Prerequisites

There are no specific prerequisites.

Customizing Path

Materials Management • External Services Management • Define Attributes of System Messages

Settings or Recommended Settings

Define how each message is to be output. The following options are available to you (see Figure 8.10):

- No message
- W: Warning
- E: Error message
- S: Result message
- P: Dialog box
- I: Information
- X: Program termination (a little joke among developers)

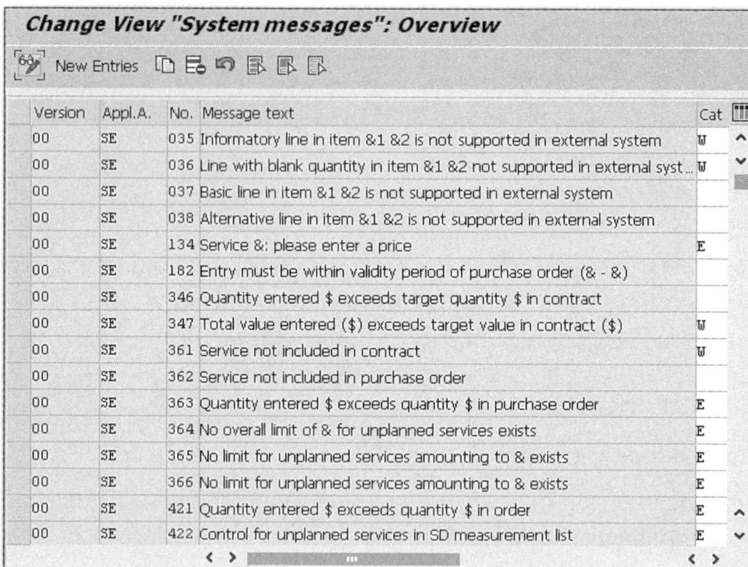

Change View "System messages": Overview

New Entries

Version	Appl.A.	No.	Message text	Cat
00	SE	035	Informatory line in item &1 &2 is not supported in external system	W
00	SE	036	Line with blank quantity in item &1 &2 not supported in external syst...	W
00	SE	037	Basic line in item &1 &2 is not supported in external system	
00	SE	038	Alternative line in item &1 &2 is not supported in external system	
00	SE	134	Service &: please enter a price	E
00	SE	182	Entry must be within validity period of purchase order (& - &)	
00	SE	346	Quantity entered $ exceeds target quantity $ in contract	
00	SE	347	Total value entered ($) exceeds target value in contract ($)	W
00	SE	361	Service not included in contract	W
00	SE	362	Service not included in purchase order	
00	SE	363	Quantity entered $ exceeds quantity $ in purchase order	E
00	SE	364	No overall limit of & for unplanned services exists	E
00	SE	365	No limit for unplanned services amounting to & exists	E
00	SE	366	No limit for unplanned services amounting to & exists	E
00	SE	421	Quantity entered $ exceeds quantity $ in order	E
00	SE	422	Control for unplanned services in SD measurement list	E

Figure 8.10 External Processing with Service Specifications: Messages

Source Determination and Default Values

You can use this Customizing function to configure source determination at the client or purchasing organization level and to define the following values:

- The unit of measure of the service at the item level
- The material group of the service at the item level
- The line number increment for the service specification
- Update of conditions in the service master record with the conditions in the purchasing documents

The settings at the purchasing organization level have priority over the settings at the client level.

Prerequisites

You must maintain the purchasing organizations and the default values (e.g., the material group or unit of measure) beforehand.

Customizing Path

Materials Management • External Services Management • Source Determination and Default Values • For Client or For the Purchasing Organizations

Settings

The Customizing function contains the following settings (see Figure 8.11):

- Define your default values for Line No. Increment in Serv. Specifications, Default Material Group at Item Level, and Unit of Measure at Item Level.
- If you set the Automatic PO Generation for Service Requisitions indicator, the system can automatically generate purchase orders from those purchase requisitions to which the vendor is assigned as a source of supply (e.g., in Transaction ME59N).
- The Condition Update Indicators allow the system to update the conditions in the service master record with the service conditions in the purchasing document (purchase order or quotation) by default. The user can accept or reject this update on a line-by-line basis.
- The settings in the Source Determination screen area support you in assigning contracts to the purchase order (e.g., when using the material group).

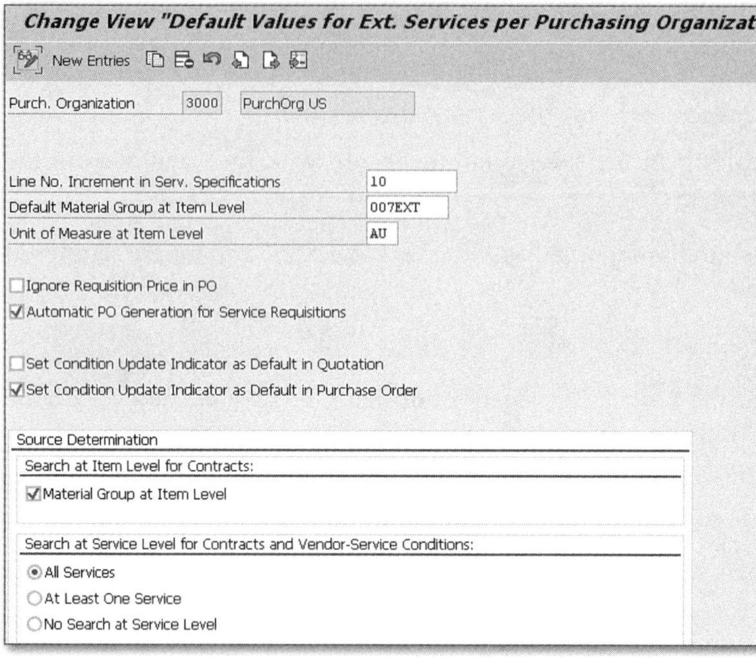

Figure 8.11 External Processing with Service Specifications: Default Values

Define Release Procedure for Service Entry Sheet

The service entry sheet is subject to a release procedure. You use this Customizing function to define the release procedure.

Prerequisites

You must create a class with class type 032 (e.g., FRG_LEIS) and assign the relevant release characteristics (e.g., purchasing organization and value).

Customizing Path

MATERIALS MANAGEMENT • EXTERNAL SERVICES MANAGEMENT • DEFINE RELEASE PROCEDURE FOR SERVICE ENTRY SHEET

Settings or Recommended Settings

The Customizing function DEFINE RELEASE PROCEDURE FOR SERVICE ENTRY SHEET contains the following subfunctions:

- In the first subfunction, create a Release Group for object category 3 (Service Entry Sheet). Assign the class to this release group.
- Use the next subfunction to create your Release Codes for the release group. You can use these codes to define those positions involved in the release procedure (e.g., 01 = accounting, 02 = technology).
- Use the Release Indicators subfunction to define those statuses that flag a service entry sheet as released or not released (e.g., N = not released, B = Released by accountant, and T = Released by technical employee).

The Release Strategy subfunction comprises several other subfunctions (see Figure 8.12):

- Define which Release Strategy applies to which Release Group. Here, specify which positions (release codes) are involved in a strategy.
- For Release Prerequisites, define the order in which the individual positions must be released.
- For Release Statuses, define which status a service entry sheet has after certain positions have been released.

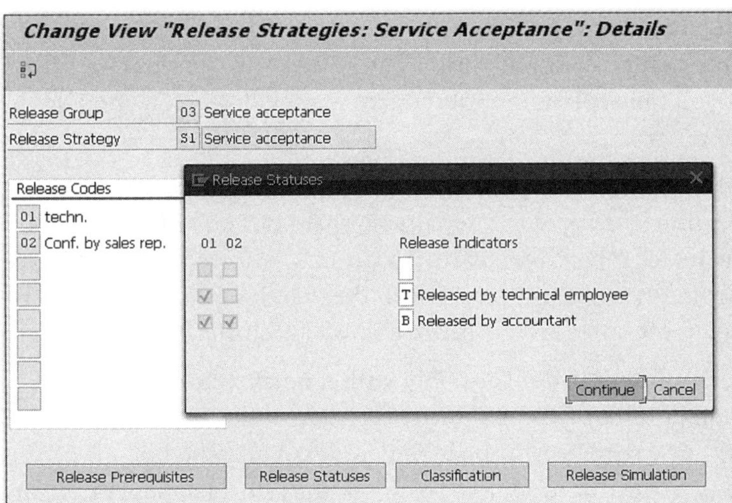

Figure 8.12 External Processing with Service Specifications: Release Procedure

The release procedure for purchasing documents isn't discussed further here. Instead, refer to SAP Note 207490, which provides a detailed description of the release procedure and the necessary Customizing settings.

8.2 Refurbishment

The refurbishment business process is characterized by the fact that repairable spares are held in stock (e.g., to ensure the availability of a particular asset). There are different statuses for repairable spares (e.g., New, Operational Again, and Faulty). Faulty parts are refurbished by your own staff or external staff. In other words, they are restored to an operational status.

As a prerequisite, you must manage repairable spares with different accounting values in the warehouse. When you refurbish a repairable spare, it has a higher value than it would have had in a faulty status. The repairable spares are either managed as material only or as individual units only (material serial numbers).

You use order refurbishment processing to process repairable spares (see Figure 8.13):

- **Procurement of repairable spares**
  You store repairable spares for certain critical and high-value components used in an asset so that you can replace the components immediately if a breakdown occurs.
- **Withdrawal of intact repairable spares and return of faulty repairable spares**
  If a material (single unit) managed as a repairable spare is faulty in an asset, you replace it with an intact repairable spare. To do this, you remove the faulty repairable spare from the asset and return it to the warehouse. You then withdraw an intact repairable spare from the warehouse and install it on the asset.
- **Creation of a refurbishment order (possibly with a notification)**
  As soon as the number of faulty repairable parts in the warehouse has reached a certain amount, you create a refurbishment order. You schedule all of the requisite operations, materials, tools, and so on, for the refurbishment of repairable spares. If necessary, you can create a refurbishment notification before the refurbishment order, either to trigger the refurbishment process or to ensure a complete history.

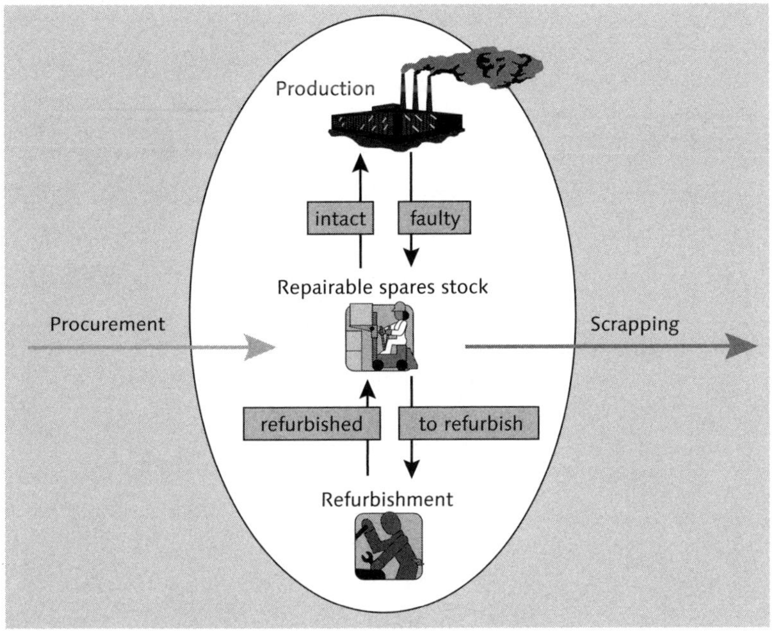

Figure 8.13 Refurbishment: Process

- **Withdrawal from the warehouse**
  The employees responsible for the refurbishment withdraw the faulty repairable spares from the warehouse, including all other materials scheduled in the order and needed for the refurbishment.
- **Performing the refurbishment**
  You carry out the refurbishment. You can post completion confirmations for internal services, goods receipts, or service entry sheets for external material or external processing.
- **Return to warehouse**
  You use a goods receipt to return the refurbished repairable spares to the warehouse. In the case of repairable spares that can't be refurbished, you cancel the reservation and post a scrapping.
- **Scrapping**
  If faulty repairable spares can no longer be refurbished, you scrap them. Don't forget to also post a goods issue for this situation.

To ensure that the process works as described here, you must make some Customizing settings.

8.2.1 Material Master

To use the material master of the repairable spare for the refurbishment, you define the *valuation category* in the accounting data at the plant level and then create several *valuation types* for the valuation category (see Figure 8.14).

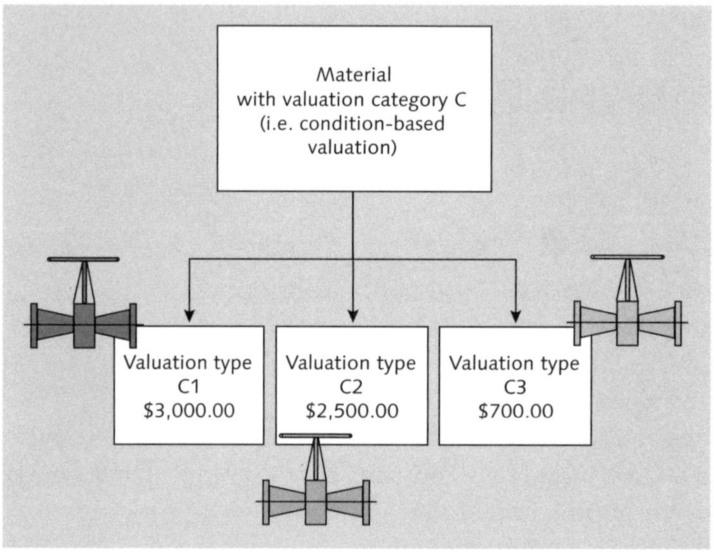

Figure 8.14 Refurbishment: Valuation Category and Valuation Type

In the following example, the valuation types are to be defined for valuation category C (Status):

- C1 = As new
- C2 = Refurbished and operational again
- C3 = Faulty and no longer operational

To be able to perform a split valuation for a material master record, you must first activate split valuation in Customizing and then configure it (i.e., define valuation categories and valuation types).

Activate Split Valuation

You use this Customizing function to configure whether a split material valuation is permitted in your enterprise. If you allow split valuation in principle, this doesn't mean that you need to perform split valuation for every single material. On the contrary, when you create the material master record, you decide whether you want to perform split valuation for the material by entering a valuation category in the material master and then creating several valuation types.

Prerequisites
There are no specific prerequisites.

Customizing Path
MATERIALS MANAGEMENT • VALUATION AND ACCOUNT ASSIGNMENT • SPLIT VALUATION • ACTIVATE SPLIT VALUATION

Transaction
OMWO

Settings
Activate split valuation by setting the SPLIT MATERIAL VALUATION ACTIVE indicator (see Figure 8.15).

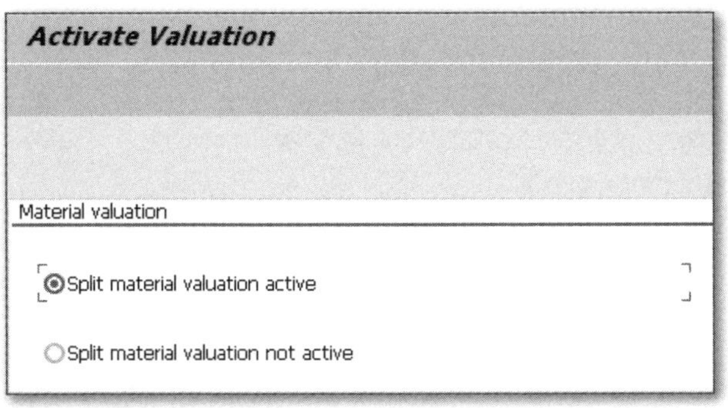

Figure 8.15 Refurbishment: Activate Split Valuation

Configure Split Valuation

You use this Customizing function to define split valuation in detail. Here, you define the following (see Figure 8.16):

- Valuation types you want to differentiate between (*global valuation types*)
- Valuation categories you want to differentiate between (*global valuation categories*)
- Global valuation types you want to *activate* for which global valuation categories
- Global valuation categories you want to use *locally* (i.e., for each plant), which automatically gives rise to *local valuation types*

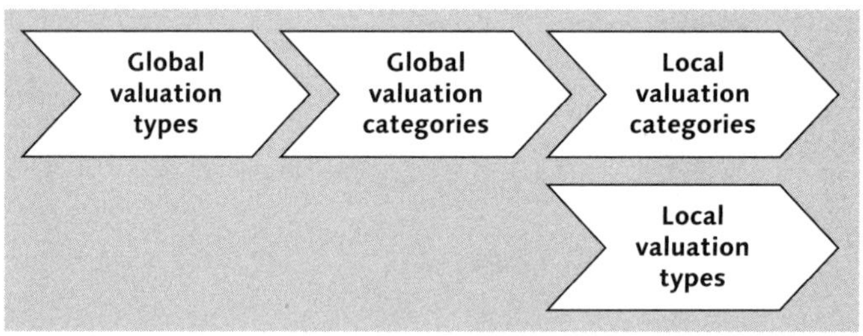

Figure 8.16 Refurbishment: Global and Local Valuation Categories and Valuation Types

Prerequisites
You must define the account category references and plants beforehand. You must also activate split valuation.

Customizing Path
MATERIALS MANAGEMENT • VALUATION AND ACCOUNT ASSIGNMENT • SPLIT VALUATION • CONFIGURE SPLIT VALUATION

Transaction
OMWC

Settings
When you call the Customizing function, you obtain the following subfunctions:

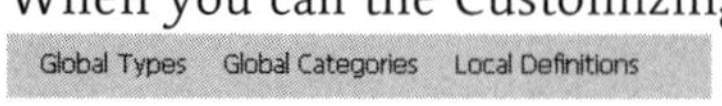

In this sequence, it's also best to work with split material valuation. First, define which valuation types you want to use (e.g., global valuation types, in Figure 8.17).

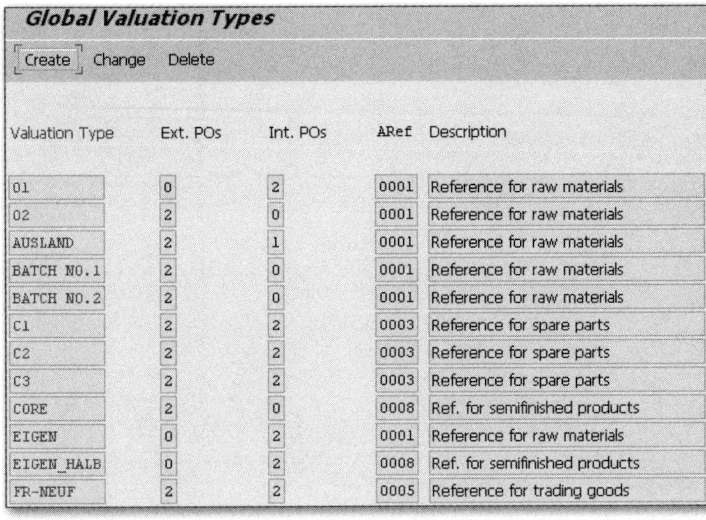

Valuation Type	Ext. POs	Int. POs	ARef	Description
01	0	2	0001	Reference for raw materials
02	2	0	0001	Reference for raw materials
AUSLAND	2	1	0001	Reference for raw materials
BATCH NO.1	2	0	0001	Reference for raw materials
BATCH NO.2	2	0	0001	Reference for raw materials
C1	2	2	0003	Reference for spare parts
C2	2	2	0003	Reference for spare parts
C3	2	2	0003	Reference for spare parts
CORE	2	0	0008	Ref. for semifinished products
EIGEN	0	2	0001	Reference for raw materials
EIGEN_HALB	0	2	0008	Ref. for semifinished products
FR-NEUF	2	2	0005	Reference for trading goods

Figure 8.17 Refurbishment: Global Valuation Types

In particular, define the following:

- Ext. POs
  The valuation type is permitted for external purchase orders.
- Int. POs
  The valuation type is permitted for refurbishment orders.

Valuation Type for Internal Purchase Orders [+]

To enable you to use a valuation type for refurbishment orders, you must set the Int. POs to 2 (Allowed).

Then, define which valuation categories you want to use (e.g., global valuation categories in Figure 8.18).

Use the Types • Cat. function to make an assignment in relation to which global valuation types are permitted for which global valuation categories. You do this by activating the valuation types for each valuation category (see Figure 8.19).

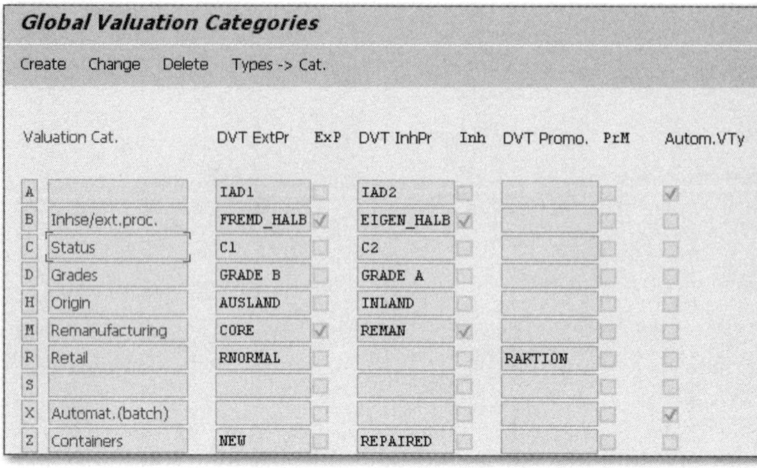

Figure 8.18 Refurbishment: Global Valuation Categories

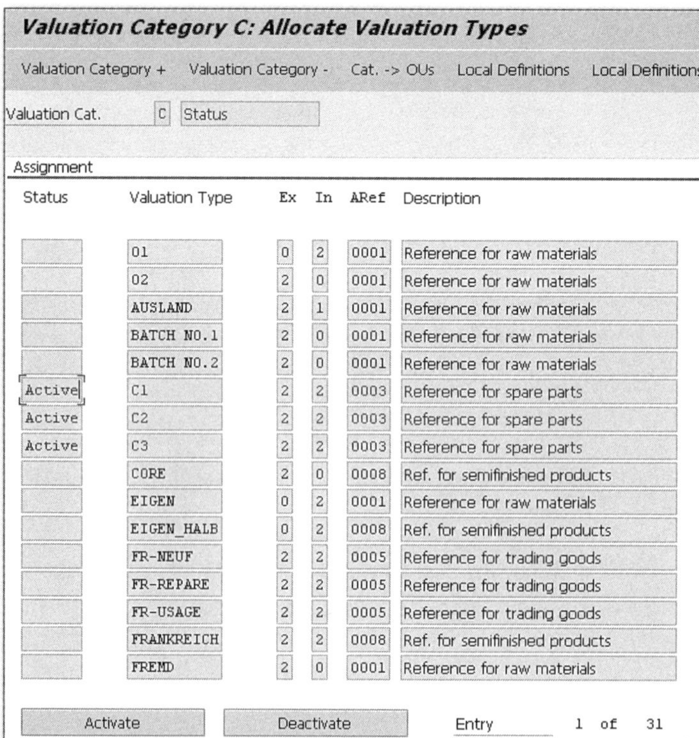

Figure 8.19 Refurbishment: Assignment of Global Valuation Types to Global Valuation Categories

Finally, use the LOCAL DEFINITIONS option to define the global valuation categories as local valuation categories. To do this, activate the relevant valuation type for the plant (see Figure 8.20).

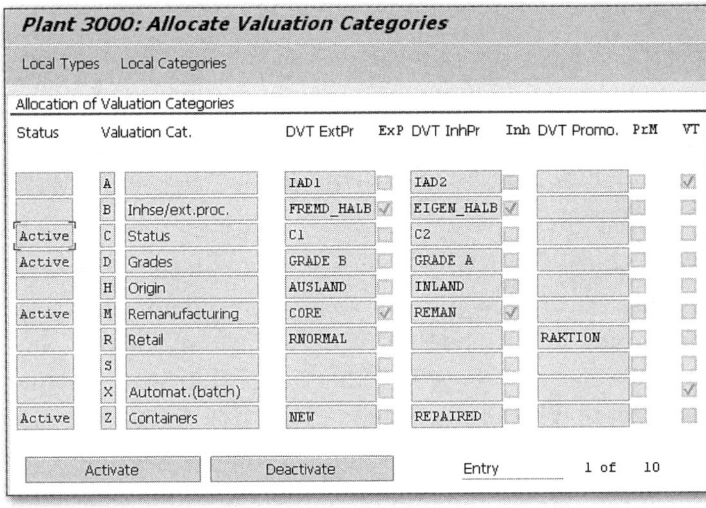

Figure 8.20 Refurbishment: Activate Local Valuation Categories

Activating the global valuation category for the plant and assigning the global valuation types to the global valuation categories causes the system to automatically create the local valuation types permitted for the plant (see Figure 8.21). In other words, you don't need to activate or assign these separately.

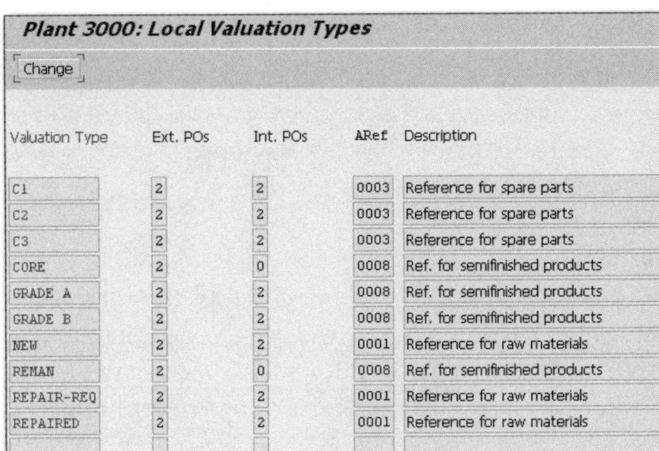

Figure 8.21 Refurbishment: Local Valuation Types

8.2.2 Refurbishment Notifications

You can use a refurbishment notification to trigger the refurbishment process. The refurbishment notification is optional, however. You use refurbishment notifications in the following scenarios:

- You attach great importance to having complete documentation in the history. In other words, you want to use notifications and orders to document the actions of the plant maintenance department.
- You always want to use a notification as a first step, and therefore as a trigger, for your maintenance processes.

If you use refurbishment notifications, we recommend that you use a separate notification type to distinguish refurbishment notifications from other maintenance notifications and to control them separately. However, the Customizing functions and settings are essentially the same as those for other notifications (see Chapter 6, Section 6.1). The only exception is the screen type object (ScrnType Object).

Overview of Notification Type

Within this Customizing function, you use the subfunction Screen Areas in Notification Header to define the notification header and reference objects you want to use for each notification type.

Prerequisites

You must define the notification types themselves and the screen layout for these notification types beforehand.

Customizing Path

Plant Maintenance and Customer Service • Maintenance and Service Processing • Maintenance and Service Notifications • Overview of Notification Types • Subfunction Screen Areas in Notification Header

Settings or Recommended Settings

In the screen type header (Screen type hdr), always enter "H100" (Header maintenance notification) unless you've defined your own headers (see Figure 8.22). The Header maintenance notification contains information such as the notification type, notification number, short text, status, and assigned order number.

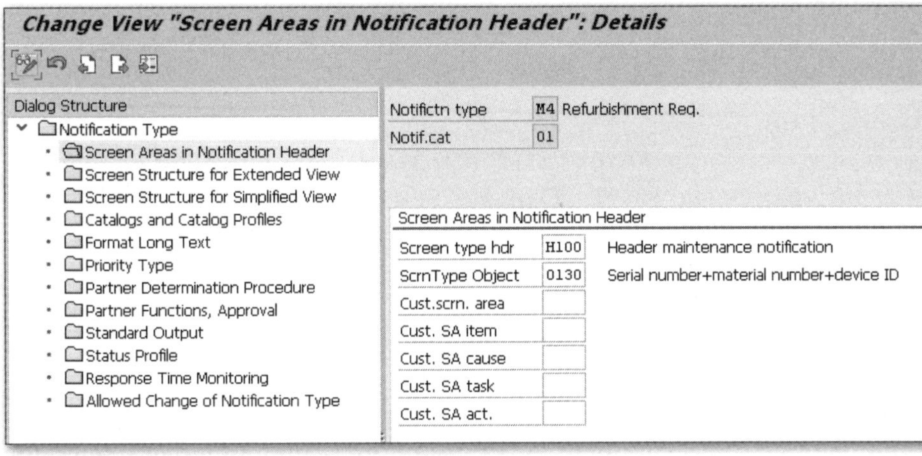

Figure 8.22 Refurbishment: Refurbishment Notification

For ScrnType Object, choose a setting that contains the material number as the refurbishment process always refers to a material number. Screen type 0130 (serial number + material number + device ID) is available for this purpose.

8.2.3 Refurbishment Orders

Next, we'll describe some additional Customizing functions as well as some special features within the Customizing settings in relation to an order type for the refurbishment process. All other Customizing functions and settings are the same as those for other orders (see Chapter 6, Section 6.2).

Configure Order Types

You use this Customizing function to define the order type and some basic parameters.

Separate Order Type for Refurbishment [+]

You need a *separate* order type for the refurbishment process. You can't use this order type for "normal" (i.e., not refurbishment) maintenance processes. Similarly, you can't use a normal order type for a refurbishment process.

Prerequisites

Because order types are used by plant maintenance in SAP EAM as well as by other user departments and their corresponding SAP applications, you must consult with your colleagues from the other user departments on the naming convention and number ranges for order types.

Customizing Path

Plant Maintenance and Customer Service • Maintenance and Service Processing • Maintenance and Service Orders • Functions and Settings for Order Types • Configure Order Types

Transaction

OIDA

Settings or Recommended Settings

The special feature of this Customizing function in relation to order types is the settlement profile (see Figure 8.23). This settlement profile *must* differ from the settlement profile for normal order types so that the orders can be settled on the basis of a material number. Further details will be provided in the next section on the Customizing function Maintain Settlement Profiles.

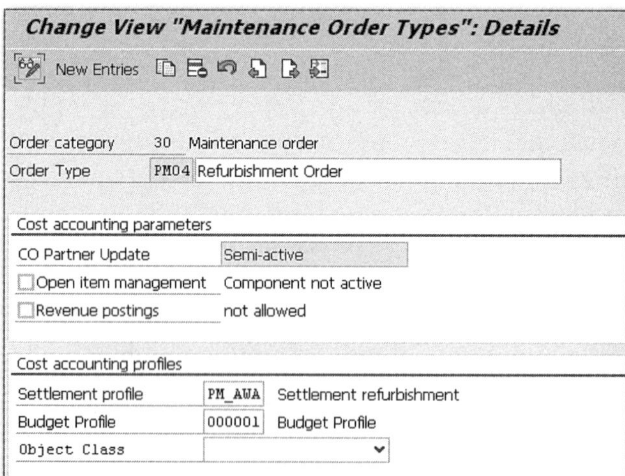

Figure 8.23 Refurbishment: Order Type

Maintain Settlement Profiles

You use this Customizing function to define the allocation structure, permitted account assignment objects, and other settlement parameters.

Prerequisites

You must define the allocation structures and document types for the settlement documents beforehand.

Customizing Path

PLANT MAINTENANCE AND CUSTOMER SERVICE • MAINTENANCE AND SERVICE PROCESSING • BASIC SETTINGS • GENERAL ORDER SETTLEMENT • MAINTAIN SETTLEMENT PROFILES

Settings or Recommended Settings

Figure 8.24 shows our recommendations for the settlement profile for refurbishment orders.

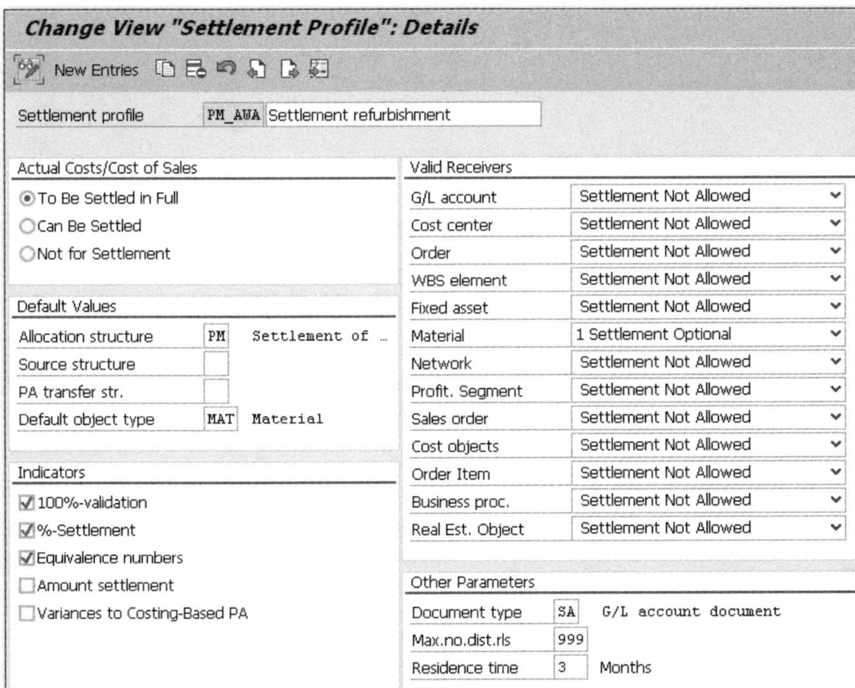

Figure 8.24 Refurbishment: Settlement Profile

[+]

Separate Settlement Profile for Refurbishment Orders

In contrast to the settlement profile for normal order types, the settlement profile for refurbishment orders has two special features:

- In the VALID RECEIVERS section, you *must* set the account assignment object MATERIAL to 1 SETTLEMENT OPTIONAL.
- You should enter "MAT" (MATERIAL) as the DEFAULT OBJECT TYPE.

Indicate Order Types for Refurbishment Processing

You use this Customizing function to identify an order type as a refurbishment order type. You're then unable to use this order type for normal maintenance processes.

Prerequisites

You must define the order type beforehand.

Customizing Path

PLANT MAINTENANCE AND CUSTOMER SERVICE • MAINTENANCE AND SERVICE PROCESSING • MAINTENANCE AND SERVICE ORDERS • FUNCTIONS AND SETTINGS FOR ORDER TYPES • INDICATE ORDER TYPES FOR REFURBISHMENT PROCESSING

Settings or Recommended Settings

Select the order type(s) that you want to use for the refurbishment (see Figure 8.25).

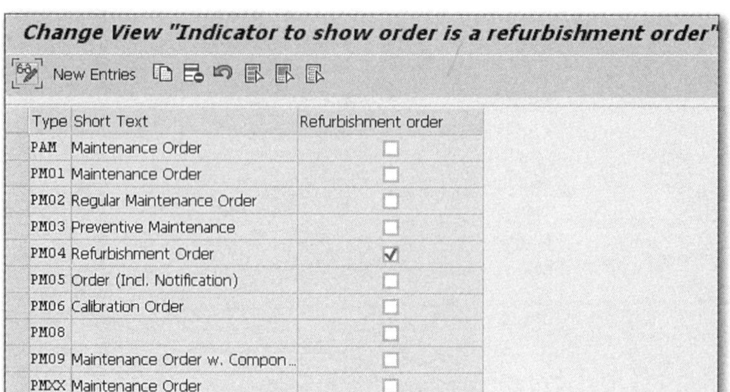

Type	Short Text	Refurbishment order
PAM	Maintenance Order	☐
PM01	Maintenance Order	☐
PM02	Regular Maintenance Order	☐
PM03	Preventive Maintenance	☐
PM04	Refurbishment Order	☑
PM05	Order (Incl. Notification)	☐
PM06	Calibration Order	☐
PM08		☐
PM09	Maintenance Order w. Compon...	☐
PMXX	Maintenance Order	☐

Figure 8.25 Refurbishment: Indicate Order Type

[+]

Indicate Order Type for Refurbishment

To be able to use an order type for refurbishments, you must identify this order type as a refurbishment order type. You're then unable to use this order type for normal maintenance processes.

Assign Notification Types to Order Types

If you've also configured a notification type for the refurbishment, you should use this Customizing function to specify that the order type for the refurbishment is proposed from the notification type for the refurbishment if an order is generated from the notification.

Prerequisites

You must define the notification types and order types beforehand.

Customizing Path

Plant Maintenance and Customer Service • Maintenance and Service Processing • Maintenance and Service Notifications • Notification Creation • Notification Types • Assign Notification Types to Order Types

Settings or Recommended Settings

For the notification type, enter the order type that you've earmarked for the refurbishment (see Figure 8.26).

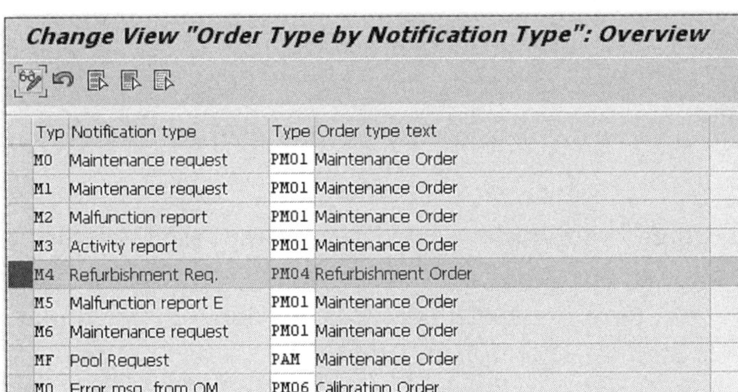

Figure 8.26 Refurbishment: Order Type and Notification Type

Define Documentation for Goods Movements for the Order

You use this Customizing function to define, for each order type and plant, which material movements are to be recorded. You can then view these in the order (either in the document flow or in the documented goods movements).

Prerequisites

You must define the order type and plant beforehand.

Customizing Path

Plant Maintenance and Customer Service • Maintenance and Service Processing • Maintenance and Service Orders • Functions and Settings for Order Types • Goods Movements for Order • Define Documentation for Goods Movements for the Order

Settings or Recommended Settings

You can document the following goods movements for the maintenance orders separately (see Figure 8.27):

- PurchOrder
  Goods receipts for the purchase order.
- PlGoodsIss
  Planned goods issues; that is, goods issues for a reservation.
- UnplGdsIss
  Unplanned goods issues; that is, goods issues without a reservation.
- GR refrbshmnt
  Goods receipts for refurbishment orders.

You can only set the latter for those order types that are identified as an order type for the refurbishment.

You also require further settings for this order type. However, they aren't discussed here because they are the same settings associated with normal order types. Instead, refer to Chapter 6, Section 6.2, for order-specific settings; Section 6.3, for completion confirmations; and Chapter 4, for generic functions.

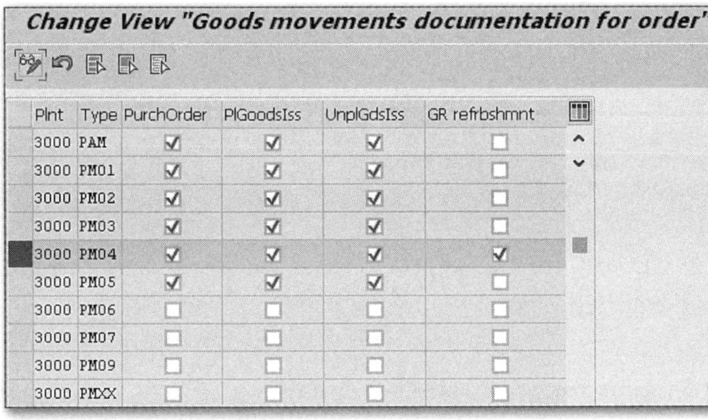

Change View "Goods movements documentation for order"

Plnt	Type	PurchOrder	PlGoodsIss	UnplGdsIss	GR refrbshmnt
3000	PAM	☑	☑	☑	☐
3000	PM01	☑	☑	☑	☐
3000	PM02	☑	☑	☑	☐
3000	PM03	☑	☑	☑	☐
3000	PM04	☑	☑	☑	☑
3000	PM05	☑	☑	☑	☐
3000	PM06	☐	☐	☐	☐
3000	PM07	☐	☐	☐	☐
3000	PM09	☐	☐	☐	☐
3000	PMXX	☐	☐	☐	☐

Figure 8.27 Refurbishment: Goods Movements Documentation

8.2.4 Material Requirements Planning

You can now also generate refurbishment orders in material requirements planning (MRP); specifically, in the stock/requirements list (Transaction MD04).

The business background is as follows: Whenever the quantity of operational parts falls below the reorder point, but nonoperational parts are in stock, MRP automatically generates planned orders. Previously, it was only possible to convert planned orders into purchase requisitions for purchasing, production orders in discrete manufacturing, or process orders in process manufacturing. Now, however, you can also convert the automatically generated planned orders into refurbishment orders to ensure that you always have the required quantity of operational parts.

Business Function

Before you can convert a planned order into a refurbishment order, you must install Discrete Industries and Mill Products (DIMP) 6.0 and activate the business functions LOG_EAM_ROTSUB, LOG_EAM_ROTSUB_2 and LOG_MM_SERNO.

Material Master

To be able to convert planned orders into refurbishment orders, you must assign a spare part class code (in the field of the same name) on the Basic Data 2 screen in the material master and define it as a Repairable spare part with CMM (code 2) or as a Repairable spare part without CMM (code 6) (see Figure 8.28).

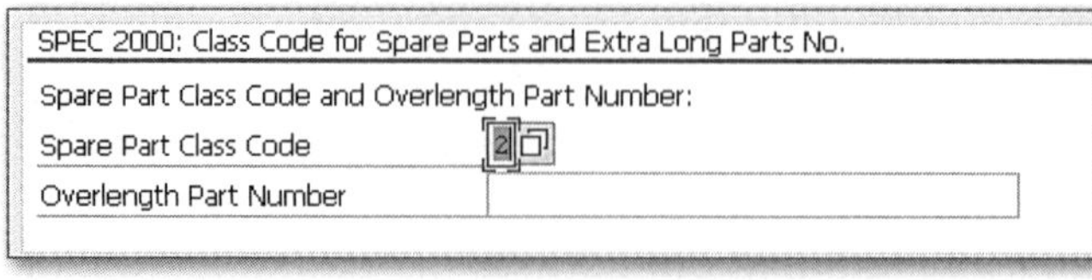

Figure 8.28 Refurbishment: Spare Part Class Code

If this subscreen doesn't exist, you must configure the layout of the data screens for the material master as described next.

Define Structure of Data Screens for Each Screen Sequence

You use this Customizing function to specify a screen sequence and to assign the necessary main screens and their subscreens.

Prerequisites

There are no specific prerequisites.

Customizing Path

Logistics - General • Material Master • Configuring the Material Master • Define Structure of Data Screens for Each Screen Sequence

Transaction

OMWO

Settings or Recommended Settings

Proceed as follows:

- Select the Screen sequence that you want to use for your refurbishment materials.
- Select the Data screen on which you want to position the spare part class code.
- Choose Subscreens to position the subscreen for the spare part class code there. To do this, enter the Program "SAPLADRT21" and Screen number "2000" (see Figure 8.29).

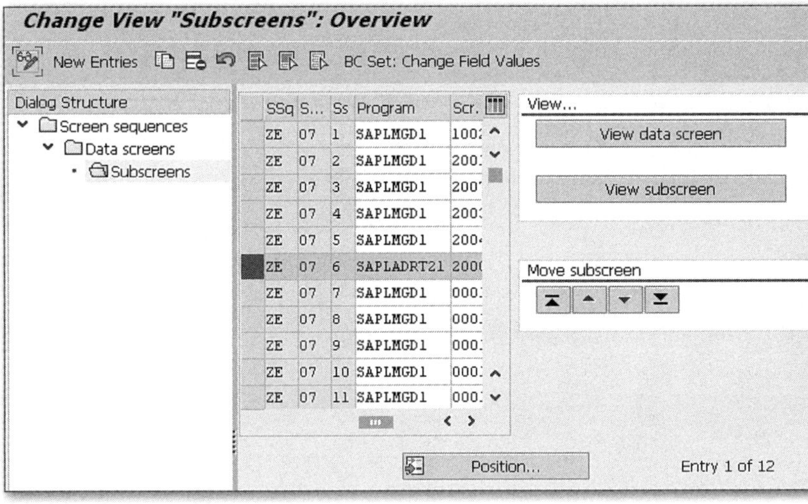

Figure 8.29 Refurbishment: Material Master for Requirements Planning

Then, in the material requirements/stock list (Transaction MD04), you can convert a planned order into a refurbishment order.

These Customizing settings should now enable you to execute the refurbishment process described in the book *Plant Maintenance with SAP—Practical Guide.*[1]

8.3 Subcontracting

The *subcontracting* business process, also known as subcontracting for MRO (maintenance, repair, and overhaul), describes the process associated with having a piece of equipment (i.e., a material serial number) repaired by a service provider. In contrast to the processes used for external processing, the object to be repaired or maintained is sent to the service provider, processed there, and then returned.

The subcontracting process comprises the following steps (see Figure 8.30):

1. You have a faulty part that requires repair and is to be processed by a service provider. You therefore create a maintenance order with a material serial number and a normal order type ❶. In this order, you enter a subcontracting

1 See K. Liebstückel, *Plant Maintenance with SAP—Practical Guide* (3rd edition, SAP PRESS, 2014, pp. 262–273).

operation (for which the SUBCONTRACTING indicator is set) and assign a material (for which the REFURBISHMENT indicator is set) to this operation.

2. The SAP system now uses the maintenance order to create a purchase requisition for the external repair or maintenance service with a subcontracting item (item category L) and the material serial number ❷.
3. You convert the purchase requisition into a purchase order for the external repair or maintenance service ❸. The purchase order item is identified as a subcontracting item (item category L) and is assigned the part that is to be returned following repair (material serial number).

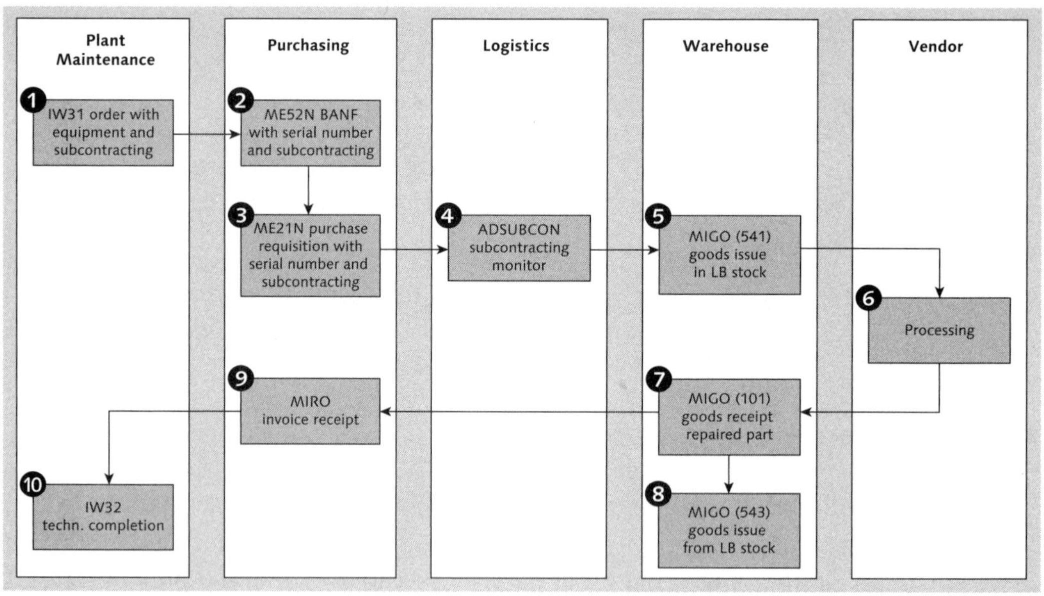

Figure 8.30 Subcontracting: Process

4. You use an outbound delivery to send the part to be repaired to the subcontractor ❹.
5. The parts provided are managed as stock of material provided to the vendor (subcontracting stock) ❺. The provision represents a transfer posting from unrestricted-use stock to stock of material provided to the vendor (subcontracting stock).
6. The subcontractor repairs, modifies, replaces, or exchanges the faulty part and returns the serviceable part ❻.

7. For the part delivered, you create a goods receipt posting that refers to the subcontracting item in the purchase order ❼.
8. For the components, a goods issue is posted from the subcontracting stock ❽.
9. You receive an invoice for the service provided by the subcontractor ❾.
10. You complete the maintenance order ❿.

Business Function

Before you can use subcontracting, you must install DIMP 6.0 and activate the business functions LOG_EAM_ROTSUB, LOG_EAM_ROTSUB_2 and LOG_MM_SERNO.

You must then configure the following Customizing settings.

Define Serial Number Profiles

You use this Customizing function to define serial number profiles. You must assign a material serial number to each piece of equipment that you want to refurbish by means of a subcontracting process. In the material master record, you must assign a serial number profile to this material at the plant level.

Prerequisites
There are no special prerequisites.

Customizing Path
PLANT MAINTENANCE AND CUSTOMER SERVICE • MASTER DATA IN PLANT MAINTENANCE AND CUSTOMER SERVICE • TECHNICAL OBJECTS • SERIAL NUMBER MANAGEMENT • DEFINE SERIAL NUMBER PROFILES

Transaction
OIS2

Settings
You must assign at least the following three serializing procedures to each of the serial number profiles that you want to use for the subcontracting process (see Figure 8.31):

- MMSL
  You can enter a goods receipt and goods issue for the material serial number (steps ❹, ❺, ❼, and ❽ in Figure 8.30).
- POSL
  You can manage serial numbers in purchase orders (step ❸ in Figure 8.30).
- PRSL
  You can manage serial numbers in purchase requisitions (step ❹ in Figure 8.30).

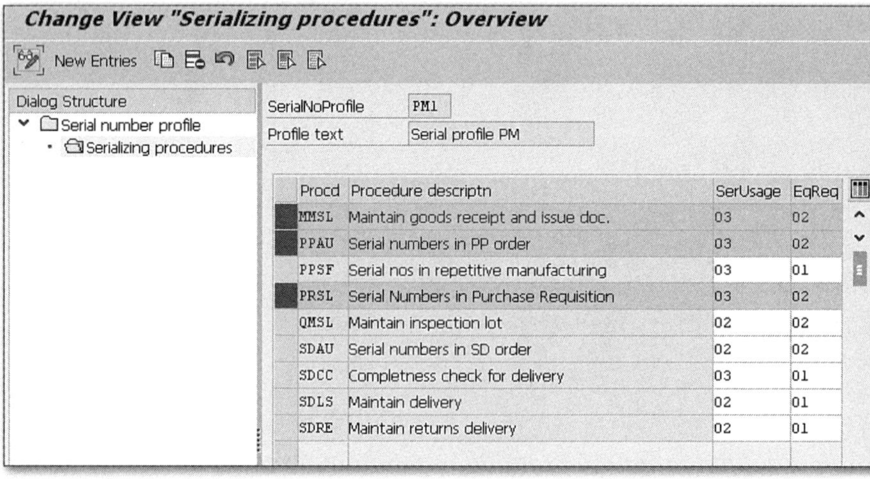

Figure 8.31 Subcontracting: Serial Number Profile

Purchase Requisition – Define Document Types

You use this Customizing function to manage the document types and item categories permitted for purchase requisitions.

Prerequisites

You must maintain the serial number profiles beforehand.

Customizing Path

Materials Management • Purchasing • Purchase Requisition • Define Document Types

Settings

For the document type that you want to use for the subcontracting process (e.g., NB = Standard Purchase Requisition), call item category L (Subcontracting), and define the serial number profile that you want to use (see Figure 8.32). You can then assign a material serial number to a purchase requisition (step ❷ in Figure 8.30).

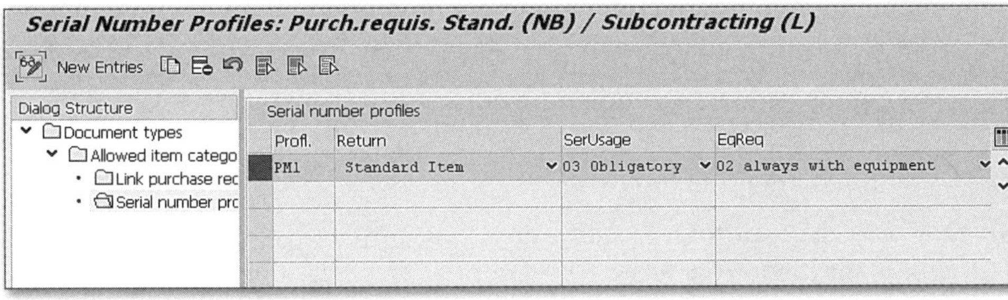

Figure 8.32 Subcontracting: Serial Number Profile in Purchase Requisitions

Purchase Order – Define Document Types

You use this Customizing function to manage the document types and item categories permitted for purchase orders.

Prerequisites

You must maintain the serial number profiles beforehand.

Customizing Path

Materials Management • Purchasing • Purchase Order • Define Document Types

Settings

For the document type that you want to use for the subcontracting process (e.g., NB = Standard Purchase Order), call item category L (Subcontracting), and define the serial number profile that you want to use (see Figure 8.33). You can then assign a material serial number to a purchase order (step ❸ in Figure 8.30).

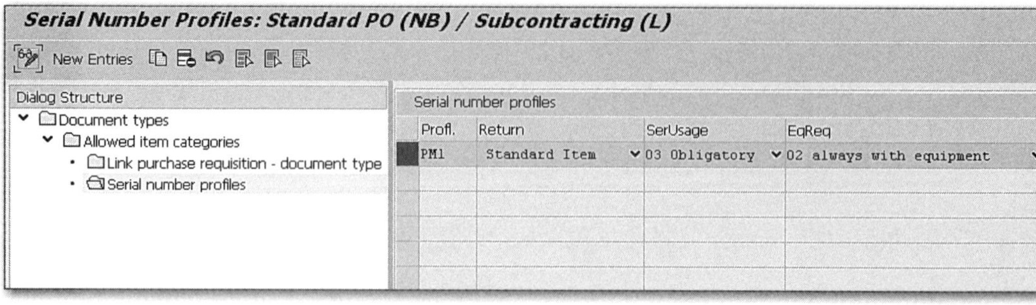

Figure 8.33 Subcontracting: Serial Number Profile in Purchase Orders

Define Material Provision Indicators

To provide a material for subcontracting, you must assign a material provision indicator to the item in the order. This indicator then flags this item as a refurbishment material (step ❶ in Figure 8.30).

You use this Customizing function to define material provision indicators.

Prerequisites

There are no specific prerequisites.

Customizing Path

Plant Maintenance and Customer Service • Master Data in Plant Maintenance and Customer Service • Bills of Material • Item Data • Define Material Provision Indicators

Transaction

OICO

Settings

Define a material provision indicator for which you set the indicator A&D MP (Aerospace & Defense material provision)[2] (see Figure 8.34).

2 Note: Subcontracting was originally developed for the SAP Aerospace and Defense industry solution, but was later made available for general use.

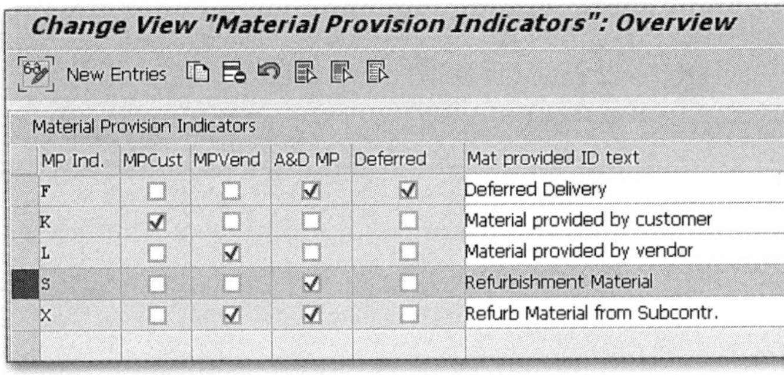

MP Ind.	MPCust	MPVend	A&D MP	Deferred	Mat provided ID text
F	☐	☐	☑	☑	Deferred Delivery
K	☑	☐	☐	☐	Material provided by customer
L	☐	☑	☐	☐	Material provided by vendor
S	☐	☐	☑	☐	Refurbishment Material
X	☐	☑	☑	☐	Refurb Material from Subcontr.

Figure 8.34 Subcontracting: Material Provision Indicator

These Customizing settings should now enable you to execute the subcontracting process described in *Plant Maintenance with SAP—Practical Guide.*[3]

8.4 Calibration of Test/Measurement Equipment

In many companies, test/measurement equipment, such as weighing machines, gauges or calipers, is used for quality inspections in the intermediate and final checking of products and equipment. To ensure that the test equipment being used always meets the specified performance criteria, it's regularly checked and calibrated. You can use the functions of test/measurement equipment management to perform the following activities:

- Manage equipment
- Plan and schedule inspections
- Execute orders and inspection lots for processing calibration inspections on equipment

Figure 8.35 provides an overview of the objects and process associated with test/measurement equipment management:

- Test/measurement equipment is managed as equipment.
- An inspection plan is created for the test/measurement equipment.
- The master inspection characteristics describe the properties that must be measured (e.g., visual inspections and linear measurements).

3 See K. Liebstückel, *Plant Maintenance with SAP—Practical Guide* (3rd edition, SAP PRESS, 2014, pp. 273–280).

- The sampling procedures describe the type and scope of the sample.
- The inspection plan is incorporated into a maintenance plan.
- The maintenance plan generates an order and an inspection lot.
- Each inspection involves processes from inspection lot management (e.g., results recording and the usage decision) and processes from order management (e.g., time confirmation and technical completion confirmation).

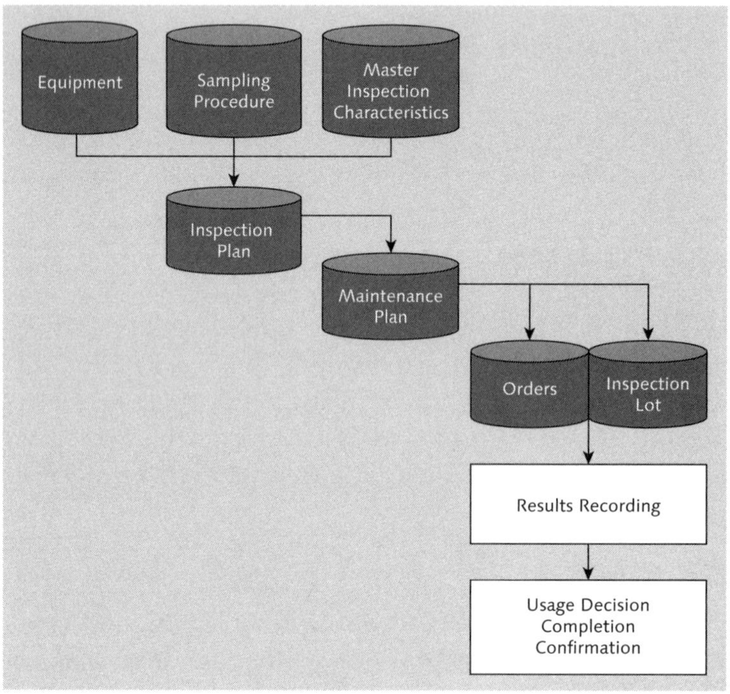

Figure 8.35 Test/Measurement Equipment Management: Overview

In the next section, we'll cover which Customizing settings are required or recommended to implement the test/measurement equipment management process, separated into Plant Maintenance and Quality Management.

[+]

PM/QM Coupling

Because Customizing functions from PM and QM are required to calibrate test/measurement equipment, we also speak of *PM/QM coupling*.

8.4.1 Customizing Functions for Plant Maintenance

The Customizing functions for plant maintenance include specific settings for the following objects:

- Equipment
- Notifications
- Orders
- Maintenance plans

Set View Profiles for Technical Objects

You can use this Customizing function to define the layout of screens associated with technical objects. You determine the following:

- Number of tab pages
- Screen group positions on tab pages
- Name of the tab page
- Icons assigned to the tab page

Separate View Profile for Test/Measurement Equipment [+]

To distinguish test/measurement equipment from other pieces of equipment, and therefore to make it independent of these, define a separate view profile for test/measurement equipment.

Prerequisites

There are no special prerequisites.

Customizing Path

Plant Maintenance and Customer Service • Master Data in Plant Maintenance and Customer Service • Technical Objects • General Data • Set View Profiles for Technical Objects

Settings

First, define a name for the view profile, and make it valid for equipment (see Figure 8.36).

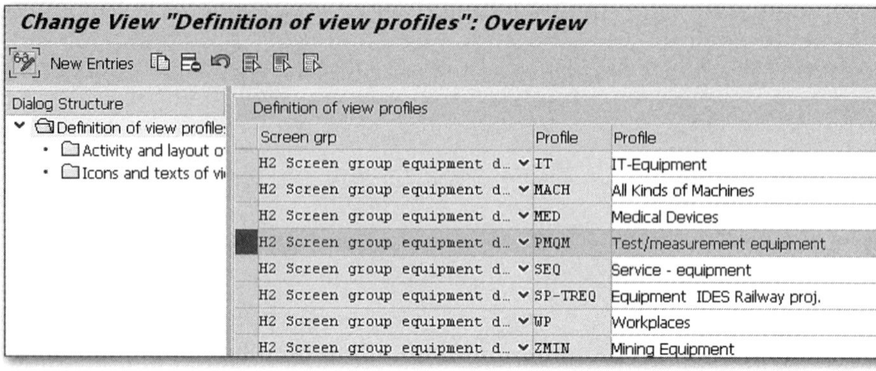

Figure 8.36 Test/Measurement Equipment Management: View Profile for Equipment

Then, use the ACTIVITY AND LAYOUT OF VIEWS subfunction to assign the relevant screen areas to the tab pages (see Figure 8.37).

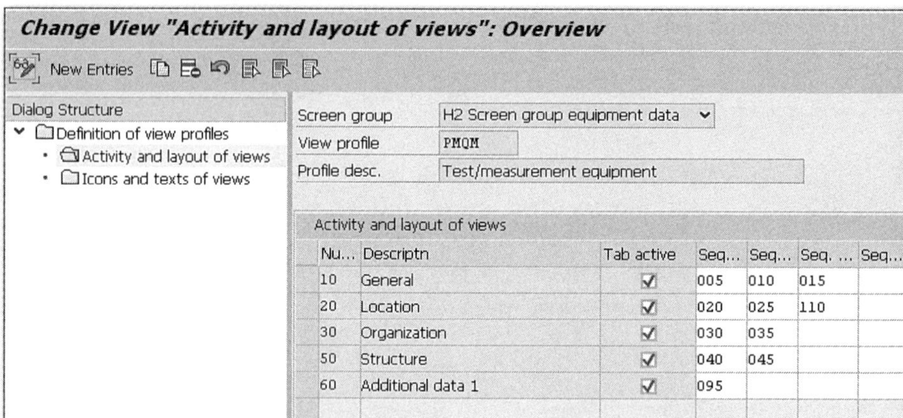

Figure 8.37 Test/Measurement Equipment Management: View Profile for Equipment

Finally, you can use the ICONS AND TEXTS OF VIEWS subfunction to define your own titles for the tab pages and to assign an icon.

Maintain Equipment Category

You use this Customizing function to define equipment categories. This enables you to control which basic properties you want a corresponding equipment master record to have. You can use control properties to define the number of equipment categories you require.

Separate Equipment Category for Test/Measurement Equipment [+]

To ensure that you can execute the business process for calibrating test/measurement equipment, you require a separate equipment category, which you must assign to the PRODUCTION RESOURCE/TOOL REFERENCE CATEGORY.

Prerequisites

If you want to assign object information keys and a view profile to the equipment categories, you must define these beforehand.

Customizing Path

PLANT MAINTENANCE AND CUSTOMER SERVICE • MASTER DATA IN PLANT MAINTENANCE AND CUSTOMER SERVICE • TECHNICAL OBJECTS • EQUIPMENT • EQUIPMENT CATEGORIES • MAINTAIN EQUIPMENT CATEGORY

Settings

Figure 8.38 shows an equipment category for test/measurement equipment management.

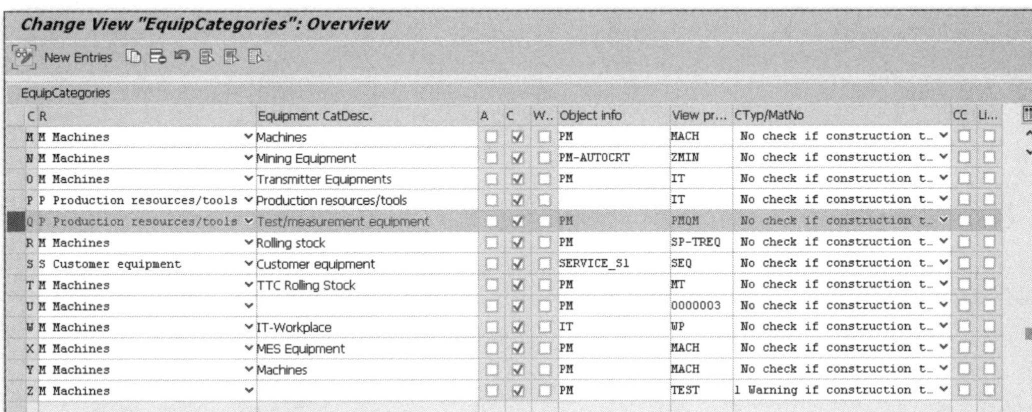

Change View "EquipCategories": Overview

New Entries

EquipCategories

C	R	Equipment CatDesc.	A	C	W..	Object info	View pr...	CTyp/MatNo	CC	Li...
M	M Machines	Machines		☑		PM	MACH	No check if construction t...		
N	M Machines	Mining Equipment		☑		PM-AUTOCRT	ZMIN	No check if construction t...		
O	M Machines	Transmitter Equipments		☑		PM	IT	No check if construction t...		
P	P Production resources/tools	Production resources/tools		☑			IT	No check if construction t...		
Q	P Production resources/tools	Test/measurement equipment		☑		PM	PMQM	No check if construction t...		
R	M Machines	Rolling stock		☑		PM	SP-TREQ	No check if construction t...		
S	S Customer equipment	Customer equipment		☑		SERVICE_S1	SEQ	No check if construction t...		
T	M Machines	TTC Rolling Stock		☑		PM	MT	No check if construction t...		
U	M Machines			☑		PM	0000003	No check if construction t...		
W	M Machines	IT-Workplace		☑		IT	WP	No check if construction t...		
X	M Machines	MES Equipment		☑		PM	MACH	No check if construction t...		
Y	M Machines	Machines		☑		PM	MACH	No check if construction t...		
Z	M Machines			☑		PM	TEST	1 Warning if construction t...		

Figure 8.38 Test/Measurement Equipment Management: Equipment Category

[+]

Special Settings in the Equipment Category

From a test/measurement management perspective, you should make the following settings in the equipment category:

- Assign a REFERENCE CATEGORY. In particular, this is used to define history-relevant fields in an equipment master record. This must be P (PRODUCTION RESOURCES/TOOLS) for equipment categories associated with test/measurement management.
- Assign the VIEW PROFILE that you've defined for test/measurement equipment.

Define Additional Business Views for Equipment Categories

In this work step, you can define additional views for each equipment category. The views represent additional tab pages within the piece of equipment.

Prerequisites

You must define the equipment categories beforehand.

Customizing Path

PLANT MAINTENANCE AND CUSTOMER SERVICE • MASTER DATA IN PLANT MAINTENANCE AND CUSTOMER SERVICE • TECHNICAL OBJECTS • EQUIPMENT • EQUIPMENT CATEGORIES • DEFINE ADDITIONAL BUSINESS VIEWS FOR EQUIPMENT CATEGORIES

Settings

Figure 8.39 shows you which additional views you can activate.

[+]

Production/Resource Tool Must Be Selected

To be able to perform the business process for calibrating test/measurement equipment, you must set the PRODUCTION RESOURCES/TOOLS view to SELECTION. This business process doesn't work for equipment categories for which the PRODUCTION RESOURCES/TOOLS view is blank.

If you want to place your test/measurement equipment in storage, it must also have a material number and serial number. You must therefore activate the SD view (serial data).

In the latter case, you still need to assign a serial number profile to the material number.

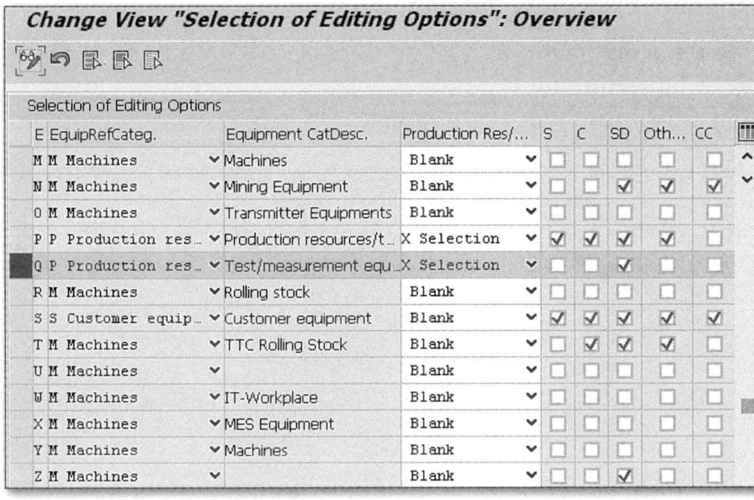

Figure 8.39 Test/Measurement Equipment Management – Equipment: Additional Views

Define Serial Number Profiles

You use this Customizing function to define serial number profiles. In the material master, you must assign a serial number profile at the plant level to each *material* that you want to serialize. This means that for each plant, you can assign a separate serial number profile to a material. A material may require serial numbers in one plant but not in another.

Prerequisites

There are no special prerequisites.

Customizing Path

Plant Maintenance and Customer Service • Master Data in Plant Maintenance and Customer Service • Technical Objects • Serial Number Management • Define Serial Number Profiles

Transaction

OIS2

Settings

On the overview screen, define a profile name and profile description (see Figure 8.40).

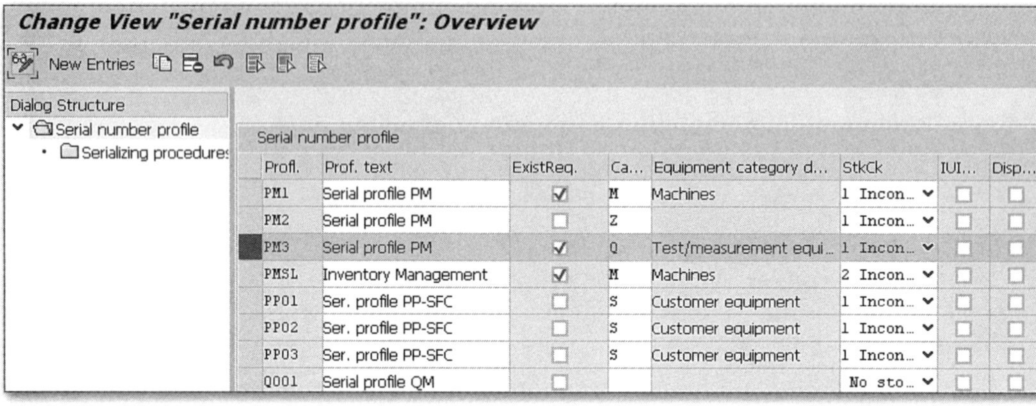

Figure 8.40 Test/Measurement Equipment Management: Serial Number Profile

If, for your test/measurement equipment, you want the system to automatically create serial numbers for a business transaction, such as the goods receipt, specify the equipment category that you've earmarked for your test/measurement equipment in the CATEGORY column. Then, define the SERIALIZING PROCEDURES that are permitted for each of the serial number profiles (see Figure 8.41) on the detail screen.

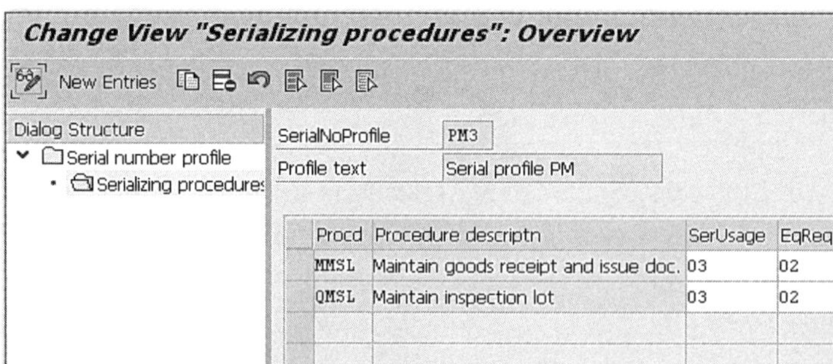

Figure 8.41 Test/Measurement Equipment Management: Serial Number Profile

Test/Measurement Equipment with a Specific Serial Number Profile [+]

You must assign the following serializing procedures to an equipment category that you want to use for the business process for calibrating test/measurement equipment:

- MMSL: MAINTAIN GOODS RECEIPT AND ISSUE DOC.
- QMSL: MAINTAIN INSPECTION LOT.

You also require other settings for this equipment category. However, they are the same as the settings associated with other equipment categories, so we won't discuss them here. Instead, refer to Chapter 3, Section 3.2, for information about equipment-specific settings, and refer to Chapter 4 for information about generic functions.

Configure Order Types

You use this Customizing function to define the order type and some basic parameters.

Separate Order Type for Test/Measurement Equipment Management [+]

You require a separate order type for the business process for calibrating test/measurement equipment. You can't use this order type for normal maintenance processes. Similarly, you can't use a normal order type for test/measurement equipment management.

Prerequisites

Because order types are by plant maintenance as well as by other user departments and their corresponding SAP applications, consult with your colleagues from the other user departments on the naming convention and number ranges for order types.

Customizing Path

PLANT MAINTENANCE AND CUSTOMER SERVICE • MAINTENANCE AND SERVICE PROCESSING • MAINTENANCE AND SERVICE ORDERS • FUNCTIONS AND SETTINGS FOR ORDER TYPES • CONFIGURE ORDER TYPES

Transaction

OIDA

Settings or Recommended Settings

When you use this Customizing function, make sure that you use the correct reference object. If you manage your test/measurement equipment as equipment, use screen reference object 0160 (Equipment). If, however, you also manage the equipment as material and serial numbers, you should choose screen reference object 0170 (equipment, material, and serial number) (see Figure 8.42).

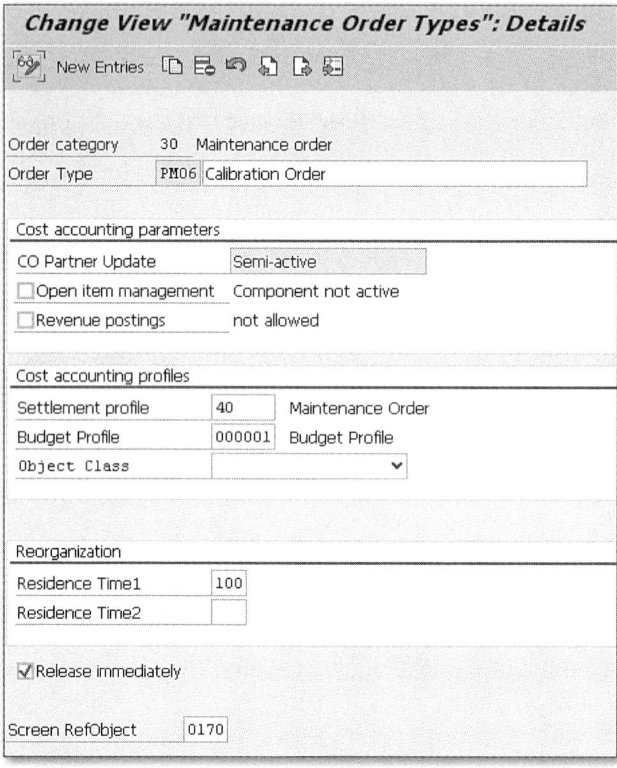

Figure 8.42 Test/Measurement Equipment Management: Order Type

You also require further settings for this order type. However, they are the same as the settings associated with normal order types, so they aren't covered here. Instead, refer to Chapter 6, Section 6.2, for order-specific settings; Section 6.3, for completion confirmations; and Chapter 4, for generic functions.

Calibration Notification Is Optional [+]

You don't necessarily require a notification for the business process for calibrating test/measurement equipment, but you can use one.

If you want to use a notification with the business process for calibrating test/measurement equipment, note the following special features.

Define Notification Types

You use this Customizing function to define the notifications types that you require. The functions described previously should be taken into consideration when defining the number of notification types you need, as well as which particular notification types you want to use.

Separate Notification Type for Calibration [+]

If you decide to use a notification with the business process for calibrating test/measurement equipment, you should define a separate notification type to distinguish it from other notification types and to control it independently.

Prerequisites

You must define the catalog profile beforehand.

Customizing Path

PLANT MAINTENANCE AND CUSTOMER SERVICE • MAINTENANCE AND SERVICE PROCESSING • MAINTENANCE AND SERVICE NOTIFICATIONS • NOTIFICATION CREATION • NOTIFICATION TYPES • DEFINE NOTIFICATION TYPES

Settings or Recommended Settings

A special feature of this notification type is the assignment of a catalog profile, which is tailored to the interests of the business process for calibrating test/measurement equipment (see Figure 8.43).

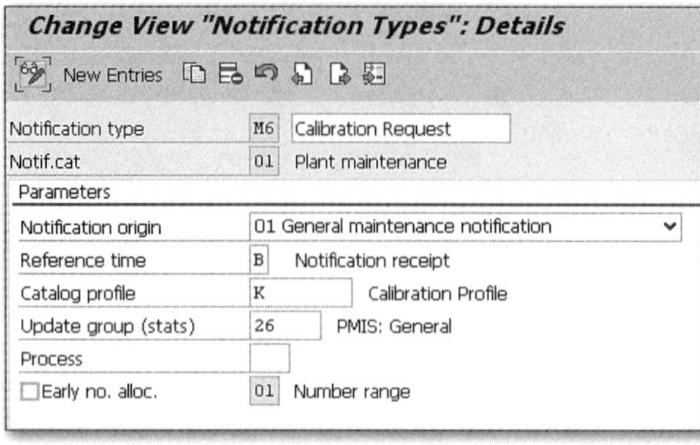

Figure 8.43 Test/Measurement Equipment Management: Notification Type

Maintain Catalogs

You use this Customizing function to maintain the catalogs, code groups, and codes that you want to use.

[+]

Catalog 9 for Calibration Notification

When you configure codes and code groups for test/measurement equipment management, we recommend that you use catalog 9 (defect types) because this can be addressed directly in Customizing for QM.

Prerequisites

There are no special prerequisites. However, you must consult with your colleagues from QM and SAP Customer Service (SAP CS) on the naming convention for catalogs because they also use catalogs.

Customizing Path

Plant Maintenance and Customer Service • Maintenance and Service Processing • Maintenance and Service Notifications • Notification Creation • Notification Content • Maintain Catalogs

Transaction

QS41

Settings or Recommended Settings

In the catalog for defect types, configure a code group and codes that you can use to document exceeding upper limits and shortfall lower limits for target values (see Figure 8.44).

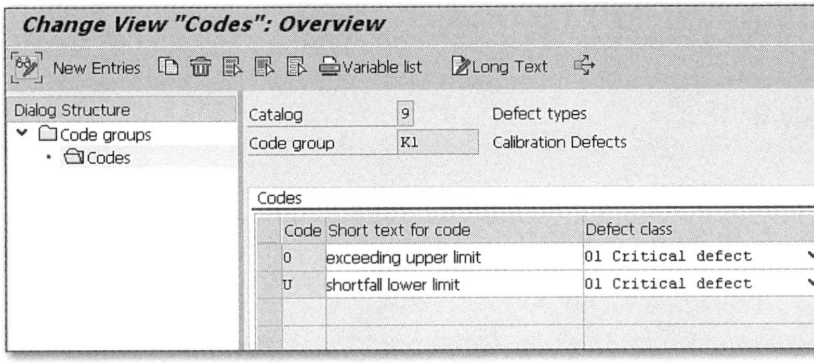

Figure 8.44 Test/Measurement Equipment Management: Catalog for Defect Types

Define Catalog Profile

To enable you to call the defect codes in the calibration notification in a useful manner, we recommend that you configure a catalog profile.

Prerequisites

You must define the catalog types, code groups, and codes beforehand. If you also want to assign a class, you must use Transaction CT04 to define the necessary characteristics and then Transaction CL02 to create a class of the class type 015.

Customizing Path

Plant Maintenance and Customer Service • Maintenance and Service Processing • Maintenance and Service Notifications • Notification Creation • Notification Content • Define Catalog Profile

Transaction

OQN6

Settings or Recommended Settings

Figure 8.45 shows the assignment of a code group from CATALOG type 9 to a CATALOG PROFILE.

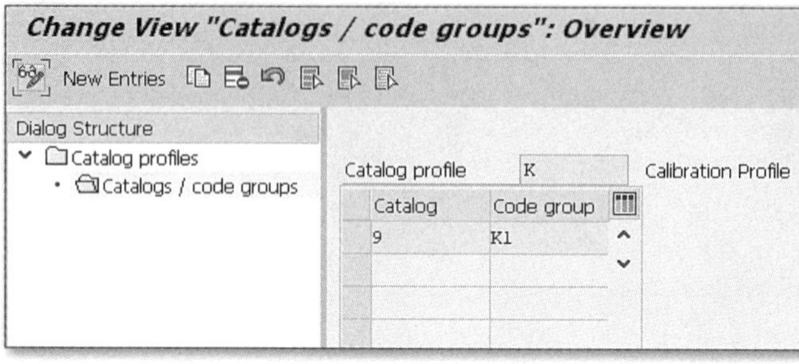

Figure 8.45 Test/Measurement Equipment Management: Catalog Profile

Change Catalogs and Catalog Profile for Notification Type

You use this Customizing function to assign your chosen catalogs to your notification type for calibrating test/measurement equipment.

Prerequisites

You must define the catalog types, catalog profiles, and notification types beforehand.

Customizing Path

PLANT MAINTENANCE AND CUSTOMER SERVICE • MAINTENANCE AND SERVICE PROCESSING • MAINTENANCE AND SERVICE NOTIFICATIONS • NOTIFICATION CREATION • NOTIFICATION CONTENT • CHANGE CATALOGS AND CATALOG PROFILE FOR NOTIFICATION TYPE

Settings or Recommended Settings

Assign the catalog profile, and enter catalog "9" in the PROBLEMS category (see Figure 8.46).

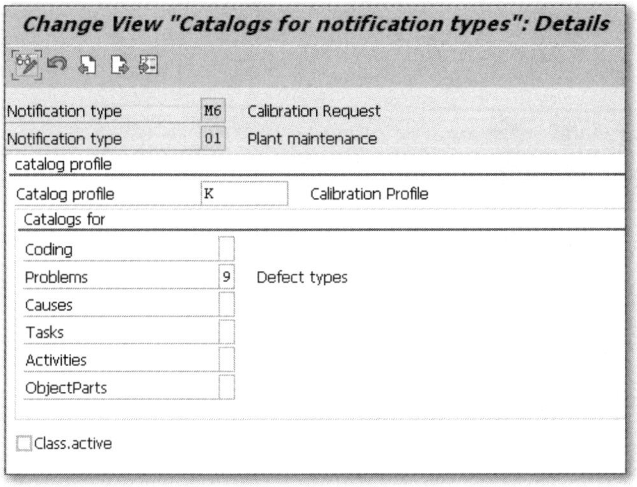

Figure 8.46 Test/Measurement Equipment Management: Assignment of Catalog Profile and Catalog

Overview of Notification Type

Within this Customizing function, you use the subfunction SCREEN AREAS IN NOTIFICATION HEADER to define the notification header and reference objects you want to use for each notification type.

Prerequisites

You must define the notification types themselves and the screen layout for these notification types beforehand. In particular, one of the tab pages must contain the entry "Subscreen 005 = Reference Object".

Customizing Path

PLANT MAINTENANCE AND CUSTOMER SERVICE • MAINTENANCE AND SERVICE PROCESSING • MAINTENANCE AND SERVICE NOTIFICATIONS • OVERVIEW OF NOTIFICATION TYPES • SUBFUNCTION SCREEN AREAS IN NOTIFICATION HEADER

Settings or Recommended Settings

In SCREEN TYPE HDR (screen type header), always enter "H100" (HEADER MAINTENANCE NOTIFICATION) unless you've defined your own headers. The HEADER MAINTENANCE NOTIFICATION contains information such as the notification type, notification number, short text, status, and assigned order number.

In ScrnType Object, assign either 0150 (Equipment Only) or 0170 (Equipment + Serial No. + Material No.), as shown in Figure 8.47.

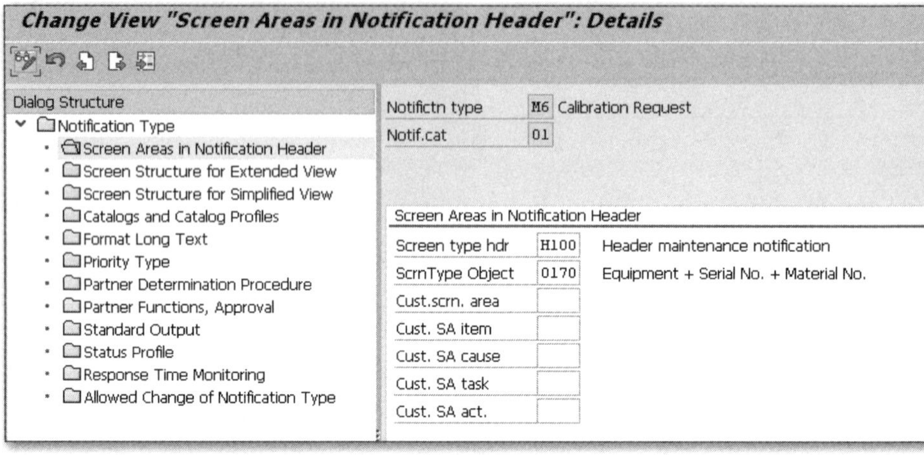

Figure 8.47 Test/Measurement Equipment Management: Screen Areas in Notification Header

Define Notification and Order Integration

Generally, you don't manually create orders for calibrating test/measurement equipment. Instead, they are automatically generated as a call in relation to maintenance plans. If you also want to use a notification for an order, it makes sense for the system to automatically create the notification for the order. You use this Customizing function to achieve this.

Prerequisites

You must define the order types and notification types beforehand.

Customizing Path

Plant Maintenance and Customer Service • Maintenance and Service Processing • Maintenance and Service Orders • Functions and Settings for Order Types • Define Notification- and Order Integration

Settings or Recommended Settings

To enable the system to automatically create a calibration notification for a calibration order, set the indicator in the NOTIFCTN column, and enter the notification type that you want to create (see Figure 8.48).

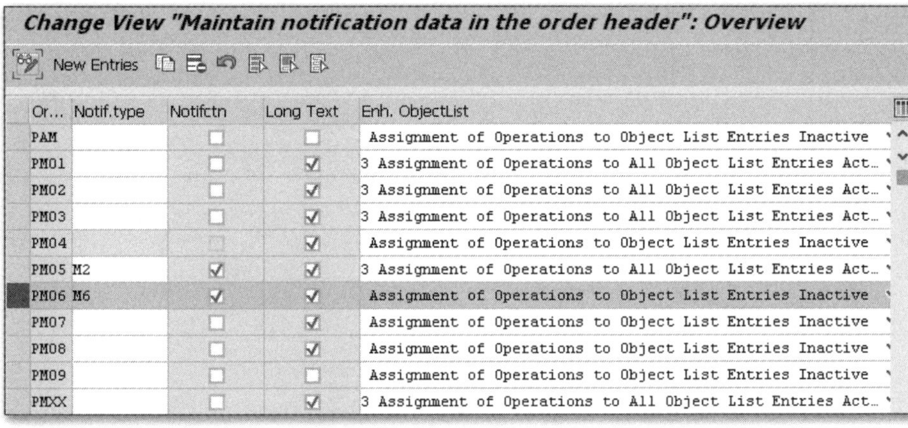

Or...	Notif.type	Notifctn	Long Text	Enh. ObjectList
PAM		☐	☐	Assignment of Operations to Object List Entries Inactive
PM01		☐	☑	3 Assignment of Operations to All Object List Entries Act…
PM02		☐	☑	3 Assignment of Operations to All Object List Entries Act…
PM03		☐	☑	3 Assignment of Operations to All Object List Entries Act…
PM04		☐	☑	Assignment of Operations to Object List Entries Inactive
PM05	M2	☑	☑	3 Assignment of Operations to All Object List Entries Act…
PM06	M6	☑	☑	Assignment of Operations to Object List Entries Inactive
PM07		☐	☑	Assignment of Operations to Object List Entries Inactive
PM08		☐	☑	Assignment of Operations to Object List Entries Inactive
PM09		☐	☐	Assignment of Operations to Object List Entries Inactive
PMXX		☐	☑	3 Assignment of Operations to All Object List Entries Act…

Figure 8.48 Test/Measurement Equipment Management: Integration of a Notification Type and Order Type

You also require further settings for this notification type, but they aren't discussed here because they are the same as the settings associated with normal notification types. Instead, refer to Chapter 6, Section 6.1, for information about notification-specific settings, and refer to Chapter 4 for information about generic functions.

Set Maintenance Plan Categories

You use this Customizing function to define the control parameters for the maintenance plans (the maintenance call object or reference object, among others).

Separate Maintenance Plan Category for Calibration [+]

You should define a separate maintenance plan category for calibrating test/measurement equipment to distinguish it from other maintenance plans and to control it independently.

Prerequisites

There are no special prerequisites.

Customizing Path

PLANT MAINTENANCE AND CUSTOMER SERVICE • MAINTENANCE PLANS, WORK CENTERS, TASK LISTS AND PRTS • MAINTENANCE PLANS • SET MAINTENANCE PLAN CATEGORIES

Settings or Recommended Settings

Figure 8.49 shows the settings for a maintenance plan category for test/measurement equipment management.

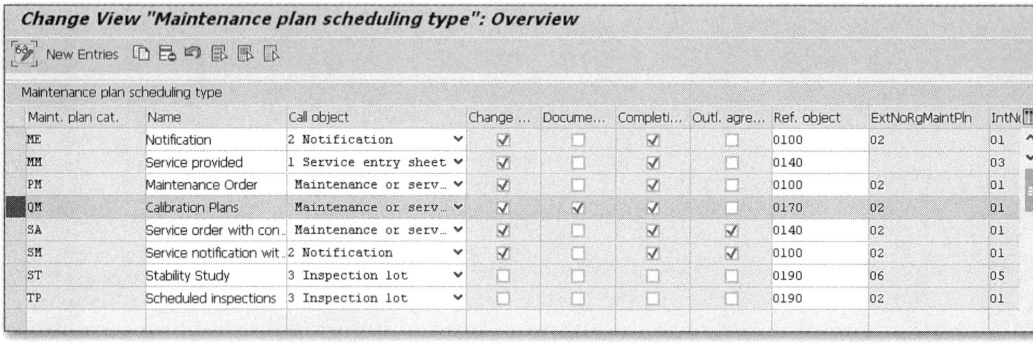

Maint. plan cat.	Name	Call object	Change ...	Docume...	Completi...	Outl. agre...	Ref. object	ExtNoRgMaintPln	IntN
ME	Notification	2 Notification	☑	☐	☑	☐	0100	02	01
MM	Service provided	1 Service entry sheet	☑	☐	☑	☐	0140		03
PM	Maintenance Order	Maintenance or serv...	☑	☐	☑	☐	0100	02	01
QM	Calibration Plans	Maintenance or serv...	☑	☑	☑	☐	0170	02	01
SA	Service order with con...	Maintenance or serv...	☑	☐	☑	☑	0140	02	01
SM	Service notification wit...	2 Notification	☑	☐	☑	☑	0100	02	01
ST	Stability Study	3 Inspection lot	☐	☐	☐	☐	0190	06	05
TP	Scheduled inspections	3 Inspection lot	☐	☐	☐	☐	0190	02	01

Figure 8.49 Test/Measurement Equipment Management: Maintenance Plan Category

[+]

Settings for the Maintenance Plan Category for Test/Measurement Equipment Management

The following settings are important for test/measurement equipment management:

- ▶ The most important setting is the CALL OBJECT. For the business process for calibrating test/measurement equipment, choose the call object MAINTENANCE OR SERVICE ORDER (and not the call object INSPECTION LOT, which you may think makes more sense here).
- ▶ In the REF. OBJECT column, assign "0150" (EQUIPMENT ONLY) or "0170" (EQUIPMENT + SERIAL NO. + MATERIAL NO.).

You also require further settings for this maintenance plan category, but we won't discuss them here because they are the same as the settings for other maintenance plan categories. Instead, refer to Chapter 7 for maintenance plan-specific settings and Chapter 4 for information about generic functions.

8.4.2 Customizing Functions for Quality Management

The Customizing functions for test/measurement equipment management comprise the following phases:

- Quality planning
- Quality inspection
- Quality notification

Define SPC Criteria

Generally, in quality planning, you use this Customizing function to define the criteria according to which the inspection results for control charts are grouped together and the processes are assigned. The statistical process control (SPC) criteria define whether individual or global control charts are managed for different orders, work centers, materials, manufacturers, or customers.

For quality planning within test/measurement equipment management, in particular, you use this Customizing function to control the inspection characteristics that you assign to the operation in an inspection plan. Here, it's necessary to distinguish between two different types of inspection characteristics:

- **Master inspection characteristics**
  Inspection characteristics whose properties you initially assign as a master record and then assign to the operation in the inspection plan.
- **Task list characteristics**
  Inspection characteristics that you assign directly to the operation and whose properties you define directly in the inspection plan.

Prerequisites

You must define the task list usages (see Chapter 3, Section 3.4) beforehand.

Customizing Path

QUALITY MANAGEMENT • QUALITY PLANNING • BASIC DATA • SAMPLE, SPC • STATISTICAL PROCESS CONTROL • DEFINE SPC CRITERIA

Settings or Recommended Settings

You require two SPC criteria:

- You use an SPC criterion to control the inspection characteristics that you define directly in the inspection plan (such as SPC CRITERION 410 in Figure 8.50). You must assign the function module QRKS_CHARACTERISTIC_PM_OBJECT to this plan.

- You use the other SPC criterion to control the inspection characteristics that you initially create as master inspection characteristics and then assign to the inspection plan (such as SPC CRITERION 420 in Figure 8.51). You must assign the function module QRKS_MASTER_CHAR_PM_OBJECT to this plan.

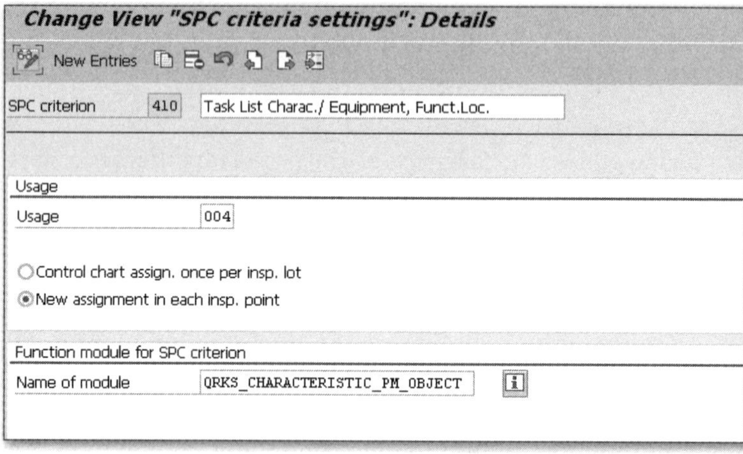

Figure 8.50 Test/Measurement Equipment Management – SPC Criteria: Task List Characteristic

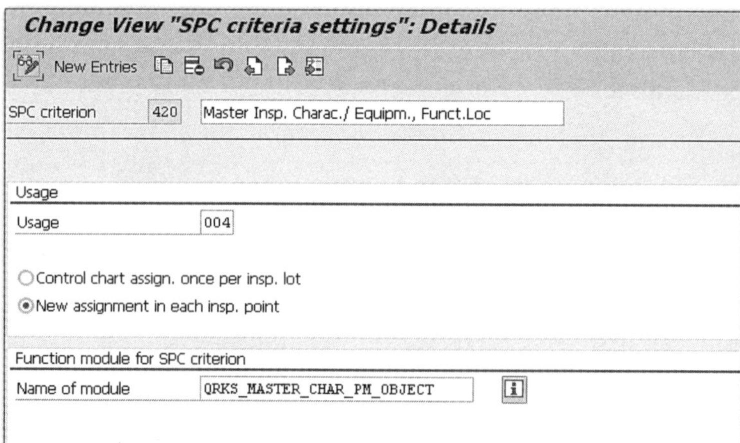

Figure 8.51 Test/Measurement Equipment Management – SPC Criteria: Master Inspection Characteristic

In addition, both SPC criteria must be assigned task list USAGE 004 (maintenance task lists) or 009 (all task list types) so that you can use them in general maintenance task lists or equipment task lists.

Furthermore, the NEW ASSIGNMENT IN EACH INSP. POINT radio button must be set for both SPC criteria because it's used to control the quality control charts for technical objects.

Define Identifier for Inspection Points

You use this Customizing function to define the inspection points and their attributes. You can use inspection points in several application areas. The following *inspection point types* are available for selection:

- **Freely defined inspection points**
  You use this inspection point type, for example, if you create *inspection lots for a production order* and want to carry out several inspections for each operation during the production process, or if you want to identify partial quantities in production.
- **Inspection points to identify samples**
  You use this inspection point type if you want to *manage samples* and record inspection results for *physical samples*.
- **Inspection points to identify maintenance objects**
  You use this inspection point type if you want to carry out calibration inspections and record their results.

Prerequisites

You must define the order types and usage decisions beforehand.

Customizing Path

QUALITY MANAGEMENT • QUALITY PLANNING • INSPECTION PLANNING • GENERAL • DEFINE IDENTIFIER FOR INSPECTION POINTS

Settings or Recommended Settings

Define an inspection point and assign inspection point type 1 (INSPECTION OF EQUIPMENT) to this inspection point as shown in Figure 8.52.

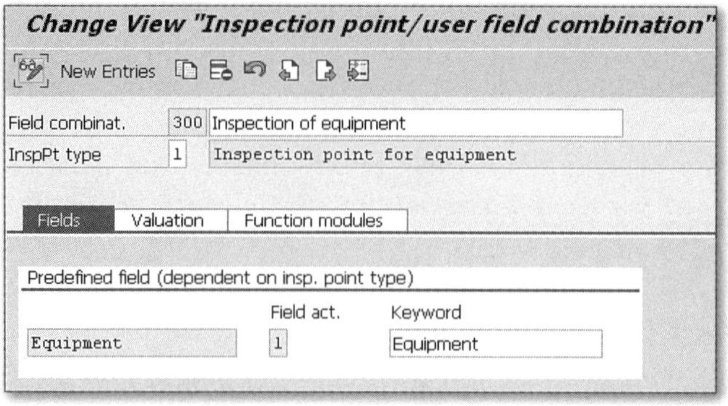

Figure 8.52 Test/Measurement Equipment Management: Inspection Point for Equipment

For this inspection point, define how you want to deal with the usage decision (see Figure 8.53):

- For the plant, use the SELECTED SET to define the code group that is to be used for the equipment usage decision.
- Define the code that you want to use if the inspection lot for the piece of equipment is to be rejected (DEFAULT WHEN AT LEAST ONE CHAR. IS REJECTED).
- Define the code that you want to use if the inspection lot for the piece of equipment is to be accepted (DEFAULT WHEN ALL CHARS. ARE ACCEPTED).

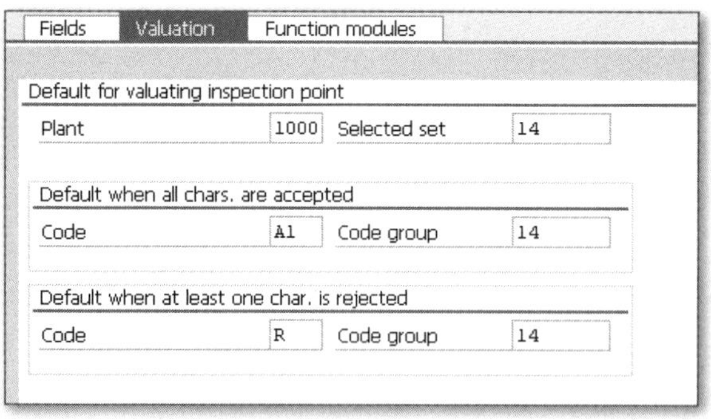

Figure 8.53 Test/Measurement Equipment Management – Inspection Point for Equipment: Valuation

To define the catalog for usage decisions, along with its code groups and codes, see the Customizing function MAINTAIN CATALOGS FOR USAGE DECISIONS.

One Inspection Point for Each Plant [+]

Because the default values for the usage decision are selected on a plant-specific basis, we recommend that you create a separate inspection point for each plant in which you calibrate test/measurement equipment.

Define Control Keys for Inspection Operations

In Chapter 3, Section 3.4, we discussed the function and use of control keys. You can use the *control key* of an operation to define which business functions you want to execute or how you want to handle the *operation* with regard to costing, printing, completion confirmations, scheduling, and so on.

Control Keys with Inspection Characteristics [+]

From a calibration of test/measurement equipment perspective, the control key has the following important property: INSP.CHAR. REQUIRED (inspection characteristic required). You must therefore assign a control key with this particular property to all operations that are to trigger a calibration inspection in the form of an inspection lot.

Prerequisites

There are no technical prerequisites. Because, however, the control key is used not only in Plant Maintenance but also in Production, QM, and Project Processing, you should consult with your colleagues from these areas on the naming convention for the control key.

Customizing Path

QUALITY MANAGEMENT • QUALITY PLANNING • INSPECTION PLANNING • OPERATION • DEFINE CONTROL KEYS FOR INSPECTION OPERATIONS

Settings or Recommended Settings

Configure one or more control keys for which the INSP.CHAR. REQUIRED indicator is set (see Figure 8.54).

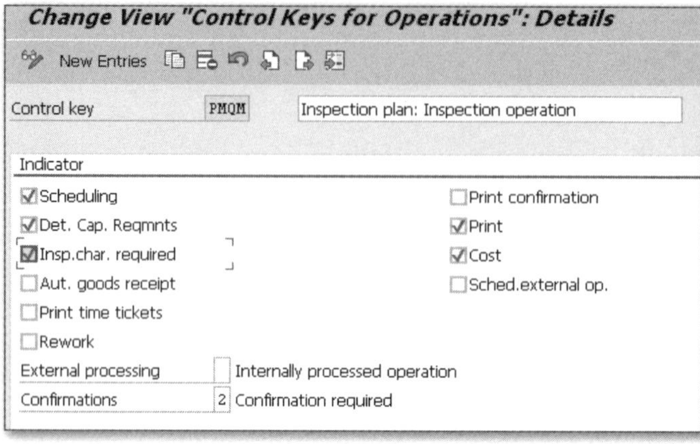

Figure 8.54 Test/Measurement Equipment Management: Control Key

[+]

Separate Control Key for Test/Measurement Equipment Management

To distinguish test/measurement equipment management not only from the other control keys in plant maintenance, but also from all other control keys that generate inspection lots, we recommend that you define a separate control key solely for use with the inspection lots associated with test/measurement equipment management.

Maintain Inspection Types

Just as the order type is at the heart of maintenance task processing and can be used to significantly influence maintenance processes, the inspection type is at the heart of inspection task processing and can be used to significantly influence inspection processes (see Figure 8.55). For example, you can control which order type in plant maintenance is to be linked with which inspection type in QM.

You use this Customizing function to define your inspection types. This Customizing function is the basis for many other Customizing functions.

Prerequisites

You must define the status profile, QM order types, catalog for usage decisions, and notification types beforehand.

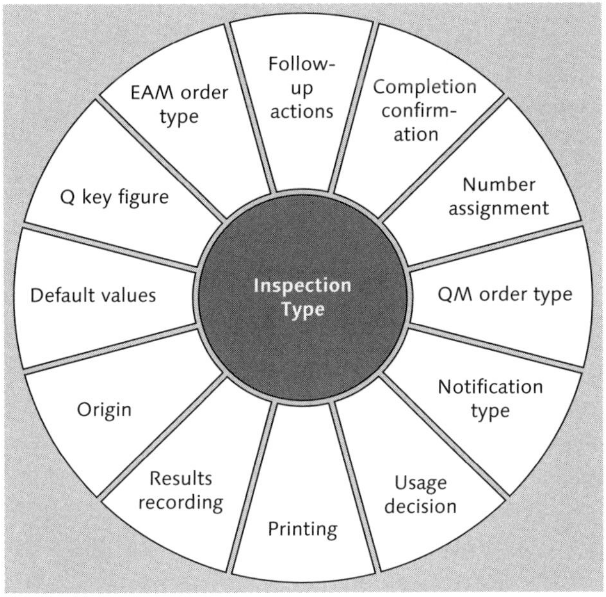

Figure 8.55 Test/Measurement Equipment Management – Inspection Type: Overview

Customizing Path

Quality Management • Quality Inspection • Inspection Lot Creation • Maintain Inspection Types

Settings or Recommended Settings

Figure 8.56 shows you the settings for an inspection type.

[+]

Separate Inspection Type for Plant Maintenance

You require a separate inspection type for calibrating test/measurement equipment. You can't use other inspection types (e.g., incoming inspections or in-process inspections in production).

The following settings are important from a test/measurement equipment management perspective:

- The UD selected set for defining the usage decisions in relation to accepting or rejecting the inspection lots
- The Notification type that is to be created if an inspection lot is rejected

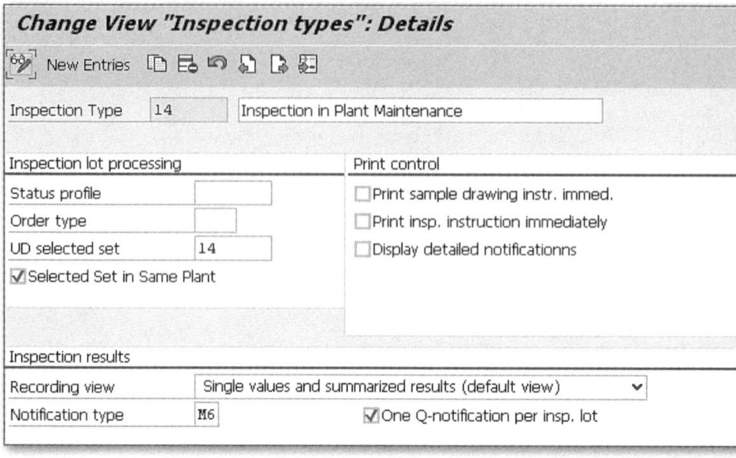

Figure 8.56 Test/Measurement Equipment Management: Maintain Inspection Types

For information about the notification type, see the Customizing function DEFINE NOTIFICATION TYPES in Section 8.4.1. For information about defining the catalog for usage decisions, along with its code groups and codes, see the Customizing function MAINTAIN CATALOGS FOR USAGE DECISIONS in this section.

Maintain Inspection Lot Origins and Assign Inspection Types

You can define inspection types yourself. However, each inspection type must refer to an inspection lot origin. You can assign an inspection lot origin to several inspection types. The inspection lot origin is predefined by SAP and can't be changed. You use this Customizing function to assign your inspection type to an inspection lot origin. The inspection lot origin for plant maintenance is 14.

Prerequisites

You must define the inspection types beforehand.

Customizing Path

QUALITY MANAGEMENT • QUALITY INSPECTION • INSPECTION LOT CREATION • MAINTAIN INSPECTION LOT ORIGINS AND ASSIGN INSPECTION TYPES

Settings or Recommended Settings

In the dialog structure below the ORIGIN subfunction, the standard SAP delivery contains entry 14 PLANT MAINTENANCE (see Figure 8.57).

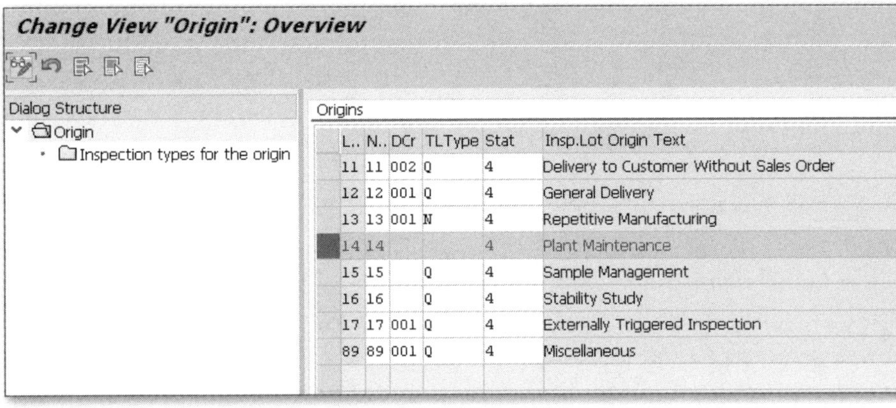

Figure 8.57 Test/Measurement Equipment Management: Inspection Lot Origin

In the dialog structure below the INSPECTION TYPES FOR THE ORIGIN subfunction, you must assign your inspection type(s) for calibrating test/measurement equipment to entry 14 (see Figure 8.58).

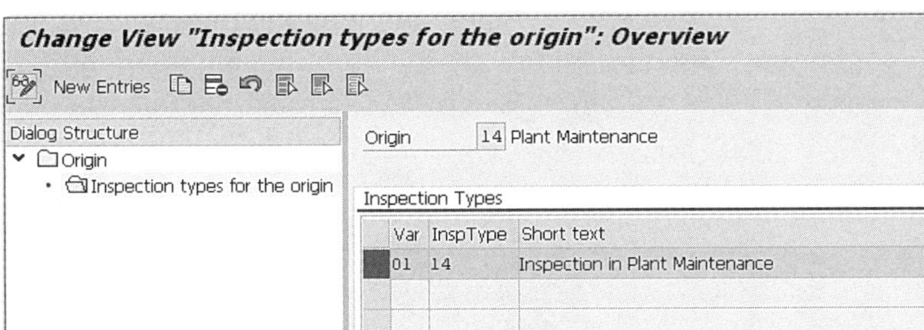

Figure 8.58 Test/Measurement Equipment Management: Assign Inspection Type to Inspection Lot Origin

Assign Origin 14 [+]

It's imperative that you assign origin 14 to all of your inspection types for calibrating test/measurement equipment management. Otherwise, you can't use these inspection types for test/measurement equipment management.

Define Default Values for Inspection Type

You can use this Customizing function to define default values for the inspection types.

Prerequisites

You must define the inspection types beforehand.

Customizing Path

Quality Management • Quality Inspection • Inspection Lot Creation • Define Default Values for Inspection Type

Settings or Recommended Settings

The following settings make sense for inspection types for test/measurement equipment management (see Figure 8.59):

- The only way to inspect test/measurement equipment is by means of a task list (general maintenance task list or equipment task list). Therefore, the Inspect by task list indicator is preselected and can't be changed.
- Because you generally want to record the results for test/measurement equipment management, you should set the Record characteristic results indicator. You can then perform characteristic-based results recording for the inspection lot.

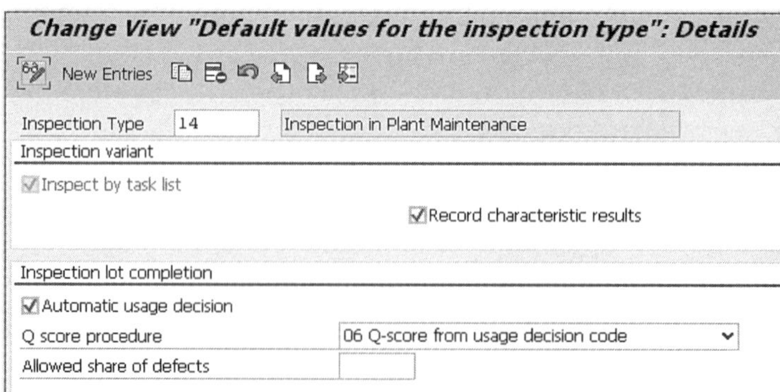

Figure 8.59 Test/Measurement Equipment Management: Default Values for Inspection Type

- If you want to make manual usage decisions only, you should not set the AUTOMATIC USAGE DECISION indicator. However, select this indicator if you want to permit automatic usage decisions in addition to manual usage decisions.
- Use the Q SCORE PROCEDURE to define how the quality code for the inspection lot is to be determined. In the case of test/measurement equipment management, this is usually the usage decision code.

Define Inspections in Plant Maintenance

This Customizing function is *the* central element of PM/QM coupling. Here, you link the order types that you've defined for test/measurement equipment management to the corresponding inspection types for each plant that you've defined. Afterward, if you create an order in this plant and for this order type (in general, this is automatically created from the maintenance plans), the system automatically generates an inspection lot with the specified inspection type as a result of linking the order types.

Prerequisites

You must define the plants, order types, and inspection types beforehand.

Customizing Path

QUALITY MANAGEMENT • QUALITY INSPECTION • INSPECTION LOT CREATION • DEFINE INSPECTIONS IN PLANT MAINTENANCE

Settings or Recommended Settings

Figure 8.60 shows you how to assign, for each plant, an inspection type to an order type. For information about configuring the order type, see the Customizing function CONFIGURE ORDER TYPES in Section 8.4.1.

Assign Inspection Types to Order Types [+]

Make sure that you assign your inspection types fully and correctly to the order types for each plant. If you don't ensure this assignment, there is no PM/QM coupling. Furthermore, the system doesn't create any inspection lots for the orders, and you can't calibrate the test/measurement equipment.

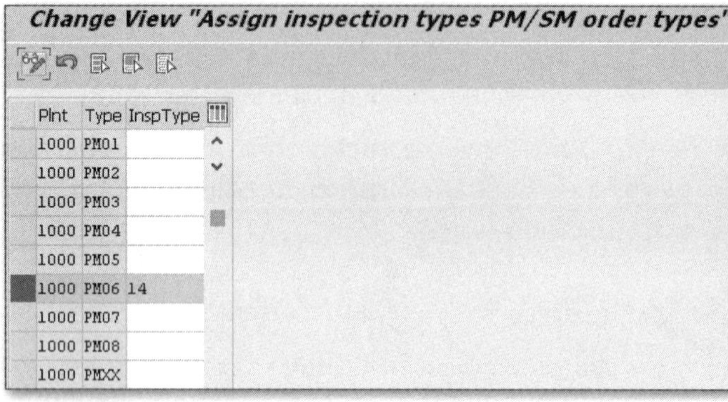

Figure 8.60 Test/Measurement Equipment Management: Assign Inspection Type to Order Type

Maintain Catalogs for Usage Decisions

You use this Customizing function to define the usage decisions for the inspection lots for test/measurement equipment management (in particular, acceptance and rejection).

Prerequisites

You must define the follow-up actions beforehand.

Customizing Path

Quality Management • Quality Inspection • Inspection Lot Completion • Maintain Catalogs for Usage Decisions

Transaction

- QS41 for editing code groups and codes
- QS51 for maintaining selected sets

Settings or Recommended Settings

This Customizing function comprises three interrelated subfunctions:

1. Use the Check catalog type subfunction to define a number and description for the usage decision catalog. In the standard SAP delivery, this is catalog 3 (see Figure 8.61), which you can also use in general.

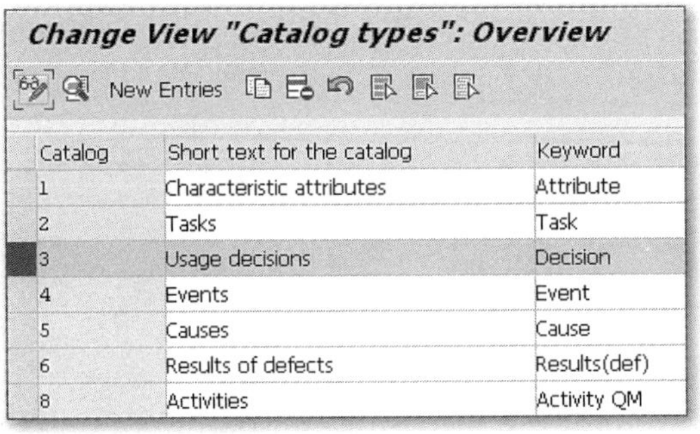

Figure 8.61 Test/Measurement Equipment Management: Catalog for Usage Decisions

2. In the next step, use the EDIT CODE GROUPS AND CODES subfunction to define, for the usage decision catalog, one or more code groups for test/measurement equipment management (see Figure 8.62).

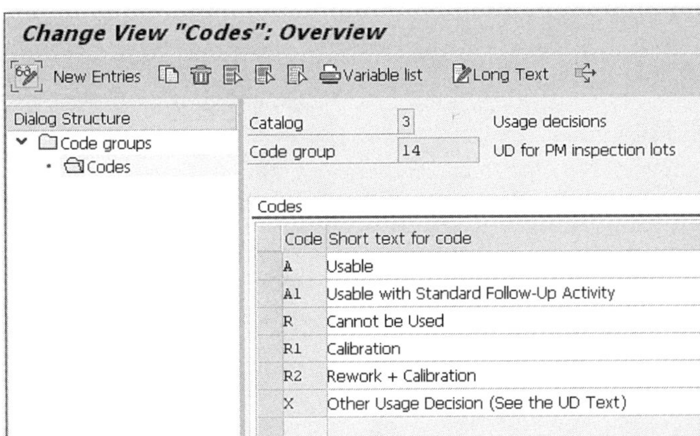

Figure 8.62 Test/Measurement Equipment Management: Code Group for Usage Decisions

3. In the third and final step, use the EDIT SELECTED SETS subfunction to define the codes in question and their consequences for each plant (see Figure 8.63):
   - In the VALUATION CODE column, define whether the code will result in the inspection lot being accepted or rejected.
   - In the QUALITY SCORE column, refine this Yes/No decision further on a scale of 0 to 100 (e.g., 0 = Not OK, 100 = OK, 66 = OK with Restrictions).
   - In the FOLLOW-UP ACTION column, define whether and, if so, which follow-up actions are to be triggered (e.g., lock the piece of equipment or technically

complete the order). For more information, see the Customizing function DEFINE FOLLOW-UP ACTION in this section.

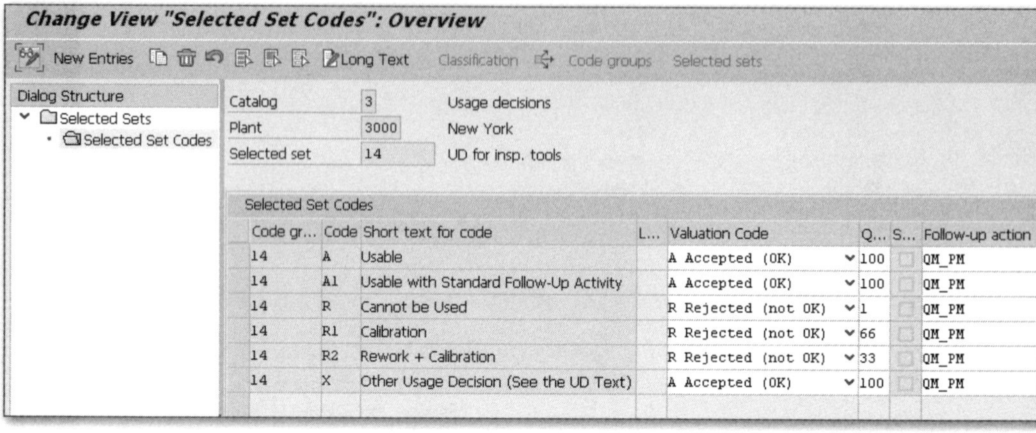

Figure 8.63 Test/Measurement Equipment Management: Valuation for Usage Decisions

[+]

At Least One Code for Acceptance and One Code for Rejection

To ensure that the inspection lots for test/measurement equipment management are completed with a useful usage decision, you need at least one code for accepting the inspection lot and one code for rejecting the inspection lot.

You must also assign these codes to each plant in which you want to implement test/measurement equipment management.

Define Follow-Up Action

You use this Customizing function to define the follow-up actions that the system is to automatically trigger when a usage decision is made.

Prerequisites

There are no specific prerequisites.

Customizing Path

QUALITY MANAGEMENT • QUALITY INSPECTION • INSPECTION LOT COMPLETION • DEFINE FOLLOW-UP ACTION

Settings or Recommended Settings

In the dialog structure, create a group of FOLLOW-UP ACTIONS for test/measurement equipment management (see Figure 8.64).

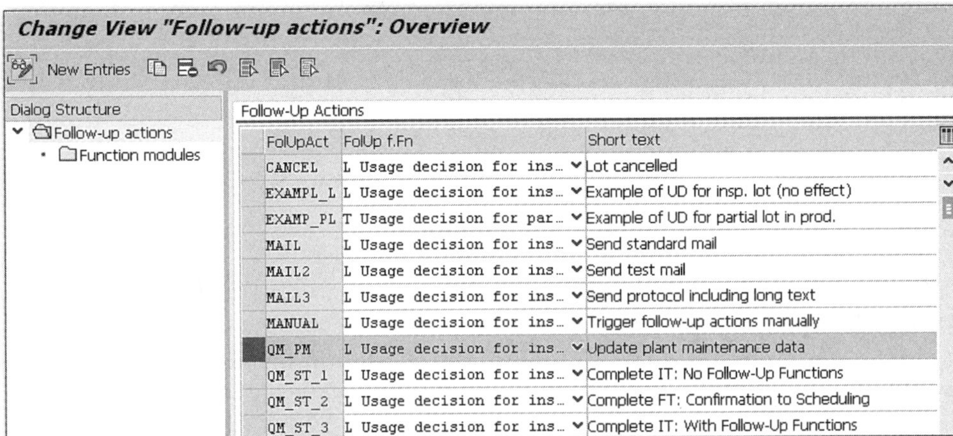

Figure 8.64 Test/Measurement Equipment Management – Follow-Up Actions: Overview

[+]

L = Usage Decision for Inspection Lot

It's important to assign the function L to the test/measurement equipment group so that the follow-up actions are controlled on the basis of the usage decision for the inspection lot.

In the next step, use the FUNCTION MODULES subfunction to define which follow-up actions are to be triggered within the group. From a test/measurement equipment management perspective, the following three follow-up actions are of interest (see Figure 8.65):

- The function module QFOA_OBJECT_STATUS_SET activates the follow-up action SET EQUIPMENT STATUS/CHANGE CYCLE MODIFICATION FACTOR. When you make the usage decision, the system displays a dialog box (see Figure 8.66) in which you can change the status of the piece of equipment from READYFORUSE to NOT READYFORUSE or vice versa, and you can change the CYCLE MODIFICFACTOR (cycle modification factor) of the maintenance plan.
- The function module QFOA_ORDER_TECHNICAL_COMPLETE activates the follow-up action TECHNICALLY COMPLETE PM ORDER. When you make the usage

decision, the system technically completes the order automatically (in the background).

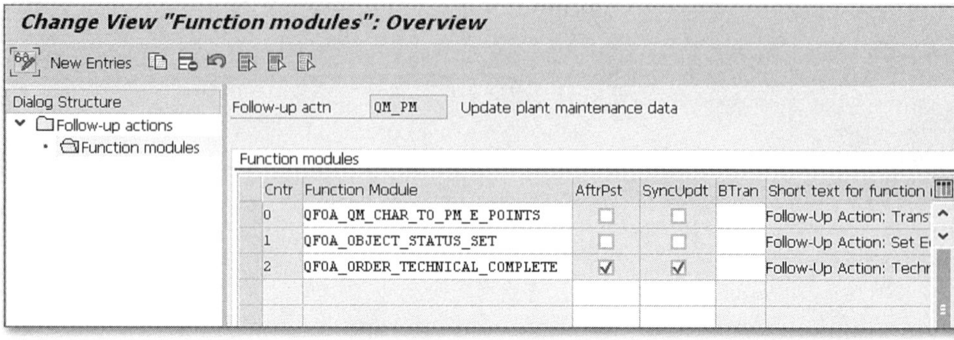

Figure 8.65 Test/Measurement Equipment Management – Follow-Up Actions: Detail Screen

- The function module QFOA_QM_CHAR_TO_PM_E_POINTS activates the follow-up action TRANSFER OF QM CHARACTERISTIC RESULTS TO PM MEASURING POINTS. You can use this follow-up action to transfer inspection results to equipment as measurement documents. Quantitative inspection characteristics (measurement readings) and qualitative inspection characteristics (characteristics with catalog attributes) are taken into consideration here. A class characteristic is used to establish a link between the master inspection characteristics and the measuring points.

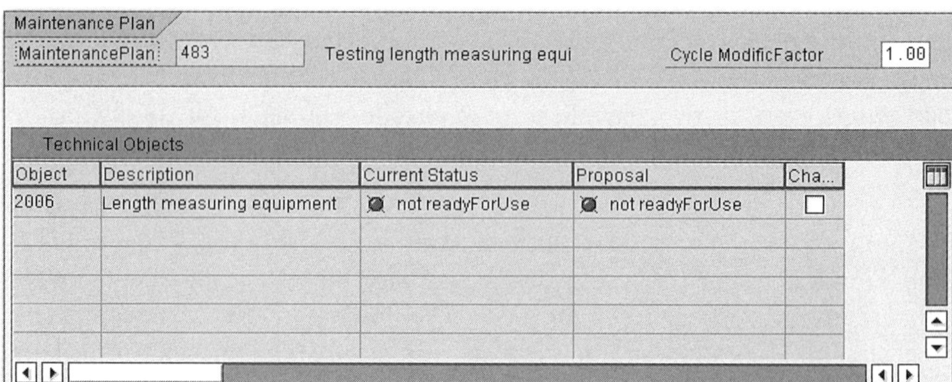

Figure 8.66 Test/Measurement Equipment Management: Set Status of Follow-Up Action

[+]

Activate Follow-Up Actions

Don't neglect to use the option to automatically trigger follow-up actions in the background. In particular, the following options are very helpful:

- The option to declare a piece of equipment as ready for use or not ready for use
- The option to technically complete orders automatically

Plan Automatic Usage Decision

You use this Customizing function to plan the process for the automatic usage decision for inspection lots. The inspection lots must satisfy the following conditions:

- All characteristics have been completed.
- None of the characteristics have been rejected.
- No defects have been recorded, and no quality notifications have been created.

The system then waits until the waiting time defined in this Customizing function has elapsed. During this period, you can still intervene in the automatic process. The system records the automatic usage decision in the long text.

Prerequisites

You must have set the AUTOMATIC USAGE DECISION indicator in the Customizing function DEFINE DEFAULT VALUES FOR INSPECTION TYPE.

Furthermore, you must define the plants, recording configurations, and usage decisions beforehand.

Customizing Path

QUALITY MANAGEMENT • QUALITY INSPECTION • INSPECTION LOT COMPLETION • PLAN AUTOMATIC USAGE DECISION

Settings or Recommended Settings

Many of the settings available in this Customizing function don't apply to the inspection lots for test/measurement equipment management. Only the following settings are relevant:

- On the Results recording tab page (see Figure 8.67), define the Recording configuration by setting it to 1 Automatic Close. For more information, refer to the Customizing function Define Recording Configuration.

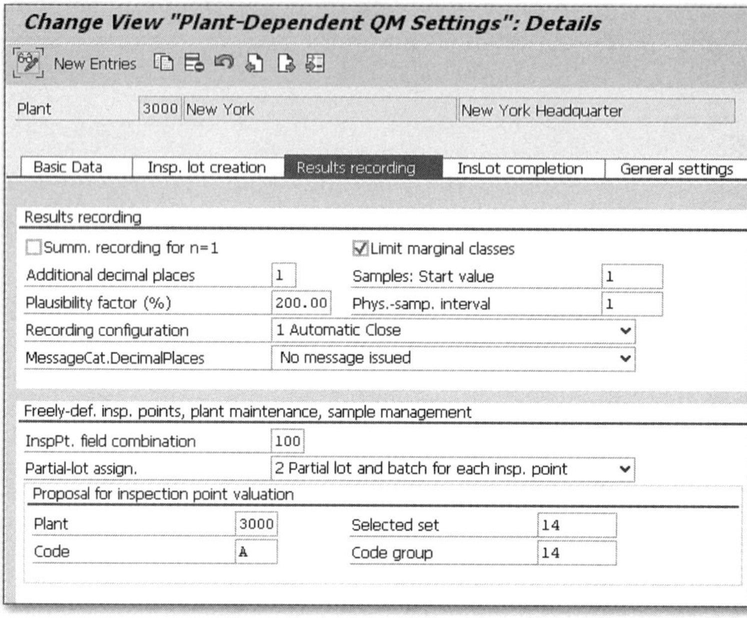

Figure 8.67 Test/Measurement Equipment Management – Automatic Usage Decision: Results Recording

- The same tab page contains the Proposal for inspection point valuation field group. Here, define which code the system is to automatically set for the usage decision, if the aforementioned prerequisites are fulfilled. Usually, this is the usage decision Acceptance.
- The InsLot completion tab page (see Figure 8.68) contains the Waiting time in hours and minutes. Here, specify how much time must have elapsed since the inspection lot was created before the system can make the automatic usage decision.

[+]

Recommendations for the Automatic Usage Decision

- Set the Recording Configuration option to Automatic Close.
- Enter a usage decision for acceptance.
- Schedule a sufficient waiting time since the creation of the inspection lot.

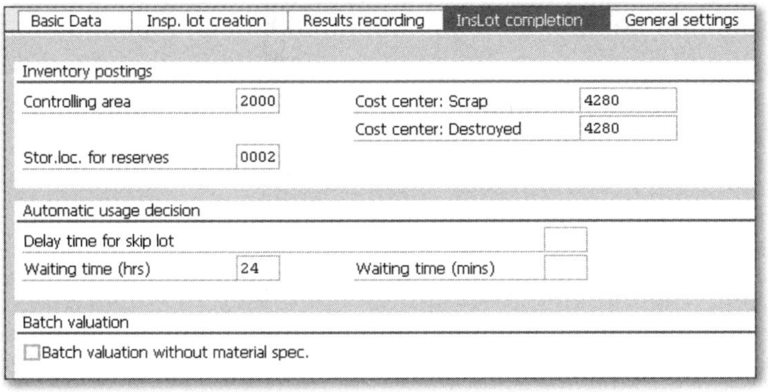

Figure 8.68 Test/Measurement Equipment Management – Automatic Usage Decision: Inspection Lot Completion

Define Recording Configuration

You use this Customizing function to define different background functions for recording characteristic inspection results. If certain prerequisites are fulfilled, you can prompt the system to complete certain functions in the background when it records inspection results.

Prerequisites

There are no specific prerequisites.

Customizing Path

Quality Management • Quality Inspection • Results Recording • Define Recording Configuration

Settings or Recommended Settings

Figure 8.69 shows you the recording configurations contained in the standard SAP delivery.

Then, on the detail screen, define which functions are triggered. For the recording configuration recommended in the automatic usage decision (1 = Automatic Close), these are, for example:

- Close When Choosing Continue
- Close When Navigating
- Close When Saving

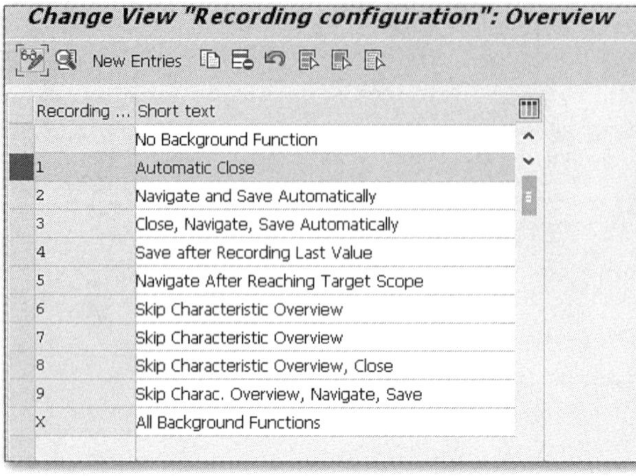

Recording ...	Short text
	No Background Function
1	Automatic Close
2	Navigate and Save Automatically
3	Close, Navigate, Save Automatically
4	Save after Recording Last Value
5	Navigate After Reaching Target Scope
6	Skip Characteristic Overview
7	Skip Characteristic Overview
8	Skip Characteristic Overview, Close
9	Skip Charac. Overview, Navigate, Save
X	All Background Functions

Figure 8.69 Test/Measurement Equipment Management – Recording Configuration: Overview

[+]

Recommendation for Recording Configuration

Retain the standard SAP settings. If, however, you want to change them, read the F1 help for the individual functions carefully.

These Customizing settings should now enable you to execute the business process for calibrating test/measurement equipment described in *Plant Maintenance with SAP—Practical Guide.*[4]

8.5 Pool Asset Management

You can use *Pool Asset Management* (PAM) to manage objects that are contained in a pool and from which they can be borrowed for a certain length of time. This can include the following objects:

- Fleet objects
- IT equipment (notebooks, projectors, etc.)
- Cell phones
- Tools
- Other objects

4 See K. Liebstückel, *Plant Maintenance with SAP—Practical Guide* (3rd edition, SAP PRESS, 2014, pp. 344–355).

These objects are returned to the pool after they have been borrowed, and the service is charged to the cost object (e.g., the cost center). Figure 8.70 shows the complete PAM process, using the example of a pool of fleet objects. PAM works in the same way for other types of pools.

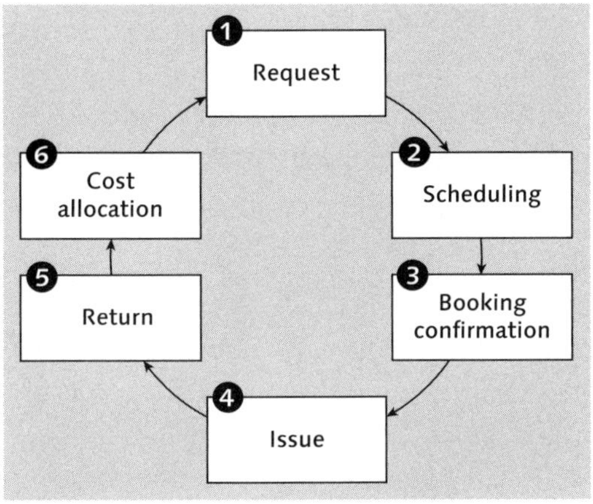

Figure 8.70 Pool Asset Management: Process

Here, PAM comprises the following steps:

- **Request**
  An employee enters a request for a fleet object in the system. He creates a PAM notification with information about the trip, driver, and settlement ❶. For this, you configure a separate notification type, a separate partner determination procedure, and the equipment characteristics of the object being requested in Customizing.
- **Vehicle scheduling**
  A vehicle scheduler assigns a fleet object to the request ❷. You configure the content and layout of the planning board in Customizing.
- **Booking confirmation**
  The employee who requested the fleet object receives an automatic booking confirmation by email, the content and layout of which is determined in Customizing ❸.
- **Fleet object issue**
  A miles reading is taken from the fleet object, the time at which it's required is recorded, and the fleet object is then issued ❹.

- **Fleet object return**
  The employee returns the fleet object, and the miles reading and date are recorded ❺.
- **Cost allocation**
  The costs associated with the employee's use of the fleet object are determined and allocated to the account assignment object ❻. The system creates a PAM order for the allocation in the background. This order contains the piece of equipment as the reference object and the settlement cost center for the PAM notification as the settlement rule. For this, you configure the order type required for the settlement as well as the activity types that you require in Customizing.

Business Function

Before you can use PAM, you must activate the business function `LOG_EAM_PAM`. Some of the following Customizing functions for configuring the business process for PAM don't become available until this business function is activated.

Overview of Notification Type

You use a PAM requirement notification to trigger the business process for PAM. You need a separate notification type here.

This Customizing function provides central access to a wide range of settings for a notification type. For the PAM notification, you need the following:

- The Screen Areas in Notification Header subfunction to define the notification header and reference object
- The Screen Structure for Extended View subfunction to define the layout
- The Partner Determination Procedure subfunction to define the partners

Prerequisites
You must define the notification types beforehand.

Customizing Path
Plant Maintenance and Customer Service • Maintenance and Service Processing • Maintenance and Service Notifications • Overview of Notification Type

Settings or Recommended Settings

Use the Screen Areas in Notification Header subfunction to determine the header and reference object for the notification, and make the following settings (see Figure 8.71):

- In Screen type hdr, always enter "H100" (Header maintenance notification), even for the PAM requirement notification.
- In ScrnType Object, choose 0150 (Equipment only) for the PAM notification because the objects in a PAM pool must be pieces of equipment.

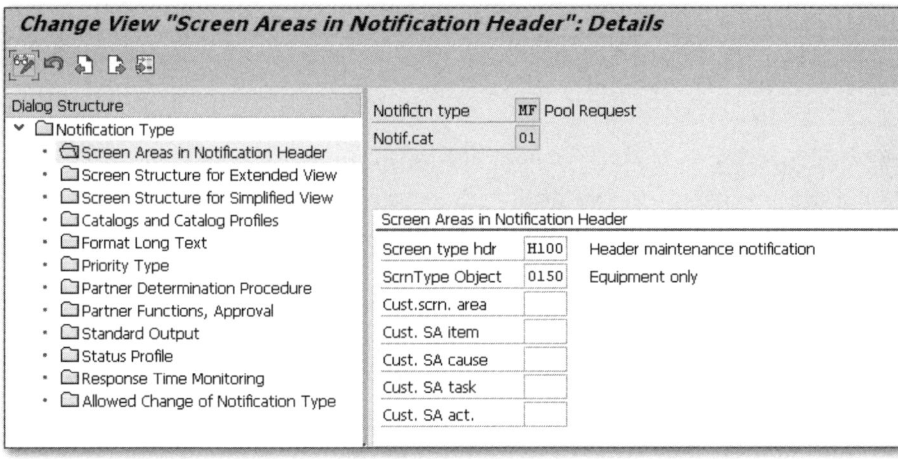

Figure 8.71 Pool Asset Management: Notification Header

Use the Screen Structure for Extended View subfunction to define the layout of the PAM notification. You require at least two tab pages (Figure 8.72):

- One of the tab pages must be 10\TAB23. SAP has predefined this tab page for a PAM notification. It contains information about the trip data, equipment characteristics, driver, and settlement.
- A second tab page should then contain subscreen 065 Partner Overview (Table Control). Use the Partner subfunction to define the partners involved in the PAM notification. The tab pages 10\TAB01, 10\TAB02, and 10\TAB03 are suitable here.

You can include additional tab pages (e.g., Location or Date Overview) if necessary.

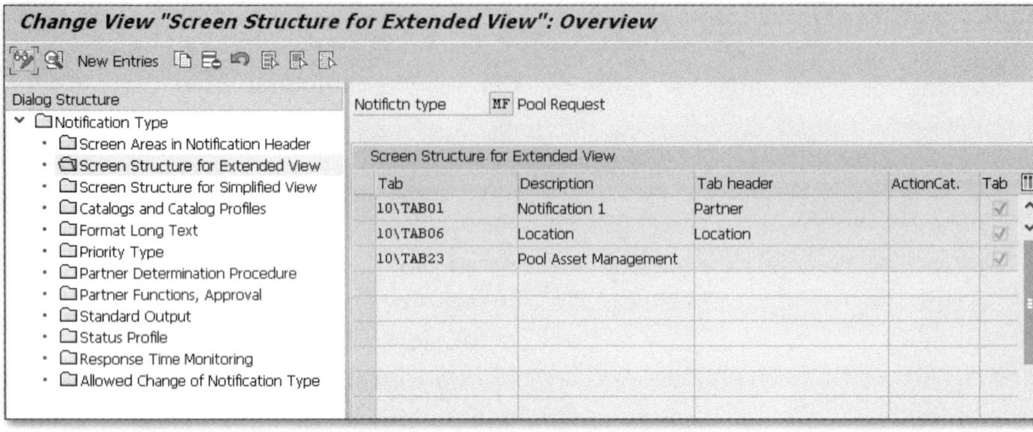

Figure 8.72 Pool Asset Management: Notification Tab Pages

Use the PARTNER DETERMINATION PROCEDURE subfunction to assign the partner determination procedure to the notification type (see Figure 8.73). The partner determination procedure that you use for the PAM notification should contain at least one partner function for the object user and one partner function for the requester. The partner functions should have either partner type PE (personnel) or US (SAP user). Later, these partners will receive emails containing the reservation confirmation, among other things. For more detailed information about these terms and about configuring a partner determination procedure, see Section 4.7.

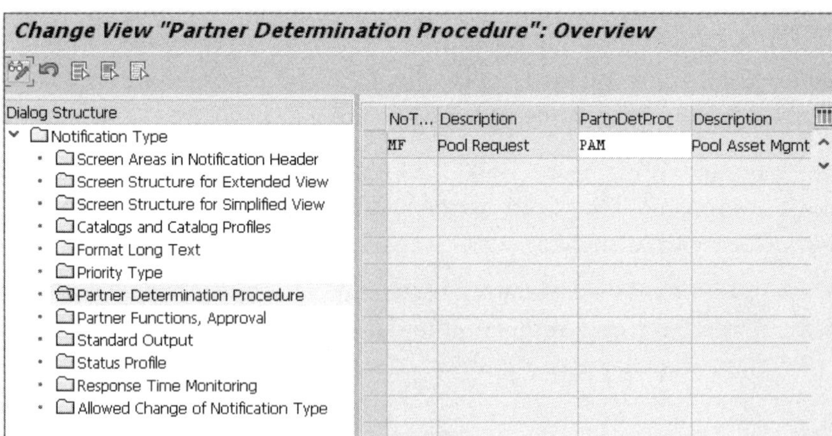

Figure 8.73 Pool Asset Management: Notification for Partner Determination Procedure

[+]

Recommendations for the PAM Notification Type

- One tab page must be 10\TAB23 so that you can report the PAM requirement data.
- The assigned partner types should be either PE or US so that an email containing the reservation confirmation can be sent later.

Basic Settings for Pool Asset Management

You use this Customizing function to make the basic settings for PAM.

Prerequisites

You must define the order type, class, and confirmation text beforehand.

Customizing Path

PLANT MAINTENANCE AND CUSTOMER SERVICE • MAINTENANCE AND SERVICE PROCESSING • POOL ASSET MANAGEMENT • BASIC SETTINGS FOR POOL ASSET MANAGEMENT

Settings or Recommended Settings

In this Customizing function, you can make the following settings (see Figure 8.74):

- The fixed value FL POOL ASSET MANAGEMENT is always preselected for the APPLICATION.
- For RESERVATION TXT, enter the SAPscript form that is to be used to send the notification (email) to the requester in the event of a reservation or cancellation. Use Transaction SE71 to maintain the SAPscript form itself and to incorporate this form into print control (see the next Customizing function for print control).
- For ORDTY. INTALLOC, enter the "PAM" order type for the internal allocation. This order type determines the internal allocations for object usage.
- For CLASS TYPE (fixed value: PAM) and CLASS, specify the class for class type PAM. This refers to the PAM-relevant characteristics of the object (piece of equipment) and the request (maintenance notification).

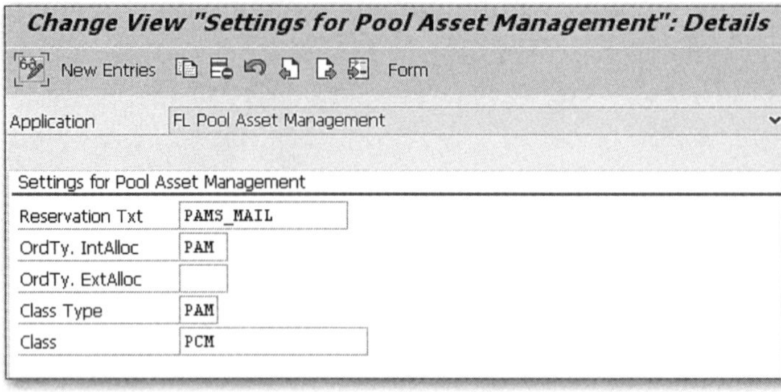

Figure 8.74 Pool Asset Management: Basic Settings

[+]

You Can Assign Only One Class

One weakness of PAM is the fact that you can assign only one class (e.g., for fleet objects). Even though other pools (e.g., tools, cranes, and notebooks) require other classes with other characteristics, at present, PAM can only be used for one pool and its full scope of functions (i.e., including the equipment characteristics).

Define Shop Papers, Forms, and Output Programs

To enable an email containing the reservation confirmation to be output or sent, you must incorporate this form into normal print control.

Prerequisites

You must develop the output program and the form itself beforehand.

Customizing Path

Plant Maintenance and Customer Service • Maintenance and Service Processing • Maintenance and Service Notifications • Notification Processing • Notification Print Control • Define Shop Papers, Forms, and Output Programs

Transaction

OIDA

Settings or Recommended Settings

This Customizing function has two subfunctions: DEFINE SHOP PAPERS and DEFINE SHOP PAPERS FOR NOTIFICATION TYPE. However, you need only the first subfunction. In this subfunction, define the following (see Figure 8.75):

- SHOP PAPER: A four-digit number (max.) for the shop paper.
- The report name and the form routine used for print control.
- The SAPscript form used as a basis for designing the pages. The form contains different elements that are used to control the layout of the individual pages and to show design information for the texts to be output on the individual pages.

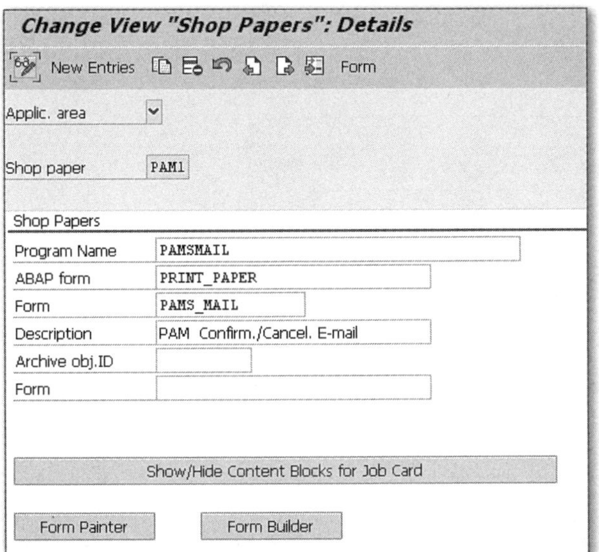

Figure 8.75 Pool Asset Management: Reservation Email

Create Classes and Characteristics (CL02/CT04)

In the Customizing function BASIC SETTINGS FOR POOL ASSET MANAGEMENT, you assign a class to the PAM notification type. You use this class so that you can specify the equipment characteristics of the equipment to be borrowed in the PAM notification. For a transport fleet, this might include the following characteristics, for example:

- Fleet object category
- Number of persons
- Transmission type
- Winter equipment

Here, use Transaction CT04 to create the requisite characteristics. Then, use Transaction CL02 to integrate them into a class of the class type PAM (see Figure 8.76).

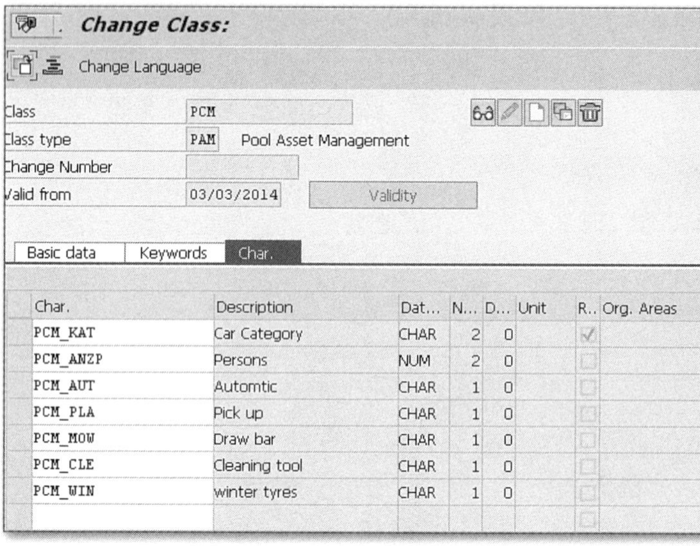

Figure 8.76 Pool Asset Management: Pool Class

Edit Relationships for Characteristics and Fields

You can use this Customizing function to edit the relationships for characteristics and fields. You can define a maximum of 12 equipment characteristics, the object category (used for settlement determination), and the number of persons (in the case of a fleet object) that are to be displayed in the planning board.

Prerequisites

You must define the class with the characteristics beforehand and then assign this class to PAM in the Customizing function BASIC SETTINGS FOR POOL ASSET MANAGEMENT.

Customizing Path

Plant Maintenance and Customer Service • Maintenance and Service Processing • Pool Asset Management • Edit Relationships for Characteristics and Fields

Settings or Recommended Settings

Specify the properties of the respective characteristic (see Figure 8.77):

- Equipment
  Determines whether it concerns a normal equipment characteristic.
- No.Persons
  Determines whether the characteristic represents the number of persons.
- PAM Type
  Determines whether the characteristic defines the fleet object category.

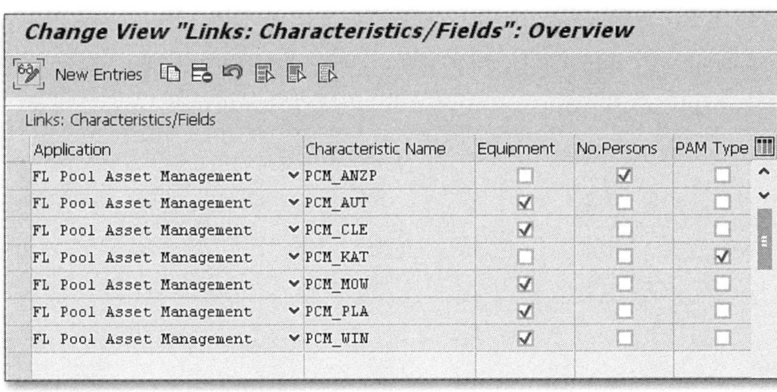

Figure 8.77 Pool Asset Management: Relationships between Characteristics and Fields

Edit Object Processing

You can use this Customizing function to define which objects are to be considered in the planning board. Even though the Customizing function already knows the order type, the current version of PAM takes only notifications into account. In other words, it doesn't take orders into account.

Prerequisites

You must define the notification types beforehand.

Customizing Path

PLANT MAINTENANCE AND CUSTOMER SERVICE • MAINTENANCE AND SERVICE PROCESSING • POOL ASSET MANAGEMENT • EDIT OBJECT PROCESSING

Settings or Recommended Settings

You can make the following settings for each notification type (see Figure 8.78):

- REQ. ROLE
  Use the REQUESTER partner function to specify the partner function for the request. The user obtains information about the piece of equipment that has been reserved or the cancellation of a request.
- USER ROLE
  Use the USER partner function to specify the partner function to be used (e.g., a driver in the case of a fleet object).
- DIR. SCHED
  Use the direct scheduling option to control the direct transfer of the notification to the planning board without the need for manual scheduling (e.g., for notifications from maintenance plans or for malfunction reports).

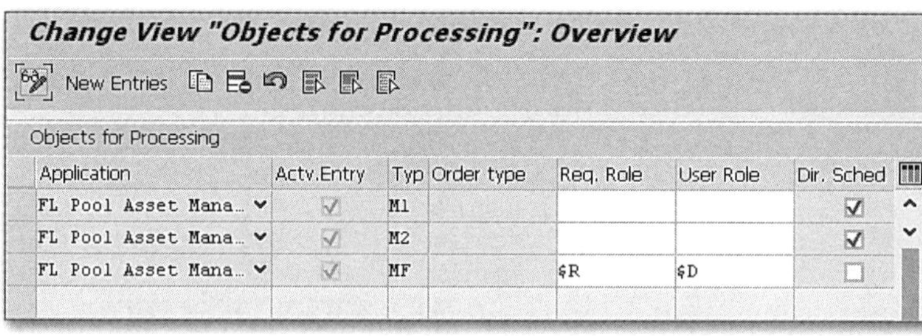

Figure 8.78 Pool Asset Management: Objects

Define Settings for Planning Board

You use this Customizing function to make some basic settings for the graphical planning board.

Prerequisites

You must define the calendar and graphic profiles beforehand.

Customizing Path

Plant Maintenance and Customer Service • Maintenance and Service Processing • Pool Asset Management • Define Settings for Planning Board

Settings or Recommended Settings

The following settings are available to you (see Figure 8.79):

- In the calendar fields, specify which Factory Calendar and which Holiday Calendar you want to use.
- In the Group, Name, and Index columns, specify which graphic profile you want to use. For information about the settings for the graphic profile, see Chapter 6, Section 6.2.3.
- According to SAP, the Number and Char01 fields contain system settings that can't be changed (and therefore could have been omitted from this screen).

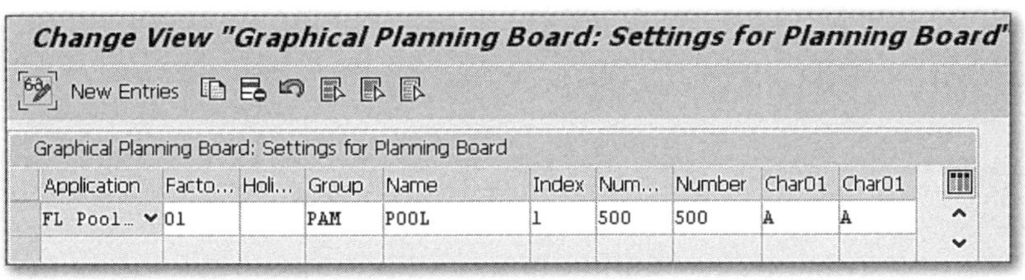

Figure 8.79 Pool Asset Management – Graphical Planning Board: Basic Settings

Edit Display Variants of Planning Board

You use this Customizing function to define display variants for the graphical planning board (Transaction PAM03); that is, which fields are to be displayed in the upper section of the planning board (fields relating to the equipment) and which fields are to be displayed in the lower section of the planning board (fields relating to the requests) (see Figure 8.80).

Prerequisites

There are no special prerequisites.

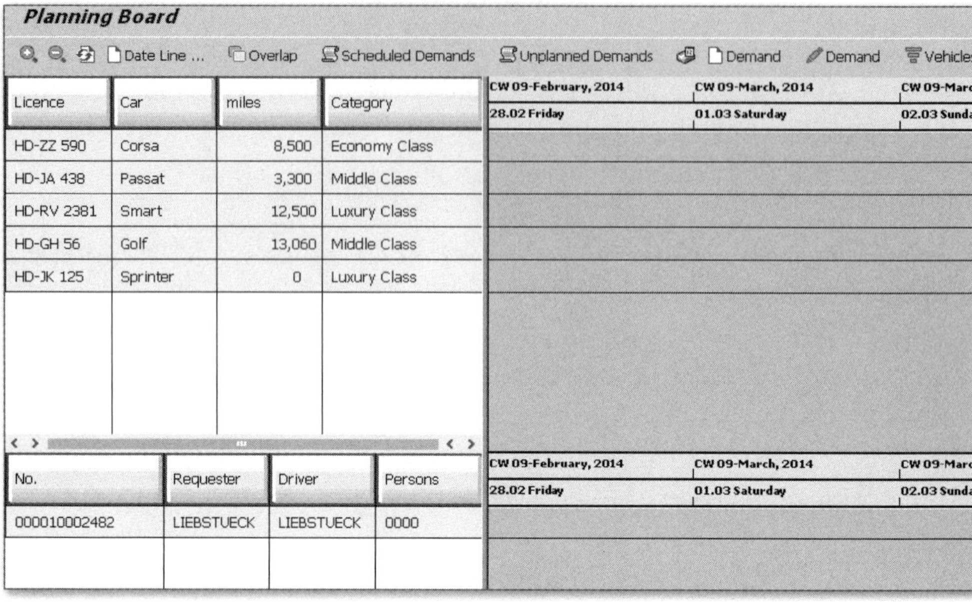

Figure 8.80 Pool Asset Management – Graphical Planning Board: Layout

Customizing Path

Plant Maintenance and Customer Service • Maintenance and Service Processing • Pool Asset Management • Edit Display Variants of Planning Board

Settings or Recommended Settings

For the application FL Pool Asset Management, you need to specify the following (see Figure 8.81):

- Enter a name for the Display Variant.
- Use ChartID 1 to define the fields in the upper section of the planning board (i.e., the fields relating to the equipment). Use ChartID 2 to define the fields in the lower section of the planning board (i.e., the fields relating to the requests).
- In the Field Name, you can choose from 33 fields.

Note

In the upper section, the equipment number is displayed for the field ROWKEY_ELEMENT while, in the lower section, the notification number is displayed for this field. The sequence numbers 0 aren't displayed.

- You can define the FIELD TEXT yourself.

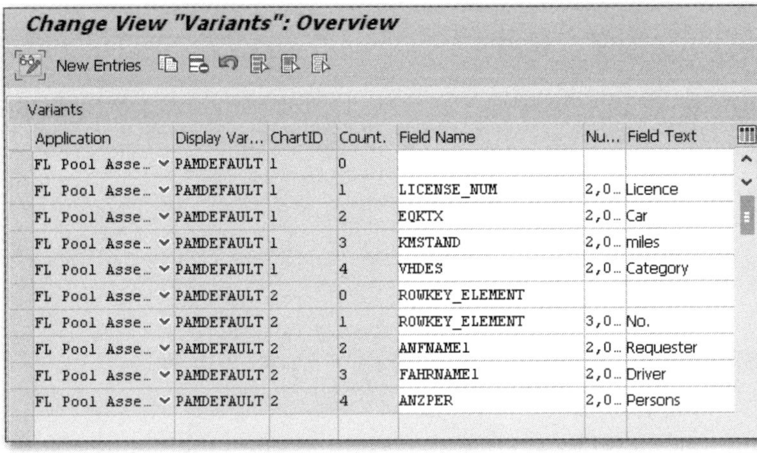

Change View "Variants": Overview

New Entries

Variants

Application	Display Var...	ChartID	Count.	Field Name	Nu...	Field Text
FL Pool Asse...	PAMDEFAULT	1	0			
FL Pool Asse...	PAMDEFAULT	1	1	LICENSE_NUM	2,0...	Licence
FL Pool Asse...	PAMDEFAULT	1	2	EQKTX	2,0...	Car
FL Pool Asse...	PAMDEFAULT	1	3	KMSTAND	2,0...	miles
FL Pool Asse...	PAMDEFAULT	1	4	VHDES	2,0...	Category
FL Pool Asse...	PAMDEFAULT	2	0	ROWKEY_ELEMENT		
FL Pool Asse...	PAMDEFAULT	2	1	ROWKEY_ELEMENT	3,0...	No.
FL Pool Asse...	PAMDEFAULT	2	2	ANFNAME1	2,0...	Requester
FL Pool Asse...	PAMDEFAULT	2	3	FAHRNAME1	2,0...	Driver
FL Pool Asse...	PAMDEFAULT	2	4	ANZPER	2,0...	Persons

Figure 8.81 Pool Asset Management – Graphical Planning Board: Layout

Maintain Status

You use this function to define how the bars that represent the requested elements are to be displayed in the planning board (in particular, which color is to be assigned to each status).

Prerequisites

There are no specific prerequisites.

Customizing Path

Unfortunately, table maintenance for PAMS_VSTAI has not been integrated into Customizing. Instead, you use Transaction SM30 to maintain this table.

Settings or Recommended Settings

For the application FL POOL ASSET MANAGEMENT, you need to specify the following (see Figure 8.82):

- Define a STATUS.
- Assign a COLOR to your status (e.g., 01 = yellow, 03 = blue, 04 = green, and 06 = red).

- Assign a split type (column Split Ty.) to your status:
  - A: Absence
    The object is unavailable due to other tasks.
  - B: Reqmt
    This status represents a requirement.
  - E: Completed Split
    The requirement is completed.
  - S: Productive Split
    The object is currently in use.

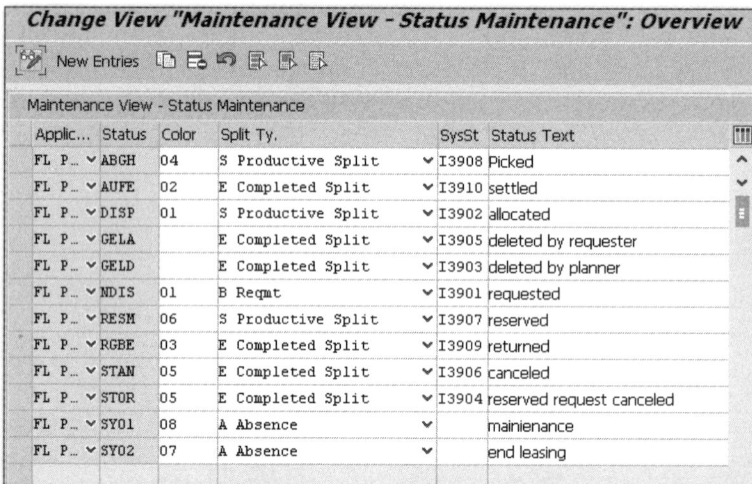

Applic...	Status	Color	Split Ty.	SysSt	Status Text
FL P...	ABGH	04	S Productive Split	I3908	Picked
FL P...	AUFE	02	E Completed Split	I3910	settled
FL P...	DISP	01	S Productive Split	I3902	allocated
FL P...	GELA		E Completed Split	I3905	deleted by requester
FL P...	GELD		E Completed Split	I3903	deleted by planner
FL P...	NDIS	01	B Reqmt	I3901	requested
FL P...	RESM	06	S Productive Split	I3907	reserved
FL P...	RGBE	03	E Completed Split	I3909	returned
FL P...	STAN	05	E Completed Split	I3906	canceled
FL P...	STOR	05	E Completed Split	I3904	reserved request canceled
FL P...	SY01	08	A Absence		mainienance
FL P...	SY02	07	A Absence		end leasing

Figure 8.82 Pool Asset Management – Graphical Planning Board: Status

- Link your status to a system status (column SysSt). The following system statuses are useful:
  - I3901: NDIS (requested)
  - I3902: DISP (allocated)
  - I3903: GELD (deleted by planner)
  - I3904: STOr (request canceled)
  - I3905: GELA (deleted by requester)

- I3906: STAN (CANCELLED)
- I3907: RESM (RESERVED)
- I3908: ABGH (PICKED)
- I3909: RGBE (RETURNED)
- I3910: AUFE (SETTLED)

- Assign a STATUS TEXT to your status.

Define Pool Categories

You use this Customizing function to define pool categories. You'll need these categories to allocate the scheduled requirements later.

Prerequisites

There are no specific prerequisites.

Customizing Path

PLANT MAINTENANCE AND CUSTOMER SERVICE • MAINTENANCE AND SERVICE PROCESSING • POOL ASSET MANAGEMENT • DEFINE POOL CATEGORIES

Settings or Recommended Settings

Define the pool categories and their descriptions (see Figure 8.83).

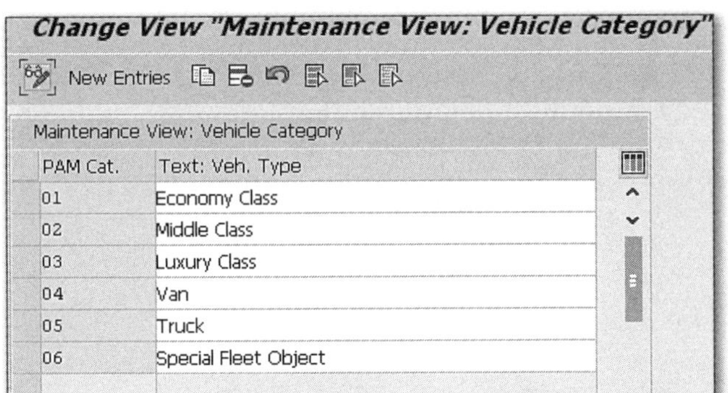

Figure 8.83 Pool Asset Management: Pool Category

Define Activity Types for Pool Categories

You use this Customizing function to define, for each controlling area and pool category, which activity types are to be used for the various allocation frequencies (e.g., for a day rate).

Prerequisites

You must define the controlling areas, PAM categories, and activity types beforehand. You create activity types in Transaction KL01.

Customizing Path

Plant Maintenance and Customer Service • Maintenance and Service Processing • Pool Asset Management • Define Service Types for Pool Categories

Settings or Recommended Settings

The following cost units are available to you (see Figure 8.84):

- RentalPr.
  For the rental price, the costs associated with the borrowing time are calculated on the basis of the unit of measurement for the activity type (hours, days).
- PrAdd.Unit
  For the price/unit, the activity-dependent costs (e.g., mileage when using a fleet object pool) are determined on the basis of the unit of measurement for the activity type.

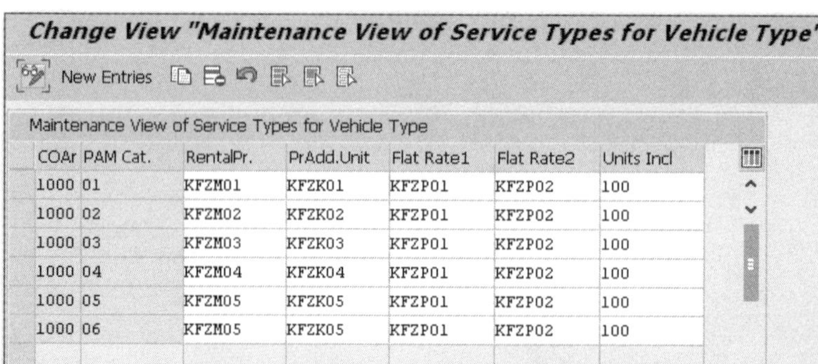
Change View "Maintenance View of Service Types for Vehicle Type"

New Entries

Maintenance View of Service Types for Vehicle Type

COAr	PAM Cat.	RentalPr.	PrAdd.Unit	Flat Rate1	Flat Rate2	Units Incl
1000	01	KFZM01	KFZK01	KFZP01	KFZP02	100
1000	02	KFZM02	KFZK02	KFZP01	KFZP02	100
1000	03	KFZM03	KFZK03	KFZP01	KFZP02	100
1000	04	KFZM04	KFZK04	KFZP01	KFZP02	100
1000	05	KFZM05	KFZK05	KFZP01	KFZP02	100
1000	06	KFZM05	KFZK05	KFZP01	KFZP02	100

Figure 8.84 Pool Asset Management: Activity Types for the Category

- Flat Rate1 and Flat Rate2
  These are used to allocate flat-rate costs (e.g., cleaning costs or refueling costs).
- Units Incl
  This is used to specify whether there are free miles for the cost unit PRAdd.Unit (e.g., free miles when using a fleet object pool).

Use Transaction KP26 to maintain the prices for each fiscal year (see Figure 8.85).

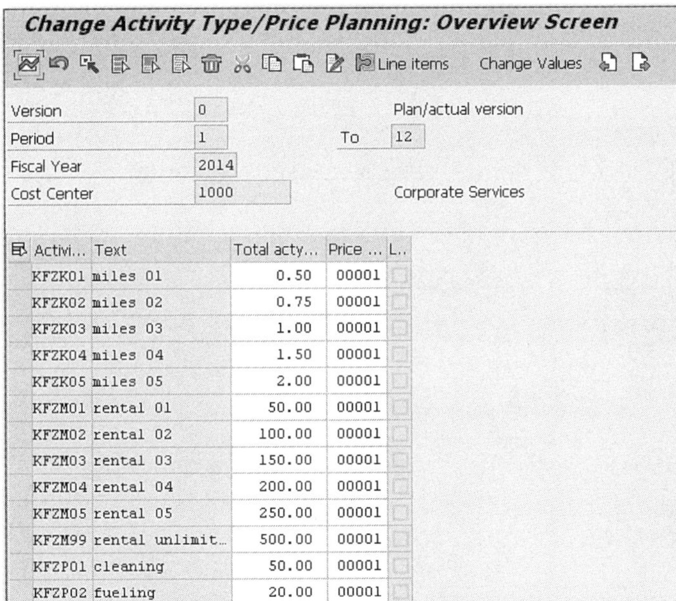

Activi...	Text	Total acty...	Price ...	L...
KFZK01	miles 01	0.50	00001	
KFZK02	miles 02	0.75	00001	
KFZK03	miles 03	1.00	00001	
KFZK04	miles 04	1.50	00001	
KFZK05	miles 05	2.00	00001	
KFZM01	rental 01	50.00	00001	
KFZM02	rental 02	100.00	00001	
KFZM03	rental 03	150.00	00001	
KFZM04	rental 04	200.00	00001	
KFZM05	rental 05	250.00	00001	
KFZM99	rental unlimit...	500.00	00001	
KFZP01	cleaning	50.00	00001	
KFZP02	fueling	20.00	00001	

Figure 8.85 Pool Asset Management: Activity Type Planning

Configure Order Types

If you settle a PAM requirement, the system creates an order in the background, which is used to settle the costs to the account assignment object in the PAM notification (order, cost center). You use this Customizing function to define the order type to be used for this purpose and to define some basic parameters.

[+]

Separate Order Type for Pool Asset Management

You should define a separate order type for the business process for PAM. You should not use this order type for normal maintenance processes. Similarly, you should not use a normal order type for this business process.

Prerequisites

Because order types are used not only by plant maintenance but also by other user departments and their corresponding SAP applications, you must consult with your colleagues from the other user departments on the naming convention and number ranges for order types.

Customizing Path

Plant Maintenance and Customer Service • Maintenance and Service Processing • Maintenance and Service Orders • Functions and Settings for Order Types • Configure Order Types

Settings or Recommended Settings

The settings for the PAM order type are the same as the settings for normal order types (see Figure 8.86).

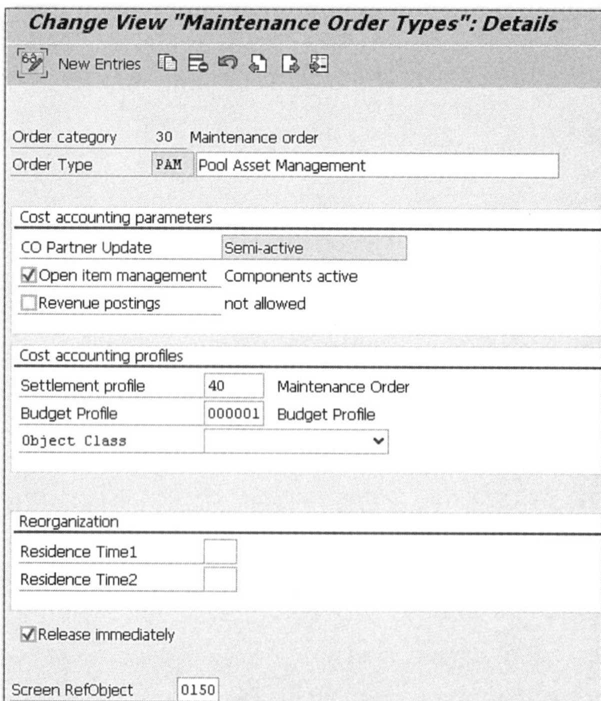

Figure 8.86 Pool Asset Management: Order Type

These Customizing settings should now enable you to execute the PAM business process described in *Plant Maintenance with SAP—Practical Guide*.[5]

8.6 Maintenance Event Builder

You can use the *Maintenance Event Builder* (MEB) to plan smaller maintenance projects in the form of individual work packages. You start the MEB in Transaction WPS1. The MEB is technically a workbench that supports you when you perform the following tasks:

- Check the work list of notifications (i.e., the backlog).
- Bundle the notifications into revisions.
- Create orders from the notifications.
- Assign the tasks.
- Display different information such as open work requirements, due dates, orders, and so on.
- Check the availability situation of resources.

If you want to process maintenance tasks within the MEB, you need to use revisions. Revisions have start and end dates and use status management (e.g., CREATED, RELEASED, ASSIGNMENTS EXIST, etc.). You create revisions either explicitly in Transaction IWR1 or implicitly in the MEB.

Business Function

Before you can use the MEB, you must activate the business functions LOG_EAM_POM and LOG_EAM_POM_2.

At the heart of the MEB is the *revision* to which all notifications and orders that need to be executed are assigned (see Figure 8.87).

Maintain Revision Type

Each revision must be assigned to a revision type. You use this Customizing function to define revision types. In particular, you determine whether you want to use the revision type in the MEB or in normal maintenance processing.

5 See K. Liebstückel, *Plant Maintenance with SAP—Practical Guide* (3rd edition, SAP PRESS, 2014, pp. 356–363).

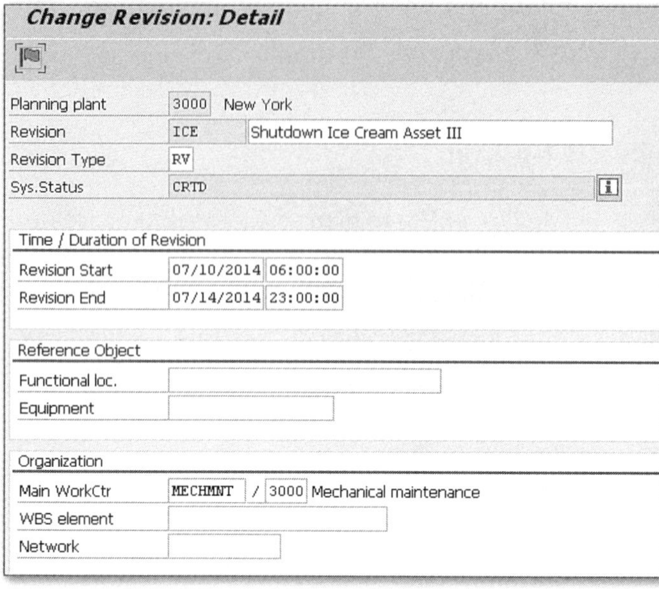

Figure 8.87 Maintenance Event Builder: Revision

Prerequisites

There are no specific prerequisites.

Customizing Path

Plant Maintenance and Customer Service • Maintenance and Service Processing • Maintenance Event Builder • Maintain Revision Type

Transaction

DIWPSC4

Settings or Recommended Settings

Assign the following revision type to a revision that you want to use in normal maintenance processing: Not WPS-Controlled (see Figure 8.88). You can then enter these revisions in notifications and orders. However, you can't use these revisions in the MEB.

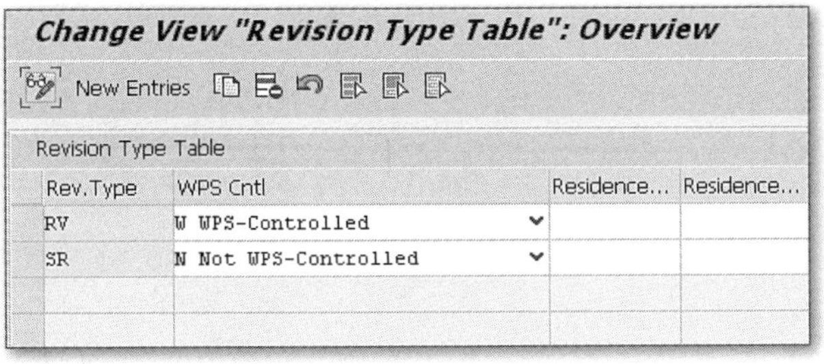

Figure 8.88 Maintenance Event Builder: Revision Type

The opposite behavior is associated with WPS-CONTROLLED revision types; that is, you can only use them in the MEB (not in normal maintenance processing).

This Customizing setting should now enable you to use the MEB as described in *Plant Maintenance with SAP—Practical Guide.*[6]

8.7 Shift Reports and Shift Notes

You use shift notes and shift reports to document events that occur during a shift. In a *shift note*, you enter information about an event (e.g., comments, times or objects). Shift notes can contain the following information, for example:

- General notes (e.g., shift interruptions, power outages)
- Breakdown documentation (e.g., a turning machine breakdown from/to)
- Suggestions for improvement (e.g., "Reduce machine speed by 10% because ...")
- Notes on employees (e.g., "Mr. Henry finished one hour early because...")
- Notes on material usage (e.g., "Casing should be clamped with a maximum pressure of 10 bar")
- Comments on the use of tools (e.g., "Hand brace 9700 isn't suitable for material T-B400")

A *shift report* is a PDF document generated at the end of a shift by the person responsible for the shift. It comprises the shift notes that have been recorded as well as other documents. If necessary, a digital signature can be used to sign a shift report. A shift report may include the following elements:

6 See K. Liebstückel, *Plant Maintenance with SAP—Practical Guide* (3rd edition, SAP PRESS, 2014, pp. 370–374).

- Shift notes
- Production activities
- Completion confirmations
- Goods movements
- Maintenance notifications
- Quality notifications
- Maintenance orders
- Measurement documents
- Graphical analyses

Business Functions

Before you can use shift notes and shift reports, you must activate the business functions LOG_PP_SRN_CONF and LOG_PP_SRN_02.

The following Customizing functions are available for adjusting shift notes and shift reports to meet your requirements.

Overview of Notification Type

From a technical data perspective, shift notes are notifications that *must* be created with a separate notification type. This Customizing function provides central access to a wide range of settings for a notification type. For a shift note, you require the following:

- The SCREEN AREAS IN NOTIFICATION HEADER subfunction to define the notification header and reference object
- The SCREEN STRUCTURE FOR EXTENDED VIEW subfunction to define the layout

Prerequisites

You must define the notification types beforehand. The notification type for shift notes *must* be assigned notification category 05 (General notification).

Customizing Path

PLANT MAINTENANCE AND CUSTOMER SERVICE • MAINTENANCE AND SERVICE PROCESSING • MAINTENANCE AND SERVICE NOTIFICATIONS • OVERVIEW OF NOTIFICATION TYPE

Settings or Recommended Settings

Use the SCREEN AREAS IN NOTIFICATION HEADER subfunction to determine the notification header. In SCREEN TYPE HDR, enter "O700" (GENERAL NOTIFICATION) for the shift note type (see Figure 8.89).

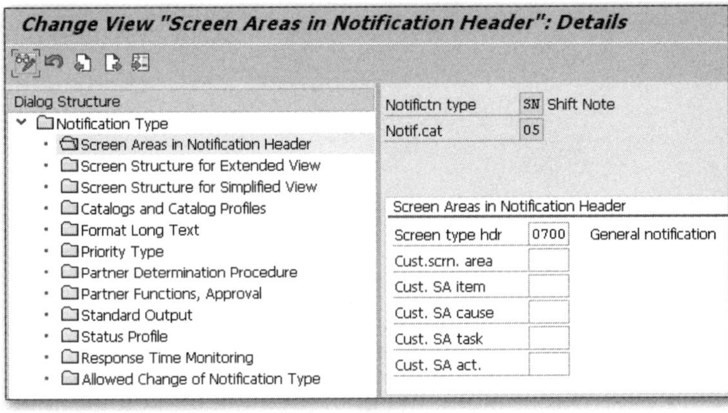

Figure 8.89 Shift Note – Notification Type: Screen Area for Notification Header

Use the SCREEN STRUCTURE FOR EXTENDED VIEW subfunction to define the layout of the shift note. You require a single tab page (see Figure 8.90). On this tab page, enter SCREEN AREA "130" (SHIFT NOTE).

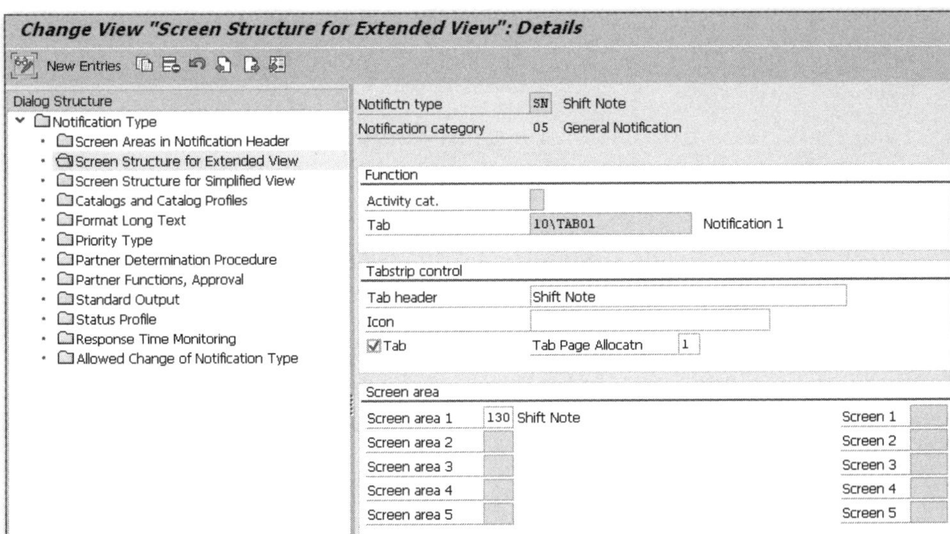

Figure 8.90 Shift Note – Notification Type: Screen Layout

Make Settings for Shift Note Type

This Customizing function is *the* central function for configuring shift notes. Here, you define the main properties of your shift notes (e.g., whether you want to assign reference objects or categories to the shift note).

Prerequisites

You must define the notification types and the catalog for the categories beforehand.

Customizing Path

Plant Maintenance and Customer Service • Maintenance and Service Processing • Shift Reports/Notes • Settings for Shift Notes • Make Settings for Shift Note Type

Settings or Recommended Settings

The following settings are available to you (see Figure 8.91):

- Set the Note Number indicator if you want the system to always display the shift note number in the processing transactions associated with the shift note.
- Set the Origin indicator if you want the system to always display the origin in the processing transactions associated with the shift note. If the origin is Work Center/Hierarchy, the system displays the number and short text of the work center as well as the hierarchy indicator and plant. If the origin is Functional Location/Equipment, the system displays the number and short text of the functional location as well as the number and short text of the piece of equipment.
- If you set the Event indicator, you can record the time or period in which the event took place in the shift note.
- If you set the Category indicator, you can assign a category to the shift note. In the Catalog type and Code group fields, specify the categories that you want to access. Use the Customizing function Maintain Catalogs to maintain the categories themselves.
- If you set the Reference Obj. indicator, you can enter a reference object (e.g., a piece of equipment or functional location) in the shift note. Use the Reference Objects subfunction to define which reference objects you want to use.

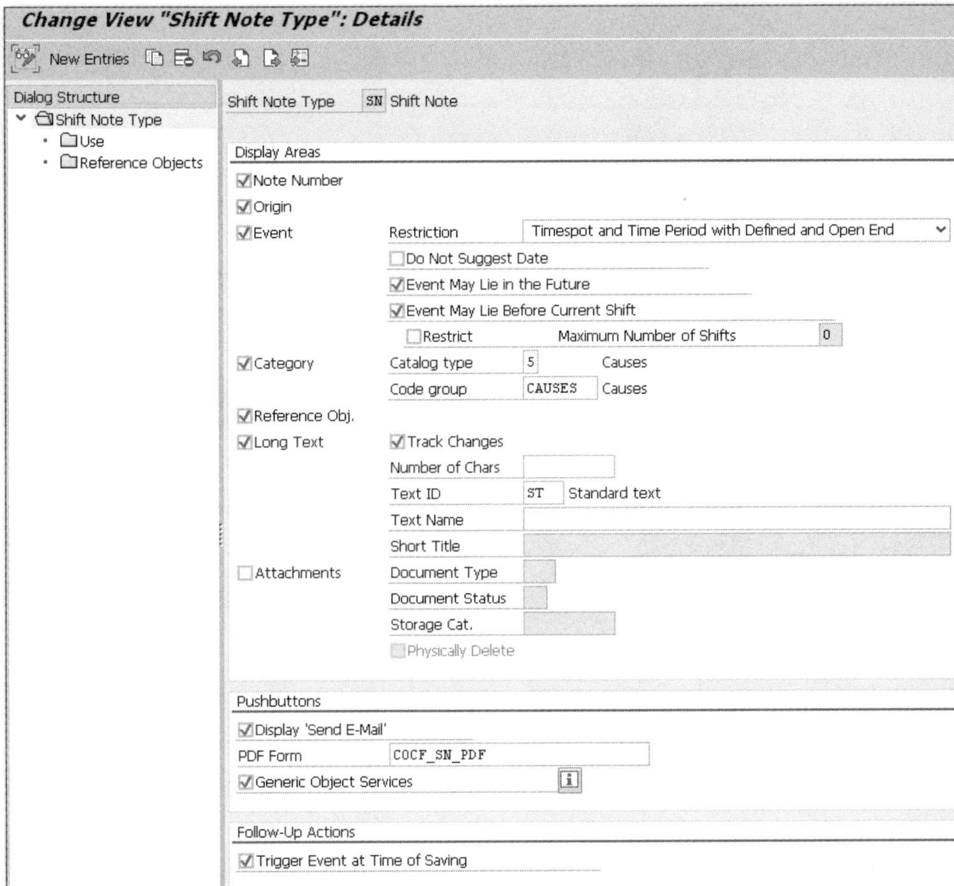

Figure 8.91 Shift Note: Shift Note Type

- If you set the LONG TEXT indicator, an OCX window for the long text is displayed in the shift note.
- If you also set the TRACK CHANGES indicator, the system logs who made which changes to the long text and when these changes were made.
- If you set the ATTACHMENTS indicator, the system displays the ATTACHMENTS area in the lower half of the shift note. You can then add documents to the shift note as attachments. Then, you must specify a DOCUMENT TYPE and storage category (STORAGE CAT.) for the documents (see Chapter 4, Section 4.8). In the Customizing function SCREEN STRUCTURE FOR EXTENDED VIEW, you must assign subscreen "131" in addition to subscreen "130".

- Set the DISPLAY 'SEND E-MAIL' indicator if you want the system to display the function key Send in the processing transactions associated with the shift note. Use the PDF FORM to define the layout of the email.
- If you set the GENERIC OBJECT SERVICES indicator, the system displays the icon and dropdown menu for generic object services in the processing transactions associated with the shift note.

Use the USE subfunction to define whether you want to create independent shift notes or whether the shift notes are to be displayed as a screen area within a normal notification (see Figure 8.92). If you want to display it in a notification, you can define whether the ORIGIN is to be WORK CENTER/HIERARCHY or FUNCTIONAL LOCATION/EQUIPMENT.

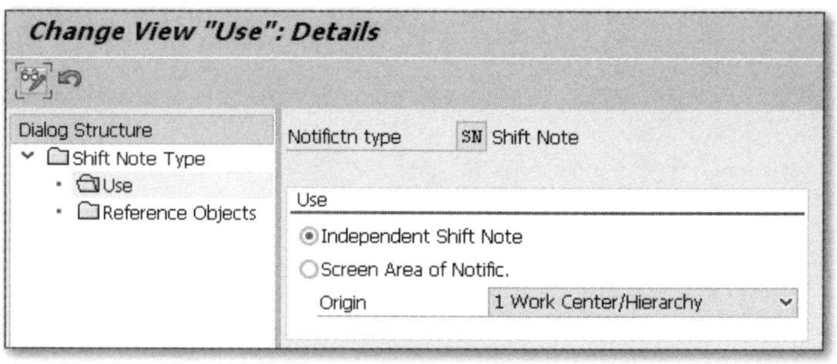

Figure 8.92 Shift Note – Shift Note Type: Use

Use the REFERENCE OBJECTS subfunction to define what you consider to be important reference objects for your shift note (see Figure 8.93). You can choose from the following reference objects:

- BUS1001002: BATCH
- EQUI: EQUIPMENT
- BUS2005: PRODUCTION ORDER
- BUS2116: PRODUCTION ORDER CONFIRMATION
- BUS1038: PRODUCTION RESOURCE/TOOL
- BUS2007: MAINTENANCE ORDER
- BUS2128: MAINTENANCE ORDER CONFIRMATION
- BUS2038: MAINTENANCE NOTIFICATION

- BUS1001: Material
- BUS2092: Measurement document
- SAP_65106: Measuring point
- BUS2004: Planned order
- BUS2016: Process order confirmation
- BUS0001: Process Order
- BUS2045: Inspection lot
- BUS2078: Quality notification
- COCF_SR: Shift report
- COCF_SN: Shift note
- BUS0010: Functional Location

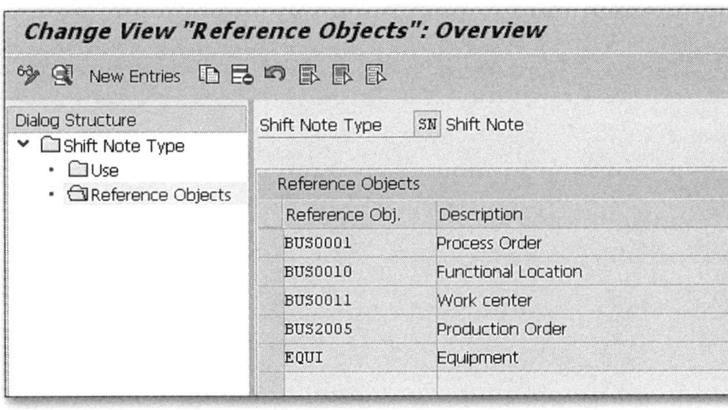

Figure 8.93 Shift Note – Shift Note Type: Reference Objects

Maintain Catalogs

The shift note categories are defined by a catalog's code group. You use this Customizing function to maintain the catalogs and their associated code groups and codes.

Prerequisites

There are no special prerequisites.

Customizing Path

PLANT MAINTENANCE AND CUSTOMER SERVICE • MAINTENANCE AND SERVICE PROCESSING • MAINTENANCE AND SERVICE NOTIFICATIONS • NOTIFICATION CREATION • NOTIFICATION CONTENT • MAINTAIN CATALOGS

Transaction

QS41

Settings or Recommended Settings

You decide which catalog to use for the shift note categories. You can define your own catalog, or you can use an existing one. Figure 8.94 shows the code group categories that were defined in the cause of damage catalog.

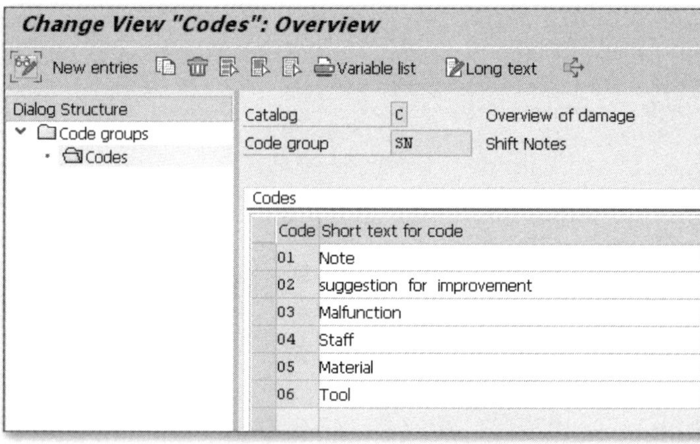

Figure 8.94 Shift Note: Category

Define Shift Report Types

This Customizing function is *the* central function for configuring shift reports. Here, you define how your shift reports are to look after they are generated and which data they should contain.

Prerequisites

If you want to use digital signatures, you must establish a signature strategy beforehand.

Customizing Path

PLANT MAINTENANCE AND CUSTOMER SERVICE • MAINTENANCE AND SERVICE PROCESSING • SHIFT REPORTS/NOTES • SETTINGS FOR SHIFT REPORTS • DEFINE SHIFT REPORT TYPES

Settings or Recommended Settings

Use the SHIFT REPORT TYPE subfunction (see Figure 8.95) to define the shift report types you require. If you require different shift reports that are to contain different structures and content, as a result of the different requirements of the technical objects or work centers, you must create several shift report types. In this subfunction, also define the following properties for each shift report type:

- Use the USAGE indicator to control whether the shift report type is to be generated later for WORK CENTER/HIERARCHY or for FUNCTIONAL LOCATION/EQUIPMENT.
- Use the W/O GAPS indicator to control whether or not a shift report must start at the time when the preceding shift report ends.
- If you set the PROTECT DOCUMENT indicator, the system uses encryption to protect the PDF document against changes.
- If you set the CHECK indicator, the system checks whether the report has been changed since the last time it was saved, each time you call the shift report.
- Use the SIGNATURE indicator to define whether digital signatures are required when saving and changing the shift report. If so, specify the relevant digital signature strategy in the SIGSTRAT column.
- Use the COMPLETE indicator to specify whether a new shift report can only be created if all preceding shift reports have been completed. Shift reports that require a digital signature aren't completed until they have been signed. In other words, the next shift report can't be created until the preceding shift report has been signed. Shift reports that don't require a signature are completed immediately after they are generated.
- Set the E-MAIL indicator if you want the system to display the function key Send in the processing transactions associated with the shift report.

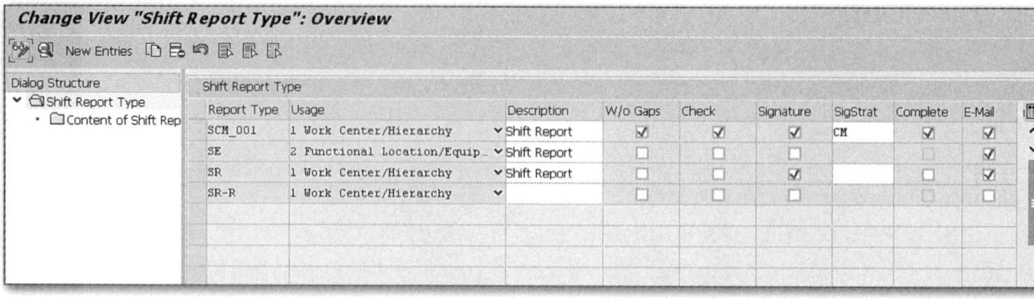

Figure 8.95 Shift Report Type: Overview

Use the CONTENT OF SHIFT REPORT subfunction to assign a class and PDF form (see Figure 8.96):

- SAP has provided the standard class `CL_COCF_SR_PDF_EAM` for this purpose.
- The standard PDF form `COCF_SR_PDF_LAYOUT_EAM` is assigned to this standard class.

A separate Customizing, which you can call via the CUSTOMIZING function key, is predefined for the standard class. Here, specify those sections of the shift report that are to comprise the PDF document.

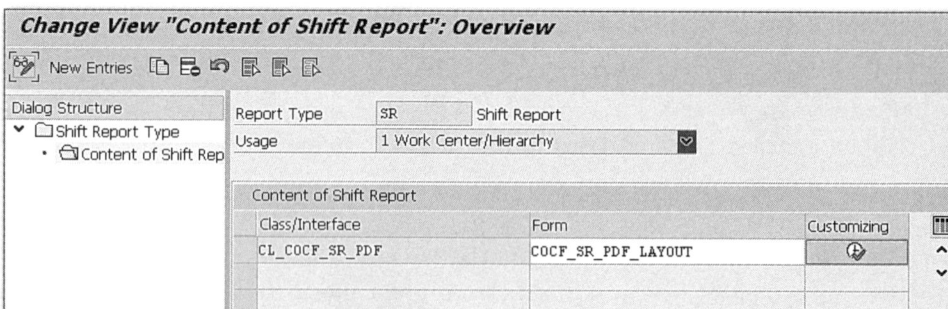

Figure 8.96 Shift Report Type: Detail Screen

The following sections are available to you (see Figure 8.97):

- Cover sheet
- List of shift notes
- List of completion confirmations
- List of aggregated quantities

- List of aggregated activities
- List of maintenance notifications
- List of quality notifications
- Customer-specific data
- Graphical analyses

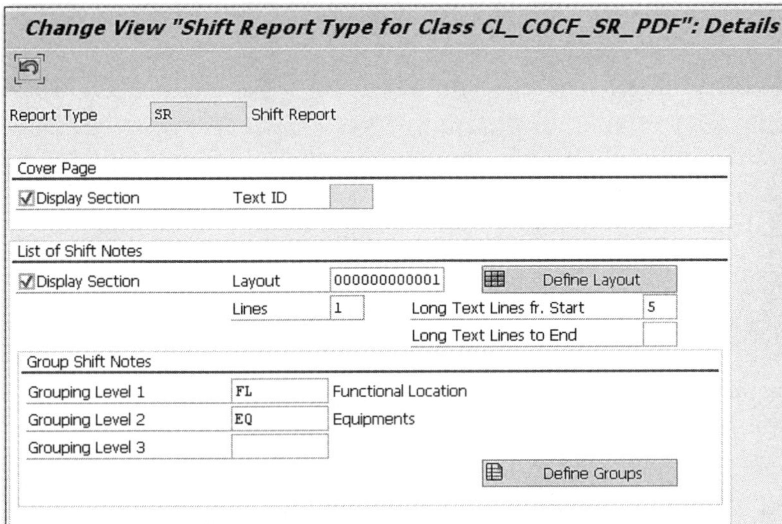

Figure 8.97 Shift Report Type: Content

For each section, you have the option to use the DEFINE LAYOUT function key to define those fields that are to be displayed in the relevant section. In Chapter 4, Section 4.10, we described how to define such layouts. You can also define multiline displays (LINES field) and display long texts (LONG TEXT LINES fields).

A special feature of the LIST OF SHIFT NOTES section is that you can group the shift notes in the shift report according to various criteria. These grouping levels enable you to obtain sorted shift notes in the shift report. For example, you want to sort the list of shift notes according to the functional location and then, within a functional location, according to the equipment number. To do this, use the DEFINE GROUPS function key to call the SHIFT NOTE GROUP subfunction. Then, use the SHIFT NOTE GROUP subfunction (see Figure 8.98) to define the following two classification criteria: FUNCTIONAL LOCATION and EQUIPMENTS.

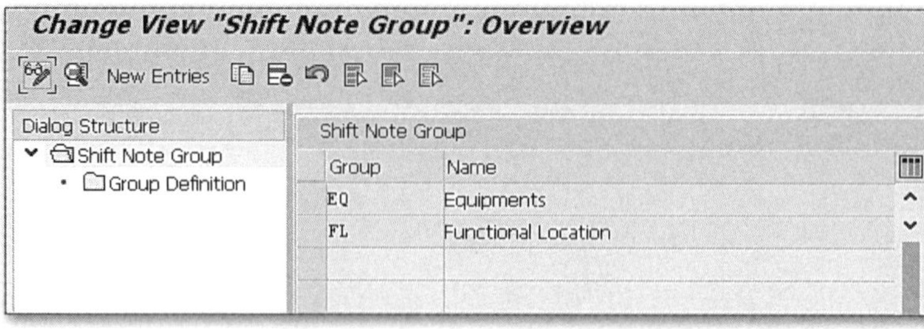

Figure 8.98 Shift Report Type – Shift Note Group: Overview

For each classification criterion, define, on the detail screen (see Figure 8.99), which field is to determine the grouping and sorting. Here, you can choose from almost 200 fields in the F4 help. You also have the option to define a layout, which overrides the general layout for shift notes.

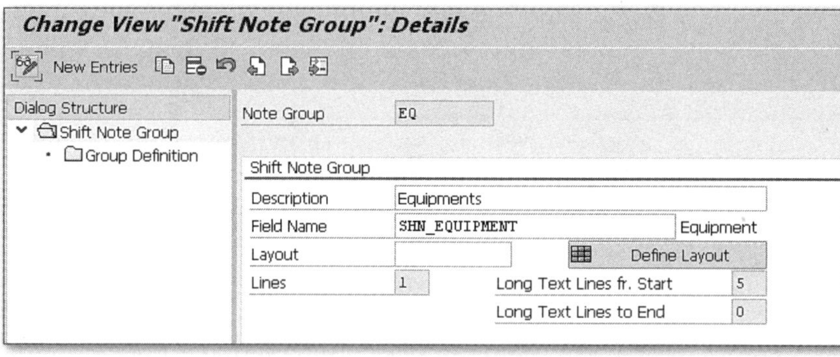

Figure 8.99 Shift Report Category – Shift Note Group: Detail Screen

This Customizing setting should now enable you to use shift notes and shift reports as described in *Plant Maintenance with SAP—Practical Guide*.[7]

7 See K. Liebstückel, *Plant Maintenance with SAP—Practical Guide* (3rd edition, SAP PRESS, 2014, pp. 238–245).

Although there is widespread prejudice that usability isn't a major feature of the SAP system, this chapter shows you the many different options within the SAP system for improving usability.

9 Usability

The book, *Plant Maintenance with SAP—Practical Guide*, introduced the different aspects of usability in terms of the SAP system.[1] The book described the following:

- The term *usability* defined in accordance with Part 110 of DIN EN ISO 9241
- The extent to which the SAP system complies with and/or violates the seven principles of DIN EN ISO 9241
- How to measure usability, primarily by means of the "processing time" key figure
- The need to implement organizational measures to achieve user acceptance, in addition to technical usability measures within the system
- Why user acceptance and usability play such an important role in plant maintenance, more so than in commercial departments, for example
- The many different options within the SAP system for improving usability
- The laboratory test that proved the processing time of business processes can improve by up to four times if you implement these measures to improve usability

In this book, we'll explore these options to improve usability in greater detail, and you'll learn *how* you should define each of these options. These options can be divided into the following three categories (these categories are listed in the order you should check their use in your enterprise):

- **Category 1**
  The user's own options.

1 See K. Liebstückel, *Plant Maintenance with SAP—Practical Guide* (3rd edition, SAP PRESS, 2014, pp. 583–627).

- **Category 2**
  The IT department's nonprogramming options.
- **Category 3**
  The IT department's programming options.

Table 9.1 provides an overview of the measures that fall into each of these categories.

Category	Who?	What?
1	User	▸ General user parameters ▸ Maintenance-specific user parameters ▸ Role menus and favorites menus ▸ List variants ▸ Personalized input helps ▸ Table controls ▸ Buttons and key combinations ▸ SAP NetWeaver Business Client (NWBC)
2	IT department (nonprogramming options)	▸ Transaction variants ▸ Customizing ▸ Action box ▸ GuiXT ▸ SAP Screen Personas
3	IT department (programming options)	▸ Upstream transactions ▸ Web interface ▸ Easy Web transactions ▸ SAP NetWeaver Portal ▸ Customer exits ▸ Business Application Programming Interfaces (BAPI) ▸ Classic Business Add-Ins (BAdIs) ▸ Enhancement points ▸ Workflows

Table 9.1 Options for Improving Usability

Next, we'll discuss each of these options in detail. You'll learn not only *what* you can do, but *how* you should do it.

9.1 Category 1: The User's Own Options

Let's start with the easiest options to implement, namely those parameters and settings that the user can change.

9.1.1 General User Parameters

From a user's point of view, many entries required by the SAP system remain unchanged over a period of time. This includes information such as the following:

- The user belongs to a specific work center.
- The work center is, in turn, assigned to a specific cost center.
- The user works in a particular plant.
- The user is responsible for a particular functional location.

The use of *user parameters* to prepopulate fields contributes greatly to improving usability. You can access this preassignment in Transaction SU3 or via the menu path SYSTEM • USER DEFAULTS • OWN DATA (see Figure 9.1). Once defined, these fields are then automatically prepopulated with the assigned values when business processes are processed.

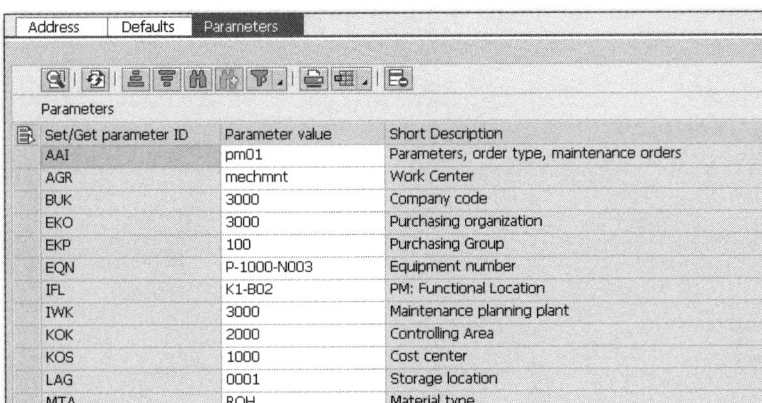

Set/Get parameter ID	Parameter value	Short Description
AAI	pm01	Parameters, order type, maintenance orders
AGR	mechmnt	Work Center
BUK	3000	Company code
EKO	3000	Purchasing organization
EKP	100	Purchasing Group
EQN	P-1000-N003	Equipment number
IFL	K1-B02	PM: Functional Location
IWK	3000	Maintenance planning plant
KOK	2000	Controlling Area
KOS	1000	Cost center
LAG	0001	Storage location
MTA	ROH	Material type

Figure 9.1 General User Parameters

From a maintenance perspective, the following parameters are used frequently:

- AAI: Order Type
- AAR: Work Center Type

- AGR: Work Center
- BUK: Company Code
- CSA: BOM Application
- CSV: BOM Usage
- EKG: Purchasing Group
- EKO: Purchasing Organization
- EQN: Equipment Number
- EQT: Equipment Category
- IFL: Functional Location
- IMD: Measurement Document
- IME: Measurement Recording List
- IP1: Measuring Point Category
- IPT: Measuring Point
- IRL: Reference Functional Location
- ISR: Structure Indicator for Functional Location
- IWK: Maintenance Planning Plant
- KOK: Controlling Area
- KOS: Cost Center
- LAG: Storage Location
- LGP: Storage Bin
- LIS: CATS Profile for List Display
- MAT: Material Number
- MPL: Maintenance Plan
- MPS: Maintenance Item
- MTA: Material Type
- PER: Personnel Number
- PIN: PP Profile/Default Values
- PLN: Task List Group
- QMR: Notification Type
- Q_ALV_GRID_INACTIVE: Deactivates the ALV Grid

- SWK: Maintenance Plant
- VAP: Main Work Center
- WAT: Maintenance Plan Category
- WGR: Material Group
- WRK: Plant

[+]

The F1 Help Supports You in Finding the Parameter IDs

You can save processing time by using parameters to prepopulate fields. You can generally view the relevant parameter ID in the Technical Info of the F1 help for a field.

9.1.2 Maintenance-Specific User Parameters

In addition to the general user parameters, you can also call maintenance-specific user parameters within the notification via EXTRAS • SETTING • CONTROL/DEFAULT VALUES and, within the order, via EXTRAS • SETTINGS • DEFAULT VALUES.

A total of five tab pages are available there. On the GENERAL tab page, you can define personal default values for the order type, notification type, planning plant, planner group, and main work center (see Figure 9.2). These parameters have priority over the general user parameters in Figure 9.1. In other words, if you also enter values for the general user parameters, these values are overwritten by the maintenance-specific user parameters.

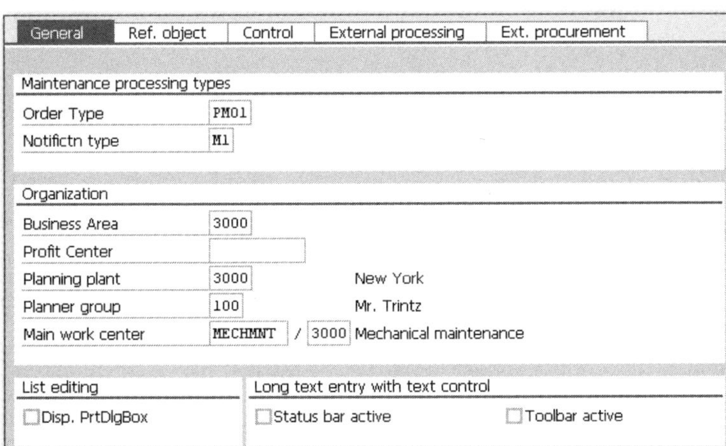

Figure 9.2 Maintenance-Specific User Parameters: Default Values

On the REF. OBJECT tab page (see Figure 9.3), you can define a scenario for the reference object.

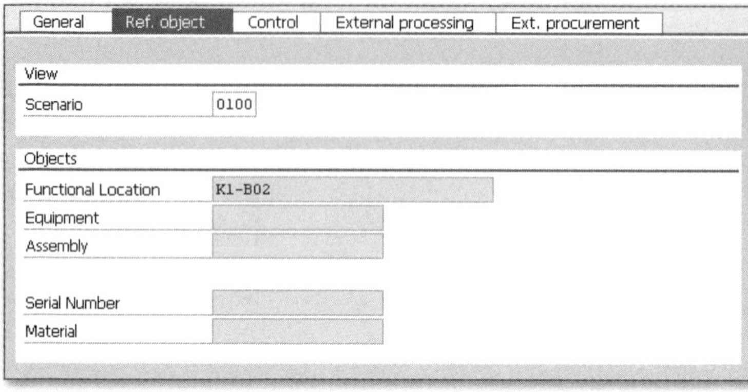

Figure 9.3 Maintenance-Specific User Parameters: Reference Object

The following scenarios are available to you:

- 0100: Functional location + piece of equipment + assembly
- 0110: Equipment + assembly
- 0120: Functional location (30-digit) + equipment + assembly
- 0130: Serial number + material number + device ID
- 0140: Without a reference object
- 0150: Equipment only
- 0160: Functional location only
- 0170: Equipment + serial number + material number
- 0180: Functional location 1:1 + equipment + assembly
- 0190: Physical sample

If you use the Customizing function OVERVIEW OF NOTIFICATION TYPE to assign a scenario to the notification type (see Chapter 6, Section 6.1.1) and the Customizing function CONFIGURE ORDER TYPES to assign a scenario to the order type (see Section 6.2.1), the maintenance-specific user parameters override these Customizing settings for the notification type and order type. If, however, you use the Customizing function SET MAINTENANCE PLAN CATEGORIES to assign a scenario to a maintenance plan category (see Chapter 7, Section 7.2), this isn't overridden.

Here, you can also assign a specific technical object, which is also then proposed in the notification and order.

The CONTROL tab page is available in the order only. Here, you have the option to influence different dialog boxes (see Figure 9.4):

- You can influence the SELECTION DIALOG in the FUNCTION: PUT ORDER IN PROCESS area and therefore override the default setting DISPLAY DIALOG.
- You can influence PRINT CONTROL and therefore override the default setting PRINT WITH DIALOG.
- You can influence the selection dialog MAINTAIN SETTLEMENT RULE, which overrides the default setting from the Customizing function SETTLEMENT RULE: DEFINE TIME AND CREATION OF DISTRIBUTION RULE (see Chapter 6, Section 6.2.7).
- You can suppress the dialog box for technical completion of the order (NO DIALOG COMPLETION indicator). Then, the order is always completed using the default values (e.g., the current date and time).

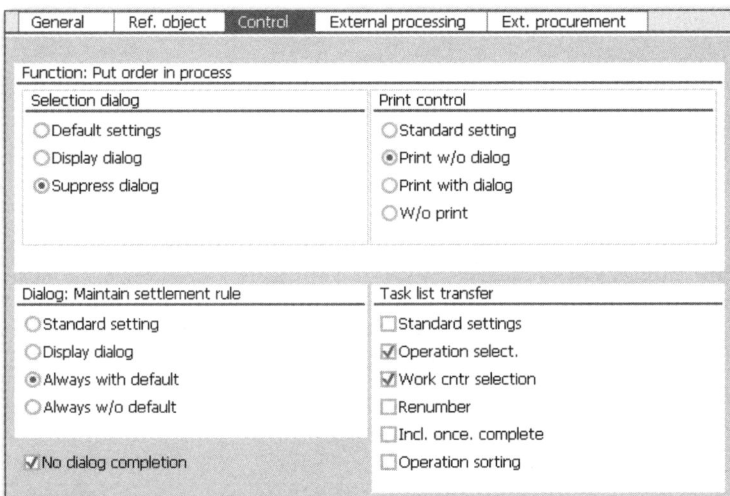

Figure 9.4 Maintenance-Specific User Parameters: Control

In addition to these default values, you can use this tab page to define how task lists are to be transferred to orders:

- OPERATION SELECT.
  Determines whether a dialog box for selecting operations is to appear.

- WORK CNTR SELECTION
  Determines whether a dialog box for selecting work centers is to appear.
- RENUMBER
  Determines whether the operations selected are to be renumbered consecutively in intervals of 10.
- INCL. ONCE. COMPLETE
  Determines whether a task list can be fully integrated once only (max.).
- OPERATION SORTING
  Determines whether the sequence of operations is the same as in the task list.

Then, use these settings to override the default settings from the Customizing function DEFAULT VALUES FOR TASK LIST DATA AND PROFILE ASSIGNMENTS (see Chapter 6, Section 6.2.9).

The EXTERNAL PROCESSING (see Figure 9.5) and EXT. PROCUREMENT (see Figure 9.6) tab pages are available in the order only.

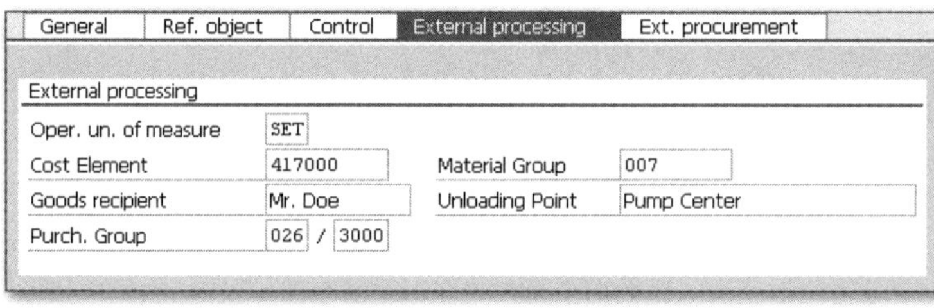

Figure 9.5 Maintenance-Specific User Parameters: External Processing

Here, define personal default values for the procurement of services and materials (e.g., for the purchasing group or material group).

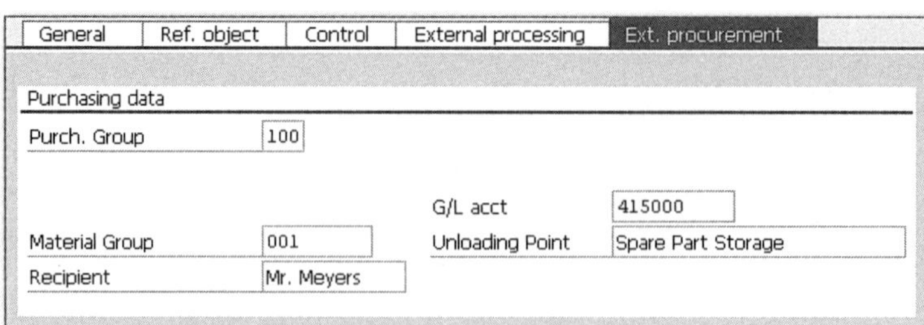

Figure 9.6 Maintenance-Specific User Parameters: External Procurement

These personal default values override the default values from the Customizing function CREATE DEFAULT VALUES FOR EXTERNAL PROCUREMENT (see Chapter 6, Section 6.2.2), which are assigned to the plant and order type.

The ACTION BOX tab page (see Figure 9.7) is available in the notification only.

Figure 9.7 Maintenance-Specific User Parameters: Action Box

If you set the ACTION BOX: TABLE indicator, the actions that you assigned to the notification type (Customizing function DEFINE ACTION BOX; see Chapter 6, Section 6.1.3) aren't displayed as a sidebar but as a table in the notification header (see Figure 9.8).

Figure 9.8 Action Box as a Table

The maintenance-specific user parameters are stored in Table TPMUS. You can use Transaction SE16 to display this table (see Figure 9.9) and make the following changes:

- Choose to make a new entry.
- Choose to change an existing entry.

Central Administration via TPMUS [+]

In other words, you can also manage the maintenance-specific user parameters centrally for your end users. You can also use the Windows key combinations Ctrl+Y, Ctrl+C, and Ctrl+V to copy entries from one user to another.

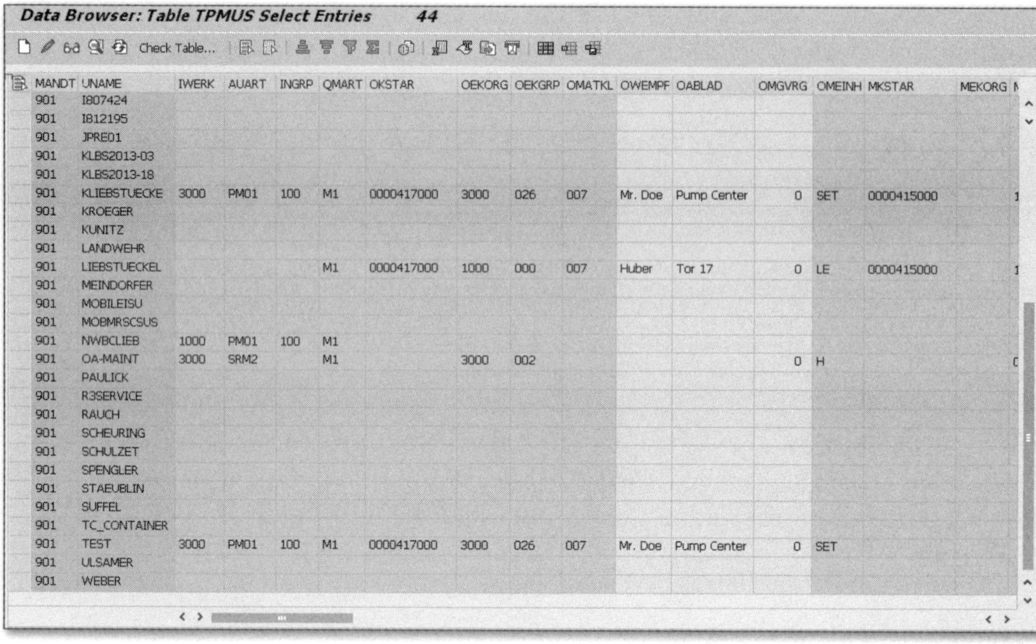

Data Browser: Table TPMUS Select Entries 44

MANDT	UNAME	IWERK	AUART	INGRP	QMART	OKSTAR	OEKORG	OEKGRP	OMATKL	OWEMPF	OABLAD	OMGVRG	OMEINH	MKSTAR	MEKORG	M
901	I807424															
901	I812195															
901	JPRE01															
901	KLBS2013-03															
901	KLBS2013-18															
901	KLIEBSTUECKE	3000	PM01	100	M1	0000417000	3000	026	007	Mr. Doe	Pump Center	0	SET	0000415000		1
901	KROEGER															
901	KUNITZ															
901	LANDWEHR															
901	LIEBSTUECKEL				M1	0000417000	1000	000	007	Huber	Tor 17	0	LE	0000415000		1
901	MEINDORFER															
901	MOBILEISU															
901	MOBMRSCSUS															
901	NWBCLIEB	1000	PM01	100	M1											
901	OA-MAINT	3000	SRM2		M1		3000	002				0	H			C
901	PAULICK															
901	R3SERVICE															
901	RAUCH															
901	SCHEURING															
901	SCHULZET															
901	SPENGLER															
901	STAEUBLIN															
901	SUFFEL															
901	TC_CONTAINER															
901	TEST	3000	PM01	100	M1	0000417000	3000	026	007	Mr. Doe	Pump Center	0	SET			
901	ULSAMER															
901	WEBER															

Figure 9.9 Maintenance-Specific User Parameters: Table TPMUS

[+]

Use Maintenance-Specific Parameters

Using maintenance-specific parameters to prepopulate fields reduces the processing time and enables you to control processing steps.

9.1.3 Roles and Favorites

You can use Transaction PFCG to create role-based menus. They have a much simpler structure than the standard SAP menu, which, in plant maintenance, can consist of up to seven levels. Starting a transaction from a role menu is, therefore, much faster than starting it from a standard SAP menu. If a role menu is assigned to you, the system automatically displays this menu on the initial screen (see Figure 9.10).

A further simplification is provided by favorites, where the user adds only those transactions he requires (see Figure 9.11). Favorites menus can be single-level or multilevel menus. Starting a transaction from a favorites menu is generally much faster than starting it from a standard SAP menu.

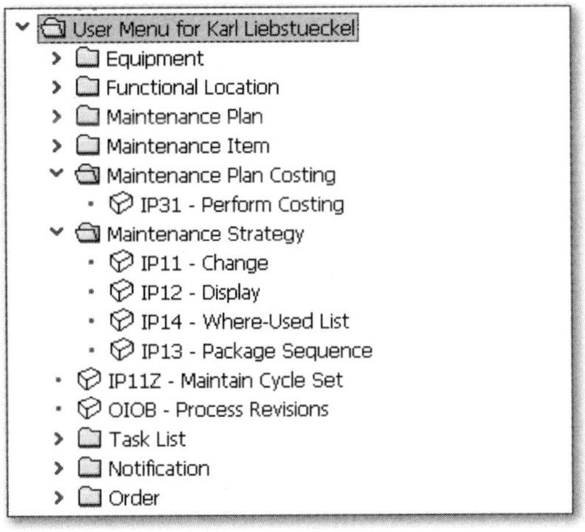

Figure 9.10 Role Menu

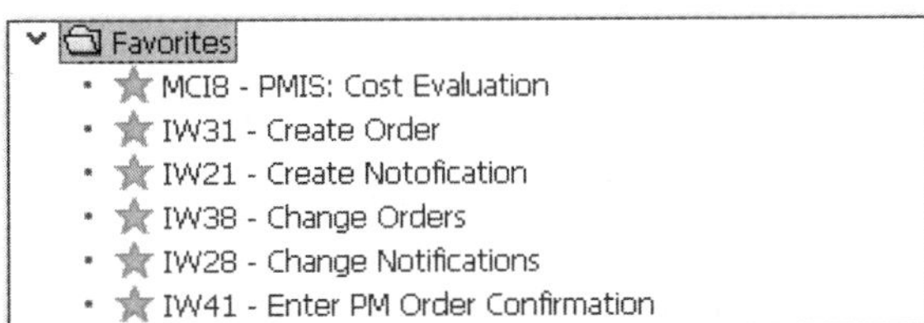

Figure 9.11 Favorites Menu

Note the following restriction: Each user creates his own favorites menu. In other words, a user can't directly use the favorites menu of his colleagues. However, you can use FAVORITES • DOWNLOAD TO PC or UPLOAD FROM PC to copy favorites menus to another user.

Use Roles and Favorites Menus [+]

Role-based menus and favorites reduce the launch time of transactions. Favorites menus can be copied to other users by using the download and upload functions.

9.1.4 List Variants

Configuring selection and display variants in SAP List Viewer is another way to make a considerable contribution toward improving usability and increasing user acceptance. SAP List Viewer doesn't present information to you as a rigid list. Instead, it makes it possible to adapt the list flexibly to your own information requirements. All lists available in SAP EAM have been converted to this technology.

The following lists are available to you in SAP EAM. In each case, the relevant transactions are provided in parentheses:

- List of functional locations (IL05, IL06, IH06)
- List of reference functional locations (IL15, IH07)
- List of equipment (IE05, IH08)
- List of fleet objects (IE36, IE37)
- List of object links and object network (IN15, IN16, IN18, IN19)
- List of measurement documents (IK17, IK18)
- List of material serial numbers (IQ08)
- List of materials (IH09)
- List of measuring points (IK07, IK08)
- List of reference measuring points (IK07R, IK08R)
- List of notifications (IW28, IW29)
- List of measures (IW66, IW67)
- List of activities (IW64, IW65)
- List of notification items (IW68, IW69)
- List of orders (IW38, IW39)
- List of order operations (IW37, IW49)
- Combined order/operation list (IW37N, IW49N)
- List of components (IW3K, IW3L)
- List of permits (IPM2, IPM3)
- List of completion confirmations (IW47)
- List of goods movements (IW3M)
- List of maintenance plans (IP15, IP16)
- List of maintenance items (IP17, IP18)
- List of maintenance dates (IP24)

- List of task lists (IA08, IA09)
- List of shift notes (SHN4, ISHN4)
- List of shift reports (SHR4, ISHR4)

In the next section, we'll use the example of the order list (Transaction IW38) to introduce SAP List Viewer. However, the information provided can be applied to any of the aforementioned lists.

Selection

When you call a list, you see a selection screen containing all of the selection options. This extends over two to four pages, depending on the list and the screen resolution. Experience shows that you'll rarely need the options listed on the second or subsequent pages. Therefore, the first step should be to create a selection variant.

Create a Selection Variant [+]

Define a selection variant whose selection conditions span a maximum of one page. All selection variants that you'll create in the future should also follow this basic rule.

To create a selection variant, choose [save icon] to save the selection screen. Then, on the next screen, you should make extensive and purposeful use of the option to hide the selection conditions (see Figure 9.12).

Selection Scr...	Field name	Type	Protect field	Hide field
1,000	Leading order	S	☐	☑
1,000	Superior order	S	☐	☑
1,000	Planning plant	S	☐	☐
1,000	Priority	S	☐	☐
1,000	Entered by	S	☐	☐
1,000	Created on	S	☐	☑
1,000	Status inclusive	S	☐	☐
1,000	Status exclusive	S	☐	☑
1,000	Description	S	☐	☑
1,000	Last changed by	S	☐	☑
1,000	Change date for ord...	S	☐	☐

Figure 9.12 Hiding a Field

Then, choose to call your list variants, which are displayed on the initial screen of every list. In this way, you can display a list of selection variants and then choose the list variant that you need. This list can become quite extensive over time, however, so follow this next practical tip:

[+]

Default Variant U_BENUTZERNAME

The selection variant that you use most often should be named U_ followed by your SAP user name (see Figure 9.13). This selection variant is then automatically displayed when you call the list.

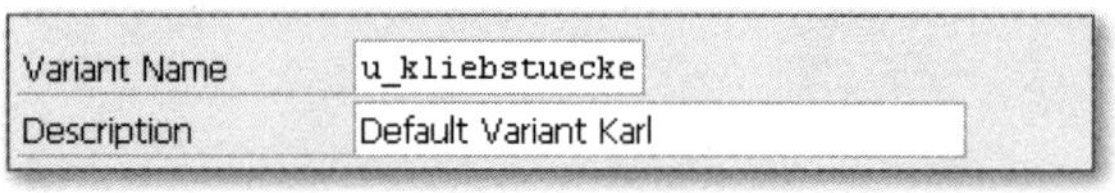

Figure 9.13 Variant Name

You can use the following selection options for each selection field:

- **Single value**
  You can search for a single value (e.g., order type PM01) or multiple single values (e.g., order types PM01, PM05, and PM10).
- **Interval**
  You can search for an interval (e.g., order type PM01 – PM05) or multiple intervals.
- **Wild card search**
  You can perform a wild card search (e.g., order type PM*, which selects all order types that begin with PM).
- **Exclusion of elements**
  You can exclude values and intervals (e.g., not order types PM03 and PM05 – PM08).
- **Clipboard**
  You can paste in search values from the Windows clipboard.

The definition of selection variables for the *dynamic date calculation* is particularly helpful, whereby, depending on the relevant current date, the from/to date selection is calculated dynamically on the basis of the selection option selected (see Figure 9.14).

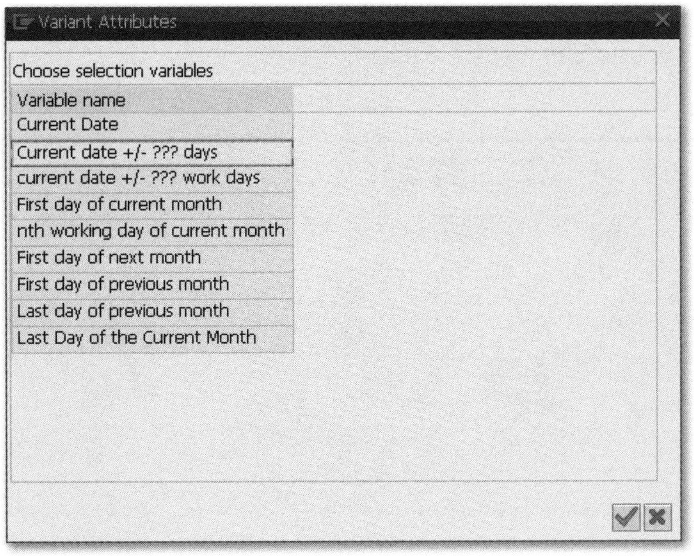

Figure 9.14 Date Variable

Dynamic Date Calculation [+]

You can use the Dynamic Date Calculation selection variables to determine the selection date of the list dynamically.

Let's take a look at the following example: On September 1, you configured the list with the selection *Start date equals current date –180 days and +60 days*. Therefore, on September 1, the system selects data from March 4 to October 31.

Activate the Monitor [+]

Some lists (not all) allow you to activate a monitor. Depending on the parameters selected (e.g., basic date), the list entries are then red, yellow, or green (e.g., red for overdue orders, yellow if the start date has been reached but the finish date has not been reached yet, and green for future orders).

List Display

If you now start the list, you obtain an initial basic list created on the basis of the selections and settings made (see Figure 9.15).

Change PM Orders: List of Orders

Order | Operations | Subtotal

S	Order	Type	Basic start date	Description	Equipment	Functional Loc.	Σ TotalPlnndCosts	Σ Total actual costs
	820340	PM01	07/10/2013	Pump defect, production downtime	10002520		2,039.82	0.00
	820360		07/29/2013	renewal pump station	P-3000-N001	ICB-M2-M1-01	5,880.75	2,271.25
	820365		07/08/2013	Modification Pump Station	P-3000-N001	ICB-M2-M1-01	7,590.00	1,437.50
	820366		07/15/2013	opeation settlement	P-3000-N001	ICB-M2-M1-01	9,487.50	977.50
	820368		07/29/2013	opeation settlement	P-3000-N001	ICB-M2-M1-01	7,590.00	1,150.00
	820369		07/10/2013	Production Pump Precision 100	P-3000-N001	ICB-M2-M1-01	6,325.00	1,725.00
	820371		07/10/2013	renewal pump station	P-3000-N001	ICB-M2-M1-01	12,650.00	0.00
	820372		07/25/2013	Production Pump Precision 100	P-3000-N001	ICB-M2-M1-01	18,975.00	0.00
	820373		07/20/2013	Pump defect, production downtime	P-3000-N001	ICB-M2-M1-01	18,212.50	0.00
	820446		07/13/2013	external inspection	P-3000-N001	ICB-M2-M1-01	1,800.00	1,500.00
	820447		07/17/2013	construction changes		C1-B01	22,836.50	0.00
	820459		07/17/2013	Maintenance Inspection for Airplane	10004170		25.30	0.00
	820500		07/14/2013	Counter based MP pump set 2		C1-B01-2	126.50	0.00
	820560		07/30/2013	Modification Pump Station	P-3000-N001	ICB-M2-M1-01	5,562.50	0.00
	820561		07/27/2013	renewal gas station	P-3000-N001	ICB-M2-M1-01	0.00	6,900.00
	820563		07/28/2013	standing order for cleaning activities		C1	0.00	0.00
	820564		07/30/2013	immediate repair	P-3000-N001	ICB-M2-M1-01	0.00	0.00
		PM01					▪ 119,101.37	▪ 15,961.25
	819899	PM02	07/15/2013	Preventive maintenance plan for Pump	10006757	SU-SC-AU-OP	0.00	0.00
	819900		07/15/2013	Preventive maintenance plan for Pump	10006757	SU-SC-AU-OP	0.00	0.00
	819901		07/15/2013	Preventive Maintenance of Pump	10006782	SU-SC-MP-P7-E1	0.00	0.00
	819902		07/15/2013	Preventive Maintenance of Pump	10006782	SU-SC-MP-P7-E1	0.00	0.00
	819903		07/15/2013	Preventive Maintenance of Pump	10006782	SU-SC-MP-P7-E1	0.00	0.00
	819904		07/15/2013	Preventive Maintenance of Pump	10006782	SU-SC-MP-P7-E1	0.00	0.00
	819905		07/15/2013	Preventive Maintenance of Pump	10006782	SU-SC-MP-P7-E1	0.00	0.00
	819906		07/15/2013	Preventive Maintenance of Pump	10006782	SU-SC-MP-P7-E1	0.00	0.00
	820490		07/12/2013	Preventive Maintenance of Pump	10006782	SU-SC-MP-P7-E1	632.50	0.00
	820547		07/16/2013	performace based pump inspection	P-3000-N001	ICB-M2-M1-01	126.50	0.00
		PM02					▪ 759.00	▪ 0.00
							▪▪ 119,860.37	▪▪ 15,961.25

Figure 9.15 List in SAP List Viewer

You now have the following options for adjusting the layout of the list to your own requirements:

- **Show and hide fields**
  You can show additional fields or hide the fields already displayed. Almost all the fields of the object to be analyzed are available for selection. For functional locations and equipment, you can also show fields from their classifications.
- **Sort**
  You can sort by a single criterion (e.g., by date) or by several criteria (e.g., by cost center and then, within the cost center, by date).
- **Calculate totals**
  For value and quantity fields (e.g., actual costs), you can calculate totals and display subtotals (e.g., for each order type).

- **Change column width**
  You can optimize the column width.
- **Search**
  You can search, within a list, for a certain term (e.g., leaks). This is an especially useful function for large lists.
- **Filter**
  You can filter a displayed list (e.g., display only entries with the system status REL).
- **Perform ABC analysis**
  You can use a key figure to perform an activity-based costing (ABC) analysis (e.g., an ABC analysis of orders relative to actual costs).
- **Graphical display**
  You can generate a graphical display (e.g., a bar chart with the number of orders for each order type).
- **Use display variants**
  You can save your settings as a display variant.

[+]

Presetting for Display Variant

You should mark the most frequently used display variant as a presetting.

Further Processing

If you now have the list in the form you want, the following options are available to you for further processing:

- **Details**
  You can select a certain line and call the database object (e.g., a certain order so that you change the date in that order).
- **Mass processing**
  You can select several lines and perform the same function for all of the lines selected (e.g., print all orders selected or assign the same date to all orders selected).
- **Mass change**
  In conjunction with list processing, there is a MASS CHANGE function for requests

and notifications, which you can use to change, in a single step, practically any field in all of the objects selected.

- **Send**
  You can send the list via SAP mail.

[+]

No Email Button?

If you can't find a button for sending a list (e.g.,), use the function LIST • SAVE • OFFICE.

- **Download**
  You can save the list in any current Office format and process it further there (e.g., download the order list to Excel to display it there using pivot functions).

[+]

Schedule Lists as a Periodic Job

You can also schedule lists as a periodic job, which then runs automatically at regular intervals and performs certain follow-up functions (e.g., an email containing the maintenance plan orders for the coming week can be sent to the production managers).

9.1.5 Personalizing Input Help

An F4 help usually displays all possible entries. Often, however, this list of all entries is very long (e.g., material groups, object types, or cause of damage) even though the individual user generally requires only a certain number of entries from this list.

Here, you now have the option to define a *personal value list*. To do this, call the F4 help in the corresponding field (e.g., MATERIAL GROUP), select the entries you require, and choose to assign them to your personal value list.

If you now call the F4 help for this field, the system automatically displays your personal value list instead of the complete list (see Figure 9.16). You can choose to return to the complete value list at any time.

[+]

Use Personal Value Lists

In a F4 help, a personal value list saves you time in relation to searching for the correct entry.

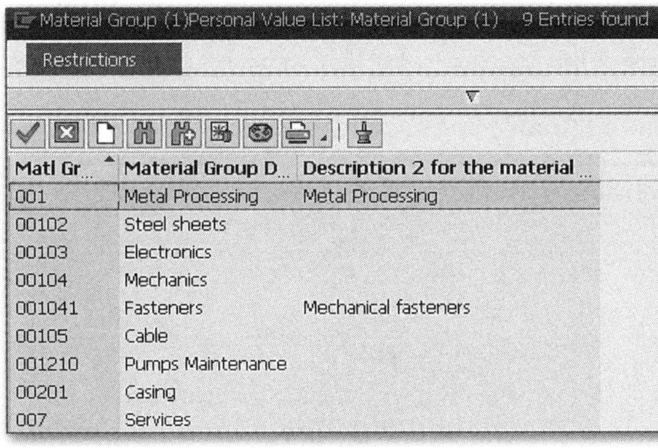

Figure 9.16 Personal Value List

9.1.6 Buttons and Key Combinations

Since the mouse was introduced as a control element for computers, opinions have been divided as to whether operation is faster with or without a mouse. All users have to answer this question for themselves. If you prefer to use the keyboard instead of the mouse, key combinations and buttons can be helpful.

Create Your Own Key Combinations [+]

Provide important and generally valid key combinations to those users who prefer to use the keyboard.

Examples, not only in the area of plant maintenance, include the following:

- You can use the F11 key to save a document. This function corresponds to the disk icon in the system toolbar.
- Another popular trick is the use of the key combination F4+Enter in date and time fields. This copies the current date or time into these fields.
- In general, you can use the F4 key to call a value list.

9.1.7 Table Controls

Table controls are a largely unknown and unused option that end users can use to adjust the SAP system to their requirements. In the past, SAP screens were pre-programmed. In other words, each field occupied a permanent position on the screen. In the meantime, however, many screens have been provided with table controls, which you can use to define your own screen layout and field sequence.

Examples of table controls in plant maintenance and related areas are provided here:

- Operation list (see Figure 9.17), component list, service specification, and object list in the order

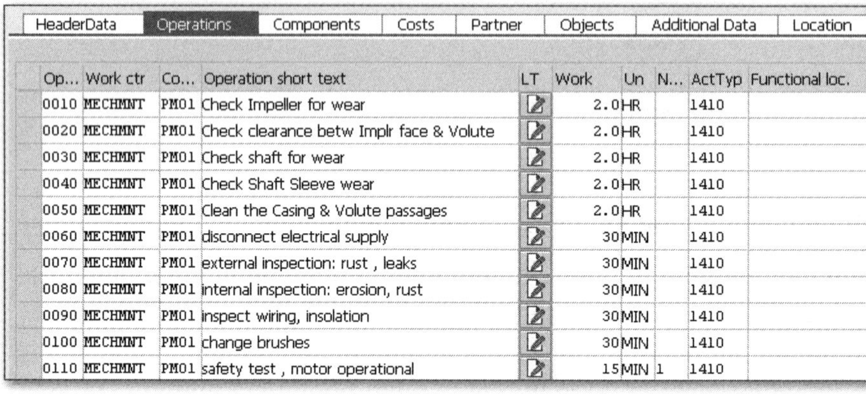

HeaderData | Operations | Components | Costs | Partner | Objects | Additional Data | Location

Op...	Work ctr	Co...	Operation short text	LT	Work	Un	N...	ActTyp	Functional loc.
0010	MECHMNT	PM01	Check Impeller for wear		2.0	HR		1410	
0020	MECHMNT	PM01	Check clearance betw Implr face & Volute		2.0	HR		1410	
0030	MECHMNT	PM01	Check shaft for wear		2.0	HR		1410	
0040	MECHMNT	PM01	Check Shaft Sleeve wear		2.0	HR		1410	
0050	MECHMNT	PM01	Clean the Casing & Volute passages		2.0	HR		1410	
0060	MECHMNT	PM01	disconnect electrical supply		30	MIN		1410	
0070	MECHMNT	PM01	external inspection: rust , leaks		30	MIN		1410	
0080	MECHMNT	PM01	internal inspection: erosion, rust		30	MIN		1410	
0090	MECHMNT	PM01	inspect wiring, insolation		30	MIN		1410	
0100	MECHMNT	PM01	change brushes		30	MIN		1410	
0110	MECHMNT	PM01	safety test , motor operational		15	MIN	1	1410	

Figure 9.17 Table Control: Operation List

- All partner overviews
- Item, cause, measure, and activity overview in the notification
- Time confirmation, measurement recording, goods movements, and activities in the overall completion confirmation
- Item overview in the maintenance plan
- Operation overview, component overview, maintenance packages, and service specification in the task list
- Collective confirmation
- Item overview for material withdrawal
- Item overview in purchase requisitions
- Item overview in purchase order

The following settings are available for all table controls:

- You can use drag and drop to arrange the columns in the order in which you require them most often.

- You can widen or narrow each column so that it accommodates the entry usually made there.
- You can hide fields that you don't need by setting the width of the column to zero.

The most important setting of all is the option to save these settings as a user-specific variant. The upper-right corner of each table control contains the icon, which you can use to access a detail screen in which you create your settings as a variant and can therefore override the default settings (see Figure 9.18).

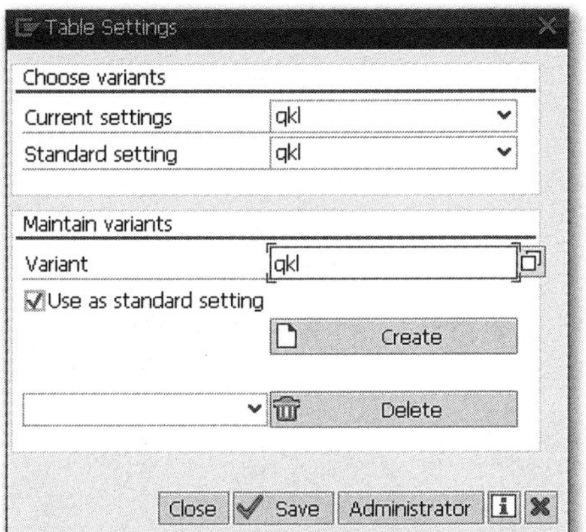

Figure 9.18 Table Settings within a Table Control

Make Extensive Use of Table Controls [+]

You can use individualized table controls and table controls stored as a variant to considerably reduce the time spent making entries and searching for entries in the screen templates, thus making a significant contribution to improving usability.

SAP ERP 6.0 with EHP6 provides the following special feature for you and your users: Previously, the table controls for the OPERATIONS overview and COMPONENTS overview contained only a selection of fields predefined by SAP. The purchasing data was sorely missed. This has now changed in EHP6 whereby you can now maintain all purchasing data (e.g., vendor, material group, ship-to party, unloading point, or purchasing group in both of the aforementioned table controls (see Figure 9.19 for the COMPONENTS overview).

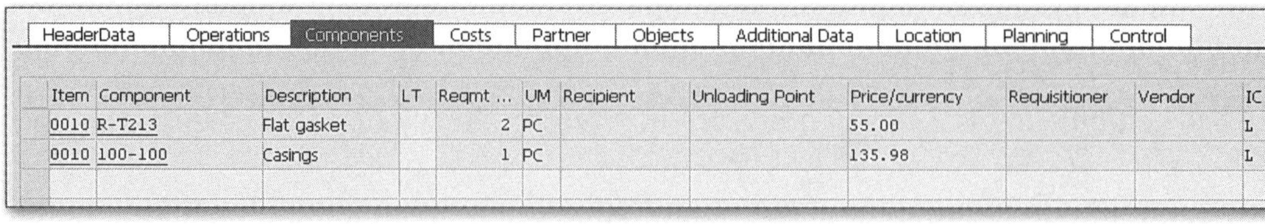

Figure 9.19 Purchasing Data in the Components Overview

As a prerequisite, you must activate the business function `LOG_EAM_CI_5`.

This concludes the description of all the options available within category 1, whereby each user can configure his own individual settings.

Let's now take a look at the options available within category 2. In most enterprises, these options are usually the responsibility of the IT department. However, the options contained in this category don't require any programming.

9.2 Category 2: The IT Department's Nonprogramming Options

As a first step, we'll introduce the option of creating any number of adjusted variants (known as *transaction variants*) for an original transaction.

9.2.1 Transaction Variants

If different user groups use a transaction to process different business transactions, it often makes sense to adjust the transaction process to the relevant business transaction or user group. For example, the process of creating an order can differ depending on the user group that creates it:

- Sold-to party or contractor
- Electricians or mechanics
- Planners or technicians

Transaction variants are suitable here.

Use Transaction Variants to Adjust Each Transaction [+]

You can create any number of transaction variants for an original transaction. In a transaction variant, you can do the following:

- Hide entire screens.
- Hide individual tab pages.
- Disable menu functions.
- Disable buttons.
- Set the field selection control for individual fields (display, required, or hide).
- Prepopulate the field content.
- Modify the column sequence, change the column width, and hide columns for table controls.
- Assign a custom name to the transaction.

Now we'll use Transaction IW31 to demonstrate how to create an adjusted transaction variant. As a first step, define a new transaction variant in Transaction SHDO. On the initial screen (see Figure 9.20), specify the original transaction in the TRANSACTION CODE field ("IW31") and enter the name of the TRANSACTION VARIANT (e.g., "Z_IW31_M" for an adjusted Transaction IW31 for the mechanical workshop).

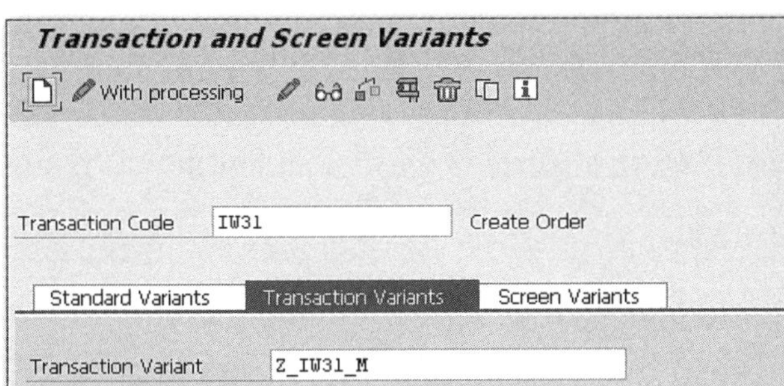

Figure 9.20 Transaction Variant

If you now choose to start the recording, the system guides you through the screens in the original transaction. Figure 9.21 shows, for example, the first screen that you see when you call Transaction IW31.

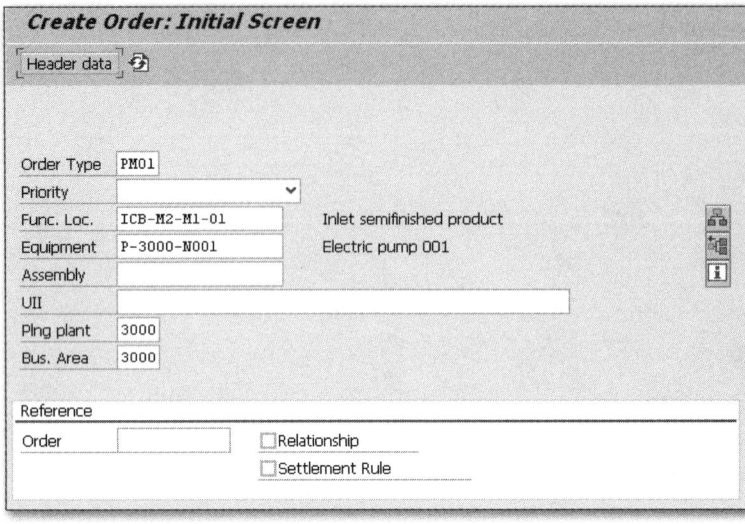

Figure 9.21 Transaction Variant: Initial Screen

Then, for each screen, specify whether or not you want it to be displayed (Figure 9.22). If you set the DO NOT DISPLAY SCREEN indicator, the screen is hidden. For each field contained in the screen, specify the following:

- W. CONTENT
  Determines whether you want to transfer the entered content.
- OUTPUT ONLY
  Determines whether you want it to be a "display only" field and not a "ready for input" field.
- INVISIBLE
  Determines whether you want to hide it.
- REQUIRED
  Determines whether you want to make it mandatory for a user to make an entry in this field.

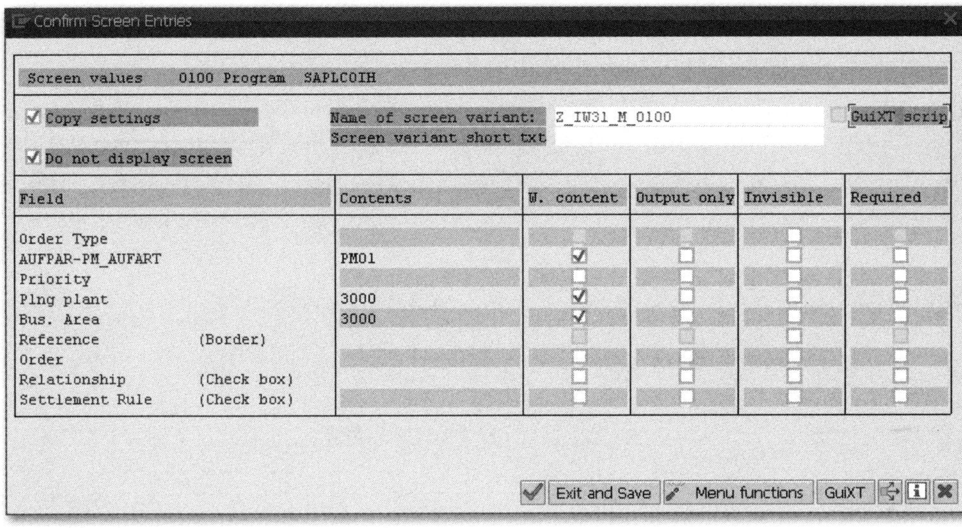

Figure 9.22 Transaction Variant: Field Settings

If the transaction has several tab pages, the system will ask you which of these tab pages are to be displayed in the transaction variant (see Figure 9.23). If you set the INVISIBLE indicator, the relevant tab page is completely hidden.

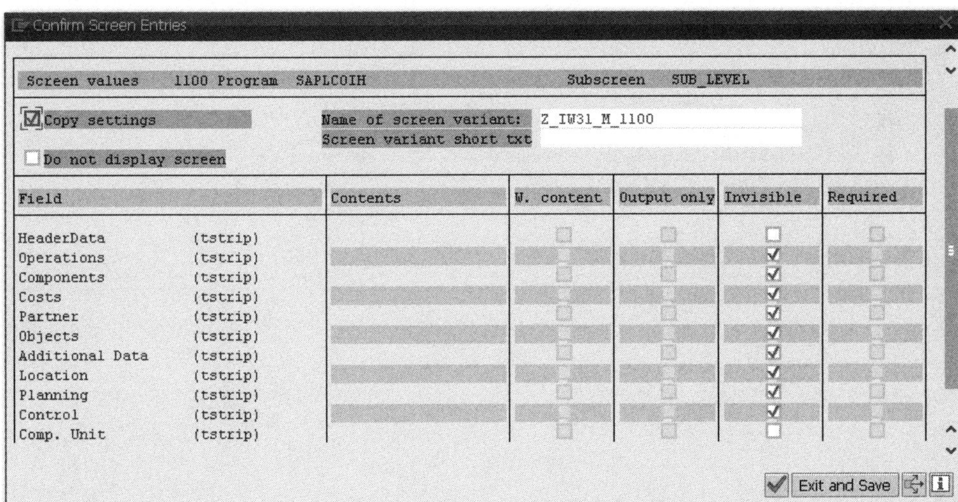

Figure 9.23 Transaction Variant: Tabstrip Settings

If you want to manipulate both the transaction menu and the function keys, you can choose Menu functions.

If you position the cursor on a menu option in the Menu Bar area, you can choose to enable or disable this option. All menu options that have been disabled are highlighted in color (see Figure 9.24). They are subsequently displayed in the menu of the transaction variant. However, they are grayed out and therefore inactive.

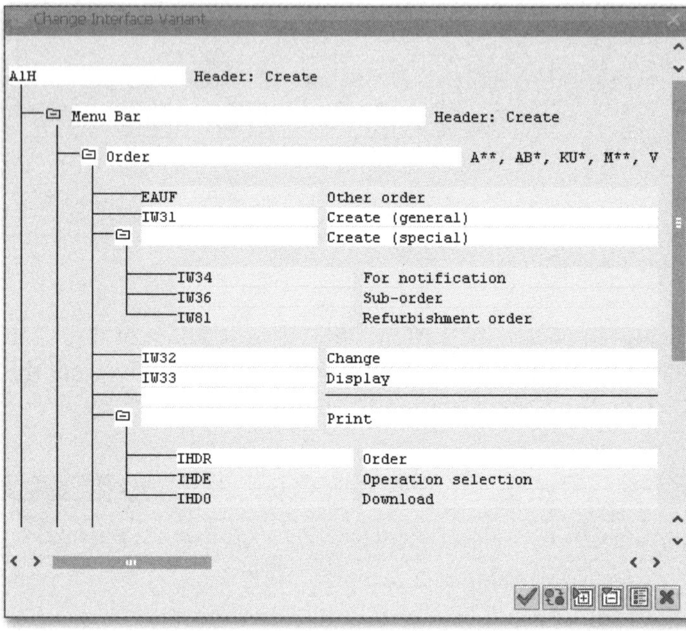

Figure 9.24 Transaction Variant: Menu Bar

If you position the cursor on a function key in the Function Key Setting area, you can choose to enable or disable this function. All function keys that have been disabled are highlighted in color (see Figure 9.25) and subsequently not displayed at all in the transaction variant.

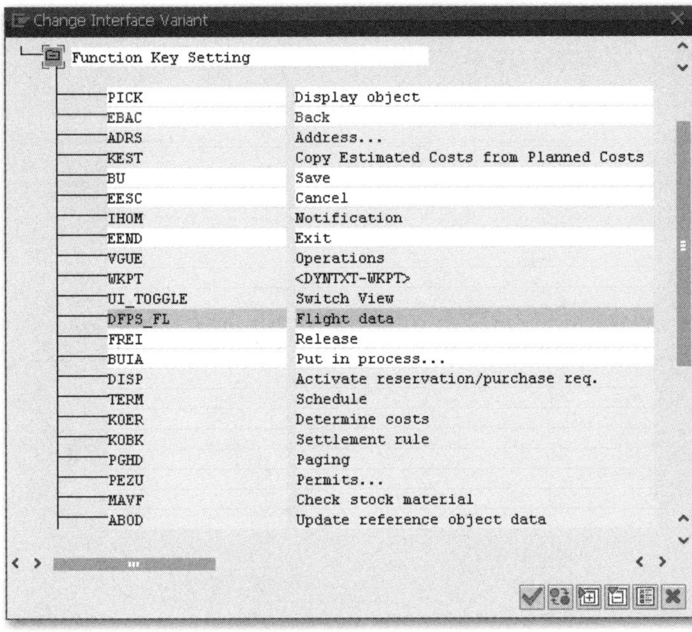

Figure 9.25 Transaction Variant: Function Keys

You can use Transaction SE93 to assign a transaction name to your transaction variant (see Figure 9.26).

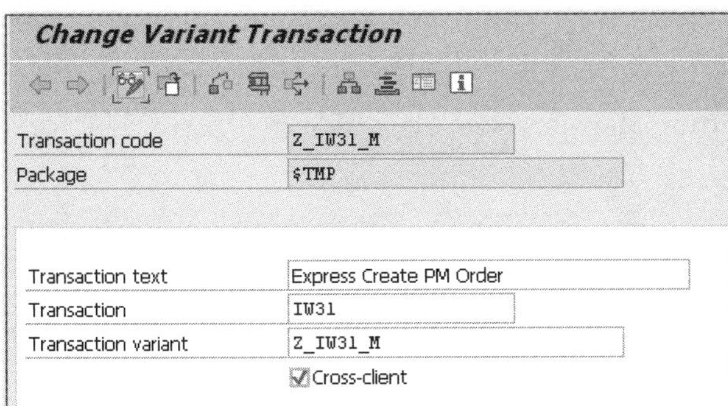

Figure 9.26 Transaction Variant: Transaction

If you now start this transaction, the original transaction is displayed in a new screen. Figure 9.27 shows you the new IW31.

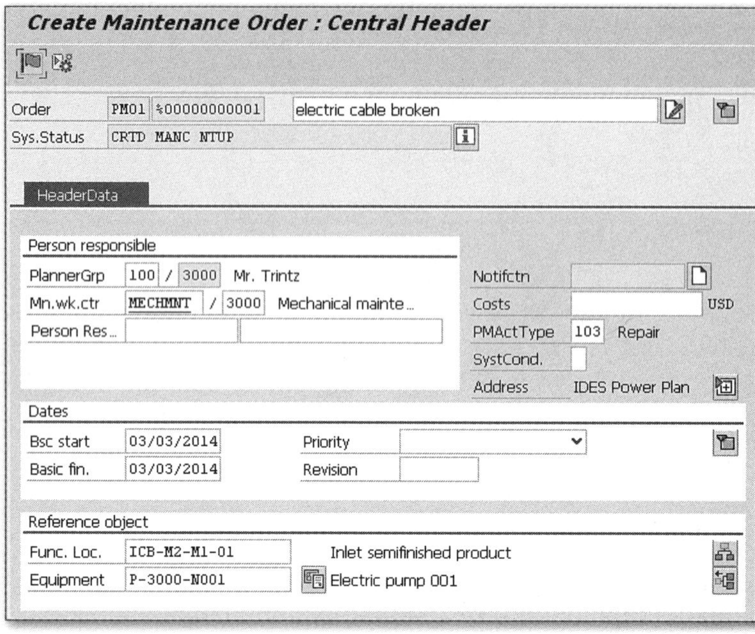

Figure 9.27 Transaction Variant: Result

To simplify the transaction, the following measures have been taken:

- The initial screen for Transaction IW31 has been skipped.
- The content of certain fields has been prepopulated (e.g., ORDER TYPE and PRIORITY).
- Apart from one tab page, all other tab pages have been hidden (e.g., OPERATIONS and MATERIAL).
- Menu options and buttons have been disabled (e.g., SETTLEMENT RULE, SCHEDULE, and PAGING).
- Subscreens have been hidden (e.g., the operation detail subscreen in the order header).
- Some fields have been hidden (e.g., ASSEMBLY).
- A unique transaction name (Z_IW31_M) has been assigned, and it can be used to call the transaction variant.

The result is a transaction whose content has been significantly reduced compared to the original transaction and contains only those screens, fields, and functions that the user needs.

9.2.2 Customizing

Customizing also provides many options in terms of improving usability. The most important of these Customizing functions are listed here, along with a description of each transaction's function if not immediately apparent from its name. The order in which these Customizing functions are listed is the order in which they occur in the SAP Reference IMG:

- Configuring the Material Master
  Adjust the screen layout for material master data.
- Material Master Field Selection
  Define required entry fields and hide fields.
- Set View Profiles For Technical Objects
  Adjust the screen layout for equipment and functional locations.
- Define Field Selection for Functional Locations
  Define required entry fields and hide fields.
- Define Field Selection for Equipment Master Record
  Define required entry fields and hide fields.
- Set Field Selection for Specific Fields in Fleet Management
  Define required entry fields and hide fields.
- Define Transaction-Based Default Values for Object Types
  Prepopulate the equipment category or functional location category.
- Define Field Selection for Maintenance Plan
  Define required entry fields and hide fields.
- Define Field Selection for Operation Data
  Define required entry fields and hide fields.
- Work Centers • Define Field Selection
  Define required entry fields and hide fields.
- Work Centers • Configure Screen Sequence for Work Center
  Adjust the screen layout for work centers.

- Task Lists • Define Profiles with Default Values
  Default values for task lists.
- Task Lists • Define Field Selection
  Define required entry fields and hide fields.
- Define Printer
  Prepopulate the values for print output.
- Maintain Settlement Profile
  Default values for account assignment.
- Overview of Notification Type • Screen Areas in Notification Header
  Define a reference object for the notification.
- Set Screen Templates for the Notification Type
  Adjust the screen layout for notifications.
- Set Field Selection for Notifications
  Define required entry fields and hide fields.
- Define Transaction Start Values
  Prepopulate the notification type and skip the initial screen.
- Assign Notification Types to Order Types
  Prepopulate the order type for the notification type.
- Define Response Monitoring
  Propose the service profile and response profile for notifications.
- Configure Order Types
  Propose a reference object for the order.
- Create Default Profiles for External Procurement
  Propose values for material and service procurement.
- Create Default Value Profiles for General Order Data
  Propose values for graphics and scheduling.
- Default Values for Maintenance Task List Data and Profile Assignments
  Propose values for external data profiles, graphic profiles, and task list transfer.
- Define Default Order Types for Maintenance Items
  Propose the order type for maintenance orders.
- Define Notification and Order Integration
  Enter the notification and order data in one screen and automatically transfer the long text from the notification to the order.

- Maintain Default Values for Control Keys for Order Types
  Propose a value for the control key.
- Default Values for Maintenance Activity Type for Each Order Type
  Propose a value for the maintenance activity type.
- Basic Order View • Define View Profiles
  Adjust the screen layout for orders.
- Define Default Values for Component Item Categories
  Propose a value for the item category for each material type.
- Activate Default Value for Current Date as Basic Date
  Define a value for prepopulating the date.
- Set Scheduling Parameters
  Activate automatic scheduling.
- Message Control
  Control whether a warning, error, or no message is to be issued.
- Define Object Information Keys
  Display object information keys automatically.
- Define Field Selection for Order Header Data (PM)
  Define required entry fields and hide fields.
- Define Field Selection for Order Operations (PM and CS)
  Define required entry fields and hide fields.
- Define Field Selection for Components (PM and CS)
  Define required entry fields and hide fields.
- Set Screen Templates for Completion Confirmation
  Adjust the screen layout for the overall completion confirmation.
- Set Field Selection for Completion Confirmation
  Define required entry fields and hide fields.
- Define Settings for Shift Note Screen Templates
  Adjust the screen layout and field selection for shift notes.
- Define Settings for Shift Note Type
  Propose values for categories.
- Basic Settings for Pool Asset Management
  Propose values for order type, class, and text.

[+]

Purposeful Customizing Improves Usability

Customizing provides many options to enhance the usability of SAP EAM, especially options for adjusting the screen layout in notifications, orders, and overall completion confirmations.

In addition, you should make specific use of field selection control, particularly for hiding information.

9.2.3 Action Box

When you process notifications, you can use the *action box* to execute follow-up actions that may make notification processing easier for you. After you've executed a follow-up action, it's documented as an activity or measure.

You can call the functions directly from the notification. Then, when you complete the function, the system automatically returns you to the notification. Examples of actions within the action box include, for example:

- Entering an internal note
- Documenting a telephone call
- Initiating a telephone call via SAPphone
- Sending an SAP mail or email
- Posting a goods movement
- Creating a repair order
- Searching the solution database for a problem or solution
- Creating a quality notification
- Generating an 8D report
- Creating a maintenance plan
- Assigning a bill of material (BOM)

As you can see, the action box concerns functions that would normally require you to exit notification processing. Therefore, the action box makes a significant contribution to simplifying a business process.

You can use the Customizing function Define Action Box to define more functions in addition to those shown in Figure 9.28.

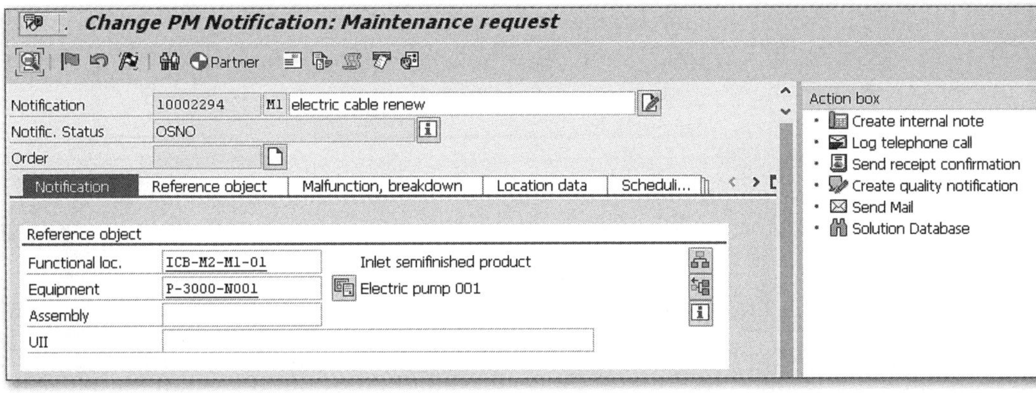

Figure 9.28 Action Box within the Notification

In Chapter 6, Section 6.1.3, you learned how to define an action box and how to assign it to a notification type.

9.2.4 GuiXT

GuiXT is an SAP GUI component that enables you to design SAP transactions according to your daily requirements without having to modify the SAP programs or screens. The following options are available:

- Prepopulate fields with values.
- Hide fields and field groups.
- Move fields.
- Add and change text.
- Add field help.
- Add new screen elements (e.g., checkboxes, buttons, graphics, and documentation).
- Adjust table controls.
- Change menu entries.
- Change field labels throughout the system.

GuiXT works with a script language that is stored as a TXT file. In a script, you can define the relevant layout in the form of individual statements for each SAP screen.

The GuiXT window is the central switching point for GuiXT and all components (see Figure 9.29). By default, the display window for screen elements and scripts is opened initially.

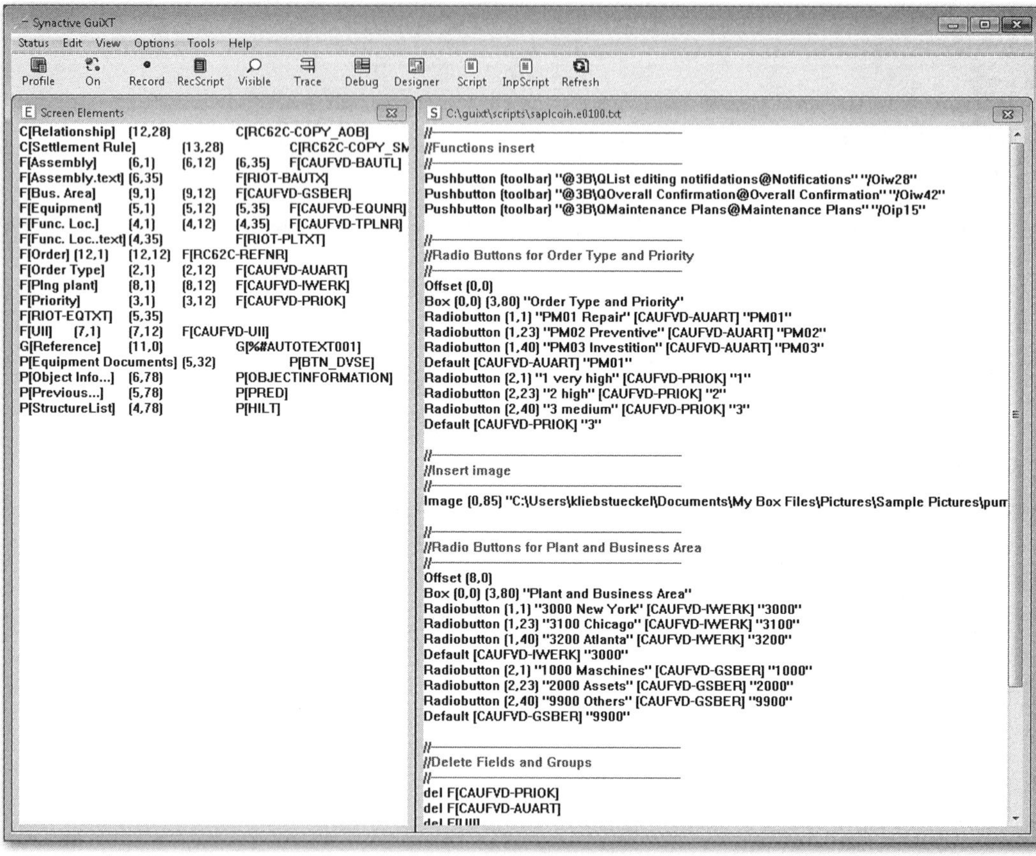

Figure 9.29 GuiXT: Control Window

Click  to access your custom-defined editor for creating scripts. In the default setting, this is the *notepad*. However, you can also use another editor. You can store the script files at different storage locations:

- As local files (e.g., at *C:\GuiXT\Scripts*)
- On central file servers (e.g., *\\demo\GuiXT\Scripts*)
- On HTTP servers (e.g., *http://www.demo.com/demo/scripts*)

- On FTP servers (e.g., *ftp://demo.com/demo/scripts*)
- In the SAP Web Repository (SAP database) (e.g., *sapwdf11.guixt.demo*)

Each of these storage options has advantages and disadvantages. For example, it's very quick and easy to save the script files as local files. However, they are then available on the local PC only.

We'll now use an example to outline the procedure and options associated with using GuiXT. We'll make some changes to the layout of the initial screen in Transaction IW31 (see Figure 9.30):

❶ In the function key toolbar, we want the user to have additional function keys for calling a list of notifications, a list of orders, and the overall completion confirmation.

❷ Due to being error prone, we don't want the user to be able to select the order type as a normal F4 selection list. Instead, we want the user to use radio buttons to choose from the order types PM01, PM02, and PM03.

❸ For priorities, we don't want the user to be able to choose the priority from a dropdown list. Otherwise, the user will always choose priority 1. Instead, we want the screen to contain radio buttons for priorities 1 – 3. We want priority 3 to be a presetting. In other words, the user must make the conscious decision to choose another priority if his order is to be assigned a higher priority.

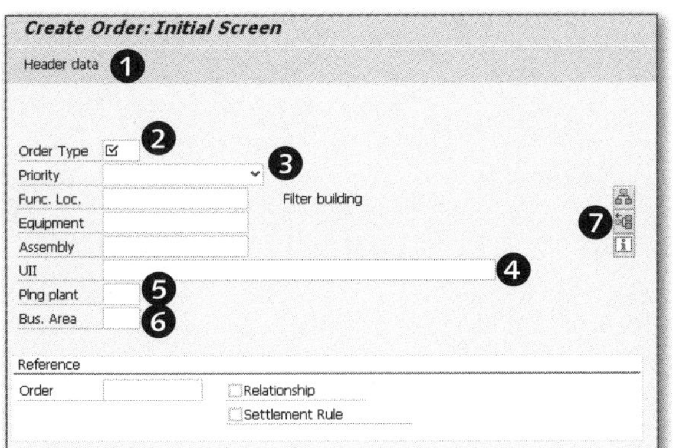

Figure 9.30 GuiXT: Standard Screen IW31

❹ Because the user doesn't require the field UII (unique parts number), we want to hide it. This can also be done using the normal field selection. The disadvantage of this, however, is that the empty line is retained. Therefore, we want to be able to move all lower fields up.

❺ Because maintenance processing occurs in planning plants 1000, 1100, and 1200 only, we want the system to display these three options only (as radio buttons).

❻ Because maintenance processing occurs in business areas 1000, 2000, and 9900 only, we want the system to display these three options only (as radio buttons).

❼ On the right-hand side of the screen, we want to display an icon that represents the user department currently using maintenance processing. Because the pump workshop is currently using maintenance processing, an image depicting a pump is to be displayed on the screen.

Listing 9.1 shows a script that satisfies the preceding requirements.

```
//-----------------------------------------------------------------
//Functions insert
//-----------------------------------------------------------------
Pushbutton (toolbar) "@3B\QList editing notifidations@Notifications" "/
Oiw28"
Pushbutton (toolbar) "@3B\QOverall Confirmation@Overall Confirmation"
"/Oiw42"
Pushbutton (toolbar) "@3B\QMaintenance Plans@Maintenance Plans" "/
Oip15"

//-----------------------------------------------------------------
//Radio Buttons for Order Type and Priority
//-----------------------------------------------------------------
Offset (0,0)
Box (0,0) (3,80) "Order Type and Priority"
Radiobutton (1,1) "PM01 Repair" [CAUFVD-AUART] "PM01"
Radiobutton (1,23) "PM02 Preventive" [CAUFVD-AUART] "PM02"
Radiobutton (1,40) "PM03 Investment" [CAUFVD-AUART] "PM03"
Default [CAUFVD-AUART] "PM01"
Radiobutton (2,1) "1 very high" [CAUFVD-PRIOK] "1"
Radiobutton (2,23) "2 high" [CAUFVD-PRIOK] "2"
Radiobutton (2,40) "3 medium" [CAUFVD-PRIOK] "3"
Default [CAUFVD-PRIOK] "3"

//-----------------------------------------------------------------
//Insert image
//-----------------------------------------------------------------
Image (0,85) "C:\Users\kliebstueckel\Documents\My Box Files\Pictures\
```

```
Sample Pictures\pumpe.jpg"

//--------------------------------------------------------------
//Radio Buttons for Plant and Business Area
//--------------------------------------------------------------
Offset (8,0)
Box (0,0) (3,80) "Plant and Business Area"
Radiobutton (1,1) "3000 New York" [CAUFVD-IWERK] "3000"
Radiobutton (1,23) "3100 Chicago" [CAUFVD-IWERK] "3100"
Radiobutton (1,40) "3200 Atlanta" [CAUFVD-IWERK] "3200"
Default [CAUFVD-IWERK] "3000"
Radiobutton (2,1) "1000 Machines" [CAUFVD-GSBER] "1000"
Radiobutton (2,23) "2000 Assets" [CAUFVD-GSBER] "2000"
Radiobutton (2,40) "9900 Others" [CAUFVD-GSBER] "9900"
Default [CAUFVD-GSBER] "9900"

//--------------------------------------------------------------
//Delete Fields and Groups
//--------------------------------------------------------------
del F[CAUFVD-PRIOK]
del F[CAUFVD-AUART]
del F[UII]
del F[CAUFVD-GSBER]
del F[CAUFVD-IWERK]
del G[Reference]

//--------------------------------------------------------------
//Positioning fields
//--------------------------------------------------------------
pos F[CAUFVD-IWERK] F[CAUFVD-BAUTL]+(1,0)
```

Listing 9.1 Script for Adjusting the Standard Screen for Transaction IW31

To activate the GuiXT script, click [icon] • GUIXT. The next time you call Transaction IW31, it will no longer have its standard layout. Instead, it will have the GuiXT-manipulated layout (see Figure 9.31).

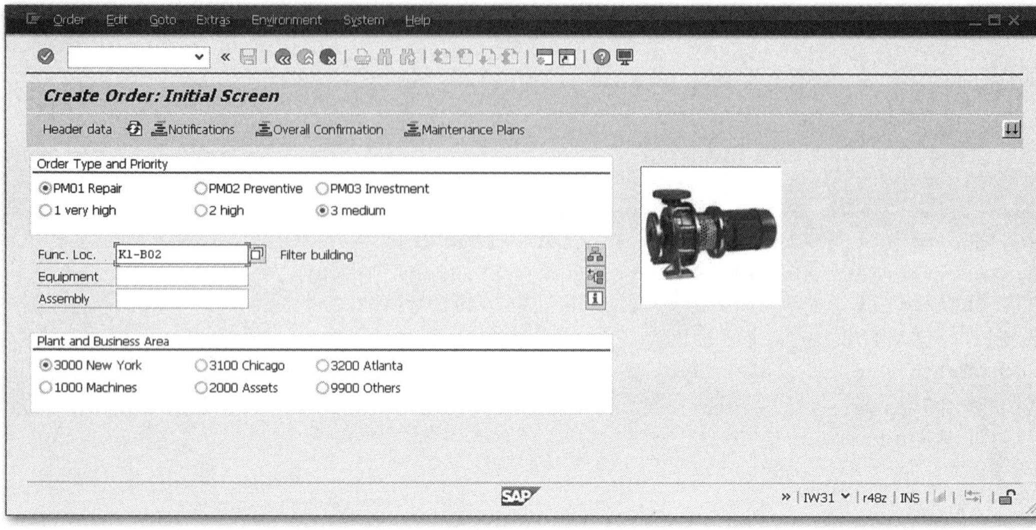

Figure 9.31 GuiXT: Manipulated Screen IW31

The Synactive website (*www.synactive.com*) provides further extensive information about GuiXT. Among other things, Synactive delivers a *PM console* for plant maintenance. This interface contains cockpit functions that provide a quick overview of the current order situation and enables you to access the most important transactions needed for your daily work. When you log on to the SAP system and call the PM console (e.g., by means of a function key or transaction), the system displays a cockpit that contains the following functions (see Figure 9.32):

- The *left block* (SEARCH) contains function keys for calling the most important transactions needed for your daily work (e.g., Transaction IH01 [Structural Display], Transaction IW38 [Order List], or Transaction IW41 [Completion Confirmation]).
- The *middle block* (MY ORDERS) provides an overview of your current order situation. It displays key figures and the number of orders in each status (e.g., overdue or confirmed [CNF]) and in each priority, both of which were assigned to the order when it was created.
- If you choose one of these pushbuttons, a list of corresponding orders is displayed in the *right block* (ORDERS) within the cockpit (e.g., which orders have priority 1 or 2 and the status OVERDUE assigned to them). You can use selection criteria such as the order type or functional location to restrict this list further.

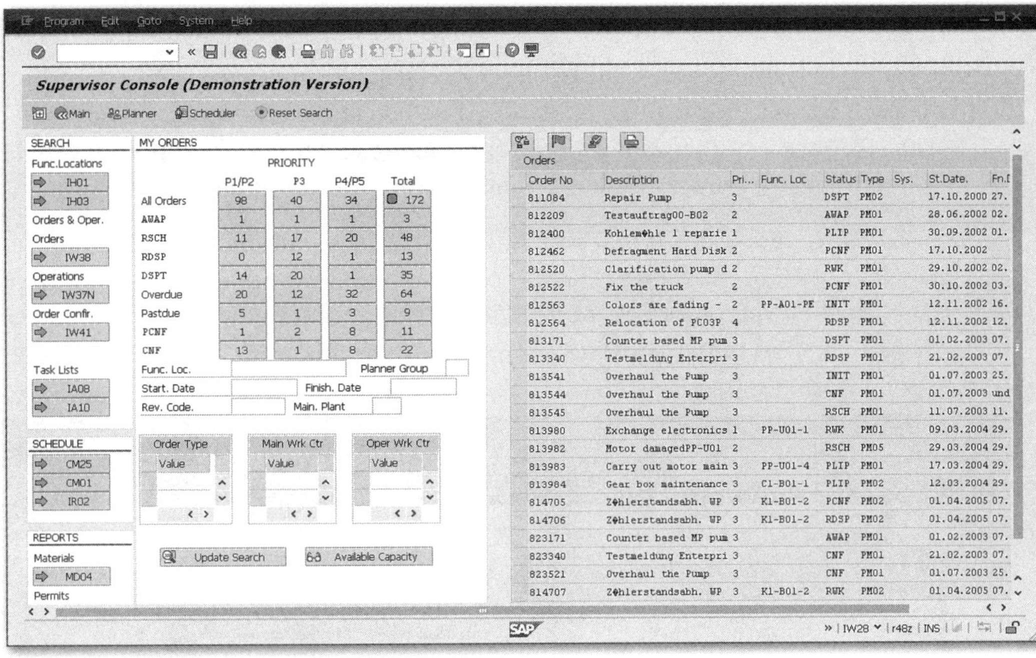

Figure 9.32 GuiXT: PM Console

The PM console is available in three variants, each of which has a slightly different appearance:

- For the maintenance planner
- For the workshop manager
- For the technician

You can use the relevant function keys to toggle between each variant.

[+]

You Can Use GuiXT to Adjust Your Screens

GuiXT can also help you adjust business transaction processing in plant maintenance to your requirements and to simplify and accelerate processing.

This now brings us to a very new technology, which was released by SAP in 2013 and can be used to change the layout of screens, namely SAP Screen Personas.

9.2.5 SAP Screen Personas

SAP Screen Personas Version 2.0 is currently available, and like GuiXT, it enables you to adjust and redesign the standard SAP screen templates to your own requirements. It comprises various functions for changing screens, thus supporting you in your endeavors to improve usability and increase user acceptance.

For example, a modified initial screen that contains the user's company logo and provides only those functions that he requires for daily work in the form of large pushbuttons and graphics (see Figure 9.33) will enjoy much greater user acceptance than the standard initial screen of the SAP access menu.

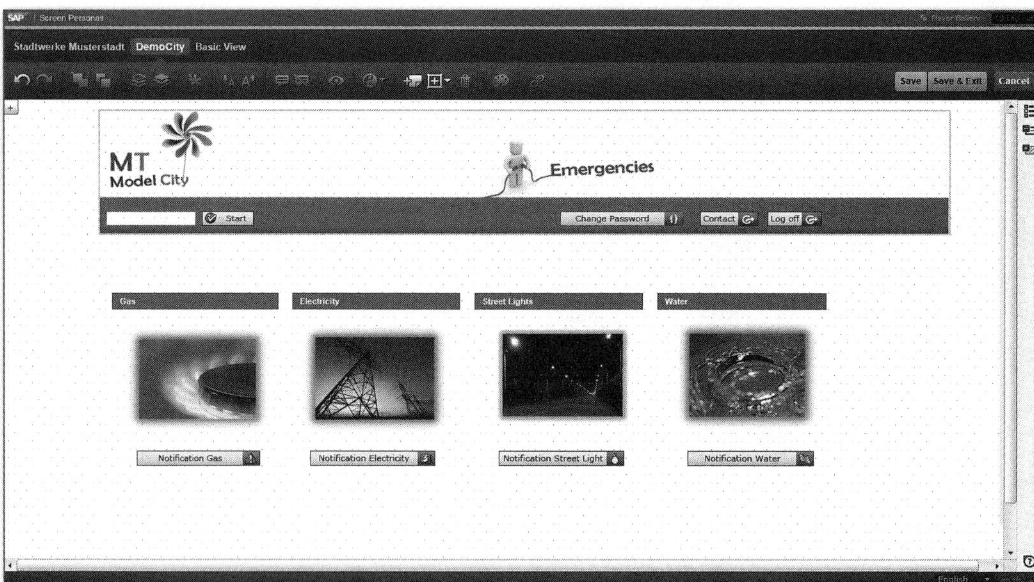

Figure 9.33 SAP Screen Personas: Modified Initial Screen

Such modified screens are known as *flavors* whereby a specific SAP transaction is personalized in a particular way. Note that a flavor is always linked to one transaction (e.g., Transaction IW31). However, you can create any number of flavors for each transaction. After you finish creating a flavor, you can release it for certain users or for all users of your SAP system.

You use a *User Interface Editor* (UI Editor) to edit the screens themselves. Here, you can edit all of the control elements of a screen:

- Titles
- Radio buttons
- Checkboxes
- Function keys
- Input fields
- Dropdown lists
- Field groups
- Texts

Then, depending on the control element, you have the following options (note, however, that not all of these options apply to all control elements):

- You can completely show or hide individual control elements or entire groups of control elements.
- You can also define new control elements (e.g., assign a URL to a pushbutton).
- You can use drag and drop to move individual control elements to the positions that you want them to occupy.
- You can change the size of the control elements.
- You can assign a new quick info to the control element.
- You can change the default value of a control element (e.g., by prepopulating an input field with a particular value).
- You can adjust the title and text of a control element.
- You can use sticky notes to add additional information to the relevant control element. Unlike quick infos, sticky notes are only displayed when the control element is selected.

Figure 9.34 shows a modified notification creation screen for Transaction IW21.

If you use the initial screen from Figure 9.33 and the modified notification creation screen from Figure 9.34, you reduce the number of clicks required by the user to process a notification from 24 clicks to just 10 clicks, all thanks to SAP Screen Personas.

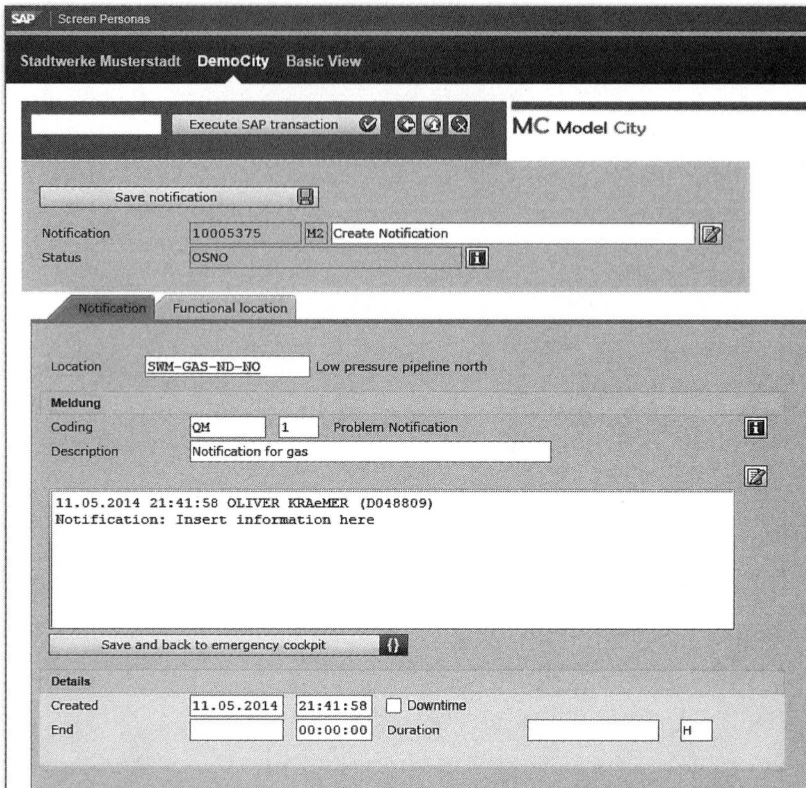

Figure 9.34 SAP Screen Personas: Modified Notification Creation Screen

In addition to individual control elements, you can also change the complete *theme* of the flavor:

- You can define your own background screens (e.g., insert logos, display context-dependent images, and define borders and color schemes).
- You can adjust the following standard SAP themes: Corbu, Tradeshow, High Contrast Black, and Blue Crystal.
- You can also change the color scheme for Corbu and Blue Crystal.

You can also integrate Google Maps, which enables you to display the location of the technical objects or orders if they have geodata (see Figure 9.35).

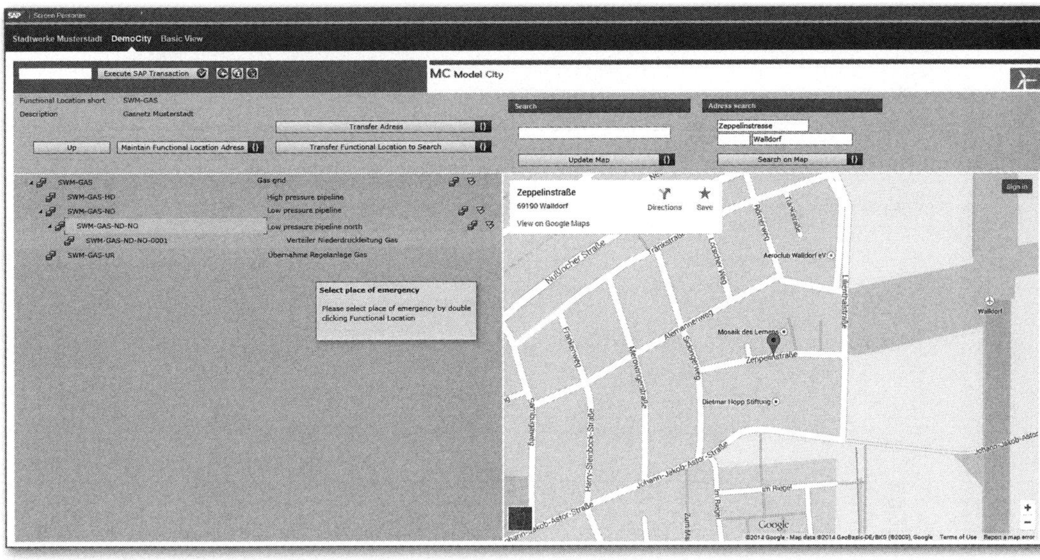

Figure 9.35 SAP Screen Personas: Modified Structural Display

Further information is available at *www.sapscreenpersonas.com* and *http://help.sap.com/personas*.

Let's now take a look at the options available within category 3 (IT with programming).

9.3 Category 3: The IT Department's Programming Options

In this section, we'll discuss IT's options for category 3.

9.3.1 Upstream Transactions

When processing business transactions, it's usually necessary to call several transactions in succession. Furthermore, within each transaction, the fields to be entered are distributed across several screens.

The basic idea of upstream transactions is to create a custom transaction containing only one or a few screens. This transaction calls the original SAP transactions in the background and transfers the data. Alternatively, the custom transaction can be used as a cockpit for controlling the original transactions and thus simplifying the process.

Let's explain this further using a real-life example: The business process for calibrating test/measurement equipment is a highly complex process that requires several SAP transactions to be called in succession. In the case of an automotive supplier who needed to manage more than 20,000 pieces of measurement equipment per plant, the standard processes were unmanageable. Therefore, the upstream Transaction ZMV01 (see Figure 9.36) was created to do the following:

- The measurement equipment is issued from the warehouse and installed on a functional location.
- The measurement equipment on the functional location is dismantled and returned to the warehouse.
- The inspection order is generated from the maintenance plan.
- The measurement results are recorded.
- The usage decision is made.
- The inspection order is confirmed.

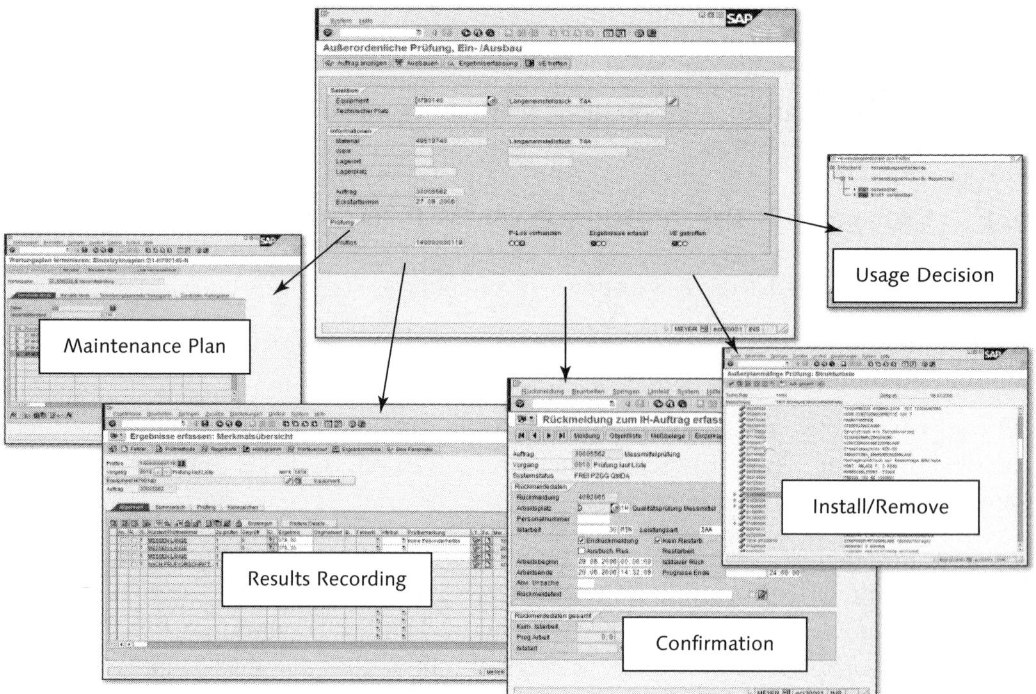

Figure 9.36 Upstream Transaction

[+]

The Programming Effort Associated with Upstream Transactions Is Very Worthwhile

An upstream transaction should substantially accelerate business transaction processing. Otherwise, it has failed in its purpose.

BAPI technology is frequently used to develop upstream transactions and other custom developments.

9.3.2 BAPIs

To simplify the development effort associated with upstream transactions and other custom developments, SAP provides a range of *Business Application Programming Interfaces* (BAPIs). BAPIs can be used for the following purposes:

- To connect SAP systems to the Internet
- To enable SAP components to communicate with one another
- To develop upstream transactions
- To connect third-party software and legacy systems to SAP systems
- To enable PC programs to be used as a *front end* for SAP systems
- To develop web applications that will communicate with the SAP system
- To enable workflow applications to communicate across system boundaries
- To enable WebFlow applications to communicate using the Internet

You can call the BAPI Explorer in Transaction BAPI (see Figure 9.37). Navigating through the hierarchy is a relatively simple way to see an overview of all available BAPIs.

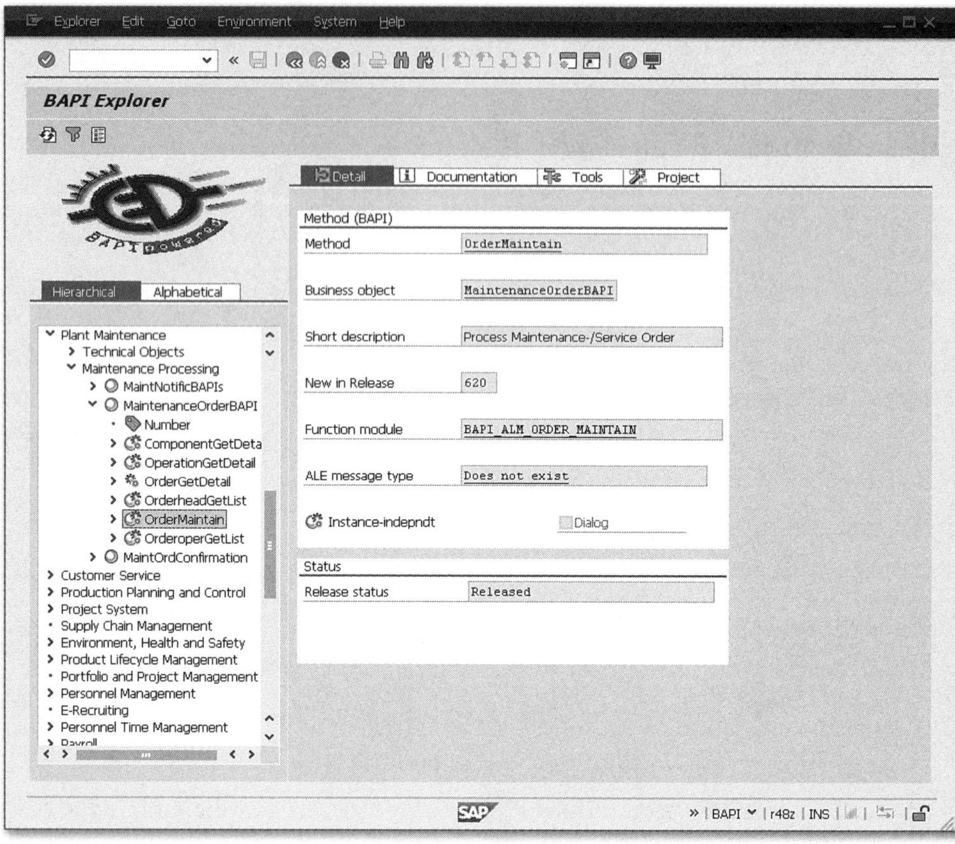

Figure 9.37 BAPI Explorer

SAP provides the following BAPIs for plant maintenance:

- **For equipment**
  The BAPI `Equipment` with methods such as `Create` to create equipment master records or `Dismantle` to dismantle equipment from functional locations or equipment hierarchies.

- **For functional locations**
  The BAPI `FunctionalLocation` with methods such as `InheritChange` to change the inheritance parameters or `StrucAssign` to replace the assignment of superior functional locations.

- **For notifications**
  The BAPI `MaintNotificBAPIs` with methods such as `NotifCreate` to create notifications or `NotifListPlangroup` to create notification lists for each planner group.
- **For orders**
  The BAPI `MaintenanceOrderBAPI` with methods such as `OrderMaintain` to create and change orders.
- **For order confirmations**
  The BAPI `MaintOrdConfirmation` with methods such as `ConfCreate` to create completion confirmations.
- **For material master records**
  The BAPI `Material` with methods such as `Edit` to change material master data or `GetMRPList` to generate the MRP list for a material.
- **For bills of material**
  The BAPI `MaterialBOM` with methods such as `CreateBomGroup` to general material BOMs.

A complete list of methods is provided in the Appendix in Section A.3. Unfortunately, there are no BAPIs for other objects such as maintenance task lists or maintenance plans.

In general, you also use BAPIs to develop web applications that will communicate with SAP systems.

9.3.3 Web Interface

As an example of a web application, we'll now introduce you to the *EAM App*, which was developed by the Institute for Design and Information Systems (IDIS) at the Würzburg-Schweinfurt University of Applied Sciences, Germany, for a well-known customer from the automotive industry.[2]

The EAM App's main functions are displayed in the main menu immediately after you log on to the app (see Figure 9.38):

2 I would like to thank Florian Wolf, Tobias Schlereth, and Daniel Schwarz, three research associates at the Institute for Design and Information Systems (IDIS), for developing the EAM App. Likewise, I would like to thank Johannes Bähr, Sebastian Büchse, and Michael Landwehr, three student assistants at the same institute, for enhancing the app and adapting it to higher education systems.

- The Express Confirmation allows you to create completion confirmations for one or more employees involved in an existing order or order operation, all of which can be done on just one screen. Here, you can also perform planned and unplanned material withdrawals as well as enter downtimes, completion confirmation texts, and causes of damage.
- The Overall Confirmation function enhances the Express Confirmation function to include the recording of measurement readings and counter readings. To provide you with a better overview, the individual functions are spread across several tab pages.
- The Technical Confirmation function contains only the technical aspects of the completion confirmation (e.g., downtimes or causes of damage). It doesn't contain any time or material confirmations.

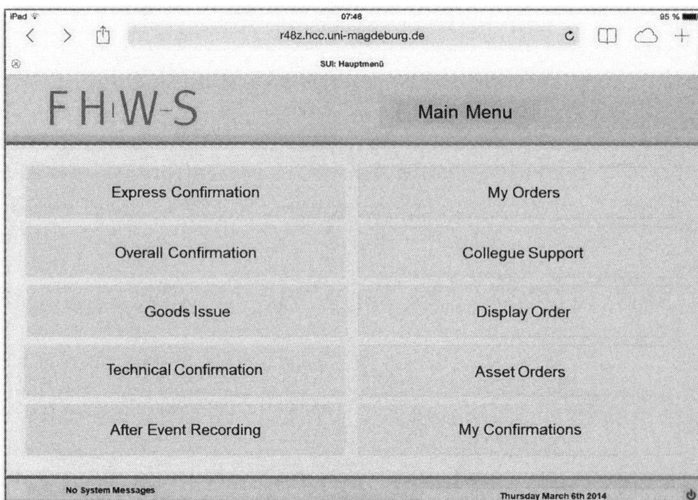

Figure 9.38 EAM App: Main Menu

- The After Event Recording function (see also Figure 9.39) is used to record work that has already been carried out. For example, you can enter an order for a technical object (functional location or piece of equipment) and perform a time confirmation for one or more employees. You can also perform planned and unplanned material withdrawals as well as specify downtimes, completion confirmation texts, and causes of damage. The following process is associated with transferring such data to the SAP system: Notification created → order created → order released → material withdrawal posted → time confirmation posted →

technical completion confirmation performed → order and notification completed.

- My Orders contains a list of operations currently open or in process for the technician.
- The Colleague Support function contains a list of operations currently open for the technician's colleagues (i.e., the operations of those personnel numbers assigned to the same work center as the technician). The technician can then assign his colleagues' orders (e.g., if a colleague is ill or hindered in his daily work for some other reason).
- The Display Order function makes it possible to display a particular order according to its order number.

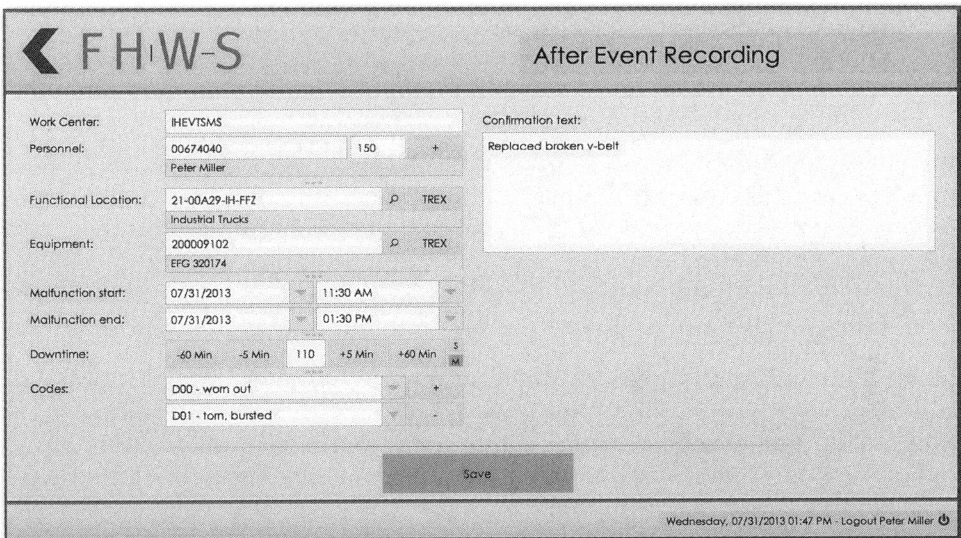

Figure 9.39 EAM App: After Event Recording

- The Asset Orders function enables you to view a list display of all open and completed operations for a functional location or a piece of equipment.
- The My Confirmations function contains the technician's completion confirmations for a fixed period (e.g., this week) or for any time interval of your choice.

In addition to pure data entry, these main functions also contain additional subfunctions such as the following:

- Display details
- Print
- Create follow-up order
- Technically complete order

In addition to these functional properties, the following technical properties of the EAM App are also worth noting:

- The EAM App was completely developed as an online scenario. In other words, it has a permanent connection via LAN, WLAN, or GSM (Global System for Mobile communications) to the backend system and is able to immediately detect wrongly entered data and output an error message to that effect.
- A BAPI is used to check the validity of the data entered (e.g., whether a valid equipment number or order number was entered).
- A BAPI is used to transfer the data to the SAP system and to post it there (e.g., the `OrderMaintain` BAPI).
- Because the EAM App was developed for HTML5, you can use it on an iPad, iPhone, and desktop PCs, industry PCs, Android smartphones, and Android tablets.
- The EAM App can be fully operated using a mouse and keyboard or by means of touchscreens (e.g., in the case of an iPad).

[+] **Web Applications Enjoy High User Acceptance**

Almost all users are familiar with web applications or apps for their private iPhone, iPad, or smartphone. For this reason, apps are also very easy to learn and use in a professional capacity and are therefore very popular among users.

9.3.4 Customer Exit

You can use *customer exits* to add your own functions to standard SAP applications without modifying the original SAP programs. In standard applications, SAP creates customer exits for certain programs, screens, and menus. Initially, these exits don't have any function. Instead, they serve as a predefined entry and exit point for adding your own additional functions to the SAP system.

You use Transaction SMOD to call customer exits. You can then view them by entering the following:

- Technical objects: ITOB*, IEQM*, ILOM*, IHCL*, CCM*
- Measuring points/counters: IMRC*
- Warranties: BG*
- Task lists: IAIH*
- Maintenance plans: IPRM*, CI*
- Notifications: QQMA*
- Orders: IWO*, CNEX*, COZF*, IREV*
- Capacity planning: COI*, CYPP*
- Completion confirmations: CMFU*, CONFPM*
- Cross-Application Time Sheets (CATS): CATS*
- Information system: MCI*
- Data transfers: IBIP*
- Graphic modules: IMSM*

By way of example, the following are five applications in which data entry was simplified and user acceptance increased significantly as a result:

- **Transfer characteristics**
  Customer exit IHCL0001 allows you to transfer the characteristic valuation from a material master record to the characteristic valuation of a piece of equipment. If you often add new equipment with reference to a material number, you can simplify the essential maintenance of the classification data by storing information such as performance, type specification, or other technical data once as a classification in the material master and then transferring these characteristics as default values to your new piece of equipment. Ideally, you then no longer have to maintain the classification.
- **Right plant, right storage location**
  Customer exit CNEX0027, which, strictly speaking, is part of SAP Project System (SAP PS) but can also be used for plant maintenance, supports you in specifying the right plant and the right storage location during material planning. This may be useful, for example, when a spare part is stored in only one plant that all other plants can access. If, within this customer exit, you then program a smart

search strategy for your enterprise, you no longer have to worry about having the right plant in material planning for the order.

- **Right task list**
  A significant improvement is also provided by customer exits IWO10021 and IWO20001. The combination of these two customer exits, for example, ensures that the right task list is derived from the damage or cause code in a malfunction report. When the order is created from the malfunction report, the correct operations are already listed in the order. This eliminates the need to find the right task list and include it in the order and therefore accelerates order processing for standardized processes in particular.
- **Right data in notification and order**
  Even though they don't make data entry any faster, customer exits that can run customized data checks when saving orders (IWO10009) or notifications (QQMA0014) help to eliminate the need for data corrections. In the notification, for example, you can immediately check whether certain combinations of breakdown and cause codes make sense. These checks help inexperienced users allay their fears about operating the system.
- **Right data in purchase requisitions**
  Customer exits COZF0001 and COZF0002 allow you to automatically prepopulate certain fields for purchase requisitions (PReq) generated within the order according to your own specifications. For example, you can prepopulate the REQUESTER field with the name of the user logged on to the system, or you can automatically enter your department acronym in the REQUIREMENT NUMBER field. This may not sound like much progress, but if a user has to enter the same department acronym 35 times for 35 nonstock items, it's easy to see why optimization becomes a priority.

The following are some more examples from professional experiences with customer projects:

- You can use customer exit IWO10004 to implement order completion checks. For example, one customer wanted to prevent orders from being technically completed if any open purchase requisitions or unconfirmed operations exist or if only some of the objects in the object list have a processing indicator.
- You can use customer exit CONFPM01 to determine your own default values when confirming an order. For example, a customer wanted to prevent the activities of certain employees from being allocated on a cost basis (indicator in

the HR master record). Now, if one of these employees enters a completion confirmation, the system proposes an activity type without a valuation.

- You can use customer exit IWO10027 to generate your own settlement rule. If you use object lists, you frequently want to proportionally allocate the resulting order costs to the cost centers involved (see Figure 9.40). You can program this customer exit in such a way that the cost centers associated with the equipment are determined from the object list and that the corresponding entries are configured in the settlement rule and assigned percentages.

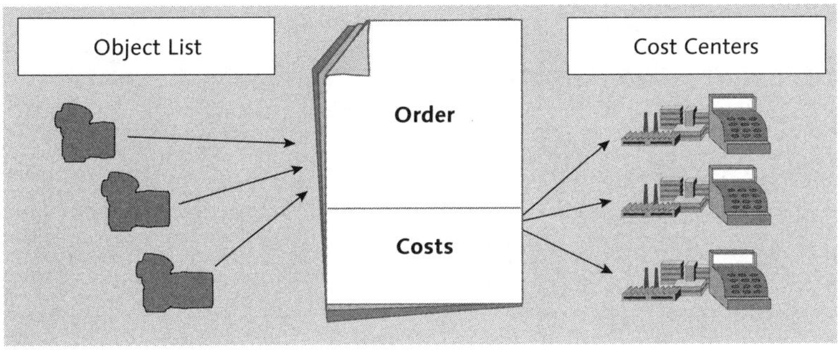

Figure 9.40 Customer Exit: Settlement Rule

- You can use customer exit IWO10012 to control priority handling within the order. For example, one customer wanted the system to use the ABC indicator for equipment to automatically determine the priority of an order (e.g., priority 1 is automatically set for "A" equipment). This customer exit is suitable for this purpose.
- One customer wanted to connect his own solution database and also wanted notifications to contain, where applicable, a reference to the solution number. Here, customer exit QQMA0001 was used to define additional fields in the notification header.

These examples clearly show the various ways in which customer exits can be used, and, ultimately, there are no limits to your imagination. More than 100 customer exits are provided for SAP EAM. A complete list of customer exits for plant maintenance is provided in the Appendix in Section A.4.

[+]

Customer Exits Simplify the Standard System and Make It More Secure

Customer exits enable you to adjust business transaction processing in plant maintenance to your requirements and to simplify and accelerate processing. However, you need to give considerable thought to whether custom programming really has the desired effect and increases user acceptance.

For the sake of completeness, we now introduce the other programming techniques for enhancing or changing the standard SAP system without having to implement a modification.

9.3.5 BAdIs and Enhancement Points

Business Add-ins (BAdIs) are used to create predefined enhancement options in SAP components. They therefore pursue the same goal as customer exits, namely adding more functions to the standard SAP system without having to implement a modification. Unlike customer exits, however, BAdIs are based on ABAP objects.

In contrast to customer exits, which support a two-tier system landscape (SAP and the customer), BAdIs support a multilevel system landscape (SAP, country versions, industry solutions, partner, customer, etc.). You can create BAdI definitions and implementations at each of these levels within the system landscape. In other words, you can implement BAdIs several times, while customer exits can only be applied once.

You use Transaction SE19, the BAdI Builder, to create BAdIs.

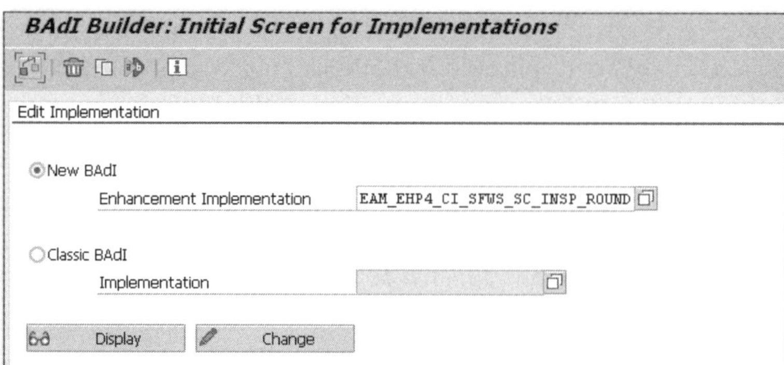

Figure 9.41 BAdI Builder

There are two types of BAdIs (see Figure 9.41):

- Classic BAdIs
- New BAdIs (also known as *enhancement points*)

The following is a list of examples that demonstrate when new BAdIs (enhancement points) can be used in SAP EAM:

- **EAM_EHP4_CI_SFWS_SC_LIST_ENH**
  To enhance the lists used in plant maintenance (e.g., to display maintenance plans, maintenance items, and maintenance packages in multilevel lists).
- **EAM_EHP4_CI_SFWS_SC_INSP_ROUND**
  To enhance inspection rounds (e.g., to activate the production resource/tool [PRT] as a measuring point).
- **EAM_EHP4_INSPECTION_ROUNDS**
  Inspections for measuring point as a PRT.
- **BADI_EAM_ITOBBAPI_SHIFTNOTE**
  Adding new fields to the BAPIs for the technical object.
- **EAM_WKCTR_SYNC_IMPL**
  Synchronization of work center for maintenance order – notification.
- **EAM_WS_ORDER_RELEASE_IMPL**
  For security-relevant enhancements when releasing orders.
- **EAM_EHP4_CI_SFWS_SC_AUTH_ENH**
  Additional authorization check for orders.
- **EAML_BADI_EAM_STRUCTURE_LIST**
  To add linear data to structure lists.
- **EAMWS_ORDER_CHECKLIST**
  Status management for the list of relevant risks in SAP EAM.
- **COCF_ES_SN_LIST**
  To define your own selection options in the list of shift notes.

The following is a list of examples that demonstrate when classic BAdIs can be used in plant maintenance:

- **EAML_PM_NOTIFICATION**
  Notifications with linear data at the header and item level.
- **IMPL_EAM_DOC_PMAUFK**
  Order documents.

- **IMPL_EAM_DOC_PMPLKO**
  Task list documents.
- **IQS_MASS_CHANGE**
  Mass change to notifications.
- **IWO_MASS_CHANGE**
  Mass change to orders.
- **PLM_CATALOG_OCI**
  Calling the catalog selection via the Open Catalog Interface (OCI).

[+]

BAdIs and Enhancement Points Simplify the Standard System and Make it More Secure

As is the case with customer exits, you can program BAdIs and enhancement points in such a way that you can adjust business transaction processing in plant maintenance to your requirements and thereby simplify and accelerate processing. However, you need to give considerable thought to whether custom programming really has the desired effect and increases user acceptance.

9.3.6 Workflow

SAP Business Workflow enables you to define business processes that aren't yet mapped in the system. These can range from simple release and approval procedures to more complex business processes such as creating a material master and coordinating the various departments involved. SAP Business Workflow is at its most efficient when a large number of workflows need to be processed repeatedly, or the business process requires a large number of processors in a precisely defined sequence.

Workflows are called and configured in the SAP menu under TOOLS • BUSINESS WORKFLOW. Figure 9.42 shows, for example, Transaction SWDD, the Workflow Builder.

From a maintenance processing perspective, the following workflows are particularly relevant:

- WS 00200065 PM: Process Notification
- WS 00200095 PM: Put Notification in Process
- WS 00200075 PM: Complete Notification

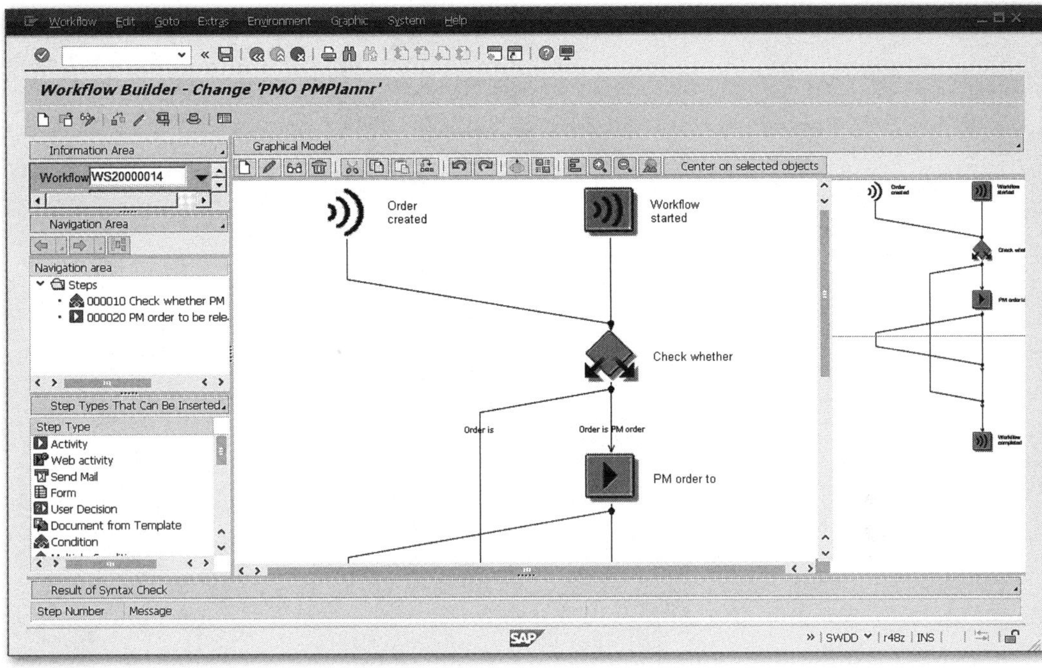

Figure 9.42 Business Workflow: Workflow Builder

- WS 00200074 PM: Complete Task
- WS 20000014 PMO (Maintenance Planner): Release Order, Technically Complete Order, among others
- WS 20000021 PMO (Order Creator): Release Order, Technically Complete Order, among others
- WS 20000031 PMO: Confirm Operation

The purpose of these scenarios is to support you in processing, monitoring, and completing newly created notifications and orders. For example, you can inform the person who created the order (order creator) when his order has been released or technically completed.

If you are entered in the workflow definition as a processor, you obtain a *work item* in your BUSINESS WORKPLACE screen (Transaction SWBP, see Figure 9.43). You can execute the work item directly (PROCESS NOTIFICATION, in this case).

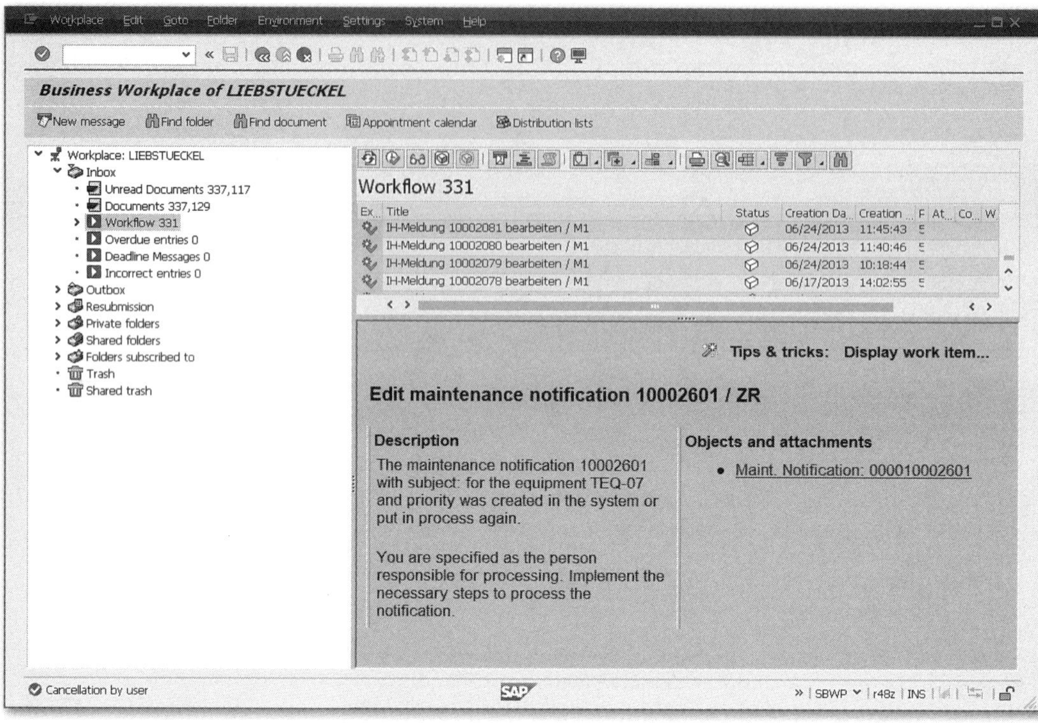

Figure 9.43 SAP Business Workflow: Work Item

Use the following Customizing functions to maintain the processor of maintenance notifications and maintenance orders:

- Set Workflow for Maintenance Notifications
  Customizing path: Plant Maintenance and Customer Service • Maintenance and Service Processing • Maintenance and Service Notifications • Notification Processing • Set Workflow for Maintenance Notifications.
- Set Workflow for Orders
  Customizing path: Plant Maintenance and Customer Service • Maintenance and Service Processing • Maintenance and Service Orders • Functions and Settings for Order Types • Set Workflow for Orders.

[+]

Recommendations for Notifications and Order Workflows

When you set up and activate workflows, your SAP system becomes an active system with a push mechanism that enables you to address all users who are to execute actions in notification and order processing but otherwise have very little to do with the SAP system (e.g., release an order or complete a task).

In particular, you can simplify task processing considerably by forwarding work items to the usual email client of the recipient (e.g., Outlook), thereby giving recipients the option to use a smartphone to process a task.

From a maintenance perspective, there is one more workflow that you can use to synchronize equipment master data in SAP EAM with the asset data in Asset Accounting, namely workflow WS00400037. For example, you can configure the system to automatically create an asset master record in the background when you create an equipment master record, and vice versa. The requisite workflow already exists in the system. All you need to do is activate some Customizing settings.

Specify Conditions for Synchronization of Master Data

You use this Customizing function to define the framework conditions in relation to whether and, if so, how equipment and assets are to be synchronized.

Prerequisites

You must define the asset class, equipment category, and object type beforehand.

Customizing Path

FINANCIAL ACCOUNTING (NEW) • ASSET ACCOUNTING • MASTER DATA • AUTOMATIC CREATION OF EQUIPMENT MASTER RECORDS • SPECIFY CONDITIONS FOR SYNCHRONIZATION OF MASTER DATA

Settings or Recommended Settings

Use this Customizing function to establish a link between the ASSET CLASS and the EQUIPCATEGORY or OBJECT TYPE (see Figure 9.44). In other words, define which equipment category or order type is to be assigned when an asset master record is created with a specific asset class.

Then, you can use EDIT ASSET MASTER RECORD and EDIT EQUIPMENT MASTER RECORD to regulate how the master records are to be synchronized in create and change modes:

- NO SYNCHRONIZATION
  When you create an asset master record, the system doesn't propose any equipment. If you manually create the related pieces of equipment at a later stage, the system doesn't synchronize the master data.
- DIRECT SYNCHRONIZATION AFTER SAVING
  When you create an asset master record, the system proposes a piece of equipment, and, when you save the asset, the system creates the corresponding equipment master record (and the assignment between both master records) immediately. Subsequent changes to the asset master record can be synchronized in the equipment master record (and vice versa).

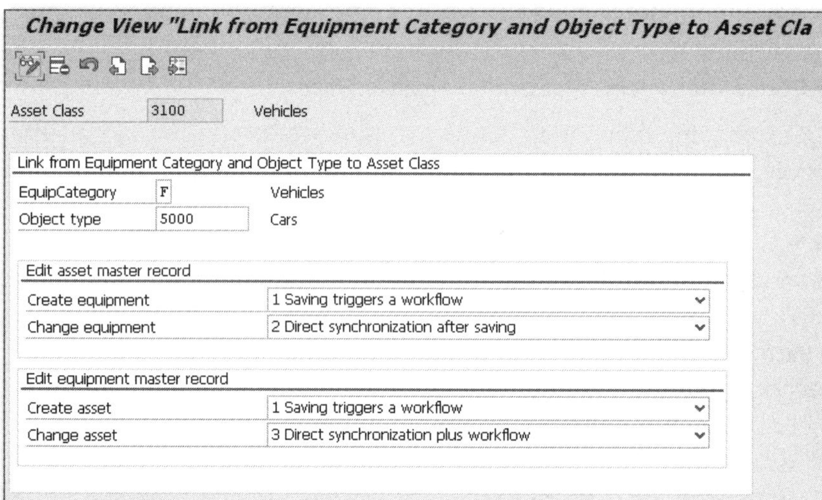

Figure 9.44 Workflow: Synchronization of Equipment and Asset

- SAVING TRIGGERS A WORKFLOW
  When you create an asset master record, the system proposes a piece of equipment, and, when you save the asset, the system triggers a specific workflow event. Depending on the attributes of the assigned workflow (e.g., a workflow to notify the person responsible), the system generates the associated equipment master record when you execute this workflow. Subsequent changes to the asset master record can be synchronized in the equipment master record (and vice versa).

- Direct synchronization plus workflow
  When you create an asset master record, the system proposes a piece of equipment, and, when you save the asset, the system generates the corresponding equipment master record (and assigns it to the asset) immediately. It also triggers a specific workflow event.

Assign Master Data Fields of Assets and Equipment

You use this Customizing function to assign, to the fields in the equipment master record, those fields in the asset master record that are to be compared when the asset master records and equipment master records are synchronized.

Prerequisites

You must activate and define the synchronization of asset master data and equipment master data beforehand.

Customizing Path

Financial Accounting (New) • Asset Accounting • Master Data • Automatic Creation of Equipment Master Records • Assign Master Data Fields of Assets and Equipment

Settings or Recommended Settings

All of the usual fields are contained in the standard SAP delivery (see Figure 9.45).

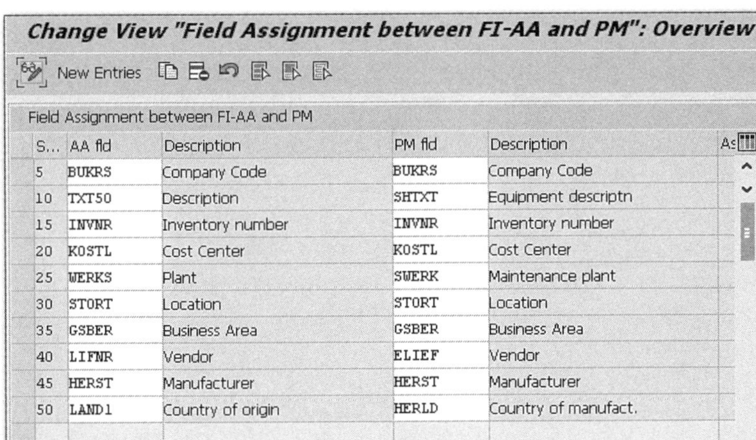

Change View "Field Assignment between FI-AA and PM": Overview

New Entries

Field Assignment between FI-AA and PM

S...	AA fld	Description	PM fld	Description	As
5	BUKRS	Company Code	BUKRS	Company Code	
10	TXT50	Description	SHTXT	Equipment descriptn	
15	INVNR	Inventory number	INVNR	Inventory number	
20	KOSTL	Cost Center	KOSTL	Cost Center	
25	WERKS	Plant	SWERK	Maintenance plant	
30	STORT	Location	STORT	Location	
35	GSBER	Business Area	GSBER	Business Area	
40	LIFNR	Vendor	ELIEF	Vendor	
45	HERST	Manufacturer	HERST	Manufacturer	
50	LAND1	Country of origin	HERLD	Country of manufact.	

Figure 9.45 Workflow: Data Synchronization

You can make other assignments by selecting additional fields in Asset Accounting and then assigning an additional field in plant maintenance to these fields. The fields available in Asset Accounting are contained in the DDIC structure SAAPM1, while the fields available in plant maintenance are contained in the DDIC structure SAAPM2.

If you want to make a more extensive field assignment, you can use customer exit AAPM0001 for this purpose. You can also use this customer exit to derive different asset classes from the fields in the equipment master record and different equipment categories from the fields in the asset master record.

[+]

Recommendations for the Asset Workflow

The workflow for synchronizing equipment and assets must be viewed somewhat critically because maintenance workers and accountants have a completely different point of view in relation to assets. In general, maintenance workers prefer a "narrow" structure in which the technical aspects of the assets are central, while accountants prefer a "broad" structure in which the taxation aspects of the assets are crucial.

Therefore, achieving a 1:1 relationship between a piece of equipment and an asset is rather unlikely.

A Project Plans and Overviews

A.1 Functional Scope of SAP EAM and Assignment of Priorities

This section contains tables that you can fill with your own dates and comments for your project planning.

You can also download this file from *www.sap-press.com*. From here, navigate to the bonus area on this book's individual page.

Legend:

- A = must be implemented in the first phase
- B = will be implemented in a later expansion phase
- C = not required, according to current knowledge

Asset Management		A	B	C
Functional locations				
Reference functional locations				
Alternative labeling				
Linear objects				
Equipment				
Serial numbers				
Classification				
BOMs				
Change management				
Equipment hierarchies				
Variant configuration				
Documents				

Asset Management		A	B	C
Permits				
Warranties				
Production resources/ tools				
Partners				
Counters				
Measuring points				
Equipment and asset synchronization				

Object	Function	A	B	C
Notification	**User status**			
	Reference objects			
	Priorities			
	Partners			
	Telephone integration			
	Paging			
	Addresses			
	Object parts			
	Damage			
	Causes of damage			
	Tasks			
	Activities			
	Notification items			
	Classification			
	Print			
	Fax			

Object	Function	A	B	C
	Download			
	Breakdowns			
	Permits			
	Response time monitoring			
	Revisions			
	Solution database			
Order	**Reference objects**			
	User status			
	Priorities			
	Partners			
	Telephone integration			
	Paging			
	Addresses			
	Print			
	Fax			
	Download			
	Permits			
	Operations			
	Scheduling			
	Relationships			
	Capacity planning			
	Capacity availability check			
	Reservation of stock material			
	Stock material availability check			
	Ordering of non-stock material			
	Catalog integration (Internet and intranet catalogs, vendor catalogs)			

Object	Function	A	B	C
	Estimated costs			
	Planned/actual costing			
	Order budgets			
	Object list			
	Production resources/tools			
	Production resources/tools availability check			
	Suborders			
	Utilization of production capacities			
Completion confirmation	**Time confirmations**			
	Technical completion confirmations			
	Goods receipts			
	Material withdrawals			
	Overhead costing			
	Order settlement			
External processing	**External processing via service specifications**			
	External processing as a purchase order			
	External processing via work centers			
	Revisions			
Preventive maintenance	**General maintenance task lists**			
	Equipment task lists			
	Functional location task lists			
	Maintenance strategies			
	Time-based maintenance plans			
	Performance-based maintenance plans			

Object	Function	A	B	C
	Single cycle plans			
	Strategy plans			
	Basic multiple counter plans			
	Enhanced multiple counter plans			
	Order maintenance call object			
	Notification maintenance call object			
	Inspection lot maintenance call object			
	"Service entry sheet" maintenance call object			
	Simulation of capacity utilization			
	Simulation of planned costs			
	Automatic deadline monitoring			
Condition-based maintenance	**PM-PCS interface**			
Refurbishment	**Refurbishment of serial numbers**			
	Refurbishment of material			
	Settlement based on standard price			
	Settlement based on moving average price			
Calibration of test/ measurement equipment	**Test/measurement equipment as equipment**			
	Inspection plans			
	Inspection maintenance plans			
	Results recording			
	Completion confirmation			
	Usage decision			

Object	Function	A	B	C
Project-based maintenance	**WBS elements**			
	Networks			
	Manual assignment			
	Automatic assignment			
	Maintenance Event Builder			

A.2 Authorization Objects in SAP EAM

Authorization Object	Description	Checked Fields
I_AER	Retroactive order entry	▸ Maintenance planning plant ▸ Order type
I_ALM_ME	Mobile Asset Management	Action
I_AUART	PM: Order type	▸ Maintenance planning plant ▸ Order type
I_BEGRP	PM: Authorization group	▸ Transaction code ▸ Authorization group for technical object
I_BETRVORG	PM: Business transaction	Business transaction
I_CONFSTOR	PM: Mass confirmation reversal	Confirmation reversal
I_ILOA	Changing location and account assignment data in orders	▸ Maintenance planning plant ▸ Order type
I_INGRP	PM: Maintenance planner group	▸ Transaction code ▸ Maintenance planning plant ▸ Planner group for customer service and plant maintenance
I_IWERK	PM: Maintenance planning plant	▸ Transaction code ▸ Maintenance planning plant

Authorization Object	Description	Checked Fields
I_KOSTL	PM: Cost centers	▶ Transaction code ▶ Controlling area ▶ Cost center
I_MASS	PM: Mass data change	▶ Mass data change for object types
I_QMEL	PM /QM: Notification types	▶ Transaction code ▶ Notification type
I_ROUT	PM: Task list	Action
I_ROUT1	PM: Task lists by maintenance planning plant, work scheduler, status	▶ Transaction code ▶ Maintenance planning plant ▶ Responsible planner group/ department ▶ Status of task list
I_SOGEN	PM: Permit	▶ Maintenance plant ▶ Permit key
I_SWERK	PM: Maintenance plant	▶ Transaction code ▶ Maintenance plant
I_TCODE	PM: Transaction code	Transaction code
I_VORG_MEL	PM /QM: Business transaction for notifications	▶ Notification type ▶ Business transaction
I_VORG_MP	PM: Business transaction for maintenance planning	▶ Maintenance plan category ▶ Business transaction
I_VORG_ORD	PM: Business transaction for orders	▶ Order type ▶ Business transaction
I_WPS_MEB	Revision planning: Maintenance Event Builder	MEB screen areas

Authorization Object	Description	Checked Fields
I_WPS_REV	Revision planning: Authorization object revision	▸ Revision type ▸ Work center ▸ Plant ▸ Revision operations for WPS-process

A.3 BAPIs for Plant Maintenance

Object	Method	Content
EQUIPMENT	CHANGE	You can use this method to change an existing piece of equipment.
	CREATE	You can use this method to create a new piece of equipment.
	CREATEBYREFERENCE	You can use this method to create a new piece of equipment with reference to an existing piece of template equipment.
	DISMANTLE	You can use this method to dismantle a piece of equipment from a functional location or a superior piece of equipment.
	DISMANTLE_IE4N	You can use this method to dismantle a piece of equipment from its superior piece of equipment or functional location. The piece of equipment must be a serial number. During dismantling, the material is transferred to the warehouse (known as a goods movement). A maintenance notification is created to document the dismantling of the piece of equipment.

Object	Method	Content
	EXCHANGE	You use this method to exchange two pieces of equipment at their installation locations. The pieces of equipment must be serial numbers. Each time a piece of equipment is dismantled or installed, a maintenance notification is created to document the dismantling or installation of the piece of equipment. Consequently, four maintenance notifications are created.
	GETCATALOGPROFILE	You can use this method to determine the catalog profile for a piece of equipment on the current date.
	GETDETAIL1	You can use this method to read an existing piece of equipment.
	GETLIST	You can use this method to select a list of equipment according to predefined selection criteria.
	GETSTATUS	You can use this method to read the system status and user status of an existing piece of equipment.
	INSTALL	You can use this method to install a piece of equipment in a functional location or to install a superior piece of equipment.
	INSTALLNONSERIALIZEDMATERIAL	You can use this method to install a non-serialized material in the BOM header of a piece of equipment without requiring a user action in Transaction IE4N. A maintenance notification is created to document the installation.

Object	Method	Content
	INSTALL_IE4N	You can use this method to install a piece of equipment on another piece of equipment or in a functional location. The piece of equipment must be a serial number. During installation, the material is transferred from the warehouse (known as a goods movement). A maintenance notification is created to document the installation of the piece of equipment.
	REMOVENONSERIALIZEDMATERIAL	You can use this method to remove a nonserialized material from the BOM header of a piece of equipment without requiring a user action in Transaction IE4N. During dismantling, the material is transferred to the warehouse (known as a goods movement).
	UPDATE	You use this method to change a piece of equipment.
FUNCTIONAL LOCATION	CHANGE	You can use this method to change the master data of a single *functional location*.
	CREATE	You can use this method to create a single *functional location*.
	GETDETAIL	You can use this method to read the master data of a single *functional location*.
	GETLIST	You can use this method to read all *functional locations* that satisfy certain selection conditions.
	GETSTATUS	You can use this method to read the system status and user status of an existing *functional location*.
	INHERITCHANGE	You can use this method to change the inheritance indicator of a *functional location*.

Object	Method	Content
	STRUCASSIGN	You can use this method to assign a superior functional location to a *functional location*.
	STRUCREPLACE	You can use this method to assign a new superior functional location to a *functional location*. This replaces the existing assignment to a superior functional location.
	STRUCUNASSIGN	You can use this method to unassign a superior *functional location* from a functional location.
MAINTNOTIFIC BAPIs	NOTIFCHANGEUSRSTAT	You can use this method to change the user status of a notification.
	NOTIFCLOSE	You can use this method to close a notification. The notification is assigned the system status NOTIFICATION COMPLETED.
	NOTIFCREATE	You can use this method to create a new notification. You can create certain fields in the notification header as well as items, causes, tasks, activities, and long texts for the notification. You can also maintain partner data.
	NOTIFDATAADD	You can use this method to add items, causes, activities, tasks, and long texts to a notification.
	NOTIFDATADELETE	You can use this method to delete items, causes, activities, tasks, and partners from a notification.
	NOTIFDATAMODIFY	You can use this method to change the header, items, causes, activities, tasks, and partners in a notification.

Object	Method	Content
	NotifGetDetail	You can use this method to display all detailed information about a notification.
	NotifListEqui	You can use this method to select notifications that are assigned to a piece of equipment.
	NotifListFuncloc	You can use this method to select notifications that are assigned to a functional location.
	NotifListPartner	You can use this method to select notifications that are assigned to a certain combination of partner functions and partners.
	NotifListPlangroup	You can use this method to select notifications that are assigned to a certain combination of planner groups and planning plants.
	NotifListSortfield	You can use this method to select notifications according to the specified sort field.
	NotifPostpone	You can use this method to postpone the notification. The notification is assigned the system status Notification postponed.
	NotifPutinprogress	You can use this method to put the notification in process. The notification is assigned the system status Notification in process.
	NotifTaskComplete	You can use this method to complete a notification task. The task is assigned the system status Task completed.
	NotifTaskRelease	You can use this method to release a notification task. The task is assigned the system status Task released.

Object	Method	Content
	NotifTaskSuccess	You can use this method to indicate that a notification task is successful. The task is assigned the system status Task successful.
MaintenanceOrderBAPI	ComponentGetDetail	You can use this method to display detailed information about an order component. Long texts are also returned.
	OperationGetDetail	You can use this method to read detailed information about an order component. Long texts are also returned.
	OrderGetDetail	You can use this method to read order information about an order.
	OrderheadGetList	You can use this method to determine a list of orders in accordance with a list of selection conditions.
	OrderMaintain	You can use this method to change orders and subobjects. You can edit the following order objects: ▶ Order header ▶ Partner ▶ User status ▶ Order operations ▶ Relationships ▶ Components ▶ Long texts for the order header, operations, and components ▶ Settlement rules ▶ Object list ▶ Link to object lists ▶ Production resources/tools ▶ Classifications of a service package

Object	Method	Content
		▸ Service lines in a service package ▸ Limits of a service package ▸ Contract limits of a service package
	OrderoperGetList	You can use this method to determine a list of operations in accordance with a list of selection conditions.
MaintOrd Confirmation	Cancel	You can use this method to cancel an order confirmation. Any goods movements that were posted together with the completion confirmation (backflushes) are automatically canceled with the completion confirmation.
	ConfCreate	You can use this method to enter time confirmations for order operations. Goods movements for a completion confirmation are determined in accordance with the standard logic for backflushes and an automatic goods receipt for the completion confirmation.
	GetDetail	You use this method to provide confirmation data for an order. Any goods movements that were posted together with the completion confirmation are also provided.
	GetList	This method delivers a list of all completion confirmations that satisfy the predefined selection criteria (order, operation, suboperation, or confirmation number).
	GetProp	You can use this method to provide default data for time confirmations for order operations.
Material	Availability	You can use this method to determine, in accordance with the ATP logic, the quantities still available for a specific material in a specific plant.

Object	Method	Content
	DISPLAY	You can use this method to display a material.
	EDIT	You can use this method to change a material.
	GETBATCHES	You can use this method to generate a list of batches for a material. You can then use input parameters to restrict this list.
	GETDETAIL	This method delivers a selection of client, plant, valuation area, and valuation type-specific detailed information about a predefined material.
	GETLIST	This method delivers a list of materials with a short text for the specified selection parameters.
	GETMRPLIST	You can use this method to select the MRP list for a material.
	GETSTOCK REQUIREMENTSLIST	You can use this method to select the current stock/requirements list for a material.
MATERIALBOM	CREATEBOMGROUP	This method is used to create a material BOM and all of its assigned objects.
	EXISTENCECHECK BOMGROUP	This method checks whether a BOM exists for a specified material, plant, and usage.

A.4 Customer Exits for Plant Maintenance

Master Data

- IEQM0001 Additional checks when installing equipment in functional location
- IEQM0002 Additional checks when defining equipment hierarchies
- IEQM0003 Additional checks prior to posting equipment

- IEQM0004 Object is permitted for contract partner (Order Maintenance contract)
- IEQM0005 Object is permitted for SD contract
- IEQM0006 Object is permitted for SD contract (maintain maintenance contract)
- IEQM0007 Check/change vendor field in equipment master
- IHCL0001 Create equipment with material: classes/characteristics template
- ILOM0001 Additional checks prior to posting a functional location
- ILOM0002 Structural check of location numbers
- ITOB0001 Customer include subscreen for functional location master data
- ITOB0002 Field changes in template
- ITOB0003 Customer include subscreen for fleet object data
- ITOB0004 Fleet identification data: checks

Measuring Points

- IMRC0001 Define field content for measuring points and measurement documents
- IMRC0002 Measuring point: menu exit for custom function
- IMRC0003 Measurement document: menu exit for custom function
- IMRC0004 Checks for new measurement document
- IMRC0005 Measuring point: exit in AUTHORITY_CHECK_IMPT

Warranties

- BG000001 Warranty check
- BG000002 Warranty check dialog box
- BG000003 Change warranty check result

Task Lists

- IAIH0001 Enhance task list header with customer-specific fields

Maintenance Plans

- IPRM0002 Determine planned dates for maintenance plan

- IPRM0003 User fields: maintenance plan
- IPRM0004 Maintenance plan/item: customer check during SAVE
- IPRM0005 Determine offset for service-based strategy plans

Notifications

- QQMA0001 User subscreen for notification header
- QQMA0008 User subscreen for notification item
- QQMA0010 User subscreen for additional data on cause
- QQMA0011 User subscreen for additional data on task
- QQMA0012 User subscreen for additional data on activity
- QQMA0014 Checks prior to saving a notification
- QQMA0015 Exit prior to calling F4 help for catalog
- QQMA0016 USER DATA function in GOTO • TASK menu
- QQMA0016 USER DATA function in GOTO • ACTIVITIES" menu
- QQMA0018 Date assignment after entry of priorities
- QQMA0019 Default partner when a notification is added
- QQMA0021 USER DATA function in GOTO menu
- QQMA0022 USER DATA" function in GOTO • CAUSE menu
- QQMA0023 QM/PM/SM: USER DATA function in GOTO • ITEM menu
- QQMA0024 Deactivation of function codes in menu
- QQMA0025 Default values when a notification is added
- QQMA0026 Authorization check when starting a notification transaction
- QQMA0027 Default values when a task is added
- QQMA0029 Changing the notification type
- QQMA0030 Checking a status change for permissibility

Orders

- COZF0001 Changing a purchase requisition for an external operation
- COZF0002 Changing a purchase requisition for an external component
- IWO10001 Creating a PM suborder

- IWO10002 Customer check for order release
- IWO10004 Customer check for order completion
- IWO10005 Customer-specific determination of profit center
- IWO10006 Exclusion of function codes
- IWO10007 Customer enhancement: permits in order
- IWO10008 Customer enhancement: determination of tax jurisdiction code
- IWO10009 Customer check during SAVE
- IWO10010 Customer enhancement for determining the WBS element
- IWO10011 Customer enhancement for component selection
- IWO10012 Priority handling in central header
- IWO10015 F4 help for user fields in operation
- IWO10016 Customer enhancement for checking operation user fields
- IWO10017 Determine external order number by customer logic
- IWO10018 User fields in order header
- IWO10020 Automatic task list integration
- IWO10021 Automatic task list transfer during order creation from notification
- IWO10022 Determine calendar
- IWO10025 Determine responsible cost center
- IWO10026 Set DO NOT EXECUTE status
- IWO10027 Generate custom-specific settlement rule
- IWO10029 Include BOM in order
- IWO10030 Prepopulate fields for results object
- IWO10031 Hide personnel number in order
- IWO10033 Customer-specific authorization check
- IWO10034 Operation status is based on capacity requirements status
- IWO20001 Task list transfer to order
- IWOC0001 Create notification: determination of reference object
- IWOC0002 Check status change permissibility
- IWOC0003 Reference object and planner group authorization check

- IWOC0004 Change single-level list processing in ALV settings
- CNEX0026 Customer enhancement for general material checks
- CNEX0027 Customer enhancement: determine plant, storage location for a component

Completion Confirmations

- CMFU0001 Define customer-specific screen layout
- CMFU0002 Set parameters for time confirmation and goods movements
- CONFPM01 Order confirmation: determine customer-specific default values
- CONFPM02 PM/SM order confirmation: customer-specific input checks 1
- CONFPM03 Order confirmation: customer-specific check after operation selection
- CONFPM04 Order confirmation: customer-specific input checks 2
- CONFPM05 Order confirmation: customer-specific amendments during save process

Information System

- MCI10001 MCI1: PMIS/QMIS update

B The Author

Dr. Karl Liebstückel is a professor of Information Management and Business Software at the Würzburg-Schweinfurt University of Applied Sciences, Germany. He has more than 25 years' experience in SAP plant maintenance. Over the course of his career, Karl Liebstückel has gained extensive practical experience in more than 70 plant maintenance projects. He was a member of the executive board of the German SAP User Group (DSAG) from 2003 to 2012, and chairman from 2007 to 2012. Furthermore, he presided over its Plant Management and Service Management work group from 2001 to 2008. Dr. Liebstückel owns his own consulting company and is the author of several books on the topic of logistics. Previously, he worked at SAP AG for 13 years in the areas of development, consulting, and training for plant maintenance and service management (last two positions held prior to leaving SAP AG: platinum consultant and global product manager). You are welcome to contact the author by email (*karl@liebstueckel.com*).

Index

A

Accelerated SAP, 23
Action box, 578
ASAP, 23
As-is analysis, 46, 62
Authorization concept, 25, 63
Authorization object, 73
Availability check, 302
Available capacity, 345

B

BAdI, 600
BAPI, 591
Basic order view, 298
Bill of material, 155
  equipment bill of material, 155, 232
  for a functional location, 155, 232
  material BOM, 155
Bill of material → see BOM, 232
BOM status, 232
BOM usage, 428
BPMN, 58
Business Add-In → see BAdI, 600
Business Application Programming Interface → see BAPI, 591
Business Blueprint, 25, 40, 60
Business processes, 243
Business Process Master List, 42
Business Process Model and Notation → see BPMN.
Business process modeling, 54

C

CAD, 22, 64
Calibration, 477
Capacity leveling, 345
Capacity planning, 344
Capacity requirement, 345
Catalog type, 263
CATS, 414, 597
Change management, 34
Chart of accounts, 77
Code group, 263
Combined order/operation list, 558
Company code, 75
Completion confirmation, 408, 597
Configure Formula Definition formula, 333
Configuring the Material Master function, 208
Confirmation reporting, 408
Construction type, 421
Content repository, 138
Control key, 100, 330, 345, 438
Controlling area, 78
Conversion module, 318
Costing, 363
Counter, 121, 597
Cross-Application Time Sheet → see CATS, 414
Cross-plant maintenance, 88
Customer exit, 596
  CNEX0027, 597
  CONFPM01, 598
  COZF0001, 598
  COZF0002, 598
  IHCL0001, 597
  IWO10004, 598
  IWO10009, 598
  IWO10012, 599
  IWO10021, 598
  IWO10027, 599
  IWO20001, 598
  QQMA0001, 599
  QQMA0014, 598
Cut-over plan, 43, 69

D

Deadline monitoring, 420
Debit cost element, 384
Default value, 426

Display variant, 563
Document, 133
Dynamic date calculation, 560, 561

E

EAM App, 593
EAM lists, 558
Electronic parts catalog, 310
Enhanced multiple counter plan, 436
Enhancement point, 601
Equipment, 154, 167
Estimated costs, 363
Event-driven process chains, 55
External processing, 439
  with service specifications, 446

F

Favorite, 556
Field selection, 142, 218
Final Preparation, 26
Fiscal year variant, 77
Flavor, 586
Fleet management, 177
Formula parameter function, 332
Function
  Activate Default Value for Current Date as Basic Date, 339
  Activate Split Valuation, 457
  Allowed Change of Notification Type, 259
  Allow Multilingual Text Maintenance for Each Equipment Category, 176
  Alternative Labeling of Functional Locations, 164
  Assign Catalog to Order Type, 313
  Assign Company Code Controlling Area, 80
  Assign Cost Elements to Value Categories, 366
  Assign Costing Parameters and Results Analysis Keys, 377
  Assign Fields to Field Selection Groups, 219
  Assign IM Assignment Key to Order Types, 392

Function (Cont.)
  Assign Master Data Fields of Assets and Equipment, 607
  Assign Network Attribute Categories to Network Types, 198
  Assign Notification Types to Order Types, 260, 467
  Assign Order Types to Maintenance Plants, 290
  Assign Secondary Screens, 214
  Assign View Profile and Equipment Categories to Fleet Object Types, 178
  Bar Chart Define Graphic Profiles, 341
  Basic Settings for Pool Asset Management, 519
  Change Catalogs and Catalog Profile for Notification Type, 268
  Configure Order Types, 287, 463, 485
  Configure Planner Group, 425
  Configure Screen Sequence for Work Center, 102
  Configure Special Functions for Maintenance Planning, 436
  Configure Split Valuation, 458
  Convert HTML Fields to SAP Fields, 313
  Convert HTML Field Values, 317
  Create Classes and Characteristics (CL02/CT04), 521
  Create Default Value Profiles for External Procurement, 324, 441
  Create Default Value Profiles for General Order Data, 404
  Create Structure Indicator for Reference Locations/Functional Locations, 157
  Create System Conditions or Operating Conditions, 398
  Default Values for Task List Data and Profile Assignments, 406, 443
  Define Access Sequence for Determining Address Data, 324
  Define Account Assignment Category and Document Type for Purchase Requisitions, 323
  Define Action Box, 275
  Define Activity Types for Pool Categories, 530

Function (Cont.)
Define Additional Business Views for Equipment Categories, 171
Define Attributes of Material Types, 205
Define Attributes of System Messages, 449
Define BOM Status, 232
Define BOM Usages, 234
Define Calculation Method for Fleet Consumption Values, 185
Define Capacity Category, 346
Define Catalog Profile, 265, 489
Define Catalogs, 311
Define Category of Functional Location, 161
Define Change Documents, Collective Purchase Requisition, MRP Relevance, 326, 444
Define Checking Control, 304
Define Checking Group, 306
Define Checking Rules, 303
Define Consumable Types, 179
Define Control Parameters for Completion Confirmations, 408
Define, Copy, Delete, Check Plant, 82
Define Default Value for Planning Indicator for Each Order Type, 291
Define Default Values for Component Item Categories, 320
Define Default Values for Inspection Type, 504
Define Default Values for Units for Operation, 340
Define Default Values for Value Categories, 367
Define Documentation for Goods Movements for the Order, 328, 468
Define Document Types, 134
Define Engine Types for Fleet Objects, 181
Define Field Selection, 417
Define Field Values for PM/PS Reference Element, 392
Define Follow-Up Action, 508
Define History-Related Fields, 174
Define Identifier for Inspection Points, 497
Define Industry Sectors and Industry-Sector-Specific Field Selection, 221

Function (Cont.)
Define Inspections in Plant Maintenance, 505
Define Installation at Functional Location, 175
Define Investment Profiles, 394
Define Item Categories, 236
Define Long Text Control for Notification Types, 256
Define LRP Type, 192
Define Material Provision Indicators, 476
Define Material Types Allowed for BOM Item, 238
Define Measuring Point Categories, 123
Define Media for Object Links, 188
Define Network Attribute Categories, 196
Define Network Group, 195
Define Network Types, 194
Define Notification- and Order Integration, 293, 492
Define Notification Types, 248, 487
Define Numbers Ranges, 114
Define Object Information Keys, 106
Define Object Types, 187
Define Offset Types, 190
Define Organizational Area for Characteristics with Linear Data, 193
Define Overall Profile in Capacity Leveling, 351
Define Overall Profiles in Capacity Evaluation, 350
Define Parameters for Subnetworks, 394
Define Partner Determination Procedure and Partner Function, 128
Define Permit Categories, 126
Define Plant Sections, 91
Define Plant-Specific Field Selection and Plant-Specific Screen Selection, 222
Define Pool Categories function, 529
Define Presetting for Free Assignment of Material, 428
Define Printer, 357
Define Priorities, 269
Define Priorities for BOM Usage, 239
Define Profiles with Default Values, 426
Define PRT Control Keys, 399
Define Recording Configuration, 513

Function (Cont.)
Define Release Procedure for Service Entry Sheet, 452
Define Relevant Fields for Assignment of IM Program, 390
Define Response Monitoring, 272
Define Scope of Check, 308
Define Selection Criteria for Alternative Determination, 240
Define Serialization Attributes for Movement Types, 230
Define Serial Number Profiles, 227, 473, 483
Define Service Category, 448
Define Settings for Planning Board, 524
Define Shift Report Types, 542
Define Shift Sequences, 347
Define Shop Papers, Forms, and Output Programs, 354, 520
Define Sort Fields for Maintenance Plan, 435
Define SPC Criteria, 495
Define Special Measurement Positions for Fleet Objects, 183
Define Status Control for Shop Papers, 362
Define Structural Display for Functional Locations, 163
Define Structure of Data Screens for Each Screen Sequence, 209
Define Structure of Data Screens for Each Screen Sequence, 470
Define Task List Usage, 424
Define Task List Usage Keys, 99
Define Transaction Start Values, 258
Define Transfer of Project or Investment Program, 389
Define Usage Types for Fleet Objects, 180
Define User Fields, 429
Define User Status, 109
Define Valid Material Types for BOM Header, 235
Define Valuation Variants, 373
Define Warranty Counters, 120
Define Warranty Types, 119
Define Workstation Application, 140
Determine Movement Types for Material Reservations, 321

Function (Cont.)
Edit Company Code, 75
Edit Display Variants of Planning Board, 525
Edit Object Processing, 523
Edit Relationships for Characteristics and Fields, 522
Global Parameters for Company Code, 77
Indicate Order Types for Investment Measures, 396
Indicate Order Types for Refurbishment Processing, 466
Maintain Allocation Structures, 381
Maintain Catalogs, 262, 541
Maintain Catalogs for Usage Decisions, 506
Maintain Catalogs function, 488
Maintain Control Keys, 100, 440, 447
Maintain Costing Sheet, 370
Maintain Costing Variants, 376
Maintain Default Values for Control Keys for Order Types, 294
Maintain Equipment Category, 169, 480
Maintain Field Selection for Data Screens, 223
Maintain Inspection Lot Origins and Assign Inspection Types, 502
Maintain Inspection Types, 500
Maintain Maintenance Planning Plant, 85
Maintain Order of Main and Additional Screens, 215
Maintain Revision Type, 533
Maintain Settlement Profiles, 384, 465
Maintain Status, 527
Maintain Storage Category, 139
Maintain Storage System, 138
Maintain Suitabilities, 430
Maintain Task List Status, 422
Maintain Value Categories, 364
Make Settings for Shift Note Type, 538
Make Settings for Units of Measurement for Monitoring of Consumption, 182
Make System Settings for Measuring Points and Measurement Documents, 122
Message Control, 400
Network Define Graphic Profiles, 343
Overview of Notification Type, 255, 462, 491, 516, 536

Function (Cont.)
Plan Automatic Usage Decision, 511
Send Shop Papers by Email, 361
Set Maintenance Plan Categories, 432, 493
Set Scheduling Parameters, 337
Set Screen Templates for Completion Confirmation, 411
Set Screen Templates for the Notification Type, 250
Set Up Data Entry Profiles, 414
Set View Profiles for Technical Objects, 479
Set View Profiles For Technical Objects, 158
Set Workflow for Maintenance Notifications, 604
Set Workflow for Orders, 604
Specify Conditions for Synchronization of Master Data, 605
Usage History Update, 173
Functional location, 153, 155

G

General maintenance task list, 421
Go-live, 70
Goods movement
documentation, 328
Google Maps, 588
GuiXT, 579
activate script, 583
PM Console, 585

I

IBIP, 65
Individual purchase order, 439
Input planning, 39
Inspection lot
maintenance call object, 433
Interval document, 122

K

Key combination, 565

L

Legacy data transfer, 25, 64
Linear asset, 189
Linear Asset Management, 189
Linear reference pattern, 192
List
activities, 558
completion confirmations, 558
components, 558
display, 199
equipment, 558
fleet objects, 558
functional locations, 558
goods movements, 558
maintenance dates, 558
maintenance items, 558
maintenance plans, 558
material, 558
material serial number, 558
measurement documents, 558
measures, 558
measuring points, 558
notification items, 558
notifications, 558
object links and object network, 558
order operations, 558
orders, 558
permits, 558
reference functional locations, 558
reference measuring points, 558
shift notes, 559
shift reports, 559
task lists, 559
variant, 557
LSMW, 65

M

Maintenance
activity type, 295
assembly, 154, 201
item, 420
notification, 245
plant, 86

Maintenance (Cont.)
preventive, 419
processing, 243
strategy, 419
Maintenance Event Builder → see MEB, 533
Maintenance plan, 420, 431, 597
multiple counter plan, 432
set categories, 432
single cycle plan, 431
special functions, 436
strategy plan, 431
type, 431
Maintenance planner group, 89, 127
Material, 154, 201, 280, 421
free assignment, 428
master, 456
planning, 302, 320
type, 201, 202
Material requirements planning, 469
Material type
Customizing, 204
Measuring point, 121, 597
MEB, 533
Mobile Asset Management, 22
Multilevel list, 150
Multiple counter plan, 432

N

Notepad, 580
Notification, 245, 533, 597
activity, 261
maintenance call object, 433
processing, 578
type, 247, 255
Number assignment, 114

O

Object information, 106
Object link, 154, 186
Object list, 279
Object networks, 194
Operate, 27
Operation, 421
Order, 279, 494, 533, 597
costs, 281
header, 279
maintenance call object, 432
material list, 280
object list, 279
operation, 280
planning indicator, 291
production resource/tool, 280
settlement, 380
settlement rule, 280
type, 284
Organizational structure, 73, 74
Organizational unit, 74

P

Paging, 62
PAM
assign class to notification type, 521
notification type, 519
Partner
determination procedure, 129
partner function, 127, 128
partner type, 127
Permit, 126
Personal value list, 565
Planning board
define objects, 523
Planning plant, 85
Plant, 82, 85
address, 83
assign company code, 84
define activity suitability, 430
maintenance plant, 86
planning plant, 85
Plant maintenance
cross-plant, 88
project tips, 32
Plant-specific maintenance, 87
PM console, 584
Pool Asset Management (PAM), 514
Pool category
activity types, 530
Production resource/tool, 280, 422
Production work center, 398

Purchase order, 439, 472
  define document types, 475
Purchase requisition, 439, 472, 598

R

Realization, 26
Reference functional location, 154, 155
Refurbishment, 454
  order, 463
Repairable spare, 454
Report RIAUFM00, 329
Results of feasibility study, 54
Risk factor, 21, 28
Role, 556
Role-based menus, 556

S

SAP0, 98
SAP Business Workflow, 602
SAP EAM
  completion confirmation, 408
SAP implementation project, 21
SAP Investment Management, 389
SAP List Viewer, 557
  further processing, 563
  list display, 561
  monitor, 561
  selection option, 560
  selection variant, 559
SAP NetWeaver Portal, 22
SAP Project Systems, 389
SAP Screen Personas, 586
Scheduling type, 330
  specify, 335
Screen sequence, 216
Selection variant, 559
Serial number, 155, 225, 454
Service entry sheet
  maintenance call object, 433
Service specification, 439, 446
Settlement rule, 280
  Define Time and Creation of Distribution Rule function, 386
Shift definition, 348
Shift note, 535
  categories, 541
  settings for type, 538
Shift report, 535
  define types, 542
Shift sequence, 349
Single cycle plan, 431
Standard value key, 97
Status management, 108
Strategy plan, 431
Structure indicator, 157
Structuring of technical systems, 153
Subcontracting, 471
Success factor, 21, 31
Suitability, 430
System message, 449
System status, 108

T

Table control, 565
Tab page, definition, 251
Task list, 419, 421, 597
TECD, 56
Test/measurement equipment, 477, 590
Test plan, 42
Transaction
  BAPI, 591
  CAC2, 417
  CM01, 346, 350
  CM33, 346
  CM34, 346
  DIWPSC4, 534
  GM01, 118
  GM04, 120
  IA08, 559
  IA09, 559
  IE05, 558
  IE36, 558
  IE37, 558
  IH07, 558
  IH08, 558
  IH09, 558
  IK07, 558
  IK07R, 558

Transaction (Cont.)
IK08, 558
IK08R, 558
IK17, 558
IK18, 558
IL05, 558
IL06, 558
IL15, 558
IN15, 558
IN16, 558
IN18, 558
IN19, 558
IP15, 558
IP16, 558
IP17, 558
IP18, 558
IP24, 558
IPM2, 558
IPM3, 558
IQ08, 558
ISHN4, 559
ISHR4, 559
IW3K, 558
IW3L, 558
IW3M, 558
IW28, 558
IW29, 558
IW37, 558
IW37N, 558
IW38, 558
IW39, 558
IW47, 558
IW49, 558
IW49N, 558
IW64, 558
IW65, 558
IW66, 558
IW67, 558
MB51, 329
OAC0, 138
OACT, 139
OIBS, 111
OICD, 234
OICG, 236
OICH, 233
OICI, 240
OICK, 237

Transaction (Cont.)
OICO, 476
OICP, 238
OICQ, 241
OIDA, 288, 464, 485, 520
OIDF, 355
OIDG, 355
OIDH, 358
OIDI, 358
OIDJ, 358
OIDS, 292
OIEN, 116
OIEZ, 175
OIK2, 366
OIL1, 423
OIL6, 426
OILJ, 429
OIMF, 273
OINM, 188
OIO4, 296
OIO5, 296
OIO6, 295
OIO9, 340
OIOD, 291
OIOF, 377
OIOI, 304
OIOR, 408
OIPK, 157
OIPU, 165
OIPV, 165
OIS2, 227, 473, 483
OITA, 395
OIW6, 148
OIWP, 163
OIZD, 99
OIZU, 103
OKP6, 376
OKP8, 374
OMS3, 221
OMS9, 223
OMSA, 222
OMSR, 220
OMT3B, 210
OMT3R, 215, 217
OMT3Z, 214
OMW0, 457, 470
OMWC, 458

Transaction (Cont.)
OP4A, 347
OP7B, 332
OPA6, 350
OPD0, 351
OPJJ, 308
OPJN, 337
OPK3, 333
OPTP, 394
OPU7, 338
OQN6, 266, 489
OVZ2, 307
OX06, 79
OX19, 80, 84
QS41, 262, 488, 506, 542
QS51, 506
SE19, 600
SHN4, 559
SHR4, 559
SMOD, 597
SU3, 549
VOP2, 129
Transaction/event chain diagram → see TECD.
Transaction variant, 568

U

UI Editor, 586
Upstream transaction, 589, 590
Usability, 61, 547
User acceptance, 61
User Interface Editor → see UI Editor, 586
User parameter, 549
maintenance-specific, 551
User status, 109

V

Valuation category, 456
Value chain diagram (VCD), 54
Vendor field selection, 132

W

Web interface, 593
Work center, 93, 127, 438
performing, 93
person as a work center, 93
responsible, 93
type, 94
Workflow, 602
Workshop, 93